621.87 SHAPI
Shapiro, Howard
Cranes and derricks /

S0-BIZ-466

DISCARD

Cranes and Derricks

Howard I. Shapiro, P.E.

Jay P. Shapiro, P.E.

Lawrence K. Shapiro, P.E.

Howard I. Shapiro & Associates
Consulting Engineers, P.C.
Valley Stream, New York

Third Edition

McGraw-Hill

New York San Francisco Washington, D.C. Auckland Bogotá
Caracas Lisbon London Madrid Mexico City Milan
Montreal New Delhi San Juan Singapore
Sydney Tokyo Toronto

Library of Congress Cataloging-in-Publication Data

Shapiro, Howard I., date
Cranes and derricks / Howard I. Shapiro, Jay P. Shapiro, Lawrence K. Shapiro.—3rd ed.
p. cm.
Rev. ed. of: Cranes and derricks. © 1991.
ISBN 0-07-057889-3
1. Cranes, derricks, etc. I. Shapiro, Jay P. II. Shapiro, Lawrence K. III. Shapiro, Howard I., date. Cranes and derricks. IV. Title. V. Title. Cranes and derricks.
TJ1363.S465 1999
621.8′7—dc21 99-16745
CIP

McGraw-Hill

A Division of The ***McGraw-Hill*** *Companies*

Copyright © 2000, 1991, 1980 by the McGraw-Hill Companies, Inc. All rights reserved. Printed in the United States of America. Except as permitted under the United States Copyright Act of 1976, no part of this publication may be reproduced or distributed in any form or by any means, or stored in a data base or retrieval system, without the prior written permission of the publisher.

3 4 5 6 7 8 9 0 IBT / IBT 9 0 9 8 7 6 5 4

ISBN 0-07-057889-3

The sponsoring editor for this book was Larry Hager and the production supervisor was Pamela Pelton. It was set in Century Schoolbook by Pro-Image Corporation.

Printed and bound by R. R. Donnelley & Sons Company.

McGraw-Hill books are available at special quantity discounts to use as premiums and sales promotions, or for use in corporate training programs. For more information, please write to the Director of Special Sales, McGraw-Hill, 11 West 19th Street, New York, NY 10011. Or contact your local bookstore.

Information contained in this work has been obtained by McGraw-Hill, Inc. from sources believed to be reliable. However, neither McGraw-Hill nor its authors guarantees the accuracy or completeness of any information published herein and neither McGraw-Hill nor its authors shall be responsible for any errors, omissions or damages arising out of use of this information. This work is published with the understanding that McGraw-Hill and its authors are supplying information but are not attempting to render engineering or other professional services. If such services are required, the assistance of an appropriate professional should be sought.

This book is printed on recycled, acid-free paper containing 50% consumer waste.

Contents

Preface to the Third Edition

The crane world is never static—that is not intended as a pun. Ever larger mobile cranes are being produced, but more importantly, innovations have come onto the scene with them. Controls and drives keep pace with microchip technology, rethinking of telescopic extension systems reduce boom weight permitting greater lengths, and new integral rapid self-assembly and disassembly arrangements are introduced. Even the notion of the load chart has been challenged by the computer revolution—separate charts for each of many configuration options result in multi-page compilations in an on-board "black box" as well as on paper.

What has not changed at all is the need for management and supervisory personnel to pay close attention to lifting operations. To do so effectively, they need to have a more complete understanding of cranes and of their strengths and weaknesses. On construction sites, and for many other applications, cranes are the key to meeting schedules and controlling costs. That is why we have written this third edition. Not only do the new technologies need to be addressed, but additional material has been developed to assist in operation planning. Our goal has always been to encourage an intelligent balance between safety and efficiency in crane use.

This third edition includes improvements to each chapter. The more significant additions are highlighted in the following paragraphs.

Chapter 2 contains new material on non-crane lifting devices, jacking towers and hydraulic gantries. We felt this material would be interesting and useful on its own, but these devices also often compete with cranes for lift assignments. Planners should therefore be conversant with them, so that rational comparisons may be made and the best equipment chosen for a project.

The wind material in Chap. 3 now includes the new wind load requirements of ASCE 7-95 which has supplanted ANSI A58.1 as the governing code. The old wind formulas and maps have been entirely superseded by this code revision which measures wind over a three-second averaging period instead of the old fastest-mile wind speed. The effect of this change in speed measurement methodology is that a new 100 mi/h (44.7 m/s) wind is *slower* than the old 100 mi/h wind. Chapter 3 offers instruction in relating the two systems.

At the end of Chap. 4 in the first edition of *Cranes and Derricks,* there was a plea for someone to devise an explicit solution to the problem of crane stability under dynamic loading. By the time the second edition was published, the authors had given up on this hope. But now a Russian engineer-friend, Anatoly Zaretsky, has come through with a solution. In the process, Professor Zaretsky has formulated a new way to conceptually consider overturning stability—a very versatile and useful way. His material has been added to Chap. 4; the chapter also now includes a study of barge-mounted cranes.

Chapter 5 has been fine tuned with more extensive discussions on crane support considerations, pick-and-carry work, and tailing operations in order to broaden planners' insights into these important matters. The authors wish to thank Mr. Ron Kohner, P.E. for his comments and observations which have served to greatly improve this chapter.

In redrafting Chap. 8, we felt an examination of the potential risks in lifting operations could help steer management and planners through their decision processes. Consideration of risk led naturally to the question of who is responsible for what among the parties involved. Significant work on that question has been done by the Specialized Carriers and Rigging Association (SC&RA), and they have graciously permitted us to use their material. Bill Smith, Director of Safety and Training for the International Union of Operating Engineers, has earned the authors' gratitude for reviewing the risk and responsibilities parts of this chapter. With his experience in the seat of a crane and in his present capacity, Bill was the right person for a reality check. The chapter and the book end with our own subjective and judgmental system for using risk assessment in determining how much planning and control is necessary for specific lifting operations.

Howard I. Shapiro, P.E.
Jay P. Shapiro, P.E.
Lawrence K. Shapiro, P.E.

Postscript

In the Preface to the first edition, the original author acknowledged the assistance of his recent-graduate sons, Jay and Lawrence. The second edition's Preface boasted, after a ten-year interval of time, that the sons were then co-authors of the book and partners in the engineering practice. Now, after nearly ten years more, the over-retirement-age original author glows with pride. His sons and co-authors are seasoned engineers with their own well established reputations and loyal client following. The technologies and methodologies they have introduced to the firm have greatly expanded the capabilities of this small office far beyond what I had ever dreamed was possible. Lastly, and with the greatest pleasure, I report that they now own and run the engineering practice, and that I work for them.

Howard I. Shapiro, P.E.

Preface to the Second Edition

There have been some important changes in the crane world during the ten years since the first edition of this book was published. Larger and more versatile telescopic mobile cranes have become commonplace, affecting both rigging and construction work, and luffing boom tower cranes are now a regular feature of the urban skyline, having replaced guy derricks, for example, at many steel erection projects. Along with those changes, however, society has become even less tolerant of accidents than in the past, and global competition now demands greater efficiency and productivity from all segments of the economy. This new edition reflects the changes in the industry, but both the first edition and this revised one address safety and efficiency in the same way with an impassioned plea for the adequate preplanning of crane operations as the primary step to achieve both goals.

A recent development in telescopic cranes, the all terrain crane, has captured a significant share of the mobile-crane market; a description of this crane type has been added to Chap. 2.

Chapter 5 has been expanded to include additional means for supporting mobile cranes, especially near building foundation walls. A shortcoming of the first edition has been rectified by inclusion of a full treatment of multiple crane lifts. This new material includes discussion of the effects of the characteristics of the load on lift planning, how load handling affects crane loading, why multiple crane lifts are far more critical than single crane lifts, and what you can do to control the risks.

Much of Chap. 6 has been rewritten to reflect current practices in tower crane installations and to amplify coverage of such subjects as exterior climbing cranes. Additionally, material on tower crane erection, jumping, and dismantling has been revised and enhanced to give far more practical information on how to plan these vital operations.

Chapter 8 has been updated with an expanded discussion of controlling lifted loads as a basic factor in crane safety. This includes consideration of operator and supervisor training as well as load and load-moment devices. Operations on rubber and "pick-and-carry" operations are also more fully discussed, and the section on tower crane accidents has been significantly enlarged. A new section on lifting personnel with cranes has been added.

Perhaps the most significant development of the last ten years is that the original author's sons are his coauthors for this new edition and his partners in a firm practicing crane engineering. Most of the new insights, new methods, and expanded coverages of this edition reflect their input. They have indeed fulfilled the original author's wish, as expressed in the first edition's dedication, by "...carry[ing] the work much further themselves."

Howard I. Shapiro, P.E.
Jay P. Shapiro, P.E.
Lawrence K. Shapiro, P.E.

Preface to the First Edition

Anyone seriously interested in cranes and derricks will be able to find pearls, in a figurative sense, between these covers, but many readers will not feel justified in opening too many oysters in their search for pearls. Although this book can be read through from end to end, readers with previous experience or familiarity with the equipment may wish to read only part of it. Since the book is addressed to an audience with a fairly broad range of interests, those interests will govern how individuals will want to read and use this book.

The first two chapters are introductory and describe the forms of equipment in use, as well as their components and accessories. These chapters are essential for beginners, but more experienced readers will find it worthwhile at least to skim through this material rather than bypass it altogether. Chapter 1 describes the mechanisms and accessories common to hoisting equipment, but more importantly—it introduces the terminology, jargon, and concepts used in the industry. The basic mathematics of hoisting devices is included. Chapter 2 deals with the equipment itself, showing how different arrangements of the basic mechanisms and components result in changes in function and capability. In addition, size range, capacity, uses, and limitations are examined.

The next two chapters deal with the operating regime, that highly technical set of conditions which affects the practical work produced by cranes and derricks and which is so intimately connected with operational safety. The mathematical treatment enables design engineers to satisfy their need to quantify the happenings in crane and derrick life, while the descriptive material gives the insights needed to appreciate the limitations and dangers associated with various equipment types. Although many readers can pass over the technical portions, at least a general understanding of the concepts in these chapters is essential for all readers.

Chapter 3 discusses both obvious and subtle static and dynamic loading effects associated with crane and derrick functional motions and delves rather deeply into the subject of wind. The effect of wind, and particularly gust action, is too often overlooked or insufficiently appreciated. Chapter 4 explores the physical concepts and the mathematics of stability against overturning, both as a static phenomenon and in a dynamic context. Overturning is the primary failure mode for the most common hoisting equipment.

The chapters that follow are type-specific in that they detail the considerations essential in making practical installations of the main equipment types. The focus of the material is the jobsite and the measures that need to be taken to assure productivity and safety. Here, again, the treatment is both mathematical and descriptive. Readers with an interest in only one equipment type can safely skip the chapters dealing with other types or, better yet, skim through them.

Chapter 5 concerns mobile crane installations and the movement of the machines both on the site and between sites. The mathematics for assuring proper crane support and for determining functional clearance is given. For those who do not need this material, it is important to know that these matters can be calculated. There is no need to resort to guesswork or rules of thumb.

Chapter 6 covers tower cranes from erection to dismantling in each of the installation configurations. Tower cranes are perhaps the least understood crane type in the United States, but their growth in physical size and frequency of use makes it necessary for all those concerned with these machines to be familiar with operational and installation concepts.

Chapter 7 is a discussion of installation details and criteria for the various derrick forms, including operational effects on the host structure. Mathematical procedures are introduced as needed.

The closing chapter, which covers safety and liability, encompasses materials applicable to all crane and derrick types. The emphasis is on preventing accidents and establishing practices that will improve productivity while reducing liability in case of accident.

If this book could be said to carry a message, it is that preplanning is a requisite for successful crane and derrick use. Preplanning will not only lead to a reduction in accidents but also measurably increase productivity and therefore directly justify its cost. During preplanning, problems are identified and solved in advance so that field crews are not left standing about waiting for someone to make a decision. With preplanning, the right equipment is sent to the jobsite, because field needs have been predetermined and capacities, sizes, etc., evaluated. And lastly, when there is preplanning, field personnel are not forced to push the machines up to and beyond their capabilities in order to get the job done. Therefore, accident potential is reduced, and the equipment can be utilized to its full productive potential.

Several people have assisted in the writing of this book by reading various portions and offering their criticisms. All but a few of their suggestions have been incorporated. Robert Del Duca, an executive of Gendelman Rigging and Trucking, Inc., who began his career on a rigging crew, has read most of the book, and his help has been invaluable. Additional critical review was provided by George A. Allin, P. E., who until his untimely death was Director of Engineering for the Harnischfeger Corporation; Erik Andersen, President of Tower Cranes of America, Inc.; Dinesh Seksaria, formerly with Lima Division of Clark Equipment Company; and William Chieco, my colleague at Charles M. Shapiro and Sons, P.C., Consulting Engineers.

My sons, Jay P. Shapiro and Lawrence K. Shapiro, have been very much involved with this work since its inception, when they were both still students. They have been militant proponents of the clearly expressed thought.

I happily acknowledge this assistance and remain indebted to the people who were kind enough to give me their help.

Howard I. Shapiro, P.E.

Chapter

1

Basic Concepts and Components

Cranes and derricks lift and lower loads by means of ropes and pulleys and move the loads horizontally. Machines that do not perform those functions, or do them by other means, are neither cranes nor derricks by common definition. A crane is a self contained piece of equipment, whereas the prime mover or power source of a derrick is a free-standing unit separate from the hoisting structure. The derrick and its power source are brought together for each particular job.

The speeds at which load handling functions are performed, the load weights that can be handled, the heights to which loads can be lifted, and the locations of the machines while doing work are attributes that differentiate an extremely diverse range of equipment. They vary from the humblest of devices to state-of-the-art automatic machines; they are built by back yard mechanics and Fortune 500 companies. The nature of their activities expose these lifting machines to dynamic loading arising from their operating motions, but they often also support loads imparted by their operating environment, including the effects of wind, snow, and ice, or even earthquakes, and temperature extremes.

Cranes and derricks are utilized for lifting service across a diverse spectrum of operating conditions. They may be exposed to infrequent duty, as in power plant turbine house service where passive work—maintenance and testing—comprise the usual use and productive working lifts are occasional, or they may be punished by intense use such as in steel mill service where round-the-clock operations induce millions of loading cycles. The diversity in forms, environments, and operating regimes makes the selection and installation of hoisting equipment as much an art as a science.

An ever present concern whenever cranes and derricks are at work is safety. By their nature, rigging and lifting loads entails risk, and when accidents occur, they are almost always dramatic and newsworthy. Safety is an abstract notion. The term is easily used in discussion or writings, but not always easy to achieve in practice; there are few simple fixes and no panaceas. Devices intended to enhance safety may yield no benefit in actual use, or even have a contrary effect. What may improve safety under one set of circumstances, can have negative effects under another. In many situations, the most effective safety enhancements are small and subtle measures tailored to the risks associated with the particular operation. Safety is best served by time spent in planning lifting operations, thereby avoiding the need to devise procedures on the fly.

Safe crane and derrick use is rooted in good planning together with competent and responsible workers. One goal of this book is to help sharpen the awareness of lift planners to the sources of risk and direct them towards means of ameliorating risk. Another goal is to assist them in applying cranes and derricks to do useful work, to aid efficiency and productivity and to reduce costs. There is no conflict between cost reduction and safety. Well planned safety enhancements will be proportional to the risks; they will not be burdensome to field crews and, in fact, they often enhance productivity.

Several fine field references and pocket handbooks are available for crane and derrick work. This book is complementary to such books rather than competitive. The field references offer "how to" and "how not to" guidance for field crews and the handbooks provide specific information about hardware and practices, whereas *Cranes and Derricks* is a planning guide and tool. Though field personnel and supervisors will find much useful material in this book, its place is in the office and its role is to support equipment selection, installation design and prelift planning.

The presentation is addressed to engineers and to crane and derrick users and supervisors with a technical background. Materials of particular interest to engineers are set off in contrasting type. For readers who do not require the engineering material, there will be no loss of continuity if the portions in contrasting type are passed over.

1.1 Introduction

The principles used in designing and installing hoisting equipment cover the spectrum of engineering disciplines, but most particularly the fields of mechanical and structural design. Take, for example, the problem of designing a truck crane. A typical truck crane has a diesel engine with a torque converter to power its several motions and pneumatically actuated mechanical brakes to stop them. The complex structural boom, the machinery deck, and the truck frame are subjected to widely varying loads. Some of the components must be designed to criteria that limit deflections, others to criteria that control stress levels, and still others to criteria concerned with service life.

Installation design has its complexities too. Loadings and motions are in three-dimensional space and vary with time. Care is required to discover the conditions that most affect machine supports and to assure that adequate clearances or protections are provided to prevent collisions with other objects. The installation designer often must also consider the logistics of getting the equipment in place and the capabilities of the crew doing the work.

The old versus the new

Engineering first came into being as a field of practice when scientific principles started to be applied to the problems of designing machines and structures. Until then, things as varied as cathedrals and flour mills were designed and built by tradesmen; journeymen working under the supervision of a master chosen for skill, experience, and proven judgment. The engineer's tools for applying scientific principles—then and now—are the various components of mathematics. Mathematical solutions were obtained by tedious calculations, often with the help of graphical constructions, tables and simplified formulas. But the use of experience did not die out. When science proved inadequate, or mathematics or calculation too difficult or slow, rules based on experience came into play.

Rules of thumb are guides based on experience, and together with tables of data and graphs of experimental results were important tools in the engineering design office. A good rule of thumb may yield reliable results, but such is not always the case. Working with this method alone leads to uncertain design, often wasteful and occasionally inadequate. Nonetheless, rules of thumb may be useful means for deriving a preliminary or trial solution to an engineering problem.

Engineering design passed through a renaissance, starting about 1970, emerging from the era of the slide rule and the rule of thumb into the electronic age. Calculators and computers now make easy work of large complex formulas and systems of equations that had previously been all but impossible to solve. Engineers have readily taken to those devices and they have become ubiquitous fixtures on their desks. Instantaneous results allow engineers to evaluate many potential conditions or to investigate more complex situations in their search for optimal, but safe designs. As new computer software and operating systems have come on line, productivity has improved. Besides offering raw computational speed, programs at the present time utilize color graphics to show results with startling clarity.

Inexpensive "user friendly" programs now provide a potent approach to problem solving. Experts and novices alike enjoy access to an ever-expanding magnitude of computational power. It would seem that much less human engineering experience is presently needed because of the expertise residing behind those desktop screens. This is an alluring notion, as inexperienced people running off-the-shelf programs cost much less than seasoned experts. But such a naive approach to problem solving is even less valid than uncritical obedience to rules of thumb.

Inordinate reliance on an off-the-shelf analysis program poses several potential perils: the inherent limitations of the program might not be understood by the user; the "friendly face" of the screen may conceal a level of decision making that the user overlooks; and the output may look authoritative but provide no means to check the results.

The computer does not emulate real life. It approximates the real world through mathematical models intended to reflect real world behavior, but a computer just as easily produces a wrong answer as a right one. Some computer models are better constructed than others, and compromises are often made to produce user friendliness. As input data is reduced and simplified, the model must, in consequence, become simpler and perhaps further and further removed from relevance. The simple model may adequately reflect reality for many actual conditions, but not all. And the typical program will not warn users to beware when a more complex model is needed. So, the problem may not be the old metaphor "garbage in, garbage out". The real danger may be from over simplification hidden behind user friendliness. In the design office, canned programs make the junior engineer feel omnipotent, and the value of experience becomes forgotten. For the novice, a program may unlock doors to places where only the expert practitioner should tread.

Too many of us have been lulled into belief by the convincing appearance of a computer program's output, later to learn that it is fatally flawed by a modeling error or over simplified methodology. How can we be so trusting of that mystical "black box"? Verification of computer solutions is a must. The engineer's toolbox still should include graphs, charts and simplified formulas to carry out these verifications. And non-engineers should beware of programs intended to replace engineering judgment with quick "on screen" answers.

While a rule of thumb is an amalgam of experience with little or no science, a computer program offers a mathematical solution that may have no attachment to the real world. Each by itself will not always prove satisfactory as a general problem solver. Together they are part of a comprehensive set of tools that give modern engineers capabilities that their predecessors would envy.

Successful crane and derrick engineering practice requires more than analytical tools and rules of thumb. It must be a "means and methods" practice, in which knowledge of what takes place in the field is important, too. The human element is a strong factor. A capable crane and rigging installation designer is one who understands the abilities and limitations of contracting firms and their people, as well as the equipment they will use.

1.2 The Basic Hoisting Mechanism

Figure 1.1 shows a lifting device with a load attached to the lower block and the block in turn supported by two ropes, or *parts of line*, suspended from the upper block. Each rope must therefore carry half the weight of the load, thus giving the system a mechanical advantage of 2. Had the load been supported by five ropes, the mechanical advantage would have been 5. Mechanical advantage is governed by the number of ropes actually supporting the load. As parts of line are added, the force needed to raise or lower the load decreases, and load movement speed decreases as well.

The blocks contain pulleys, or *sheaves,* so that the rope is in one continuous piece from the end attached to the upper block to the winding drum. This makes the force in all parts of the rope uniform in a static system. The value of the rope load is found by dividing the weight of the lifted load by the mechanical advantage; in Fig. 1.1 the

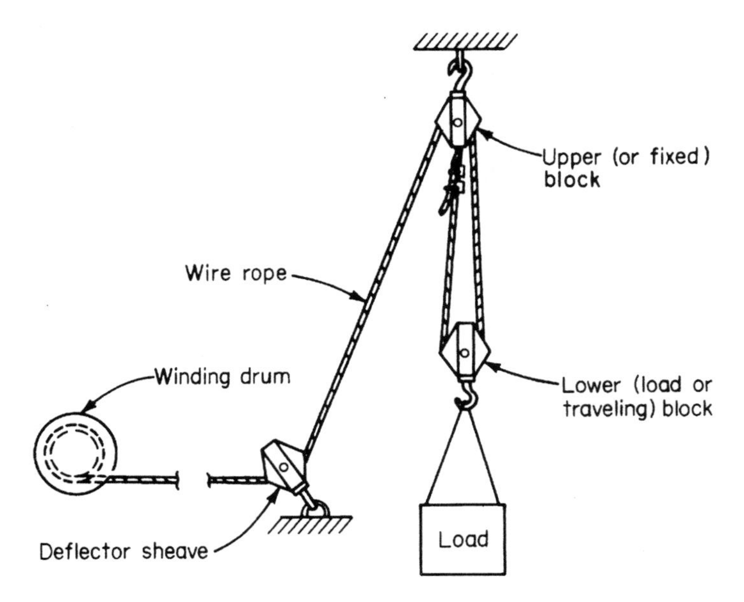

Figure 1.1 Basic hoisting mechanism.

lifted load would include the lower block. When the distance between the upper and lower blocks is great, it is necessary to include the weight of the parts of line as well.

The load in the rope is also equivalent to the force that must be generated at the winding drum in order to hold the load.

The effects of friction come into play as soon as the system is set into motion. Friction losses occur at the sheave shaft bearings and in the wire rope itself, where rope losses result when the individual wires rub together during passage over the sheave. These losses induce small differences in load between each rope segment (i.e., each section of rope from sheave to sheave). The loss coefficient can vary from a high of about 4½% of rope load for a sheave mounted on bronze bushings to a low of as little as 0.9% for a sheave on precision ball or roller bearings. An arbitrary value of 2% can be taken for sheaves on common ball or roller bearings when the rope makes a turn of 180°.

The tension in the rope at the winding drum is different when the load is raised and when it is lowered. Friction losses are responsible for this difference. When load-weighing devices that operate by reading the tension in the line to the drum are used, the variation is readily observed.

When an unloaded hook must be lowered, lowering will be resisted by friction, by the weight of the rope between the upper block and the deflector sheave, and by the inertia of the winding-drum mass. Mechanical advantage works in reverse in this case, as a mechanical disadvantage so to speak, so that the weight at the hook must exceed the rope weight times mechanical advantage plus an allowance to overcome friction and inertia. If the weight at the hook is less than the result of this calculation, the hook will not lower; for that matter, if the weight is significantly less, the hook will rise on its own and will not stop until it strikes the upper block. To prevent this action, it is necessary to have a lower block with adequate weight or to add an *overhauling weight* (overhaul ball) so that the rope will overhaul through the system. Since the overhaul weight becomes part of the dead weight of the mechanism and remains in place throughout operations, it must be taken into account in operating plans. It is part of the lifted load.

Figure 1.2 shows the basic hoisting mechanism on a working crane. In fact it illustrates two separate sets of mechanisms; the *main fall* is a multi-part line suspended from the boom tip and the *auxiliary fall* is a single part arrangement. The lower block on the main fall is provided with heavy side plates, called *cheek weights,* for overhauling while the auxiliary fall is overhauled by a cast weight sometimes called a *headache ball.*

Figure 1.2 Link-Belt model RTC-8050 rough terrain crane. Note the two wire ropes each leading from a winding drum over the upper surface of the boom to a deflector sheave at the boom tip. One rope continues to the fixed upper block and thence to the load block, while the other runs over a second deflector sheave to the auxiliary hook and ball. (*Link-Belt Construction Equipment Company*)

Friction effect on line pull can be large on systems with many parts of line or multiple sheaves between the fall and the winch. On such systems, friction should not be ignored. For a given number of sheaves and friction loss per sheave, the system loss can be calculated. Referring to Fig. 1.1, let W be the weight of the load and the lower block, P the force at the winding drum, and μ the loss coefficient. During raising, the rope between the deflector sheave and the upper block carries the force $(1 - \mu)P$, and the ropes supporting the load carry $(1 - \mu)^2P$ and $(1 - \mu)^3P$, respectively. In order to lift but not to accelerate the load we must have

$$P = \frac{W}{(1 - \mu)^2 + (1 - \mu)^3}$$

When lowering the load, the friction effect is opposite, so that the force in the rope from deflector sheave to upper block becomes $P/(1 - \mu)$. The ropes supporting the load will then experience forces of $P/(1 - \mu)^2$ and $P/(1 - \mu)^3$, and the holding force at the drum will be

$$P = \frac{W}{(1 - \mu)^{-2} + (1 - \mu)^{-3}}$$

The preceding equations can be generalized. If n is taken as the number of parts of line supporting the load and m is the number of 180° turns taken by the rope between the upper block and the drum (turning angles for each of the sheaves are added to find the number of 180° multiples), then

$$P = \frac{W}{r} \tag{1.1}$$

where for raising the load

$$r = (1 - \mu)^{m+1} + (1 - \mu)^{m+2} + \dots + (1 - \mu)^{m+n}$$

but after simplification[1]

$$r = \frac{(1 - \mu)^{m+1} - (1 - \mu)^{m+n+1}}{\mu}$$

For lowering the load

$$r = \frac{1}{(1 - \mu)^{m+1}} + \frac{1}{(1 - \mu)^{m+2}} + \dots + \frac{1}{(1 - \mu)^{m+n}}$$

which simplifies to

---

[1]For these simplified expressions, the authors wish to thank Mr. Keith Trommler of Bragg Crane and Rigging, Long Beach, California.

$$r = \frac{1 - (1 - \mu)^n}{\mu(1 - \mu)^{m+n}}$$

Example 1.1 A load of 15,700 lb (7121 kg) is to be lifted by a hoisting device with four parts of line and a lower block of 300 lb (136 kg). Neglecting friction, how much force must be developed at the drum?

When friction is neglected, drum force will be the same as rope loading. The MA (mechanical advantage) is 4 when four parts of line support the load:

$$P = \frac{15{,}700 + 300}{4} = 4000 \text{ lb } (17.79 \text{ kN}) \qquad \textit{Ans.}$$

When friction is taken into account, how much force must the drum exert to raise the load? How much to lower it? Assume that there are sheaves mounted on ordinary ball bearings and three deflector sheaves taking the rope through two 180° turns.

Using $\mu = 0.02$, $m = 2$, and $n = 4$, we get $1 - \mu = 0.98$. From Eq. (1.1), for raising a load

$$r = \frac{0.98^3 - 0.98^7}{0.02} = 3.65$$

$$P = \frac{15{,}700 + 300}{3.65} = 4384 \text{ lb } (19.50 \text{ kN}) \qquad \textit{Ans.}$$

and for lowering the load

$$r = \frac{1 - 0.98^4}{0.02 \times 0.98^6} = 4.38$$

$$P = \frac{15{,}700 + 300}{4.38} = 3653 \text{ lb } (16.25 \text{ kN}) \qquad \textit{Ans.}$$

If the rope from the deflector sheave to the upper block weighs 20 lb (9.1 kg) and 50 lb (22.7 kg) is needed to overcome drum friction, will the system overhaul when unloaded? If not, how much overhauling weight must be added?

Ignoring sheave friction, the weight that must be overcome in order to lower the load block is $20 + 50 = 70$ lb (31.7 kg). With an MA of 4, this requires $4 \times 70 = 280$ lb (127.0 kg) at the load block. But the load block weighs 300 lb (136.1 kg); the system will overhaul.

With sheave friction considered, the weight to be overcome is $50/0.98^2 = 52.1$ lb (23.6 kg) after passing through two 180° turns, plus 20 lb of rope weight, for a total of 72.1 lb (32.7 kg). For lowering, r was found to be 4.38. The weight required for overhauling is then

$$72.1 \times 4.38 = 316 \text{ lb } (143.2 \text{ kg})$$

which is greater than the lower block weight. An additional weight of at least 16 lb (7.2 kg) is needed. *Ans.*

In practice, an overhaul weight somewhat in excess of the calculated value is used. This allows for possible inaccuracy in the loss coefficient and provides the mass needed to overcome inertia and induce acceleration. When friction losses are neglected in calculations, it would be wise to increase the overhauling weight appreciably.

1.3 Drums, Hoists, and Sheaves

One or more winding drums mounted on a frame with a power plant and the necessary controls is called a *hoist, hoisting engine,* or *winch* (Fig. 1.3). The means of powering these machines and the drives and control systems employed vary widely, but a basic unit could comprise a diesel engine with gear train and two or three drums. A disconnect clutch would be provided between the engine and gear train with additional clutches, called *frictions,* mounted at each drum. A brake band is built around the periphery of one flange on each drum, which is widened for this purpose. Each drum can therefore be individually controlled. For additional safety, a ratchet-and-pawl system is included at each drum, providing a means for positively locking the drum against inadvertent spooling out of rope. When the pawl is engaged, the drum is said to be *dogged* or *dogged off.*

When ready to begin work, the operator starts the winch engine and engages the *disconnect,* or *main, clutch.* This sets the gear train in motion, but the drums will not turn with frictions disengaged. To lift a load, the operator engages the friction on the appropriate drum. As the drum starts to spool in rope, the pawl will automatically disengage from the ratchet. The engine throttle is usually set to operating speed before lifting so that load acceleration is controlled by gradual engagement of the friction in conjunction with use of the foot brake. To stop lifting, the friction is disengaged and the foot brake applied. The drum is dogged if the load must be suspended for an extended period.

To lower a load, the dog is first released by momentarily engaging the friction. Lowering is controlled by the foot brake.

Before the operator leaves the winch unattended, the load must be lowered to the ground, all drum dogs engaged, frictions and main clutch disengaged, and the engine stopped. A grounded load will pose no threat of falling, but a machine with the engine running and main clutch engaged is potentially dangerous. Any inadvertent or unauthorized engagement of a friction will raise the hook. A partial or momentary engagement will release the dog and cause the hook to fall.

Winch units designed for use in a crane are tailored to the specific needs of the crane, often including advanced control and mechanical

(a)

(b)

Figure 1.3 (*a*) Three-drum gasoline hoist with torque converter. (*b*) Detail showing ratchet-and-pawl dogging arrangement. (*Clyde Iron, a unit of AMCA International Corporation.*)

systems. Hydraulically driven winches and some electric winches can be furnished with automatic braking systems in which the brakes are normally engaged. When either power-up or power-down control signals are initiated, the application of power to the drum triggers release of the brake. These winches often are not equipped with dogs.

Modern free-standing winch units, also called *base mounted drum hoists*, can be quite sophisticated too. Some now include pneumatic, electric or hydraulic controls and hydraulic drives coupled with gasoline, diesel, or electric power plants. Powered or free-fall lowering options are available, and dogs may disengage automatically.

Because so many older winches remain in use, the typical derrick operation will still employ mechanical winches with frictions at each drum. On some older winches, lowering is a free-fall operation, retarded by the brakes. On others, lowering may be controlled by torque converter slippage or by the brakes.

Derricks installed on the roofs of tall buildings for machinery replacement or tower crane removal pose special winch problems. Lowering significant loads over long distances overheats the brakes or torque converter, necessitating cooling-off stops. Rigging contractors engaged in this work often use modified winches with oversized brake bands and transmission-oil coolers for this severe service. More information about derrick winches will be found in Chap. 7, Derrick Installations.

Tower cranes usually have electric or hydraulic winches that do not permit free-fall lowering. Powered lowering, or lowering against a torque converter, provides good load control for placing machinery on anchor bolts and similar precision work.

Hoist drums

A winding drum transmits power to the wire rope, but in many applications it also serves as a reservoir by spooling and storing rope that is in excess at the moment. Each turn of the rope around the full circumference of the drum is called a *wrap*. Rope is helically wrapped around the drum, starting at one end flange (see Fig. 1.4) and progressing toward the opposite flange. A series of wraps extending from flange to flange is referred to as a *layer*. If spooling continues after completion of a layer, the wraps proceed back toward the starting flange in a second layer. In an operating winch, when the drum is spooled full, some flange must be left projecting beyond the surface of the outermost layer of rope. Although some authorities recommend a minimum projection of ½ in (12.7 mm), one rope diameter seems like

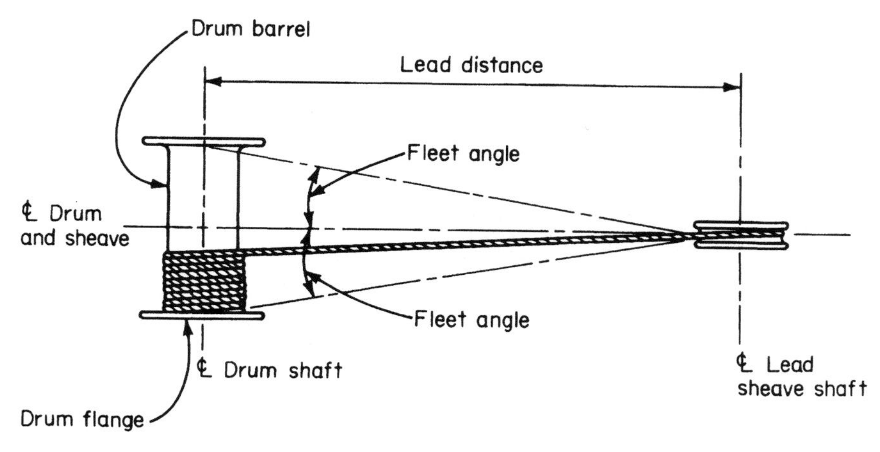

Figure 1.4 Fleet angle.

a more logical and prudent minimum. In Canada, 2 in (50 mm) is recommended.†

Flange projection above a fully loaded drum is a necessary precaution so that the rope will not slip off the drum. As the rope spools in on the uppermost layer, the rope is being forced in the direction of the flange by the previous wrap. When the flange is reached, friction between rope and flange will cause the rope to attempt to climb the flange. That friction force is quickly overcome, as the rope climbs, by the component of line tension that develops in the opposite direction. A small flange projection is therefore adequate to prevent the rope from climbing over the flange.

Another condition that requires ample flange height develops when the hoist line is relieved of load. If the unloading rate is rapid, the rope releases its strain like a mildly stretched rubber band and could jump over the flange edge.

The end of a rope is attached to the drum barrel by a socketing or clamping arrangement. During hoisting operations, as rope is spooled out, at least two full wraps must remain on the drum as a safety margin. (If unspooling continued until all the rope was paid out, the rope attachment would be subjected to a shock loading as the lowered load suddenly stopped and reversed. The rope attachments are neither as strong as the rope itself nor as capable of absorbing shock.)

†D. E. Dickie, *"Crane Handbook,"* Construction Safety Association of Ontario, Toronto, 1975, p. 11.

To assist in making first-layer wraps closely placed and uniform, drum barrels are often grooved. A good first layer is a necessity if succeeding layers are to wrap properly; the rope itself provides a groove effect for subsequent layers. Grooves are cut to suit a particular rope diameter, and grooved drums can be properly used only for that diameter of rope. Grooves may be cut helically, but better performance results from grooves cast parallel to the rope direction, in a proprietary arrangement known as the *Lebus style.*

Removable shells, called *laggings,* can be added to drum barrels that are arranged to receive them. Laggings permit drums to be adapted for rope of another size but can also be used to increase line speed by increasing barrel diameter (reducing rope storage capacity at the same time). Drums operating at very high line speeds exhibit spooling problems that increase as the number of layers increases. Lower layers tend to become loose and sloppy, causing excessive wear and premature failure. This occurs in part from rapid braking of the drum rotation, which induces greater inertia on the upper layers than on the lower. Subsequent reloading tightens the upper layers, leaving the lower layers loose. Respooling under tension, starting at the first layer, is the recommended remedy. When a hoist continues to operate with loose lower layers, wraps from one layer may fill the gaps created in a lower layer. As rope spools out under load the wraps that had become pinched in a lower layer can experience significant wear, abrasive damage and shock load when pulled free. Layers may also loosen from raising an unloaded hook with an overhaul ball that is too small.

Wear and damage can also occur at *flange points* and *cross-over points*. A flange point is where the rope contacts the drum flange as the rope starts another layer. A cross-over point is where rope on one layer contacts and crosses the rope in a preceding layer.

Single layer drums are used, where feasible, to eliminate several causes of wear and rope damage, but another feature of a single layer drum is its ability to deliver constant line speed and constant line pull. Winches equipped with single layer drums posses two additional interesting features. They can be furnished with spooling protection devices which shut down the hoist drum should a second layer develop. They can also be arranged to spool both ends of the rope, eliminating the dead end and doubling hoisting speed.

Fleet angle

To aid proper spooling and to prevent excessive wear on the rope and on drum grooves, the maximum angle at which the rope leads onto the drum, called the *fleet angle,* must be kept within controlled limits. Figure 1.4 illustrates the fleet angle, which is the angle whose tangent is one-half of the drum-barrel width divided by the distance between the shaft centerlines of the drum and lead sheave. The lead sheave is a fixed-position deflector sheave aligned with the center of the drum. Fleet angles should be no less than ½° and no more than 1½° for smooth drums or 2° for grooved drums when the lead sheave is centered on the drum. A fleet angle of 1½° requires that the lead distance be 19 ft for each foot of drum width. A 2° fleet angle requires 14⅓ ft. (In the SI, the same numbers apply but with meters as the unit.)

Sometimes it is neither practical nor possible to install the lead sheave at the required distance, and a shorter distance must be accommodated. Proper spooling can still be maintained if a pivoted block (Fig. 1.1) is used for leading to the drum. A pivoted block will lie over from side to side following the rope as it spools on the drum. When using a pivoted block, one must be sure that it is set close enough to the drum to provide at least ½° of fleet angle in the maximum-layover position. Too small a fleet angle will cause the wraps to pile up at the flange (particularly at low loading levels), whereas an adequate angle will guide the rope away from the flange.

If a still shorter lead distance is needed, the lead sheave can be mounted on a horizontal shaft that lets the sheave move laterally as the rope spools. This is called a *fleeting sheave*. The shaft length must be selected so that the minimum fleet angle is respected.

In all lead sheave arrangements, but especially when the lead is short, it is necessary to mount the sheave preceding the lead sheave so that it aligns as closely as possible with the lead sheave; this will mitigate the wear on rope, sheaves, and bearings.

Drum capacity

Wire ropes are initially made oversized. During service they elongate and reduce in diameter as they wear, seat in, or tighten up. Also, from one spooling in to another, the tightness of the wraps and rope tension both vary. These factors all affect the length of rope that can be spooled onto a drum. Such calculations should not be expected to produce exact results, but reasonably good values can be determined using the method that follows. Referring to Fig. 1.5, let us consider three cases of drum capacity, each using the equation

$$L = (D + E)EBs \tag{1.2}$$

where with U.S. customary units L is given in feet when dimensions D, E and B are in inches and the spooling factor is taken from Table 1.1 or calculated using Eq. (1.2a). Table 1.1 assumes new rope and includes a rope diameter oversize factor of 5%. Taking d as the actual

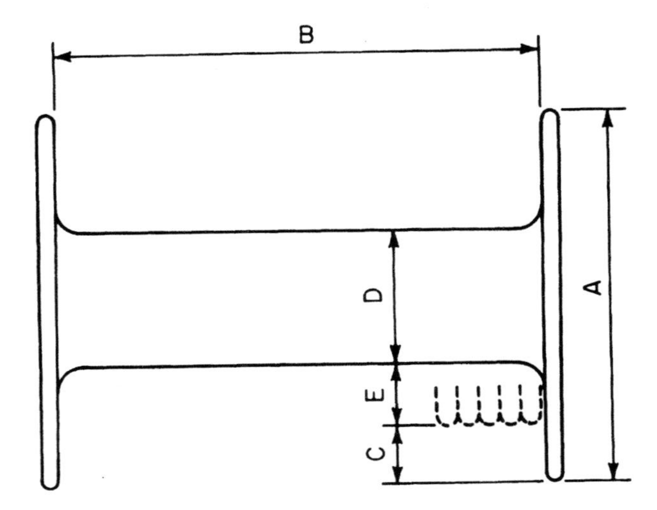

Figure 1.5 Hoist-drum dimensions.

TABLE 1.1 Spooling Factors *s* for Drum-Capacity Calculations in U.S. Customary Units

Rope diameter *d*	Spooling factor *s*	Rope diameter *d*	Spooling factor *s*	Rope diameter *d*	Spooling factor *s*	Rope diameter *d*	Spooling factor *s*
½	0.925	13⁄16	0.354	1⅜	0.127	2	0.0597
9⁄16	0.741	⅞	0.308	1½	0.107	2⅛	0.0532
⅝	0.607	1	0.239	1⅝	0.0886	2¼	0.0476
11⁄16	0.506	1⅛	0.191	1¾	0.0770	2⅜	0.0419
¾	0.428	1¼	0.152	1⅞	0.0675	2½	0.0380

rope diameter, in inches for U.S. customary units or millimeters for SI units, the following expressions for the spooling factor may be used in lieu of Table 1.1 values.

$$s = \frac{0.2618}{d^2} \text{ in U.S. customary units}$$

$$s = \frac{0.00285}{d^2} \text{ in SI units} \tag{1.2a}$$

The first case involves the maximum quantity of rope that can be stored on a drum when the hoist is not in operation. Dimension C is taken as zero and

$$E = \frac{A - D}{2} \tag{1.3}$$

which is then substituted into Eq. (1.2), giving the maximum stored length L.

The second case is used to determine the maximum quantity of rope that can be spooled onto the drum of an operating winch. C is taken as ½ in (12.7 mm) or preferably as one rope diameter, and

$$E = \frac{A - D - 2C}{2} \tag{1.4}$$

The third case is used to estimate the quantity of rope found on a drum. The dimension C is measured, and Eq. (1.4) is solved for E. When E is substituted into Eq. (1.2), the stored length is found. The above equations can be manipulated and used to solve any number of practical drum problems, as will be seen later in this chapter.

Line pull

Hoist drums are rated by *line pull* (the tension the drum is capable of applying to a rope leading onto it) and by line speed. Ratings are generally specified at the first layer but may be given for the top layer as well. As rope spools onto the drum, line pull decreases and line speed increases but available torque remains constant. If we use the dimensions of Fig. 1.5, with a rated line pull of P_r on the first layer, the rated torque T is given by

$$T = \frac{P_r(D + d)}{2}$$

for a system with nominal rope diameter d and with consistent dimensional units. The usable line pull P_u at any other drum layer is found from

$$P_u = \frac{2T}{A - 2C - d} = \frac{P_r(D + d)}{A - 2C - d} \tag{1.5}$$

If the rated first-layer line speed is given in feet per minute as V_r and drum dimensions in inches, the drum speed ω in revolutions per minute is given by

$$\omega = \frac{12V_r}{\pi(D + d)}$$

For SI units, the value 12 is replaced by the power of 10 needed for dimensional consistency. The line speed V_u at any other rope layer is

$$V_u = \frac{\omega\pi(A - 2C - d)}{12} = \frac{V_r(A - 2C - d)}{D + d} \qquad (1.6)$$

Sheaves and blocks

Sheaves are used to change the direction of travel of wire ropes. Assembled in multiples, in the form of blocks, they are able to provide almost any required mechanical advantage.

Ideally, sheaves should be mounted in exact alignment with each other, but since in practice this rarely occurs, the grooves are shaped to provide some tolerance for misalignment. In the discussion on fleet angle, it was noted that a 2° lead angle can be accommodated without difficulty, but an appreciable constant misalignment causes the rope to rub the sides of the groove, resulting in wear on the rope and sheave and shortening the useful lives of both.

Sheaves rotate about their mounting shafts on bushings or bearings. A reasonable value for friction loss at bushings can be taken as 4½%, while bearings produce losses of 1 to 2%, depending upon their quality. These losses are rough average figures for ropes making a bend of 180° over the sheave and can be reduced for smaller turning angles. Actual friction losses are a function of the style of rope, the ratio of sheave to rope diameter, and bearing type.

There is no minimum sheave or drum diameter that would prevent a hoisting mechanism from operating, but sheave and drum diameters have a direct bearing on rope life. Rope life decreases with decreasing sheave diameter, as indicated by Fig. 1.6. The minimum ratios of sheave or drum to rope diameter for cranes and derricks stipulated in American codes are rigidly fixed and do not vary with a rope-life parameter, as some European codes do. American practice requires that the winding-drum barrel diameter be no less than 18 rope diameters. While the ratio for the upper block is also 18 minimum, the lower-block ratio may not be less than 16. These ratios apply to the load-hoisting systems of construction cranes and derricks. Overhead and industrial crane practice is more conservative.

A sheave is described by the rope diameter for which it is grooved and by four other diameters: the outside diameter of its flanges; the diameter to the base of the groove, or *tread diameter;* its shaft diameter; and its *pitch diameter.* Ratios of sheave to rope diameter are given using the pitch diameter, which is the diameter to the center of the rope on the sheave, in other words, the tread plus rope diameter. These ratios are often referred to as D/d ratios.

The weight of a sheave increases with its size, as do the weight and size of its mountings. This dead-load increase reflects a direct reduction in lifting capacity; conversely, a larger crane structure will be

RELATIVE BENDING LIFE FACTORS

Rope Construction	Factor
6 x 7 or 7 x 7 Aircraft	.60
19 x 7 or 18 x 7 R.R.	.70
6 x 19 S	.80
6 x 19 W	
6 x 21 FW	
6 x 26 WS	
6 x 25B FS	.90
6 x 27H FS	
6 x 30G FS	
6 x 31V FS	
7 x 21F W	
6 x 25 FW	
6 x 31 WS	1.00
8 x 19S	
8 x 21 FW	1.10

Rope Construction	Factor
7 x 25 FW	
6 x 29 FW	
6 x 36 WS	
6 x 36 SFW	1.15
6 x 43 FWS	
7 x 31 WS	
8 x 25 FW	
6 x 41 WS	
6 x 41 SFW	1.25
6 x 49 SWS	
7 x 36 FW	
6 x 46 SFW	
6 x 46 WS	
8 x 36 WS	1.35
6 x 61 FWS	
6 x 57 SFWS	

*Note: This table, with some modifications, is based on outer wire diameter relationships.

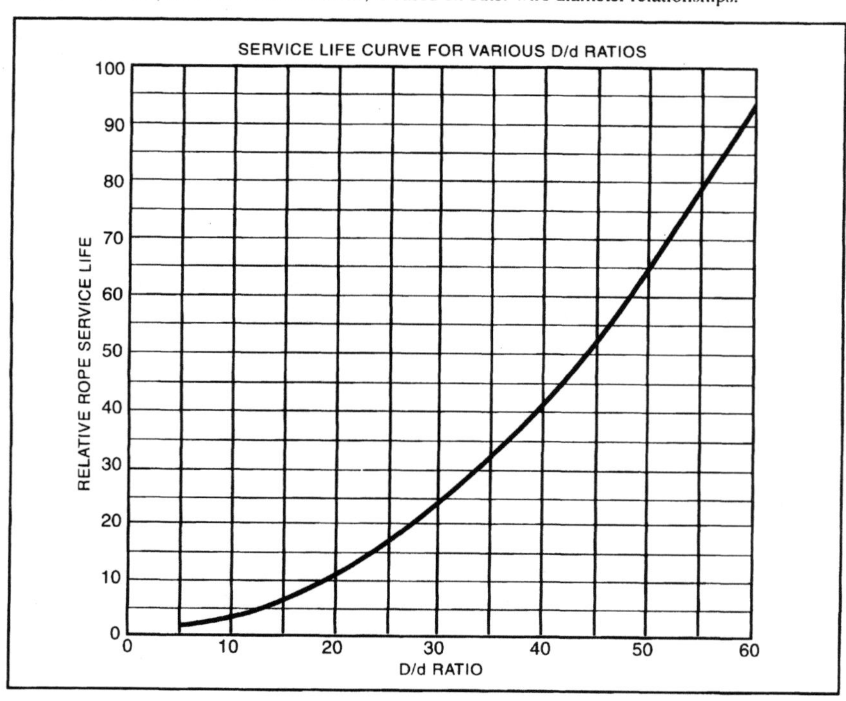

Figure 1.6 This generic curve relates relative rope service life to the ratio of sheave to rope diameter (D/d) considering only bending and tensile stresses during use. The scale is nominally the percentage of the life of a straight rope undergoing the same loading without passing over sheaves. The table of relative bending life factors permit adjustment of the curve values to different styles of rope. (*Wire Rope Technical Board Users Manual,* 3d ed., used with permission.)

needed to attain the previous capacity. For this reason, crane designers are motivated to use the smallest D/d ratios permitted by code unless particular economic considerations dictate otherwise. Assuming that adequate rope-inspection procedures will preclude actual rope breakage, only the effect on rope replacement costs need be compared with costs attributable to an increase in D/d ratio. Replacement costs could include loss of production for continuous-duty equipment, or high labor costs on machines where replacement poses difficulties.

Figure 1.6 shows that for the common range of D/d ratios, from 15 to 30, the plot is nearly linear. This gives rise to an approximate relationship between service life at any two ratios of

$$L_r = \frac{r_2 - 10}{r_1 - 10} \tag{1.7}$$

where r_1 and r_2 are the D/d ratios being compared and L_r is the ratio of the relative rope lives. As an example, if a machine with a sheave ratio of 18 has given satisfactory service and rope life in single-shift operation, what ratio would provide similar rope life (calendar) for two-shift operation? Solving Eq. (1.7) with $L_r = 2$ and $r_1 = 18$ gives a required ratio of 26. In a similar manner, three-shift operation would suggest a ratio of 34.

Every time a rope bends around a sheave, there is an episode of stress for the rope. When bending direction reverses in adjacent sheaves, the stress affects are disproportionally severe. The smaller the sheave diameter, the more severe is the stress effect. Since deterioration due to fatigue or abrasive wear determines rope life, they must be evaluated in any rational scheme of rope and sheave selection. Little is known at present of the relationships involved; rope and sheave selection is hardly a science.

Blocks are manufactured in several styles to satisfy the varying needs of hoisting service. In a general sense, they can be classified as being of the oval or diamond pattern, or of the *snatch-block* type, also known as a *gate block,* as illustrated in Fig. 1.7. Blocks can be provided with fixed or swiveling single or double hooks as well as with fixed or swiveling shackles or with bales. Snatch blocks have the advantage of permitting reeving when the end of the rope is not free.

Blocks or any other sheave-mounting arrangement should be provided with guards to prevent the rope from leaving the sheave groove. The simplest form of guard is a pin or bolt placed just clear of the edge of the sheave flange. Where appropriate, a pipe spacer on the bolt will keep the side plates at constant separation, allowing the bolt to be torqued to prevent unwanted loosening.

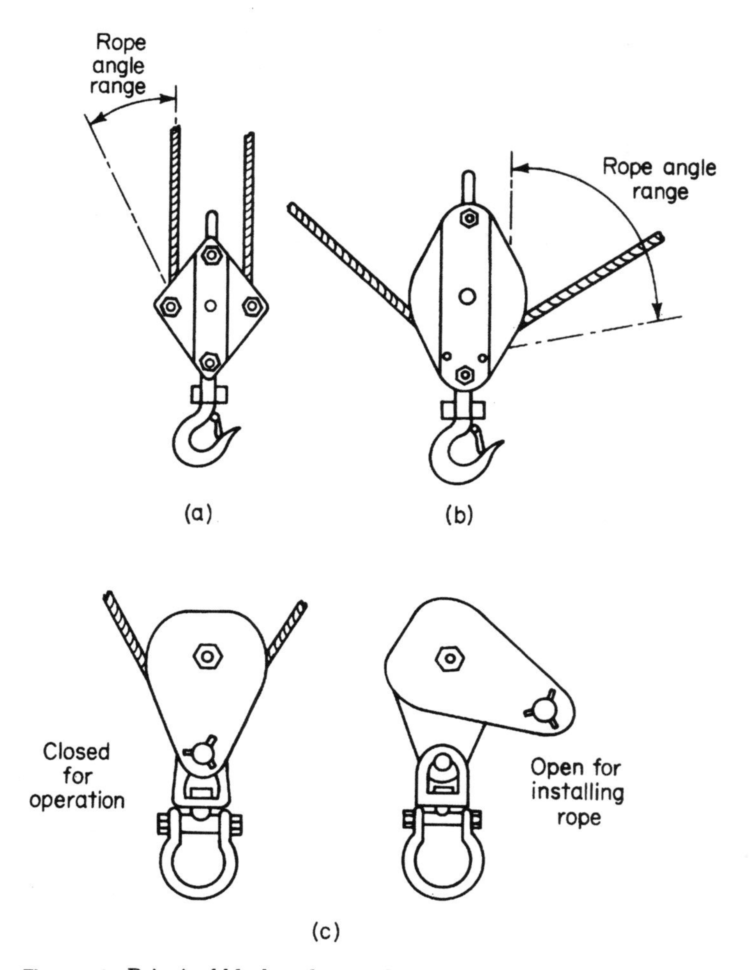

Figure 1.7 Principal block styles: (*a*) diamond pattern; (*b*) oval pattern; (*c*) snatch blocks. The rope angle ranges shown reflect the degree of confinement of the rope in the block and do not mean to imply that hoisting should be done under those conditions.

A diagram like Fig. 1.8 showing the arrangement of ropes in a hoisting system is called a *reeving diagram.* There are two ways to reeve a set of two blocks, so that parts of line are either equal to the number of sheaves or 1 greater. This depends on which block the fixed end, or *dead end,* of the rope is fastened to. For blocks with many sheaves, more complicated reeving arrangements are often used in an attempt to balance the friction loads. When more sheaves are present than are needed or used for a particular reeving, care should be exercised in

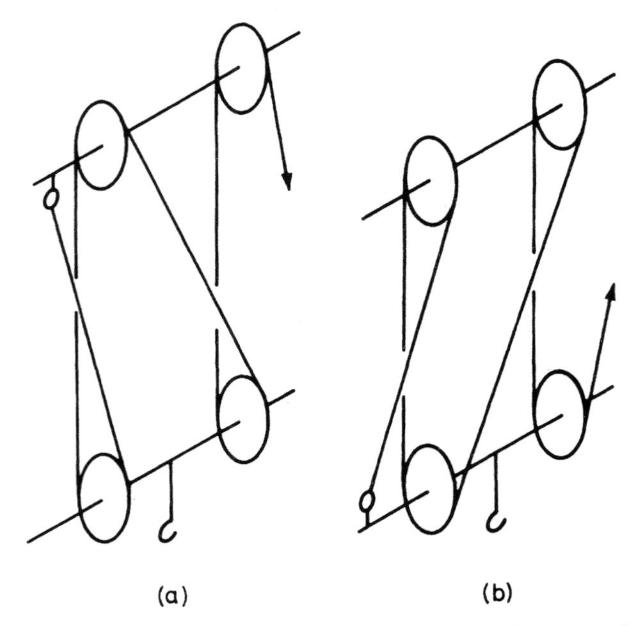

Figure 1.8 Reeving diagrams for (*a*) four and (*b*) five parts of line.

selecting the active sheaves so that the block will be balanced and the lines will be leading straight out of the grooves.

1.4 Wire Rope and Fittings

Without wire rope, there would be few cranes or derricks. Fiber rope is so limited in strength that it is capable of only marginal hoisting use, and chains are both awkward and heavy. The cold-drawn wire used for wire-rope construction has a tensile strength ranging from about 225,000 to 340,000 lb/in^2 (1550 to 2350 MPa), giving wire ropes outstanding strength-to-weight ratios and making possible the array of lifting equipment now available.

In constructing wire rope, individual wires are laid, not twisted, together into *strands* and the strands in turn are laid over a core to form the rope. The number of wires in a strand, the number of strands in a rope, and the nature of the core vary. Ropes are categorized by classes, such as 6 × 19, which give the number of strands to the rope and the nominal number of wires in the strand. The class designations have a basis in tradition rather than in absolute fact, and a 6 × 19 class rope may have anywhere from 15 to 26 wires in its strands. A particular 6 × 19 class rope is a 6 × 25 rope, which has 25 wires in each of its 6 strands. Figure 1.9 shows a few of the many styles of rope available.

Rope cores of several types are available: namely, fiber core (FC), wire strand core (WSC), or wire rope itself (independent wire-rope core, IWRC). The core acts to support the strands, holding them in position, and the wire cores add strength to the rope. Wire cores reduce flexibility, however, they generally increase resistance to crushing and bending fatigue.

Wire ropes are manufactured in several grades; improved plow steel (IPS), extra improved plow steel (EIPS or XIPS), and now extra extra improved plough steel (EEIPS or XXIPS). The older plow steel is rarely used today. Rope strengths are given in catalog listings for the size, construction, and material of the rope; they are listed in terms of breaking strength. The safety factors, or *design factors,* used with wire rope vary with rope application and will be covered later in this section.

For ordinary rope-design situations, only four rope properties are of importance:

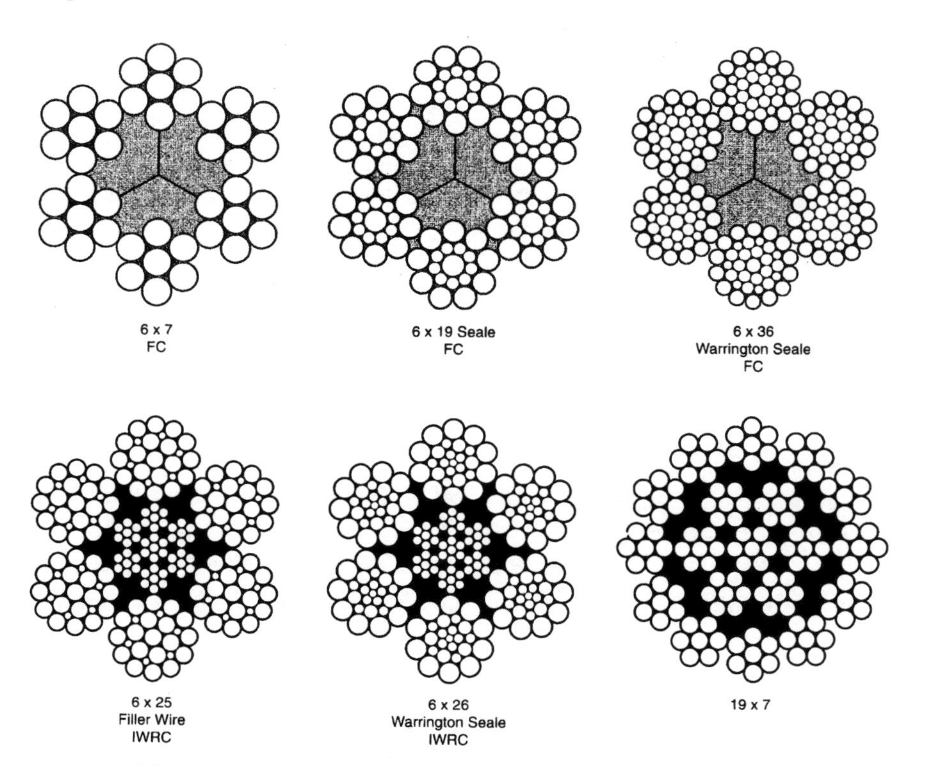

Figure 1.9 A few of the many wire rope constructions. (*Wire Rope Technical Board Users Manual,* 3d ed., used with permission.)

Strength. Controlled by size, grade, construction and core

Flexibility and fatigue resistance. Improved by strands with a large number of small wires and by preforming

Abrasion resistance. Enhanced by large outer wires or by Lang's lay construction

Crushing resistance. Best with IWRC or WSC and large outer wires

The *lay* of the rope refers to the direction of rotation of the wires and the strands. Regular-lay (right or left) ropes are made with the wires in the strands laid in one direction and the strands laid in the opposite direction, so that individual wires have the appearance of running generally parallel to the long axis of the rope. Regular-lay ropes are easy to handle, resist kinking and twisting, and are stable. In ropes with Lang's lay, wires and strands are laid in the same direction, and individual wires seem to run diagonal to the axis of the rope, offering greater surface exposure and hence greater abrasion resistance. They are more susceptible to untwisting and kinking than regular-lay ropes and have less crushing resistance. But ropes with Lang's lay are more flexible and fatigue-resistant.

When a rope is preformed, each strand is shaped to its ultimate helical form just before being laid into the rope, and the individual wires are also shaped by the process. This produces a stable rope which (1) will not unwind when a cut end is left free, (2) will return to its original form after unwinding under load, and (3) is less likely to kink or foul. Preformed rope also has improved fatigue resistance. Nearly all rope is now preformed.

The arrangement of the wires in a strand contributes to the properties of the rope. The common arrangements are:

Simple. All wires of the same size

Seale. Large wires on the outside for abrasion resistance and small wires inside for flexibility

Warrington. Alternate wires large and small to combine flexibility with abrasion resistance

Filler wire. Very small wires placed in the spaces between the wires in the inner and outer layers of wire for increased fatigue resistance

A complete specification for a particular wire rope construction could read 6 × 25 *filler wire, preformed, IPS IWRC, right-regular-lay rope,* but if not designated, right regular lay is assumed.

For derrick or general hoisting use, 6 × 19 class ropes seem to be most popular, as they offer a good balance between abrasion resistance and flexibility. Cranes employ 6 × 19 as well as 6 × 37, 8 × 19, and other rope styles to meet particular requirements.

Drums are called *overwound* or *underwound* depending on whether the rope spools onto the drum from the top or the bottom. When an *overwound* drum is seen from the rope side, if the rope attachment is at the right side, a right lay rope is required for proper spooling; left lay is needed when the attachment is at the left. The opposite is true for *underwound* drums seen from the rope side.

The nature of ropes, when loaded, is such that the strands seek to abandon their original helical shape for an elongated helix, and the rope attempts to unwind if the ends are unrestrained. The rope will stretch under initial loading until each wire is in line contact with adjacent wires. This elongation is called *constructional stretch*. Elastic strain will cause further elongation. If the rope ends are unrestrained, some unwinding, or spin, will occur, marked by end rotation and additional elongation.

The rope end can be restrained by clamping against rotation. To resist end rotation, the clamp must develop a torque reaction. If a load were to be freely supported by a single rope, the clamp would transmit the torque to the load and the load would spin unless the load itself were prevented from rotating. *Taglines,* ropes from the load to workers on the ground, are used to prevent load spin (taglines also orient and steady the load). A somewhat controversial alternative involves the use of swivels. Some authorities maintain that the unwinding of the rope that takes place when swivels are used leads to significant rope weakening and may hasten the onset of fatigue failure.

When an 8 × 19 rope is constructed with an IWRC so that the strands and IWRC are laid in opposite directions, the rope has spin-resistant characteristics at some levels of loading and reduced spin at others. Another construction, 19 × 7, has similar characteristics when it is made with the inner and outer strands laid in opposite directions. The tendency of one layer of strands to rotate one way is counteracted by the tendency of the other layer to rotate in the reverse direction, and a measure of torque balance is maintained. These ropes, and some other styles similar to them, are called *rotation-resistant* ropes.

A spinning concrete bucket, steel beam, or other load is a danger to the workers who must land it. The advantage in using rotation-resistant ropes, therefore, is that under many jobsite conditions the load will be stable and a tagline will not be necessary. The time that would otherwise be required to connect and remove taglines is instead applied to placing additional loads.

The disadvantages of rotation-resistant ropes are threefold: (1) They are more sensitive to damage from mishandling than standard ropes, (2) their torque balance generally occurs at low levels of loading, about one-eighth of breaking strength, and (3) at normal loading levels the ropes rotate somewhat, the amount of spin increasing with loading ratio.

As a rotation-resistant rope rotates, its core becomes overpowered by the outer strands and twists tighter; concurrently, the outer strands unwind. The tightening of the core tends to shorten the rope while the unwinding outer strands tend to lengthen it, causing the load on the rope to redistribute. The core becomes burdened with a disproportionate share of the load, which creates a tendency toward internal damage that is difficult to identify during ordinary visual inspections.

Special rotation-resistant ropes have been developed with proprietary designs that render them truly torque balanced. Under the full range of loading ratios experienced in practice, these ropes spin little or not at all. Without rotation, there is no load redistribution between the outer and inner portions of the rope; all wires are equally loaded. These ropes do not suffer the disadvantages of common rotation-resistant ropes.

Working loads

American practice has been simply to divide the rope-breaking strength by a safety factor, or *design factor,* and use the result as the working load. American National Standards Institute (ANSI) codes for cranes and derricks generally require a design factor of 3.5 for *running lines,* those which travel over drums or sheaves, 3.0 for *standing lines,* or guys, and 5.0 for rotation-resistant ropes, which must only be used as running lines. Derrick practice, as evidenced by some handbooks, has been to use a factor of 5 or more. High design factors have historical precedent and reflect a time when load weights were not easily determined, material properties were less closely controlled, and rigging design was done mostly by rule of thumb. Under present conditions, and with federal Occupational Safety and Health Act requirements mandating frequent rope inspection, a design factor of 5 is too high for ordinary derrick operations.

When applying the stipulated design factors, the practice is to use dead and live loads without augmenting them to include the effects of sheave friction or dynamic (impact) loading. Needless to say, an element of judgment should be applied here, as there are design situations where the combined friction and impact can be significant and demand consideration.

In European practice, rope selection is based not only on maximum loading but also on factors related to the intensity of use the rope will undergo and the number of sheaves the rope will have to pass over. Where the arrangement of the reeving is fixed and reasonable loading predictions can be made, this system is rational and can be very effective.

Slings are outside the scope of this book, but mention must be made of the necessity to use high design factors for them. Since slings are often used repeatedly under particularly severe conditions, provision must be made for their inevitable deterioration with time. Federal OSHA regulations specify sling capacities for various configurations and conditions of use.

Fittings

The fittings used to attach wire ropes to structural connections and to each other are of several kinds and serve different purposes. Of importance here are the general types of fittings and their structural efficiency.

Rope-end fittings provide the rope end with a loop, eye, pinhole, or pin for attachment. Permanent fittings, which require cutting the rope for removal, are the most efficient, usually matching the strength of the rope. The exception is hand-spliced loops which have about 80% of the rope strength in the commonly used rope sizes. The removable fitting types, wedge sockets, clip and thimble loops, and variations, also develop about 80% of rope strength. Figure 1.10 illustrates some common rope fittings. More detailed information can be found in any wire-rope handbook (see also Fig. 1.11).

Permanent rope fittings should be used wherever feasible because of their superior strength and reliability. There are, however, many purposes for which permanent fittings are impractical, the dead-end attachment of the hoist rope being an obvious example.

The general practice in the design of rope systems is to base the design on rope requirements, ignoring the efficiency of rope-end fittings. It is apparent that judgment should be used rather than blind adherence to this practice.

Shackles are used to connect a rope-end loop or eye at a structural pinhole or bale. The length of the shackle opening along its pin is usually quite large compared with the thickness of pin plate needed to develop shackle strength. This is necessary for flexible shackle use but often requires that the pin plate be built up in thickness to provide a stable mounting.

Turnbuckles are used on guy lines and offer a means for removing slack from the system. They are available with hook, loop, or jaw (shackle) end fittings.

1.5 The Basic Luffing Mechanism

The word *luffing,* like so many crane and derrick terms, has its origin in nautical usage from the days of sailing ships. Booms were rigged

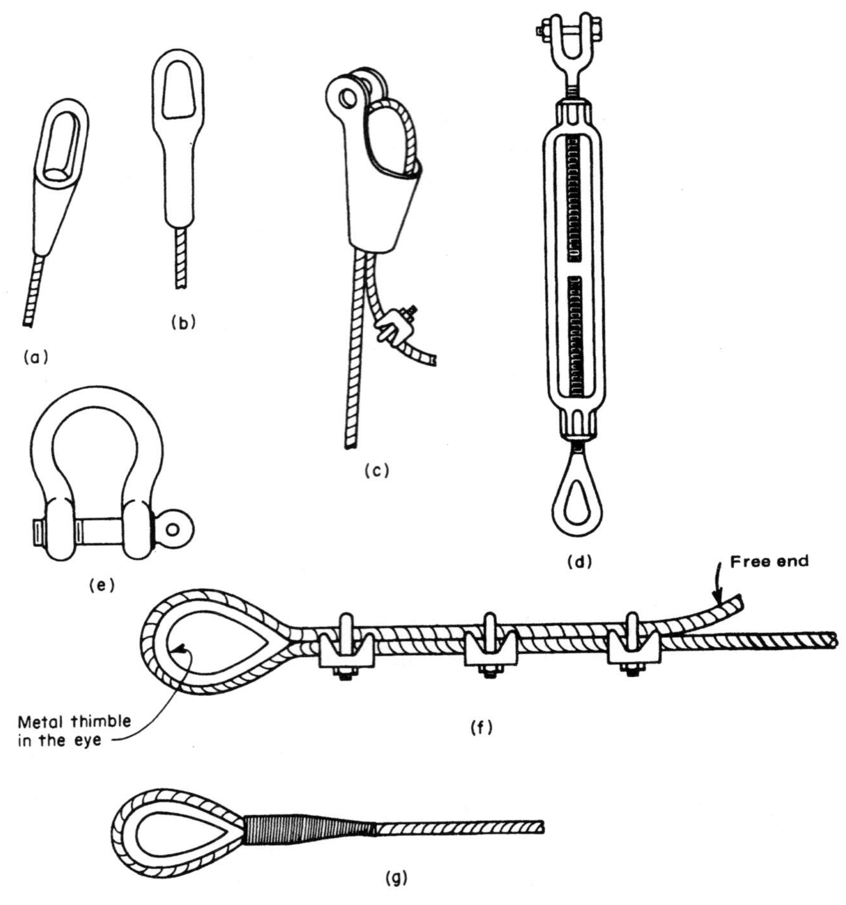

Figure 1.10 Some common wire-rope fittings: (*a*) spelter (zinced) socket (efficiency 100%); (*b*) swedged socket (efficiency 100%); (*c*) wedge socket (efficiency about 80%); (*d*) turnbuckle; (*e*) shackle; (*f*) Crosby-type clips and thimble (efficiency about 80%); (*g*) hand-spliced eye and thimble (efficiency about 80%).

to handle cargo as well as sails, and the technologies of hoisting and seamanship developed together.

As used in crane practice, *luffing* means changing the angle the main load-supporting member makes with the horizontal. Other names used for this same motion include *topping, derricking,* and *booming*. Henceforth, these terms will be used interchangeably.

Raising and lowering loads with the hoisting mechanism has been covered previously. Derricking also raises and lowers loads to a limited extent, but mainly this motion serves to move loads horizontally. Together with a *swing,* or horizontal rotating motion, luffing enables a

Figure 1.11 Riggers slinging a load in preparation for lifting. Note the overhaul ball and fittings. This photograph was taken before OSHA required the use of hard hats and other protection. (*Gendelman Rigging and Trucking, Inc.*)

hoisting device to service loads within a portion of a vertical cylindrical zone. The photographs in Figs. 1.14 and 1.15 show the two ends of a luffing system that one of the authors designed for a stiffleg derrick. Note the arrangement of pairs of blocks to allow for use of 13 parts of line.

The derricking lines always carry system dead load and suffer an increase in loading whenever a hook load is lifted. Nonetheless, the luffing-system ropes experience less severe service than do hoisting-system ropes as a rule. Luffing is a relatively slow operation. A well-planned installation will be arranged to keep luffing movements at a minimum for the sake of production efficiency. This implies that luffing ropes will undergo most of their loading sequences while static, a condition that ropes are far better able to endure than loaded passage over sheaves. For this reason, the ANSI codes specify minimum D/d ratios for luffing sheaves and drums of only 15.

Calculating the forces in the luffing lines is more involved than for the load hoist lines. Referring to Fig. 1.12, if W is taken as the weight of the load and lower load block, and if the strut is assumed to have negligible weight, the moment about the strut bottom pivot is

$$M = WL \cos \theta = WR$$

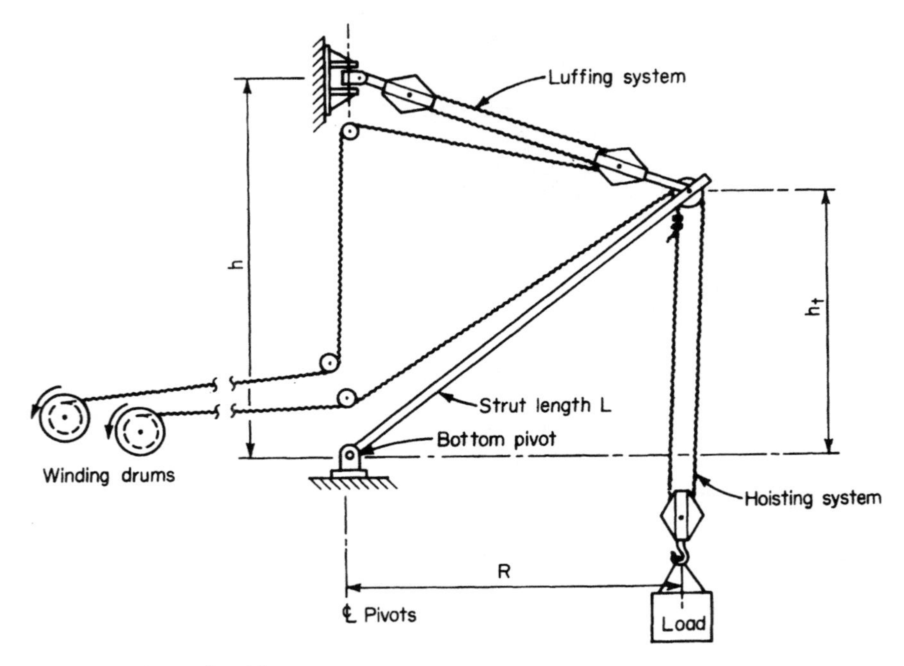

Figure 1.12 Basic derrick arrangement.

where θ is the angle the strut makes with the horizontal. In order to support that moment, the luffing lines must develop the horizontal reaction

$$T_h = \frac{M}{h} = \frac{WL \cos \theta}{h} = \frac{WR}{h}$$

so that the load in the luffing lines is given by

$$T = \frac{T_h}{R} [R^2 + (h_t - h)^2]^{1/2} = \frac{M}{Rh} [R^2 + (h_t - h)^2]^{1/2} \qquad (1.8)$$

As with the hoisting mechanism, when the topping lift is static, each part of line is equally loaded. When derricking starts, friction is introduced and the resulting losses can be taken as having the values given in Sec. 1.2.

An unloaded strut may not have enough weight to overcome resistance and derrick out when required. The overhaul weight, described in Sec. 1.2, must also serve to ensure that the system will be able to derrick out. The forces which resist derricking out and which must be overcome are the same as for the load hoist system but of course are now the analogous forces in the derricking system. They are friction at the sheaves and at the derricking drum and the weight of the vertical segment of the lead rope.

The overhaul weight calculated for the load hoist system will often be adequate for the derricking system as well; it is a question of system geometry. If the height to the derricking top pivot h is less than the strut length L, derricking should not be troublesome. Problems with derricking will decrease as h decreases. On the other hand, if h is greater than L, the potential for derricking problems increases. As with so many things in life, however, matters are not quite so simple. As h increases, the force in the derricking lines, from the load and from system dead weight, decreases. The designer must decide whether greater advantage lies in increasing h to obtain a lightly loaded derricking system (to the detriment of the hoisting system because more overhauling weight will be needed) or in the reverse choice.

When it is necessary to produce more accurate calculations, friction must be taken into account. If the line to the first luffing deflector sheave carries a force P_h, an equilibrium state must be satisfied such that

$$T = P_h + (1 - \mu)P_h + (1 - \mu)^2 P_h + \cdots + (1 - \mu)^{n-1} P_h$$

or conversely

$$P_h = \frac{T}{1 + (1 - \mu) + (1 - \mu)^2 + \cdots + (1 - \mu)^{n-1}}$$

where the angle the line to the deflector sheave makes with the derricking lines has been neglected, sheave friction loss is μ, and the number of parts of line supporting the strut is n. Adding the effect of the losses at the deflector sheaves, we find P, the force at the drum required to initiate luffing in Fig. 1.12, to be

$$P = \frac{P_h}{(1 - \mu)} = \frac{T}{(1 - \mu)^1 + (1 - \mu)^2 + \cdots + (1 - \mu)^n}$$

For luffing out, equilibrium requirements will show that

$$P = \frac{T}{(1 - \mu)^{-1} + (1 - \mu)^{-2} + \cdots + (1 - \mu)^{-n}}$$

The equations governing these operations can be generalized. If n is taken to include the line to the deflector sheave and m is the number of 180° turns over sheaves between the upper luffing block and the drum, then the luffing-line, or the boom-hoist-line, load at the drum is

$$P = \frac{T}{r} \tag{1.9}$$

where for luffing in

$$r = (1 - \mu)^m + (1 - \mu)^{m+1} + \cdots + (1 - \mu)^{m+n+1}$$

$$= \frac{(1 - \mu)^m - (1 - \mu)^{m+n}}{\mu}$$

and for luffing out

$$r = \frac{1}{(1 - \mu)^m} + \frac{1}{(1 - \mu)^{m+1}} + \cdots + \frac{1}{(1 - \mu)^{m+n-1}} = \frac{1 - (1 - \mu)^n}{\mu(1 - \mu)^{m+n-1}}$$

Although in the derivation for Eq. (1.8) the weight of the strut was ignored for simplification, in practice strut weight is significant and must be included when determining the moment M.

Example 1.2 Assume that the load-hoist-system design required a block plus overhaul weight of 600 lb (272.2 kg). With the following data:

Strut	*Luffing system*
Weight 240 lb (108.9 kg)	Distance h = 12 ft (3.66 m)
Length 20 ft (6.1 m)	Minimum R = 3.5 ft (1.07 m) (h_t is therefore 19.7 ft = 6.0 m)
CG located 11 ft (3.35 m) from bottom pivot, or at $0.55L$	n = 5 parts of line
	m = 2
	μ = 0.045 and $1 - \mu$ = 0.955
	Drum friction 50 lbs (222.4 N)
	Weight of vertical part of boom hoist line = 8 lb (3.63 kg)

(*a*) What is the static load in the luffing system? Will the strut be able to derrick out with no load on the hook?

Taking moments about the strut bottom pivot at R = 3.5 ft (1.07 m), we have

$$M = 600(3.5) + 240(0.55)(3.5) = 2562 \text{ lb} \cdot \text{ft } (3.47 \text{ kN} \cdot \text{m})$$

The load in the luffing system is found using Eq. (1.8)

$$T = \frac{2562}{3.5(12)} [3.5^2 + (19.7 - 12)^2]^{1/2} = 516 \text{ lb } (2.30 \text{ kN})$$

and from Eq. (1.9) for luffing out

$$r = \frac{1 - 0.955^5}{0.045(0.955)^6} = 6.024$$

$$P = \frac{516}{6.024} = 86 \text{ lb (381 N)}$$

but this does not reflect the weight of the vertical lead rope, so that the net force at the drum is

$$P_{net} = 86 - \frac{8}{0.955} = 78 \text{ lb (345 N)}$$

which is greater than the friction at the drum; therefore, the strut will derrick out.

(*b*) With the h distance changed to 25 ft (7.62 m) will the strut still be able to derrick out?

The moment remains the same, but

$$T = \frac{2562}{3.5(25)} [3.5^2 + (19.7 - 25)^2]^{1/2} = 186 \text{ lb (827 N)}$$

so that

$$P = \frac{186}{6.024} = 31 \text{ lb (138 N)}$$

and

$$P_{net} = 31 - \frac{8}{0.955} = 22.6 \text{ lb (101 N)}$$

which is less than the 50 lb (222.4 N) of resistance at the drum. The system will not derrick out; more overhaul weight is needed, demonstrating the effect of varying h. The required weight can be found by working backward through the same equations:

$$\text{Required } T = \left(50 + \frac{8}{0.955}\right) 6.024 = 351.7 \text{ lb (1.56 kN)}$$

$$\text{Required } M = 351.7(3.5) \frac{25}{6.35} = 4846 \text{ lb} \cdot \text{ft (6.57 kN} \cdot \text{m)}$$

$$\text{Required } W = \frac{4846 - 240(0.55)(3.5)}{3.5} = 1253 \text{ lb (568 kg)}$$

An overhaul weight in excess of 1253 lb (568 kg) is needed. Had the 8-lb (3.63-kg) lead rope been neglected, calculations would indicate an overhaul weight of 1054 lb (478 kg). This illustrates the extreme sensitivity of the overhaul problem and induces the suggestion that an overhaul weight somewhat in excess of the calculated value be used in practice. Of course, once derricking is started, the R distance, and hence the moment, increases and reduces overhauling-weight needs. But if insufficient weight is provided, the system will not start to boom out at all.

Example 1.3 Given the same data as in the previous example with h = 25 ft (7.62 m), if the strut is required to carry a load of 22,000 lb (9979 kg) at the hook in addition to an overhauling weight (including the block) of 1500 lb (680 kg), what will be the load in the derrecking system and how much line pull must the derricking winch provide for booming in? Take 20 ft (6.10 m) as the R distance.

The moment about the strut bottom pivot becomes

$$M = (22{,}000 + 1500)(20) + 240(0.55)(20)$$

$$= 472{,}640 \text{ lb} \cdot \text{ft } (640.8 \text{ kN} \cdot \text{m})$$

$$h_t = 0$$

and the load in the luffing system is then

$$T = \frac{472{,}640}{20(25)}[20^2 + (0 - 25)^2]^{1/2} = 30{,}265 \text{ lb } (134.6 \text{ kN}) \qquad \textit{Ans.}$$

The line pull needed at the drum is found from Eq. (1.9)

$$r = \frac{0.955^2 - 0.955^7}{0.045} = 4.168$$

$$P = \frac{30{,}265}{4.168} = 7260 \text{ lb } (32.30 \text{ kN}) \qquad \textit{Ans.}$$

The rope used in the luffing system must have a minimum strength margin of 3.5, as the ropes are running lines. The minimum permitted rope-breaking strength is therefore

$$\text{BS}_{\text{min}} = \frac{30{,}265}{5}\left(\frac{3.5}{2000}\right) = 10.6 \text{ tons } (94.23 \text{ kN})$$

which can be satisfied by a ½-in-diameter (12.7-mm) 6 × 19 class rope of IPS or better, but the winch must be capable of delivering a line pull of 7260 lb (32.30 kN) with whatever quantity of rope will be on the drum at R = 20 ft (6.10 m). Will line pull be adequate for the following data?

From Fig. 1.5

A = 27 in (686 mm) B = 23 in (584 mm) D = 11 in (279 mm)

First-layer line pull = 9000 lb (40.0 kN)

Spooling factor for ½-in rope s = 0.925 (SI: $s = 1.77 \times 10^{-5}$)

Initially 2000 ft (610 m) of rope was stored on the drum, and luffing-system dimensions are given in Fig. 1.13.

Assuming that the sheaves are of minimum D/d, that is, 15, the pitch diameter will be 15(½) = 7½ in (190 mm). With the strut at R = 3.5 ft (1.07 m) the total length of rope spooled out is

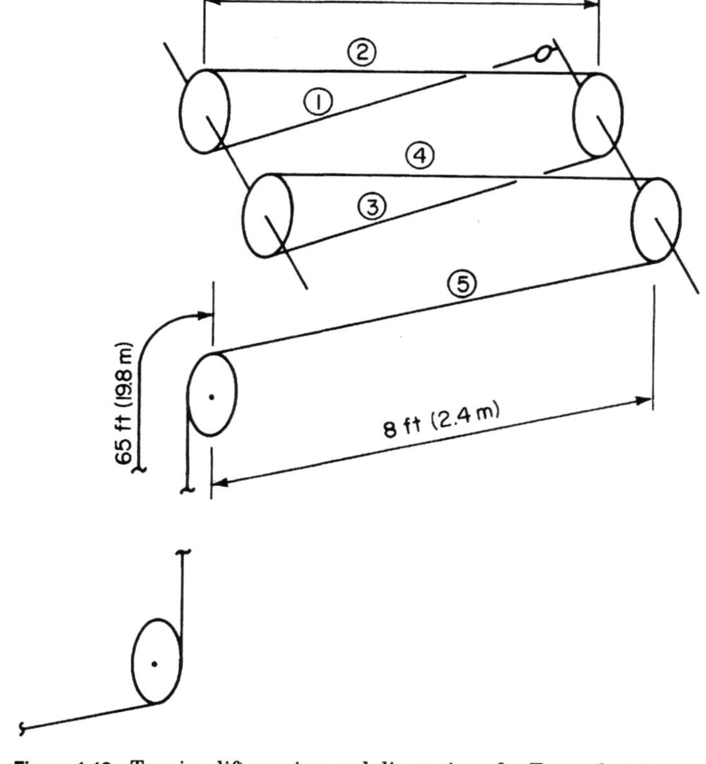

Figure 1.13 Topping-lift reeving and dimensions for Example 1.3.

$$L_r = 65 + 8 + 4(6) + 4(\tfrac{1}{2})(\pi)\,\frac{7.5}{12} = 101 \text{ ft } (30.8 \text{ m})$$

The luffing-system lengths at $R = 3.5$ ft (1.07 m) and $R = 20$ ft (6.10 m) are found from the geometry to be

$$L_{3.5} = [3.5^2 + (25 - 19.7)^2]^{1/2} = 6.35 \text{ ft } (1.94 \text{ m})$$

$$L_{20} = [20^2 + (25 - 0)^2]^{1/2} = 32.02 \text{ ft } (9.76 \text{ m})$$

When the strut is derricked out from $R = 3.5$ to $R = 20$, each part of line must increase in length by $32.02 - 6.35$, so that the total quantity of rope spooled out at $R = 20$ is

$$L_r = 101 + 5(32.02 - 6.35) = 229 \text{ ft } (69.80 \text{ m})$$

With 2000 ft (610 m) of rope on the drum initially and 229 ft (69.80 m) spooled out, Eq. (1.2) can be used to determine the radius at which the lead line will act at the drum so that usable line pull can be checked

$$2000 - 229 = (11 + E)E(23)(0.925)$$

$$E^2 + 11E - 83.24 = 0 \qquad E = 5.15 \text{ in (131 mm)}$$

In Eq. (1.5), for usable line pull, the C dimension is required

$$C = \frac{27 - 11 - 2(5.15)}{2} = 2.85 \text{ in (72.4 mm)}$$

$$P_u = \frac{9000(11 + 0.5)}{27 - 2(2.85) - 0.5}$$

$$= 4976 \text{ lb (22.13 kN)}$$

but 4976 < 7260 lb (22.13 < 32.30 kN); therefore the winch will not be capable of delivering the needed line pull unless excess rope is unloaded. Check again with 500 ft (152 m) of rope initially on the drum:

$$500 - 229 = (11 + E)(E)(23)(0.925) \qquad E = 1.06 \text{ in (26.9 mm)}$$

$$C = 6.94 \text{ in (176.3 mm)}$$

$$P_u = \frac{9000(11.5)}{27 - 2(6.94) - 0.5} = 8201 \text{ lb (36.48 kN)}$$

Since 8201 > 7260(36.48) > 32.30), line pull is sufficient.

When stored rope is not involved, the usual way of solving insufficient-line-pull problems without substituting a more powerful winch simply requires that the number of parts of line be increased or the h dimension be changed. Experimentation by varying the h dimension will show that for any R, increasing the h will cause the load in the luffing system to decrease. Further experimentation will confirm that overhauling problems increase with increasing h. Figures 1.14 and 1.15 illustrate a derrick luffing system with 13 parts of line.

1.6 The Basic Derrick

The derrick illustrated in Fig. 1.16 is of the most basic form, comprising a hoisting system, a luffing system, and a strut, or *boom*. It is called a *Chicago boom* and is used by rigging contractors to hoist machinery and equipment onto high-rise buildings, by stone contractors to hoist and place exterior stone or precast facing panels, and by other trades as well. Capacities range from less than 1 ton (900 kg) to at least as high as the 35-ton (32-t) model designed by one of the authors. Chicago booms are usually installed by mounting them to building

Figure 1.14 The mast top end of a stiffleg derrick luffing system. The two top lines are the load-hoist and derricking lead lines; they run down the center of the mast over deflector sheaves to the winch. (*Photo by Lawrence K. Shapiro.*)

Figure 1.15 The boom tip end of the luffing system of Fig. 1.14. Note the load lead line coming up at the right. It passes over a deflector sheave and through the boom tip. It then passes over the vertical sheave mounted on the derricking block and back to the mast. This arrangement was designed by one of the authors. (*Photo by Lawrence K. Shapiro.*)

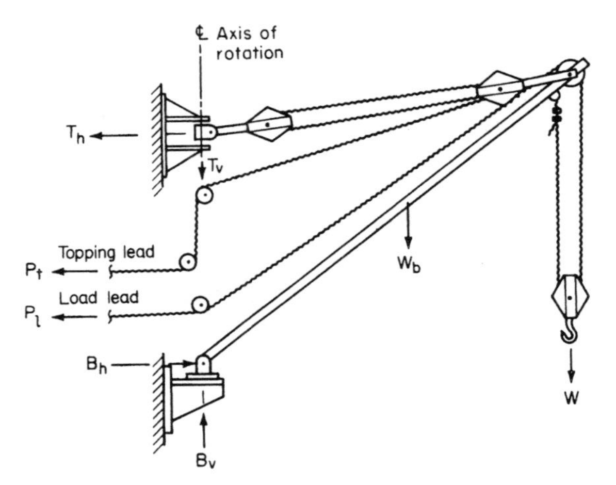

Figure 1.16 A Chicago boom derrick.

columns as in Fig. 1.17. The same arrangement of derrick becomes a guy derrick when it is mounted to a mast and a stiffleg derrick when it is fixed to a frame. The various forms of derricks and how they relate to the Chicago boom will be treated more fully in Chap. 2.

Figure 1.17 Chicago boom mounted at the fiftieth floor during construction of the McGraw-Hill Building in New York City. What appears to be an upper topping lift is a side (swing) guy. The load lead line is reeved over a sheave in the upper block and then runs under the boom. The topping lead comes off a sheave at the boom tip to a deflector sheave at the building face. Derrick and installation were designed by one of the authors.

Lateral motions

In derrick practice, the luffing system is generally called a *topping lift,* and, as shown in Fig. 1.16, the topping lift and the bottom of the strut are attached to pivots vertically in line with each other. These pivots permit the derrick to rotate, or *swing,* so that the hook traverses a path along the arc of a horizontal circle. The pivots define a vertical line called the *axis of rotation.* The horizontal distance from the axis of rotation to a plumb line through the CG of the load (hook) is called the *operating radius* (previously referred to as the R distance). The topping lift is used to change the operating radius.

The derrick configuration must be stable in both the vertical and horizontal planes. Vertical stability is ensured by the triangular arrangement of the boom, the topping lift, and the structure on which the derrick is mounted. Lateral stability is provided by guy lines attached to each side of the boom (strut) tip and running back to the structure.

The same lines that provide lateral stability are used to swing the boom. This is done by *taking up* on one guy line while *paying out* the other, a procedure usually done manually but sometimes by winches or hydraulic or electric devices.

Recognizing critical conditions

In Secs. 1.2 and 1.5, methods were developed for calculating the loads in the hoisting and luffing systems. The loads in the boom and on the supporting structure follow from them by simple statics.

The designer needs to be concerned only with those operating situations that create maximum loading conditions in the derrick and on the support structure. In the derrick of Fig. 1.16, maximum reactions for a constant hook load are easily discernable:

1. B_v is maximum at minimum radius.
2. B_h is maximum at maximum radius.
3. Topping-lift load decreases as radius decreases.
4. T_h increases as radius increases.
5. T_v is maximum downward at minimum radius and maximum upward at maximum radius.
6. P_t decreases as radius decreases.
7. P_l remains constant. Depending on geometry, this line exerts a small moment about the boom foot which can often be neglected.

If the maximum reactions are to be calculated by hand, the number of load cases can be whittled down by a rational examination. With the availability of spread sheet programs, however, it is as easy to evaluate every reaction for every case. Of course, the spread sheet should be spot checked by hand calculation.

Table 1.2 shows the reactions in the derrick of Fig. 1.18 for a constant hook load. For all parameters except axial load, the minimum and maximum radii yield the extreme values.

Resolving forces

Forces in the derrick and in the supporting structure are resolved by simple statics. The three conditions of static equilibrium

$$\Sigma H = 0 \qquad \Sigma V = 0 \qquad \Sigma M = 0$$

must be satisfied. A quick method, suitable for rough evaluation but not for design, ignores the angles made by the lead lines. In this method, using the information in Fig. 1.18 and taking moments about the boom foot, we have

$$\Sigma M = WR + W_b(8.25) - T_h h = 0$$

$$(22{,}000 + 1500)15 + 240(8.25) - T_h(10) = 0$$

from which $T_h = 35{,}450$ lb (157.7 kN) and

$$T_v = \frac{T_h(h_t - h)}{R} = \frac{35{,}450(13.23 - 10)}{15} = 7630 \text{ lb } (33.9 \text{ kN})$$

Continuing with the static equilibrium requirements, we get

TABLE 1.2 Chicago Boom Reactions for the Dimensions and Loads Given in Fig. 1.18

Operating radius, ft†	h_t, ft	T_h, lb	T_v, lb	Topping load, lb	P_t, lb	B_h, lb	B_v, lb	Boom axial load, lb
2.5	19.84	4,975	19,575	20,200	5,050	6,765	54,080	54,510
5.0	19.36	9,915	18,560	21,050	5,260	13,535	52,820	54,520
7.5	18.54	14,800	16,850	22,430	5,600	20,320	50,615	54,540
10.0	17.32	19.615	14,360	24,300	6,080	27,115	47,320	54,540
12.5	15.61	24,370	10,940	26,710	6,680	33,920	42,690	54,520
15.0	13.23	29,075	6,260	29,740	7,435	40,710	36,210	54,480
17.5	9.68	33,735	−620	33,740	8,435	47,485	26,540	54,400
20.0	0	38,290	−19,150	42,810	10,700	54,150	0	54,150

† 1 ft = 0.30480 m; 1 lb = 4.4482 N.

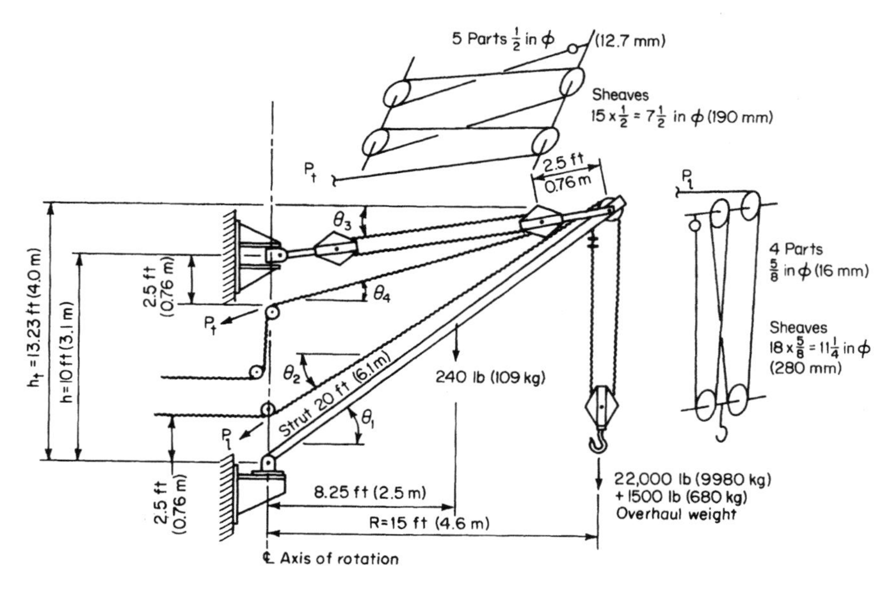

Figure 1.18 Derrick of Fig. 1.16 dimensioned for sample calculations.

$$\Sigma H = T_h - B_h = 0 \qquad T_h = B_h = 35{,}450 \text{ lb } (157.7 \text{ kN})$$

$$\Sigma V = W + T_v - B_v = 0 \qquad B_v = (22{,}000 + 1500) + 7630$$

$$= 31{,}130 \text{ lb } (138.5 \text{ kN})$$

The topping-lift load is given by

$$T = \sqrt{T_h^2 + T_v^2} = \sqrt{35{,}450^2 + 7630^2} = 36{,}260 \text{ lb } (161.3 \text{ kN})$$

But four parts are in the topping lift and one is the lead line, so that the lead-line force is

$$P_t = \frac{T}{n} = \frac{36{,}260}{5} = 7250 \text{ lb } (32.3 \text{ kN})$$

and the topping lift is left with 36,260 − 7250 = 29,010 lb (129.0 kN). The load lead-line force is found by dividing the lifted load by four parts of line, giving (22,000 + 1500)/4 = 5875 lb (26.1 kN). If the angle the lead makes with the boom is ignored, the lead-line force will act as an axial loading on the boom. The boom load will then be

$$B_a = P_l + \sqrt{B_h^2 + B_v^2} = 5875 + \sqrt{35{,}450^2 + 31{,}130^2}$$

$$= 5870 + 47{,}180 = 53{,}050 \text{ lb } (236.0 \text{ kN})$$

Note that all horizontal reactions are in the vertical plane defined by the boom and the topping lift. As the boom swings, the horizontal reactions rotate as well (Figs. 1.17 and 1.19). Structural members supporting these reactions may then be exposed to axial forces, biaxial bending, and torsion. Ways of dealing with these reactions are covered in Chap. 7.

For more accurate calculations, it is necessary to evaluate the angles made by the derrick members, as indicated in Fig. 1.18. The first three angles are readily found from the given dimensions

$$\cos\theta_1 = \frac{R}{L} = \frac{15}{20} \qquad \theta_1 = 41.41^\circ$$

$$\tan\theta_2 = \frac{h_t + 11.25/(2 \times 12) - 2.5}{R} = \frac{11.20}{15} \qquad \theta_2 = 36.74^\circ$$

$$\tan\theta_3 = \frac{h_t - h}{R} = \frac{13.23 - 10}{15} \qquad \theta_3 = 12.15^\circ$$

Figure 1.19 A 125-ft (38-m) Chicago boom mounted on a steel stack at a Consolidated Edison Company plant in New York City during alterations. Mounting provisions and installation were designed by one of the authors. (*Photo by Lawrence K. Shapiro.*)

θ_4 is somewhat more complex but can be expressed with satisfactory accuracy as

$$\tan \theta_4 = \frac{13.23 - 10 + 2.5 - 2.5 \sin \theta_3 - 7.5/(2 \times 12)}{15 - 2.5 \cos \theta_3}$$

$$= \frac{4.89}{12.56}$$

$$\theta_4 = 21.28°$$

Taking moments about the boom pivot under static conditions and noting that four parts of the topping lift go to the anchorage while the fifth is the lead gives

$$(22{,}000 + 1500)15 + 240(8.25) - (1/4)(22{,}000 + 1500)(2.5 \cos 36.74)$$
$$- (10 - 2.5)P_t \cos 21.28 - 4(10P_t) \cos 12.15 = 0$$

from which $P_t = 7435$ lb (33.1 kN) and $T_h = 4P_t \cos 12.15 = 29{,}075$ lb (129.3 kN). From the horizontal equilibrium condition

$$\Sigma H = B_h - 29{,}075 - 7435 \cos 21.28 - 5875 \cos 36.74 = 0$$

$$B_h = 40{,}710 \text{ lb } (181.1 \text{ kN})$$

The vertical requirement then gives

$$\Sigma V = B_v - 29{,}075 \tan 12.15 - 22{,}000 - 1500 - 240$$
$$-7435 \sin 21.28 - 5875 \sin 36.74 = 0$$

$$B_v = 36{,}210 \text{ lb } (161.1 \text{ kN})$$

and the boom axial load is

$$B_a = (40{,}710^2 + 36{,}210^2)^{1/2} = 54{,}480 \text{ lb } (242.3 \text{ kN})$$

In like manner, the topping lift load is

$$T = (29{,}075^2 + 6260^2)^{1/2} = 29{,}740 \text{ lb } (132.3 \text{ kN})$$

The above loadings are about $2\frac{1}{2}\%$ greater than those found using the approximate method.

Table 1.2 gives all member loads and reactions at several radii with the hook load remaining constant.

Using the same format, the reactions induced during motion of the luffing or the hoisting systems can be determined by inserting friction losses where appropriate.

1.7 Basis for Load Ratings

In crane and derrick practice, a *load rating* is the maximum allowable load for a specific radius with the crane or derrick in a particular configuration while operating under defined conditions. Although somewhat awkward, that statement sets forth meaningful and necessary terms that limit and bound the concept of load rating, a concept that may seem to be rather simple at first glance. If we consider the elements of the definition one by one, we will get a clearer understanding of the concept.

Starting with the last element, defined conditions of operation, it is obvious that if the device had been designed for a particular loading in conjunction with wind at 25 mi/h (40 km/h), operation in winds of 45 mi/h (72.4 km/h) would require a reevaluation of the load-lifting capacity. Wind *forces* have increased to about three times the design value. In like manner, speed of operation, levelness or plumbness of mounting, improper reeving, and other factors can create loadings or variations in loadings not contemplated in the design.

In Example 1.3 the load, operating radius, and boom length are the same as for the data in Table 1.2 for the last radius. Note, however, that the topping-lift load in one case is 30,265 lb (134.6 kN) including the lead line while in the other it is 42,810 lb (190.4 kN) without the lead line. The difference, aside from minor effects of accuracy of method, is brought about solely by the difference in topping-lift heights h, which are 25 and 10 ft (7.62 and 3.05 m), respectively. The boom axial loads vary as well and are 24,560 and 54,150 lb (109.2 and 240.9 kN) for the two cases. Thus, a change in derrick configuration has imposed a completely different set of member loads on the system.

The effects of change in radius are obvious from the data listed in Table 1.2.

Consequently, each element in that initial statement on load rating can be seen as conveying the concept that load ratings are valid only within the bounds of defined conditions. Although the term "load rating" almost invariably appears by itself without the essential modifiers, one should never fail to understand that specific conditions are implied whenever the term is used.

Limitations

The numerical examples given in earlier sections indicate that the load-lifting capacity of a crane or derrick can be limited by

1. Topping-lift or load-hoist rope strength
2. Available line pull at the winch

3. Structural strength of the boom
4. Structural strength of the mountings

In cranes and other derrick forms there are additional members whose strength can also limit capacity, and deflection is sometimes a limiting criteria too.

Cranes and derricks are subjected to repeated loadings and unloadings in the course of their service lives. When the combination of the number of loading cycles and the range of stress variation causes accumulative damage, there is a limit on probable life, after which fatigue failure is likely. Most lifting devices do not operate with set loads and motions but are exposed to variable working conditions. The loads and the stress variation they produce are random, and to complicate the matter further, operating radii and even machine configurations can be random variables as well. To evaluate the equipment in order to prepare a design capable of a specified service life requires the application of probabilistic concepts in place of the ordinary deterministic methods of structural analysis. Much work has been done toward applying those mathematical concepts to design problems, but much

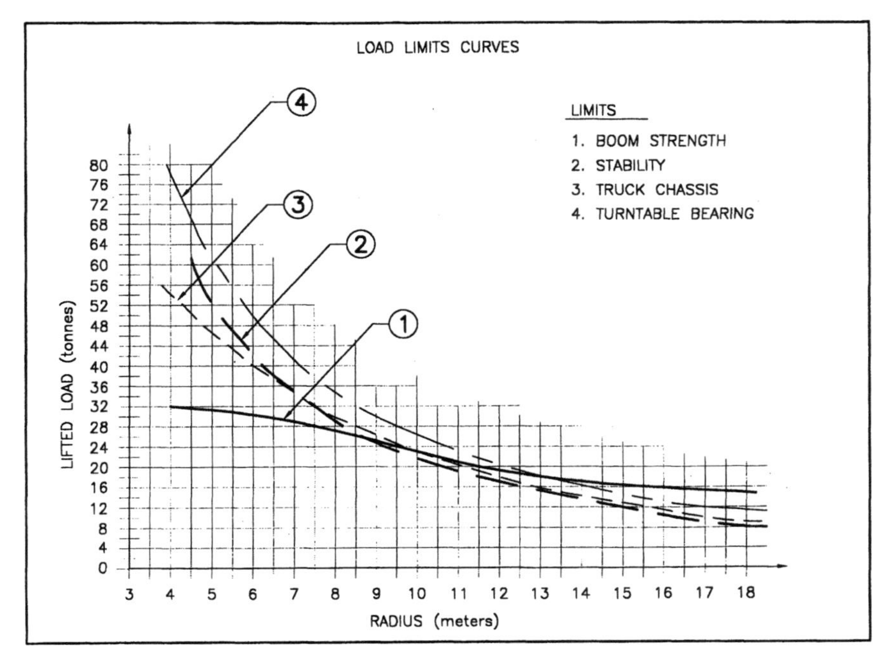

Figure 1.20 Curves of lifting capacities as limited by boom strength, crane stability, truck chassis strength, and turntable bearing strength. For each radius, the crane rated load will be the lowest value from these and other limiting factors.

practical work remains to be done before usable design-related methods will be ready for general application. Fortunately, however, most construction cranes and derricks have few, if any, components falling in the fatigue design range.

A most important factor controlling load ratings for several forms of equipment is stability against overturning. Most ratings of mobile cranes, for example, are governed by stability. Rail-mounted portal and tower cranes are often similarly limited. For still other equipment, load ratings are based on the assumption that foundations or bases will be constructed that will provide security against overturning. Fig. 1.20 presents four curves of load limits that are considered in establishing load ratings for a truck crane.

Stability against overturning

The crane shown in Fig. 1.21 can be made to overturn if a large enough load is boomed out to a great enough radius. Tipping will take place about a definite line called the *tipping fulcrum*. That line can be identified for the various forms of equipment and will be in the more complete stability studies in Chap. 4. At this point, it is sufficient to note that such a line exists and is definable. The CGs of the crane

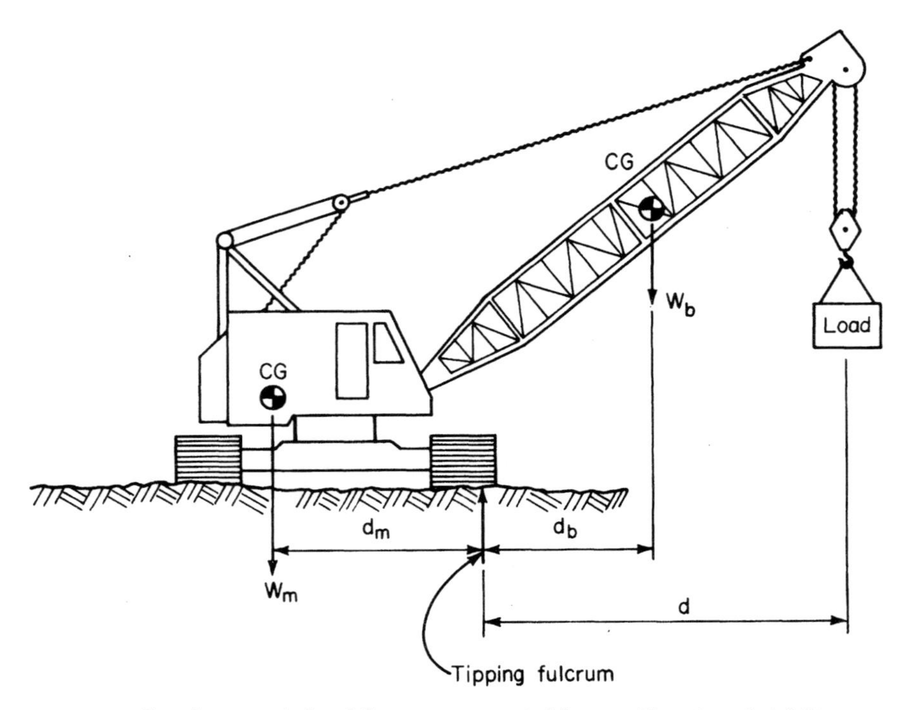

Figure 1.21 Crawler-mounted mobile crane annotated for consideration of stability.

components are well defined. Given these data, the moment of the machine weight about the tipping fulcrum can be expressed as $W_m d_m$; it is called the *machine resisting moment,* the *stabilizing moment,* or simply the *machine moment.* The moment of the load and the boom, given by $W_b d_b + Wd$, is called the *overturning moment.* If, with any one load, the radius is increased to the point where the overturning moment and the machine moment are equal, the point of incipient overturning has been reached. Any increase in the load or radius will cause the crane to fall over (Fig. 1.22). The load that produces incipient overturning is the *tipping load* at that radius.

In the United States, when stability governs, load ratings are limited to a fixed percentage of the tipping load. The percentage used for each type of equipment has been set as the result of cumulative experience over the years rather than from some logical mathematical determination. With the recent acceptance of very high-strength steels and the increasing needs at construction sites, cranes have experienced a remarkable growth in capacity and in the lengths of boom accommodated. With this growth have come situations where the old stability margins are not providing adequate reliability.

Machine moment remains constant if the machine does not swing, but as radius increases, the boom moment absorbs an increasing proportion of available moment and the tipping load decreases rapidly. As maximum radius is approached, the rated load can become rather small and the overturning moment can become 95% or more of the resisting moment. At such high percentages, cranes are quite sensitive, and a small perturbation can induce overturning. A crane weighing 400,000 lb (181 t) with ratings at 85% of the tipping load and a long-radius rated load of 3000 lb (1360 kg) will tip with the addition of only 530 lb (240 kg), or 0.13% of machine weight. That small increment of load can easily be induced by wind or by dynamic effects. An alternate approach toward stability-based ratings that makes use of a variable percentage of the tipping load has been adopted in some countries. This method is described fully in Chap. 4.

On the other hand, as radius decreases, the tipping load rises sharply, so that at radii close to the tipping fulcrum it can become so high that strength limits govern.

Figure 1.22 shows a 300-ton-capacity (272-t) truck crane during stability research experiments conducted some years ago by the Harnischfeger Corporation at a test facility. That 360,000-lb (163.3-t) machine is in a state of balance about its tipping fulcrum, the entire weight of the machine and load being supported on the two outrigger beams and pads.

Since floating cranes do not have a firm base for support, they must list under load (Fig. 1.23). List angles are therefore also limiting factors for load ratings. Stability of barge mounted cranes is also covered in Chap. 4.

Figure 1.22 P&H model 6250-TC truck crane of 300 tons (272 t) capacity in a state of balance during stability experiments. (*Harnischfeger Corporation.*)

Figure 1.23 Barge-mounted crane listing under the weight of a 75-ton (68-t) bridge section during unloading operations. Lifting accessories were designed and operations planned by one of the authors. (*Photo by Lawrence K. Shapiro.*)

Chapter

2

Crane and Derrick Configurations

In Chap. 1 the basic concepts of crane and derrick mechanisms were introduced; this chapter will make use of those concepts together with additional materials to describe various machines. Equipment form, operating motions, capabilities, advantages, and disadvantages will be discussed. The reader will become generally familiar with individual machines and with the range and diversity of equipment available. The ability of each device to satisfy construction or industrial needs will be seen in the perspective of the array of products marketed.

2.1 Introduction

The fellow who facetiously said that a camel is a horse designed by a committee obviously was not an engineer. An engineer would certainly have appreciated the fact that although both animals are beasts of burden, their operating environments and performance characteristics are in no way alike. He would have recognized that the differences are an inevitable and necessary consequence of their places of habitation and their working roles.

The same facetious person might have offered a similar observation about mechanical beasts of burden, such as cranes. Here too, evolution has led to differentiation to fit a variety of tasks and environments. The major differentiation has been between machines for construction verses those for general industry. The dividing line is rather hazy, however. For the most part, industrial machines find application in long-term permanent installations with consistent operating conditions. Construction cranes, on the other hand, are used for work-site assignments that may last from several hours to several years; these machines are likely to be exposed to a broad variety of work assignments and operating conditions. Despite these differences, a surprising number of machine types find extensive application in both construction and industry.

This book devotes particular attention to general-purpose construction machines, those that are most widely used. There is less emphasis on equipment less broadly used and little more than passing mention of highly-specialized devices. Innovative devices have been created for rigging loads in unusual circumstances. A few of these are described because of their interesting characteristics and to round out the presentation.

2.2 Derricks

Derrick (or Derick) was the name of a London hangman of about 1600. Given the similarity between a gallows and lifting devices in use at that time, it is not hard to see how our present equipment came to bear this name. But, for the sake of clarity, a definition is in order. A *derrick* is a device for raising, lowering, and/or moving loads laterally through use of a hoisting mechanism employing ropes but whose hoisting engine is not an integral part of the machine. This definition is the authors', and there are some who may not agree with it. In the ANSI Safety Code for Cranes, Derricks, Hoists, Jacks and Slings, a derrick is defined as "an apparatus consisting of a mast or equivalent member held at the end by guys or braces, with or without a boom, for use with a hoisting mechanism and operating ropes." This definition is perhaps too restricting, leaving a number of hoisting devices without a generic name. Also if the proposed definition is accepted, a crane simply becomes a similar device that has an integral hoisting engine.

Although we tend to think of derricks as comprising struts and perhaps other structural members, the author's definition could include derricks made of nothing more than ropes and fittings. They would rely on the structure on which they are mounted for rigidity and stability.

Since many types of cranes and derricks are little more than elaborations or variations on the Chicago boom discussed in Chap. 1, the Chicago boom is the first derrick to be treated in detail.

Chicago boom derrick

A Chicago boom can be mounted on a building frame during or after construction, on a tower, or on any frame. Indeed, Fig. 1.19 shows one installed on a power plant stack. When a boom or strut is assembled in the form of a Chicago boom, it can range from as little as 10 ft (3 m) to as much as 125 ft (38 m) in length, while capacities can range from a low of say ¼ ton (225 kg) to a practical upper limit of perhaps 35 tons (32 t). In the not too distant past, booms were made of wooden poles, but they are rarely used today. Short lightweight booms are easily and inexpensively made of single steel pipes fitted with the necessary attachments, but most booms are trussed, or latticed, structures of angle irons or tubing or a combination of the two. An occasional aluminum boom is seen.

Topping blocks are usually common hoisting blocks; one is fitted at the boom tip, using steel straps, and the other is mounted at the pivot fitting on the support structure with similar straps. The upper load block may consist of sheaves built into the boom head, or it may be a common block suspended on straps. Figures 1.14, 1.15, 1.17, and 2.1 show blocks with strap attachments.

Chicago booms can be used to hoist materials to a height somewhat above the boom foot, but the horizontal reach is limited by the length of the boom. In order to swing the boom, swing guys often are fitted

Figure 2.1 A 36-ft (11-m) Chicago boom installed at the McGraw-Hill Building in New York City during construction. The curved steel straps on the upper load block permit concentric mounting while allowing the derrick to boom in to very small radii. The chainfall in the foreground is for transferring loads from the derrick to the building floor. (*Photo by Lawrence K. Shapiro.*)

to the boom tip and are run laterally on each side to a point of anchorage. Wind, friction at the pivots, and the resistance in the opposite swing line must be overcome during swinging. With a manual arrangement, this requires the advantage of several parts of line and the use of fiber rope. Hand pulling through several parts is a slow operation that can take as long as several minutes to swing the boom through 90°. Where production economics justifies the expense, mechanical swing systems are used, and swing time is accordingly reduced.

In the typical installation, a two-drum winch is used to power the hoisting and topping motions, but when the work involves only lifting and swinging, the topping motion will not be needed. A fixed-rope guy line can then be installed, or to make adjustment easier and to provide flexibility, the system can be reeved to a chainfall of the type shown hanging in Fig. 2.1. In this case, only a single drum winch is needed.

Operational control is often enhanced when the winch is located at the floor level of the boom foot, particularly when the loads are to be hoisted to this floor, but the winch can be located at any level. When the winch is too large or too heavy to be lifted in the jobsite material hoist or in the elevator of an existing building, it may be necessary to use a small temporary winch to operate the derrick in order to hoist the working winch. Conversely, it may prove advisable to locate the winch on the ground.

When the winch is on the ground, the operator has direct communication with the ground crew and can have the boom and load in view at all times. When the winch is on the same floor as the boom foot and the loads are to be hoisted to that floor, the operator has direct communication with the swing and load landing crew and has the boom but not the load in view at all times. Landing the load is a critical operation, so that direct and immediate communication at landing level is a distinct advantage. Recall also that less overhauling weight is needed when the winch is at the higher elevation.

Although any structurally adequate column can be used as the mast for a Chicago boom, corner columns should generally be avoided. The authors' firm once designed an installation where booms were mounted on all four corners of a 34-story building, and the work required that they be jumped twice to higher elevations after the initial mounting. The "old timers" said that it would not work and that corner columns could not be used, but it was done and it did work. All the stone facing panels for the building were hoisted, and most were set, by those booms. Corner columns do raise formidable installation problems, however. Arrangements to permit swinging through 270° can be quite complex, as can be the fittings needed to allow the lead lines to follow the boom through the swing.

A Chicago boom is not a high-production machine. Although its first cost is very low and installation costs are often moderate, its crew size is usually large in relation to its productive capacity. Typical applications are "on again, off again" rigging work or situations where the load placement is above the reach or capacity of mobile cranes.

Guy derrick

Guy derricks at one time were the only practical means to erect steelwork on high-rise structures, but nowadays that work is mostly done by mobile or tower cranes. Guy derricks may also be used for general rigging, stone quarrying, and construction of refineries and chemical plants. A guy derrick can be described as a Chicago boom with its own integral column, called a *mast,* which is held vertical by six or more guy ropes, as shown in Fig. 2.2. The guys radiate in a horizontal circle about the derrick and are spaced as evenly as site conditions permit. A common configuration would include a mast of about 125 ft (38 m) and a boom of about 100 ft (30 m), although both larger and smaller derricks may be found. Capacities can range up to 200 tons (181 t) or more. Both boom and mast are almost invariably latticed.

The boom of a guy derrick is usually made shorter than the mast so that at close-in radii the boom can swing under the guys. This enables the derrick to work through an arc of 360°. Narrow loads, such as steel beams, are easily handled in this way, but care must be exercised to prevent loads from striking the boom. Many steel erector's derricks have boom bottom faces protected by steel plates.

When a guy derrick swings, the boom and mast move together; a pivoted fitting is provided at the mast top, and a ball-and-socket joint is mounted at the bottom. Above the top pivot is the fitting to which the guys are attached, called a *spider.* The top pivot itself is called a *gudgeon pin.* Swinging, or slewing, is accomplished by using a large horizontal wheel, called a *bull wheel,* which is fitted at the bottom of the mast. Wire rope is run around the bull wheel, and winch power is used for swinging. Alternatively, on small derricks, an arm, called a *bull pole,* can be used to provide leverage for pulling the derrick around manually. Another method, sometimes used on steel erection derricks, makes use of a handwheel and gearing so that one person at the base of the derrick can swing the boom.

The topping lift is attached to the mast just below the gudgeon pin. The topping and load lead lines come into the mast at the top and above the boom foot, respectively, and run down the center of the mast. The lead lines run into the derrick base, which is fixed in position, over sheaves, and out to the winch. Because the lead lines run through the center of rotation, they do not encumber the swing motion.

Figure 2.2 A dramatic lift by a guy derrick during steel erection at a New York City high-rise office building. Note the planks inserted to prevent guy turnbuckles from unwinding.

The mast must be held plumb; excessive lean will make swinging difficult, like continuously going uphill. The guys, of course, serve this function. Even when taken up to remove excess sag, the guys hang in a catenary shape. Under load, those opposite to the boom are stressed, causing rope elastic stretch and a reduction in catenary sag, which makes the mast lean toward the boom. In order to control mast lean, the size, construction, and initial tension of the guy rope must be taken into consideration, together with guy configuration and load levels.

Guy derricks are practical for steel erection work because it is possible for the derrick to lift itself, or *jump,* as the height of the work increases. The winch is left at base level as the derrick jumps; the winch operator works blind, receiving all instructions by signal or voice communication.

A guy derrick must be supported at the mast base, and anchorage points are required for the guys. On a building, a steel grillage is usually placed under the mast to transfer the loads to the host structure. The guys are then anchored to the building frame. Inasmuch as steelwork first proceeds with only partial bolting at the connections, analyses must be made to assure that the partially connected structure is adequate to support the derrick reactions together with structure dead weight, construction loads, and wind. This must be done for each derrick-mounting level.

Modern high-rise steel structures utilize composite construction in which the steel framing members and the concrete floor act in combination to support loads. One result of this efficient technique is smaller-sized framing. The derrick will be in place and must operate before the concrete is installed, making support much more difficult and expensive.

The initial erection of a guy derrick requires use of another lifting device such as a crane. The grillage and its supports, the guy anchorages, and the jumping operation are all cost factors that need to be considered. But in addition, in order to swing a load past a guy, the derrick must be boomed-in until the boom can pass under the guy, or alternatively the guy must be lowered and then replaced after the boom passes. This is both awkward and time-consuming; nonetheless production rates are usually acceptable for steel erection work.

Gin pole derrick

A gin pole resembles a guy derrick without a boom. Like the mast of a guy derrick, the gin pole is held in place by guys fastened to its spider. Gin poles are not intended to swing under load and are mostly used for vertical lifting without change of radius. Gin poles are either very large or rather small, with almost no role for units of intermediate size. Lifting that might be done by midsize gin poles is done more economically by other means in nearly all instances.

The heavy lift types operate at a fixed radius with a lean of only 5° or 10° from the vertical. Gin poles are primarily used for rigging work, raising heavy machinery, vessels, or structural components into place. Poles of 250 ft (76 m) or more are available with capacities ranging to about 300 tons (270 t). The practical limit to the size of gin poles results from the need to use other lifting equipment for erection.

Lighter capacity gin poles can be used at greater angles of lean, but in that case they must be fitted with a topping lift and side guys which would permit changing radius under load. These are often used for raising loads to the roof of an existing building, or for erecting a larger derrick.

The gin pole is a flexible piece of equipment that lends itself to adaptation to jobsite conditions. The authors' office has designed an installation in which a 250-ft (76-m) pole was placed in the center of a power plant building frame. The base was anchored to a concrete foundation at ground level and the pole head projected some 40 ft (12 m) above the roof framing. Four 160-ton (145-t) plate girders were to be erected at an elevation of 210 ft (64 m); the erection procedure required raising and placing the first girder, decreasing the lean for the second, rotating the pole 180°, and repeating for the next two girders. The unusual condition of having the pole within a high and substantially braced steel framework led to the idea of using only four guys, one to each building corner at roof level. Each guy was fitted at the anchorage end with a multiple-part reeving and a hand-operated winch. This arrangement permitted rapid and accurate changes in the lean between lifts as well as precise placement of the loads. The high elevation at which the guy anchorages were placed resulted in rather small guy loads considering the magnitude of the lifted loads.

We have also used small gin poles for installations at completed buildings and elsewhere. They can be placed on a roof or setback of limited area more easily than other derrick forms. Poles of 30 ft (9 m) or so with capacities of about 5 tons (4.5 t) are readily arranged with a topping lift and side guys under these conditions; however, caution is required. If the side guys are not properly placed, they can usurp the topping-lift load, causing them to fail. This can occur particularly when the side guy anchorages are not on the same axis as the boom foot pivot. Anchorages on this axis permit change of boom radius with no change in guy length. When the anchorages are either forward of the line (toward the load) or at a lower elevation, booming-in necessitates simultaneous paying out of line to the guys. If the guys are not evenly paid out, the tighter guy can pull the boom over and simply break the boom at its base. If the anchorages are behind the line or above the elevation, the guy lines must be precisely taken up during booming-in. Done unevenly, the more taut guy will destabilize the boom and transfer the topping load to the other guy, pulling the boom over. Failure is likely to ensue.

At refineries and chemical plants, gin poles are often paired and used to erect vessels, as shown in Fig. 2.3. The derrick hooks are made fast to the vessel at a point above the center of gravity. As the derricks lift, the bottom end of the vessel will rotate and must be moved laterally toward the derrick bases. To accommodate those movements, a *tailing crane* has to lift the vessel bottom and lead it in. The tailing crane is just visible in Fig. 2.3. Tailing cranes are discussed in detail in Chap. 5.

Stiffleg derrick

Like a guy derrick, a stiffleg derrick has a boom supported from a vertical mast. However, instead of guys it has two rigid inclined legs supporting the mast. Without guys for the boom to clear, the mast can be short. The *stifflegs* are often at right angles to each other but may be as little as 60° apart. The legs of the derrick shown in Fig. 2.4 are 85° apart. The leg positions dictate that the swing arc must be limited to about 30 degrees less than the angle outside of the legs.

When permanently installed at ground level, the mast bottom and stifflegs are anchored in structural foundations. For aboveground use

Figure 2.3 A pair of gin poles prepared for lifting an ammonia vessel. Double guys can be seen emanating from the spiders at the pole tops. A tailing crane is barely visible behind the right-hand pole and at the rear of the vessel. (*American Hoist and Derrick Company.*)

Figure 2.4 Stiffleg derrick with a 62-ft (19-m) boom lowering part of an air-conditioning chiller. This derrick of 35 tons (32 t) capacity is made up of short pinned components so that it can fit into elevators or jobsite material hoists. The light-capacity latticed pole in the foreground is a gin pole used for erection and disassembly. Both derrick and installation were designed by the authors' firm. (*Gendelman Rigging and Trucking, Inc.*)

and for construction work, sill members (i.e., horizontal framing elements) are added to connect the leg ends and mast bottom, forming vertical triangles that give the assembly rigidity. Needless to say, the mast bottom and leg ends must still be supported.

Stiffleg derricks are available in a wide range of sizes and capacities. For general light building work, 12-ton (11-t) and lighter models are made with booms from 30 to 50 ft (9 to 15 m) or so. At the other end of the scale are machines with capacities to 700 tons (635 t) and booms to 265 ft (80 m). For special applications, stiffleg derricks can be mounted on towers or on portal frames either fixed or traveling. The usual installation would be on the ground at industrial facilities or some construction activities and on the roof framing at high-rise building sites.

The mast, legs, sills, and boom can be of latticed construction with angle-iron or tubular members, or legs and sills can be pipes, rolled shapes, or laced channels. Typically, the mast rotates together with the boom during swinging; it has a top pivot at the joining of the legs and a pivot or ball-and-socket joint at the mast bottom. The mast base is mounted in a fixed position and houses the lead sheaves. Lead lines come down the center of the mast and out at the base sheaves, just as in a guy derrick. The erection of the mast and legs requires use of additional lifting equipment, usually a small gin pole (Fig. 2.4).

The stiffleg derrick designed by the authors' firm shown in Figs. 2.4 and 2.5 has mast and legs of pin-connected pipes in short lengths so that the parts can be lifted in a jobsite material hoist. The boom is

Figure 2.5 The derrick of Fig. 2.4 dismantling a tower crane 58 stories above the street. The very small roof area made it necessary to support and tie down both sill ends on a setback two stories below. The crane installation, the derrick installation, and the dismantling procedures were designed by the authors' firm. (*Delro Industries, Inc.*)

also in short pinned lengths. To accommodate variations in building framing dimensions from jobsite to jobsite, the derrick sills are provided with sliding anchorage fittings. The sills themselves are rolled beams in pinned segments. The mast is fixed and does not rotate; instead, the boom is stepped in a fitting that rotates about the mast on a Teflon bushing. The derrick is intended for rigging work on high-rise buildings, and it was first assembled, tested, and used on a roof some 56 stories above the street. The work entailed replacement of 20,000-lb (9100-kg) air-conditioning machines with the hoisting done at a 55-ft (17-m) radius. Swinging was arranged through use of a small-diameter bull wheel (30 in or 0.76 m) and a hand-operated winch.

During operation, as a stiffleg-derrick boom swings, mast and leg end reactions change and actually reverse. The compressive reactions, particularly under the mast, are of considerably greater magnitude than the lifted load. Derrick dimensions seldom match building framing dimensions, so that even when the mast is placed on a building column, it is usually necessary to install framing reinforcement or transfer beams at the leg ends. Tie-downs are always needed at the three mounting points to prevent uplift. A second lifting device is needed for erection so that installation of a stiffleg derrick becomes somewhat expensive. But the operating motions of a stiffleg derrick are the same as those of a fixed-position crane, except for the limited swing arc. Thus, stiffleg derricks can be viable low-cost alternatives to cranes in places where tower or portal mounting is called for, especially at such ground installations as concrete batching plants, stoneyards, and barge unloading facilities.

Other derrick forms

Gallows frame.† When two gin poles are assembled with a horizontal beam across the pole heads, the device is called a *gallows frame* and resembles the apparatus that the hangman Derrick used for his grisly work. The upper load blocks are fitted to the beam, which is arranged so that the poles are loaded concentrically. This gives the gallows frame a greater capacity than is available with a similar double gin pole; the increase can be on the order of 50%.

As with gin poles, the height of a gallows frame is limited by the equipment needed to erect it, but units with hook lifts of 200 ft (61 m)

†Riggers in the New York City area use this term to refer to another device, a four-legged frame used in plant machinery rigging.

are available with capacities of 600 tons (545 t). Shorter configurations achieve ratings as high as 1200 tons (1090 t).

A double gin pole can be used to erect a vessel taller than the poles, but because of the gallows beam, a gallows frame is limited to shorter vessels.

Column derrick. A column derrick is a modified guy derrick that is useful where loads must be lifted to a building setback or penthouse. Instead of guys, a structural frame is installed at the mast top spider and is attached to building columns or other substantial bracing means (Fig. 2.6). The frame acts as a rigid mast support system as opposed to the guys ordinarily used; because guys are somewhat slack, a guyed mast will lean toward the boom as a load is lifted. A column derrick can be installed at locations where space is limited or where there is not sufficient space or clearance for a full set of guys. Of course, there must be an adjacent structure high enough for attachment of the top frame.

A column derrick can swing freely since there are no guys in the way, and it is not necessary to utilize that time-consuming maneuver of booming-in and ducking under the guys.

Shaft hoist. This device is a derrick only under the author's definition. It is a simple device without compression members, made by suspending an upper load block in an elevator shaft or other framed shaftway by wire ropes tied to the four corner columns or to other members. The weight of the load together with the four tensioned ropes holds the block fixed in position either at the center or offset in the shaftway. An alternative form would have the block supported from beams spanning the shaft. The lead line can be run from the upper or the lower block to a snatch block and thence to the winch. The capacity of a shaft hoist is limited only by the strength of the shaft framing or the cost of reinforcing the framing. This device is inherently inexpensive for most installations. Figure 2.7 shows a shaft hoist the authors' firm designed for lowering equipment to a subcellar machinery space in an office tower.

Cathead. Another derrick under the authors' definition, the cathead, is a beam cantilevered out from a support with the upper block attached to its outer end. An alternate form has the outer end of the beam supported by ropes tied back to the structure. A cathead is used for rather lightweight loads ordinarily on the order of 5 tons (4.5 t) maximum, but capacity is limited mainly by the strength of the support framing or the cost of frame reinforcement. The typical cathead

Figure 2.6 A column derrick used to place heavy precast facing panels on a high-rise hotel under construction. This is a guy derrick without guys—a concept of A. J. McNulty & Co., Inc., designed by the authors' firm. The upper support (which replaces the guys) and the base support are cantilevered from the building frame. (*A. J. McNulty & Co., Inc., photo by Robert Weiss.*)

Figure 2.7 The bottom element of a large air-conditioning chiller ready to be lowered into a subcellar by a shaft hoist. The unit was rigged over the opening on steel beams, which will be removed by the small telescopic crane. The 30-ton (27-t) machine was one of several installed at the McGraw-Hill Building. *(Photo by Lawrence K. Shapiro.)*

is incapable of changing load radius, so that loads must be handled by *drifting*. A load is drifted when it is pulled laterally to change its horizontal position, thereby making the support ropes depart from the free vertical. The cathead is an inexpensive device, but its capabilities are limited.

When a load is suspended from a long line, drifting a short distance requires very little force and the lateral, or side, loading imposed on the derrick is slight. Both forces increase as suspension length decreases or drift increases. A load cannot be landed onto a building floor until its center of gravity is within the building. This often cannot be accomplished by drifting, and the ordinary rigger's methods of shifting loads may be too time-consuming to be economical for small loads. In such cases catheads must be modified to make it easier to land the load. They can be fitted with a trolley to support the upper block and to permit radius change and ease of landing. The trolley can be controlled by a small hand winch, a chainfall, or a come-along.

On one job, a cathead was needed to hoist large building-facing panels weighing some 7½ tons (6.8 t) to a maximum height of about 500 ft (150 m). The panels were to be transferred from the cathead to monorails for lateral distribution and positioning for placement. Because the panels had to be hoisted past panels previously set into place and were large wind-catching surfaces, hoisting had to be done well clear of the building face. The monorails, on the other hand, had to be set close to the face to simplify panel placement. The necessary radius-change capability was provided by a trolley system and hand winch. Capacity and load stability were enhanced by use of a double beam. Figure 2.8 shows the installation.

Flying strut. The flying strut is an invention of one of the authors that has been used at only one installation. It is mentioned here only to demonstrate that there are options available when jobsite conditions prevent the use of standard equipment. The options are limited only by the designer's imagination.

The flying strut was used at a hospital construction site (Fig. 7.38) after the framework was in place and the concrete floors had been

Figure 2.8 Unique cathead and monorail system for installing stone facing panels on the AT&T Long Lines Building in New York City. The cathead lifts units well clear of previously set panels and then trolleys the load in for transfer to monorails for lateral distribution and placement. Installation and equipment were designed by the authors' firm. (*A. J. McNulty, Inc.; photo by H. Bernstein Assoc., Inc.*)

poured. It was necessary to lower a series of 34-ton (31-t) chiller units and other machinery into a subcellar. Conditions dictated that the lowering be done halfway back into an alley between an existing building and the new frame. The alley was of limited width, its bottom was a broken-up surface at subcellar level, and there were beams framed across and a partial deck poured at street level. The loads had to be trucked in at street level, and there were no equipment or procedures known that could perform the work economically under the given conditions.

The flying strut seems to be complicated, but really it is not. The strut is a compression member suspended horizontally by ropes perpendicular to the building face (Fig. 7.37), half inside and half outside the building. The inboard end of the strut is held by four positioning ropes, one to the top and bottom of each of the columns on either side of the strut at the building face. Stabilizing ropes, or *preventers,* are run to interior columns as well. (The name preventer comes from their use to prevent unwanted or unexpected occurrences.) At the outboard end of the strut, the upper load block is mounted together with suspension ropes capable of supporting the load and running up several stories to column anchorages. Side preventers are also provided at the outboard end. The load block is not supported by the strut; instead, the strut is a means for positioning the block for the work. By providing take-up devices on the positioning and preventer ropes, the position and attitude of the strut can be controlled.

The strut and its suspension system need be designed only for the lateral thrust resulting from positioning the load block and can be designed to permit large-order drifting of the load as well. At the hospital job, it was necessary to lift the load with an initial drift offset of about 18 ft in 35 ft (5.5 m in 10.7 m) of vertical rise (27°) from the side of the strut. The strut rolled in its suspension ropes to accommodate the drift and proved to be a quick and easy means of handling the loads. The device was relatively inexpensive to make and install.

2.3 Mobile Cranes

Years ago, some Danish friends who are tower crane engineers expressed to one of the authors their conviction that the spirit of the old American west was still alive. Cowboys may have laid down their guns, they said, but now they use mobile cranes to kill each other instead. To Americans, that belief seems rather strange, but it is prevalent among many European crane engineers because of the stark differences in philosophy and history that separate American and European crane design concepts. Mobile cranes were born in open spaces, while the tower crane came into being amid the narrow streets and moderately tall buildings that make up European cities. Mobile cranes rely on the individual, the operator, who can do damage to the machine and its surroundings by failing to exercise due care; in the frontier spirit, destiny is in the operator's hands. Tower cranes, particularly in early days, have been shaped by the restraints and demands on the individual that are necessary for successful city living; most tower cranes have automatic controls and limits, so that the operator's essential function is to just "aim".

Contrary opinions notwithstanding, mobile cranes are not lethal weapons. In the hands of good operators, mobile cranes have been proven, worldwide, to be efficient, versatile machines with a good safety record. In the hands of really competent operators, they are even more effectual. "Competent" should be taken in the broad sense to encompass not only skill in operating the crane controls for the various motions, but to include an understanding of the machine's underlying principles, its capabilities and limitations, and ambient conditions and how they affect the load that can be safely lifted.

Crawler, truck, all terrain, and rough terrain carriers

A crane that can move freely about the jobsite under its own power without being restricted to a predetermined travel path requiring extensive preparation is a mobile crane. Movements about the jobsite are called *travel* movements, and movements from site to site are *transit* movements. All mobile cranes can travel; wheeled mobile cranes are capable of self-powered transit movements, but crawler cranes are not.

Crawler bases. The crawler assembly, upon which the rotating superstructure is mounted, is referred to as a *base* or *base mounting,* and is not, strictly speaking, a *carrier.* Wheeled crane mountings are true carriers. The base mounting includes a *carbody,* which is the central or main portion, with the swing circle fixed on top. The swing circle is the point where the upper works are joined to the base and must not only transmit the loads generated by the superstructure to the base but also serve as the bearing or roller circle for the swing motion. At the sides of the carbody, *crawler frames* are mounted which in turn hold the axles, propel drive, idlers, and tracks or crawler treads. Carbody, swing circle, side frames, axles, and treads, together with the propel mechanisms, constitute the base mounting.

Most crawler-crane propel systems are powered by the engine in the superstructure through a vertical shaft on the axis of rotation, which in turn drives horizontal shafts to sprockets and drive chains on the side frames. An alternate method makes use of hydraulic motors mounted on the side frames. Some of the newer, very large machines have a separate propel engine mounted on the base. Travel speeds from about 0.5 to 1.5 mi/h (0.8 to 2.4 km/h) are usual, smaller cranes operating at the higher speeds.

Each tread propel drive has separate clutch, brake, and lock for full independent control of travel and turning motions. The locks are engaged during operation and when parked on grades. The propel system is geared down for power, enough to "climb the side of a barn," as one manufacturer's engineer put it. The power is needed for maneuvering on soft ground and for traveling on ramps and grades. Crawler cranes can manage a 30% grade when not under load, but travel on steep slopes should be done only under controlled conditions.

Figure 1.21 shows that the spread of the crawler tracks affects the stabilizing moment; the wider the spread, the greater the lifting capacity is. The practical limitations on width when transporting cranes has led to the development of extensible crawlers. Cranes can be transported and can travel and operate in the narrow configuration, but to achieve increased capacities the crawlers are extended to a greater width. It may be necessary to add counterweight in order to gain full capacity with crawlers extended or, conversely, to remove counterweight in order to operate in the narrow configuration. Most machines with extensible crawlers are equipped to extend or retract without assistance but may require jacks, blocking, and tools (see Fig. 2.9).

The base mounting transmits crane loads to the ground. To do this without excessive increase in radius under load and without excessive change in radius during swing motions requires that the base be extremely stiff and practically free of deflection. This is accomplished by using complex castings or weldments of great weight.

Beam-type low-bed trailers are used to carry crawler cranes during road transit. For rail transit or under roadway restrictions the side frames can be removed without disassembly of the drive chains. The base mounting can also be separated from the superstructure, and all or part of the counterweight removed to reduce weight.

In a comparison of cranes of equal lifting capacity, crawler cranes offer lower cost and rental rates than do truck cranes, but their transit costs are higher. Among the largest mobile cranes are crawler cranes with heavy-lift attachments.

Truck carriers. Truck-type crane carriers should not be confused with ordinary commercial truck chassis. A crane carrier is designed and built for no other purpose than crane service. Figure 2.10 shows a

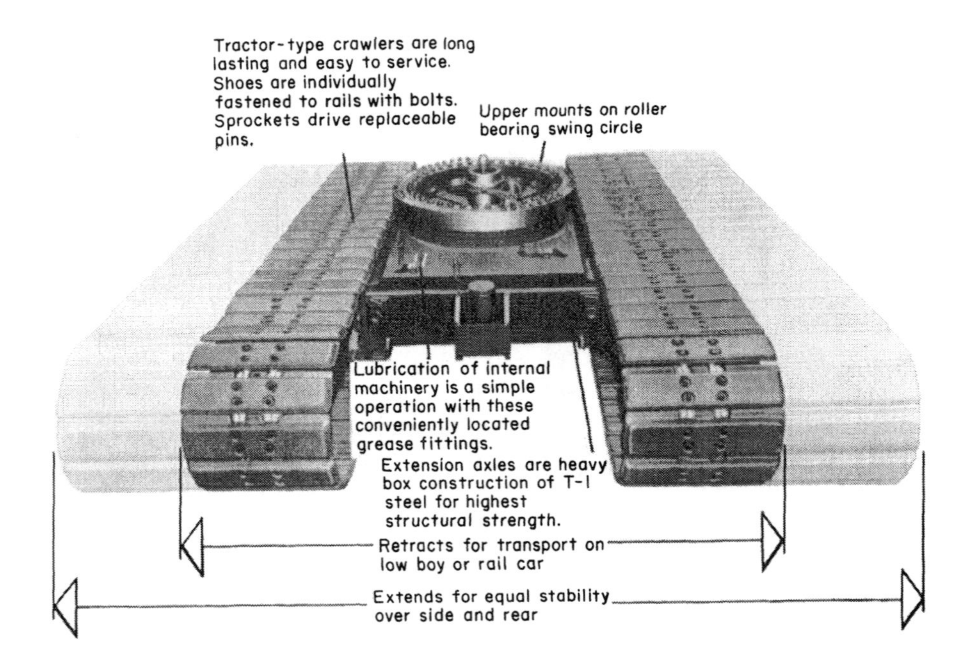

(a)

Figure 2.9 Increasing the spread of the crawler tracks adds capacity to any crane, but it makes rail or truck transit more difficult or impossible. Many larger cranes, therefore, are equipped with extensible-retractable crawlers. (*a*) Crawler base designed to extend and retract crawlers with self-contained equipment and mechanisms. (*b*) The P&H 670-WLC with crawlers in travel position, 11 ft 2 in (3.4 m) wide. (*c*) Hydraulic jacks raise the carbody and the crawler opposite the counterweight. Axles are extended by internal hydraulic rams. A drive-shaft extension is slipped into place. (*d*) With the superstructure swung 180°, the second crawler is extended and locked into position. Crawler width is now 16 ft 6 in (5.03 m). (*Harnischfeger Corporation.*)

rather light carrier, but it is still clear that this chassis is quite different from that of a standard truck. The largest units may have as many as nine axles.

Carriers must be designed for movement about the jobsite and on the highway. To accommodate those extremes, transmissions with as many as 33 forward gears are provided, and most carriers are

(b)

(c)

(d)

Figure 2.9 *(Continued)*

Figure 2.10 Carrier for a truck crane of fairly light capacity. The turning circle with roller bearing is already mounted. In the right background are chassis weldments turned opposite to their position in the parked carrier (but not for the same carrier model). (*Harnischfeger Corporation.*)

equipped with special *creeping gears* as well. The high creeping-gear ratios, above 200:1, provide so much power that the drive shaft can be twisted off if this gear is used for other than creeping purposes on reasonably good ground. On the other hand, road speeds vary from crane to crane but range from 35 to 50 mi/h (56 to 80.5 km/h). Jobsite ramps of 20 to 40% can be traveled without load, depending on the crane model and surface conditions. Brakes are designed to hold on the maximum travel grade.

A good crane carrier must have an extraordinarily stiff, torsionally rigid frame for operating radius stability, but this requires the addition of diaphragms, stiffeners, and reinforcements, which add to weight. For roadability, weight must be held to a minimum. Many important differences in the quality of carriers rests on the compromises made in satisfying those conflicting needs.

Carriers in the United States can be as wide as 13 ft (4 m), and the entire crane, in transit configuration, can weigh as much as 350,000 lb (160,000 kg). In many areas, highway load limitations do not permit such load weights or widths. Machines up to 12 ft (3.7 m) wide can be readily shipped on United States railroads, but wider loads require special routing. Often special provisions for transit must be made, as covered more fully in Chap. 5.

All terrain carriers. First appearing in Europe during the late 1970s, the all terrain crane now accounts for perhaps 60% of mobile crane sales throughout the world. The carriers used on these telescoping cantilevered-boom machines combine the high road speeds of truck carriers with the off-road capabilities of the rough terrain crane. Europe still remains the predominant marketplace for all terrain machines, but sales in the United States have been increasing.

Large-capacity multiaxle models have appeared featuring the characteristics necessary to carry the all terrain label: high road speed, off-road maneuverability, pick-and-carry ratings, and drive positions in both the chassis-mounted cab and in the operator's cab mounted on the rotating superstructure (Fig. 2.11). To achieve maneuverability, these cranes typically have all-axle drive and steering as well as crab steering. Many machines are furnished with sophisticated suspension systems that maintain equalized axle loading on uneven surfaces while the crane is in motion or is static. New models are continuing to appear, and models with more than 300-ton (272-t) capacity are available. Outriggers are provided for increasing capacity.

Rough terrain carriers. Rough terrain carriers are two-axled carriers used with telescoping cantilevered-boom cranes of up to about 100-ton (90.7-t) capacity. There is one operator-driver station that may swing with the boom (Fig. 2.12) or remain fixed. Carriers of this type are characterized by oversized tires that make travel easier under adverse site conditions. As with all terrain cranes, outriggers are provided for increasing capacity.

Rough terrain carriers perform transit moves at less than highway speeds, about 30 mi/h (48 km/h), and with some discomfort to the operator as a result of the high center of gravity and unsprung suspension. For long moves, these cranes are generally transported on low-bed trailers.

Latticed boom cranes

Latticed boom mobile cranes are highly versatile, capable of high productivity, and an amazing range of applications. They are used to load sand and gravel at concrete batching plants, erect high electric transmission towers, set vessels at refineries, load and unload ships, demolish buildings, erect steel, pour concrete, and perform a myriad of other construction and industrial tasks. The very nature of the machine, as it has evolved, permits almost unlimited use for lifting tasks. Presently, machines are available with capacities greater than 1000 tons (900 t).

Figure 2.11 A 20-ton (18-t) all terrain crane in travel mode. Characteristics of all terrain cranes include high road speed, off-road maneuverability, pick-and-carry ratings, all-wheel drive and steering, crab steering, and driving controls in both the carrier and the upper cabs. (*Grove Manufacturing Company.*)

Figure 2.12 A swing-cab rough terrain crane at an urban jobsite. Note the unusually large wheels which make maneuvering easier on uneven ground. Driving and operating functions are controlled from one station. (*Photo by Howard I. Shapiro.*)

Latticed boom cranes are mounted on either truck or crawler bases. A rotating structure, called a *superstructure* and sometimes an *upper,* is supported on the carrier or base and includes the power plant and drives, winches, counterweights, operator's station, and boom-mounting accommodations. The term *front end attachment* refers to the boom or the combination of boom and other struts that make up the lifting structure.

The most common front end attachment is the ordinary boom. It is made up of a *base* or *butt* section, any number of *inserts* of varying lengths in the center, and a *tip* section. A butt- and tip-section assembly, without inserts, is referred to as a *basic boom* and is the arrangement with which the crane's maximum lifting capacity is achieved. Hammerhead tips are used (see Fig. 2.13) typically for heavy close-up picks because they permit increased lifting height with respect to the length of boom in use. They are stubby in shape because the boom tip

Figure 2.13 A two-crane lift with short hammerhead-tipped booms. The hammerhead tips hold the load away from the boom by means of offset sheaves to permit the load to be raised clear of the booms. Both cranes are P&H 790-TC models. (*Harnischfeger Corporation.*)

sheaves are well offset from the centerline. This also offsets the hook and gives more clearance between the load and boom.

When booms are assembled to long lengths, very lightweight tips with only one or two sheaves are used. With various combinations of inserts, as specified by the manufacturer, any boom length can be assembled in increments that normally are 10 ft (3 m). With few exceptions, mobile cranes are capable of raising the maximum boom lengths assigned to them without assistance from other equipment.

Smaller cranes mount basic booms of 30 ft (9 m) or so, but for large cranes they can be 70 to 100 ft (20 to 30 m). Maximum boom lengths vary with the crane model; the longest booms in use are about 350 ft (107 m). The maximum practical boom length, for the present, has reached a plateau. Boom cross sections are now 8 ft (2.44 m) square so that an increase in cross section will require nonstandard transportation equipment and special permits in most jurisdictions. Higher-strength steels [100 kips/in^2 (690 MPa) yield-strength steels are in general use] bring diminishing cost to benefit returns should slenderness ratios increase beyond the present levels. Longer booms must therefore await technological advances or an easing of the economic impact of shipping wider booms. Long booms are shipped in sections on flat-bed trucks, but some cranes carry short booms of 100 ft (30 m) or so folded for quick jobsite erection. Many cranes of small to mid size can self-assemble their booms, however larger ones need an assist crane.

In comparison with that of a Chicago boom, the topping lift for a crane is positioned rather low, particularly on long booms, thus inducing high compressive loads in the boom. To help reduce this load, and thereby increase capacity, a *boom foot mast* is used on some cranes. The mast, shown in Fig. 2.14, serves the single purpose of raising the topping lift to reduce boom compression.

The term "topping lift" is not used in the mobile crane industry. It is replaced by two terms, one for the fixed-length ropes fastened to the boom tip, called *pendants, boom guy lines,* or *hog lines,* and the other for the running lines that produce the luffing motion, called *boom-hoist* ropes. The frames in which the boom-hoist sheaves are mounted are called the *upper* and *lower spreader.* The luffing-system components mentioned can all be seen in Figs. 2.13 and 2.14. The brakes on the boom-hoist winches are spring-set so that they automatically engage when the system is not in motion. They are also equipped with dogs to lock the winch against accidental release of the boom.

Mobile-crane load ratings, except for short booms, close radii, and deflection limitations on very long booms, are almost always governed by the ability of the machine to resist overturning. The machine weight itself provides considerable overturning resistance, but in or-

Figure 2.14 The boom foot mast on this Link-Belt truck crane serves to increase the angle between the boom and the boom guy lines, thus reducing the compressive force in the boom and increasing structural capacity. (*Link-Belt Construction Equipment Company*)

der to increase capacity counterweights are added at the rear of the upperstructure. There is a limit, however, to the maximum counterweight that can be added; the crane must have adequate resistance against tipping over backward when short booms are in use. Counterweights are detachable for transporting when road limitations require. Some cranes are rated for different counterweight options.

Crane controls function through mechanical linkages, pneumatic or hydraulic pressure, or electrical signals, depending on the size of the machine, manufacturer's preference, and control function. Power trains are geared from high to very low ranges, and torque converters or magnetic flux transmissions are common (Fig. 2.15). The elaborate

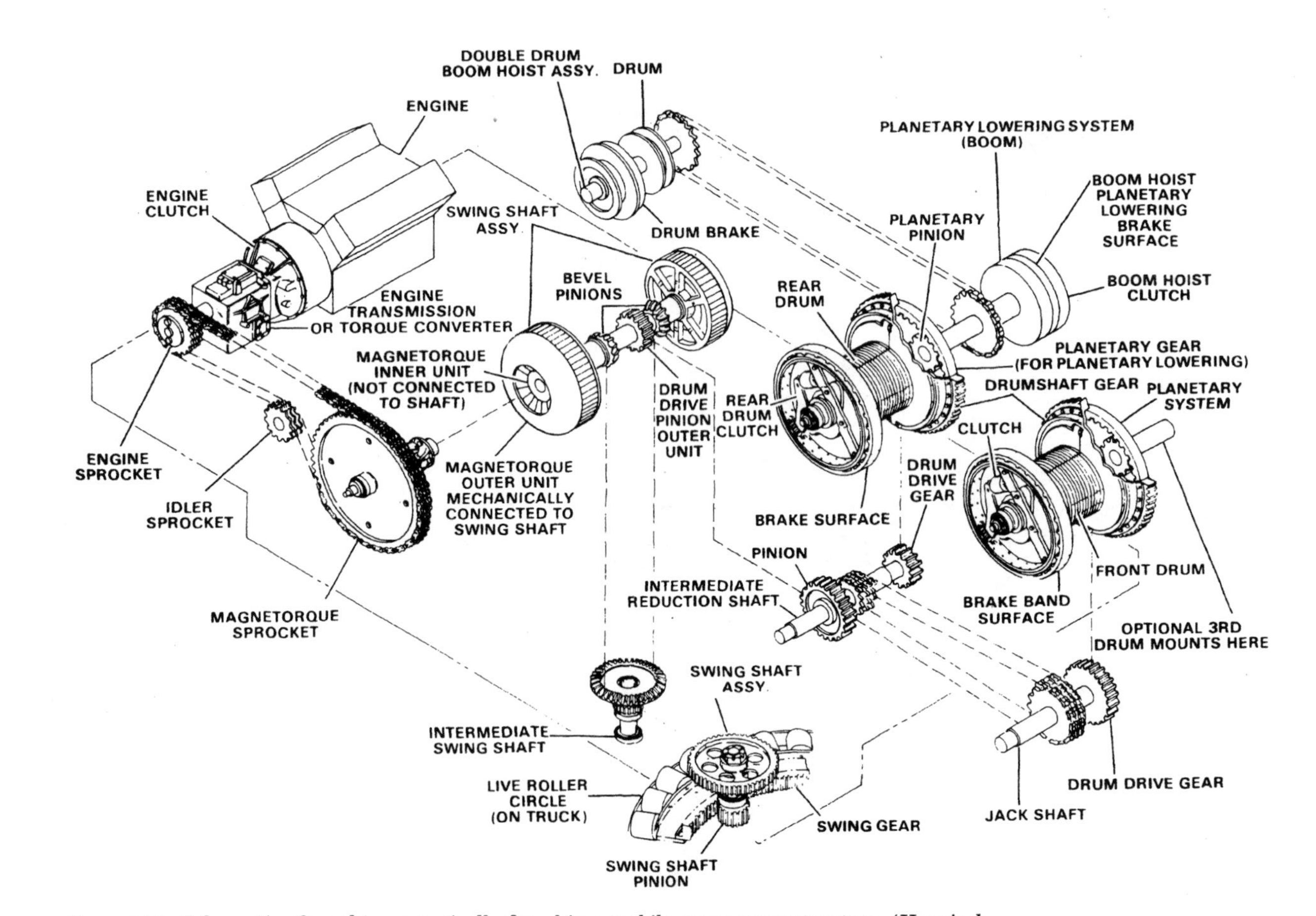

Figure 2.15 Schematic of machinery typically found in a mobile-crane upperstructure. (*Harnischfeger Corporation.*)

control and drive systems are intended to enable an operator to maintain high production rates or to crack the shell of an egg with a 100-ton load. But cranes typically have the power to produce operating speeds so extreme that they can destroy themselves. The high lift, luffing, swing, and braking speeds and accelerations attainable are not a reflection of recklessness or irresponsibility on the part of the manufacturers but a consequence of the range of operating configurations and conditions in which the crane must function. Drivers need not operate their automobiles at top speed, but the power needed to achieve top speed makes for good performance and often enhances safe operation at lower speeds. The same is true with cranes, and, as with the automobile, operator judgment and responsible behavior are necessary.

Wind must be mentioned in any discussion of latticed boom cranes because so few people involved in the use of mobile cranes seem to be aware of wind effects and limits. A friend from a crane-rental company telephoned one day in a state of confusion and consternation. One of his cranes, with long boom and jib mounted, was loading a ship at a local pier. Although the rating chart indicated that the cargo to be loaded could be taken to the required radius and then some, the crane became "tippy" at a far shorter radius. He telephoned to confirm that he was reading the chart correctly; he was. We then reviewed a series of installation and assembly questions in an effort to find the cause of the loss of capacity. Finally, when asked about wind, his response was, "It sure is windy. It's blowing so hard you can hardly stand up!" Sure enough, the wind was blowing from behind the operator, thus acting like more load on the hook. That friend, with his years of experience with mobile cranes, did not think to connect the tipping of the crane with wind.

The recommended practice for designing latticed crane booms in the United States† makes no direct provision for the wind velocity the boom must endure when the crane is out of service. Instead, it stipulates that the manufacturer shall specify the velocity at which the boom must be lowered to the ground or otherwise secured. Booms up to 200 ft (60 m) in length can be quickly lowered to the ground at almost any operating site; however, a 450-ft (137-m) boom and jib combination is another matter. Lowering such equipment at an urban jobsite when high winds threaten is a daunting task. It is true that not many cases of storm-wind crane failure are on record, but the vulnerability of those machines cannot be ignored even though only property damage is the usual outcome.

†SAE Information Report J1093, Crane Boom Systems—Analytical Procedure, Society of Automotive Engineers, Warrendale, Pa., March 1994.

A few years ago a hurricane threatened New York City with winds of over 80 mi/h (36 m/s). At the time, a crawler crane with 350 ft (107 m) of boom and a 100-ft (30-m) jib was at work on a high-rise concrete structure on Broadway about half a mile (800 m) north of Times Square. There was no way to secure that boom with confidence with such winds expected. The only viable choice for safety was to lay the boom down in coordination with the police and traffic authorities; the boom and jib crossed and completely blocked-off two crosstown streets.

Owners and users of long-boom cranes must be prepared with a plan of action in case high winds should occur at any time at each and every jobsite. Crane manufacturers should be consulted for recommendations.

Although the recommended industry practice referred to above does not include wind as a condition to be considered while the crane is in service and lifting loads, it does presume that the design factor used will accommodate wind effects. That being the case, one could assume that cranes are designed for winds of about 20 mi/h (8.9 m/s) in conjunction with rated loads unless the manufacturer's documentation states otherwise. It is reasonably safe to assume that all current domestic models can support such winds, but without allowance for wind on the lifted load. A note common to most crane rating charts requires that the operator reduce ratings (derate) for wind when necessary. A 20-mi/h wind will exert a force of only 1⅛ lb/ft^2 (54 Pa) on a flat-surfaced load; therefore only loads with appreciable sail area will require crane capacity derating. The operator is not given the means or guidance on how and under what conditions to derate, nor is it feasible to provide simple means at present. Operator judgment and experience are required.

However, a relatively quick, but approximate, jobsite method is given in Chap. 8 for use by crane operators or supervisors capable of carrying out the mathematics.

Jibs. To increase the vertical reach of a crane, jibs are used although on occasion increased horizon reach can be attained, this is an exception to the general case. A jib is a lightweight boomlike structure mounted at the boom tip, as shown in Fig. 2.16. The fixed-length *stay ropes,* or *guy lines,* hold it in place, and the backstay attachment is located on the boom or at the boom base. Jibs can be mounted parallel to the boom centerline or offset at any angle to 45° from the centerline as the manufacturer directs. Some models feature luffing jibs. Figure 2.17 shows a rather long jib, 80 ft (24.4 m), mounted on a 320-ft (97.5-m) boom at a 10° offset angle, making the jib tip some 400 ft (122 m) above ground.

A jib is a relatively low-capacity structure in comparison with the boom on which it is mounted. In particular, jibs are not very strong

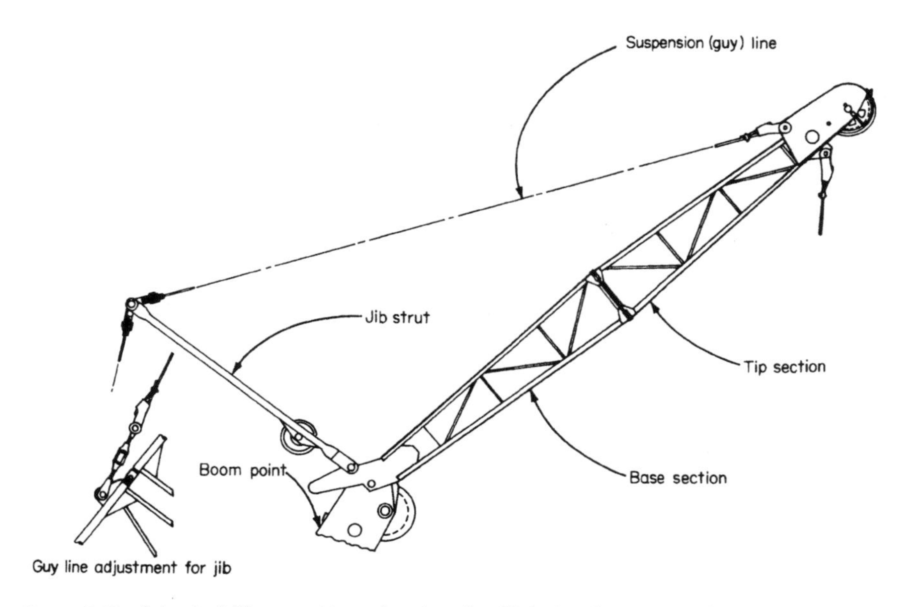

Figure 2.16 A typical jib mounting, showing the jib in its shortest configuration. The jib is offset 20° from the boom.

Figure 2.17 P&H model 6250-TC with 320 ft (97.5 m) of boom and an 80-ft (24.4-m) jib at 10° offset. Jibs are used to increase lifting height. (*Harnischfeger Corporation.*)

with respect to side loading. Jib lifts are most often made with a single part of line, although they can be rigged for two, and in some cases more, parts of line. A typical crane is arranged so that it can carry the main hook suspended from the boom tip while another hook is suspended from the jib. When that is done, the jib line is used for light duty service; then, it is usually referred to as a *whip line*.

Towers. Some mobile-crane models are provided with an optional tower *front end attachment*. It consists of a vertical mast or tower pinned at the position on the superstructure where a boom is ordinarily mounted; a luffing boom is affixed at the tower top. A jib is sometimes fitted at the end of the boom.

The tower attachment creates a hybrid between a mobile crane and a tower crane, offering the mobility and ease of assembly of the former along with the reach over obstructions afforded by the latter. One of these may be advantageous at an urban or dense industrial worksite where the crane must be placed close to a structure and still be able to raise loads over and beyond (Fig. 2.18).

Towers and booms are made up of inserts of various lengths so that a number of tower-height and boom-length combinations can be assembled. The manufacturer's instructions generally include an advice that the tower boom must be lowered and fastened to the mast when the crane is out of service (an exception is when very high winds are expected, in which case the mast must be lowered to the ground or guyed). Therefore, the mast should be longer than the boom. However, there are exceptions, in which case the crane manufacturer issues guidelines for securing a boom that cannot be fully folded down.

Most tower attachments require a small assist crane, commonly a telescopic machine, for unloading and placing tower and boom sections, and for assembling the rigging and the counterweights. Tower attachments utilize unique rigging and require many adjustments and the assembly of many special parts. Typically, assembly takes more than a standard workday for the larger models. Some models, the very tallest, may require a large assist crane to help raise the tower and boom off the ground.

These devices have moderate maximum lift capacities, but maintain capacity at long radii as compared to conventional booms. Tower attachments have proven to be productive and versatile; their popularity has substantially limited the penetration of tower cranes into the U.S. market.

Special attachments. Some mobile cranes can be set up as a guy derrick, which results in a great increase in capacity, particularly at long radii. The machine, as converted, is no longer a stability-limited device

Figure 2.18 An American 9310 crawler crane with 190 ft (58 m) of tower attachment and a 170-ft (52-m) boom being used to erect a poured-in-place high-rise concrete structure. The tall tower permits the crane to be placed close to the work while giving hook access over the entire deck. (*Photo by Jay P. Shapiro.*)

but is rated on the basis of the structural strength of the various components. The advantage over an ordinary guy derrick is in the mobility of the machinery platform between setups and in the ease of erection. Guy derrick attachments can be made for truck- or crawler-mounted cranes and can have capacities to about 600 tons (544 t) and booms to near 400 ft (122 m).

There are several proprietary systems, called *heavy lift attachments,* for increasing crawler-crane capacity. These systems require considerable ground space and heavy setup costs, but they are capable of lifts to as much as 2000 tons (1800 t) and reaches close to 700 ft (213 m) high. The American Crane Corporation's Skyhorse makes use of added counterweights on a wheeled dolly linked to the machine so that it follows during swing motions and acts as an anchorage for fixed-length backstay ropes to a very high latticed boom foot mast. The Manitowoc Engineering Company's Ringer (Fig. 2.19) utilizes a circular track surrounding the machine. The boom is stepped onto a carrier extending from the superstructure; the carrier rides on the track. Counter- weights are placed on a rearward-extending carrier that also rides on the track. A high latticed boom foot mast is also used. The Link-Belt Construction Equipment Company's heavy lift attachment also utilizes a circular path. Its boom and added counterweights

Figure 2.19 A Manitowoc 4100W crawler crane, fitted with a Ringer attachment, at work on Metrorail in Dade County, Florida. Note that the boom and high mast are stepped onto a carrier riding on the large-diameter ring which constitutes the machine's tipping fulcrum. The added concrete counterweight blocks also ride on the ring. The high mast gives the boom added structural capacity, while the additional counterweight and the placement of the tipping line well in front of the machine increase its stability. (*Manitowoc Engineering Company.*)

mount on a single frame, linked to the superstructure, that rides on the prepared path. The largest heavy-lift system is the Lampson Transi-Lift, which utilizes a boom and counterweights mounted on separate carriers with independent but coupled drives.

Telescoping cantilevered boom cranes

We have never heard anyone propose an adequate and technically acceptable shorter name for this type of crane. Other names commonly used, such as *cherry picker, zoom boom,* and *hydraulic crane,* are either undignified or inaccurate. For want of something better, *telescopic crane* will be used here.

Telescopic cranes can be mounted on any type of carrier or base, but truck, rough terrain, and the relatively new all terrain carriers predominate. A crawler base, although offered by some manufacturers, seems inconsistent with the concept of a telescopic crane, as we shall see, but a crawler base can be useful for special applications.

Although telescopic cranes have been marketed for about 30 years, new design innovations continue to appear. Self weight has always been a limiting factor for cantilevered boom structures, therefore the thrust of innovation has been to reduce weight so as to permit longer booms and greater lift capacities. Boom shapes offering improved structural efficiency have been introduced and telescoping systems have changed. Some cranes now use combinations of hydraulic cylinders and rope reevings for boom extension. Others use one cylinder to extend and retract boom sections and remotely activated pins to hold the sections in position for work. Typical conventional designs use one cylinder or rope reeving per section, and those cylinders or ropes support the section and the working loads.

Present rough terrain cranes are available in capacities from about 14 to 100 tons (12.7 to 90.7 t) and carry boom plus extensions as long as 200 ft (61 m). Telescopic truck cranes are produced up to 850 ton (770 t) capacity with telescoping booms as long as 197 ft (60 m) in length. All terrain cranes are offered within those capacity ranges. Fixed and luffing jibs and latticed extensions of various types are furnished with many telescopic cranes to provide greater hook heights and add other capabilities to these versatile machines. When jibs or extensions are mounted at an angle to the boom, this gives the crane the ability to reach over obstructions, such as a roof edge, and place the load further onto the roof than a straight member would. When a luffing jib is mounted, the telescopic boom is set at a fixed high angle, and the jib is rigged with a topping lift so that its angle may be adjusted to change radius. Luffing jibs give the crane capabilities similar to a tower attachment on a latticed boom crane.

Telescopic booms are made up of a series of rectangular or trapezoidal cross-sectioned segments (or of other symmetric shapes) fitting into each other. The largest segment, at the bottom of the boom, is called the *base,* or *base section.* The smallest in cross-section, at the top of the boom, is the *fly*, or *fly section.* In between, there can be one or more sections, which we shall call the first, second, third, etc., *midsections* (there is no agreement in the industry on boom-section terminology). At the upper end of the fly section is the *boom head,* which carries the upper load and lead sheaves.

When telescopic booms are fully powered by hydraulic cylinders, each telescoping section is coupled to an *extension cylinder*, although fly sections are often unpowered; they are extended manually or by other means and pinned in the extended position. Surprisingly, long extension cylinders may buckle under load, but upon buckling they come to rest in contact with a wall of the boom section. This "stiffens up" the cylinder enabling it to carry more load. When load is removed, the cylinders become straight again. This phenomenon explains many instances where booms cannot be extended while under load.

Some booms utilize combinations of extension cylinders and rope reevings. These tend to be lighter cranes, say 50 tons (45 t) or so in capacity. The extension cylinder drives a boom section while pulling the rope which extends another section; both sections extend together. Yet another system uses light-weight, light-capacity cylinders to extend the boom sections, but once extended the sections are locked into place with automatically inserted pins. These booms can only be extended while unloaded because the pins carry lifted load effects, not the cylinders. Some cranes utilize only one extension cylinder which sequentially extends as many sections as needed. These "extension only" cylinder schemes result in lighter weight boom assemblies; the weight reductions end up as increased lift capacity. It should be noted that these booms operate at a limited number of fixed booms lengths, as opposed to the continuously variable lengths offered by conventional telescopic booms.

For luffing, or changing the boom angle, telescopic cranes use one or two exposed hydraulic rams called *boom hoist cylinders* (Fig. 2.20). As a further means for increasing lift capacity without increasing weight, some telescopic truck cranes utilize movable counterweights. With outriggers extended, when the boom is lifted from its travel cradle, the counterweight automatically extends backwards and remains in extended position throughout lifting operations. Crane motions and the controls that activate them are shown in Fig. 2.21.

The boom sections telescope, one within the other, on slider pads or rollers. During use, particularly in dusty environments, the moving parts wear, allowing increased clearance, or *slop,* to accumulate be-

Figure 2.20 A Grove telescopic truck crane with four-section boom at work on a cooling tower. Note how the counterweight extends when the crane is in working position. (*Grove Worldwide*)

tween sections. Routine inspection and occasional replacement of bearings or pads or insertion of shims is required, as slop can become dangerous. Slop increases load radius without changing the reading of the boom-angle indicator. As slop accumulates, the margin of stability decreases and there is increased danger of tipping failure.

To increase the length of telescopic booms, an additional nonpowered section, often called a *manual insert,* can be added at the tip. When this is done, the boom head is mounted on the insert. The insert is pinned in place in either the retracted or the extended position. For further increase, a latticed extension is sometimes offered. This extension attaches to the boom head at four points and acts as a cantilevered member. When not in use, some extensions can be folded sideways and latched to the boom base section for storage. Extensions usually are mounted in-line with the boom centerline, but some can be fitted at an offset angle. Because of their light weight, load ratings may be greater on a latticed extension than on the telescopic boom alone while at the same radius. A latticed extension is mounted on the crane in Fig. 2.22.

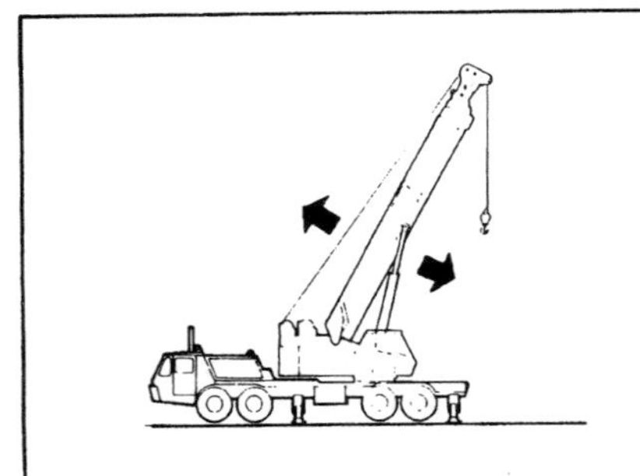

SETTING BOOM ANGLE. Position the boom at the proper angle for the load and working conditions. Refer to the rating plate for proper angle. Pull the boom hoist lever back, or depress the heel of the boom hoist pedal, to raise the boom. Push the boom hoist lever forward, or depress the toe of the boom hoist pedal, to lower the boom. Be sure to pay out line from the main and/or auxiliary winch, to prevent the hook blocks from coming in contact with the boom point.

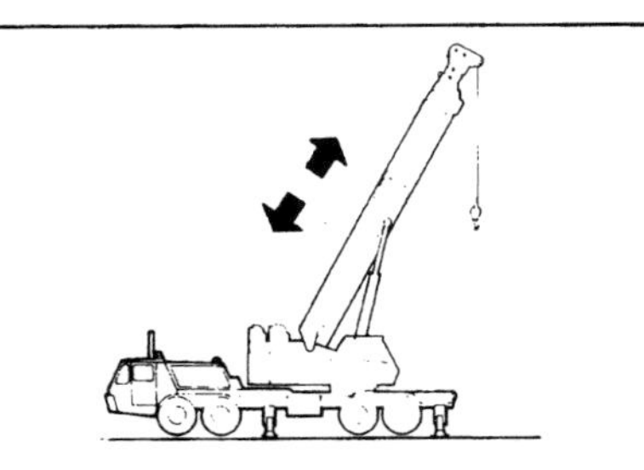

SETTING THE BOOM LENGTH. Push or pull the telescope lever to position the boom at the required length. Check to see that all boom sections are equally telescoped. If section lengths are uneven, depress and hold the individual button for the section to be telescoped. Move the telescope lever forward or back to control the movement in or out for the individual section. Be sure to pay out line from the main and/or auxiliary winch to prevent the hook block(s) from coming into contact with the boom point. Always refer to the rating plate for the proper boom length for the load being lifted.

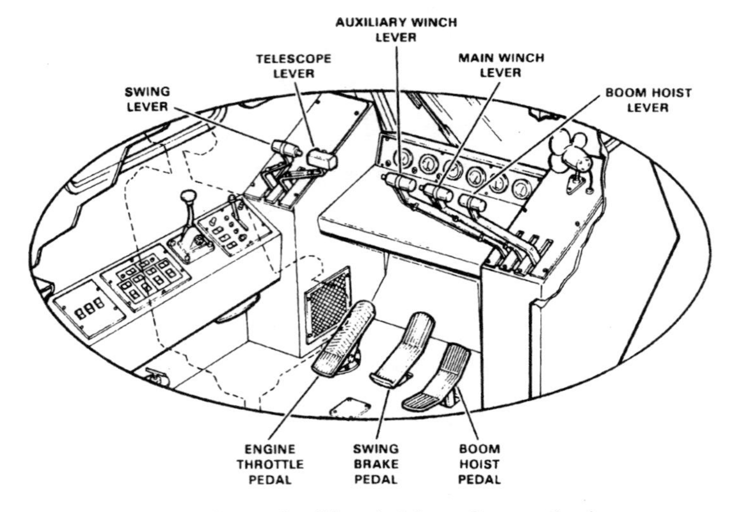

Figure 2.21 Crane operating cycle. (*Harnischfeger Corporation.*)

Latticed jibs and several styles of solid member jibs (Fig. 2.23) are offered for telescopic cranes to increase lift height. They mount either on the boom head or at the tip of the latticed extension.

Most telescopic cranes can be immediately ready for work upon arrival at the jobsite; fly sections and jibs take a little time to deploy. Only maneuvering into position, cribbing the outriggers, and leveling the crane would be required before operation. Many inserts, extensions, and jibs are arranged for rapid deployment. Thus, a great advantage of this crane style is its availability for work with little investment of time or labor. Increasingly these advantages overcome the

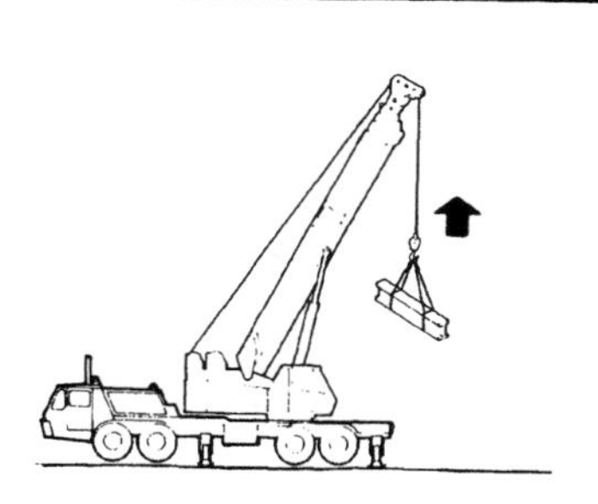

LIFTING THE LOAD. Push the main or auxiliary winch control lever forward to lower the hook block. Attach the load to the hook block. Then pull the main or auxiliary winch lever back to raise the load. Depress and hold the button on the side of the winch lever to operate the winch in high speed. The winch will operate in low speed when the button is not depressed.

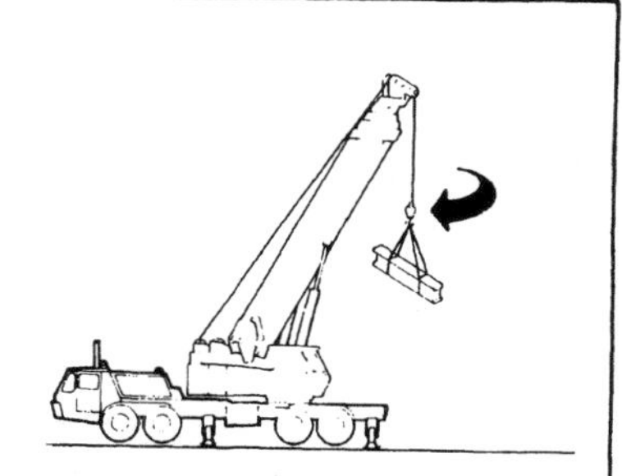

SWINGING. Push the swing lever forward to swing the upper to the left (toward the boom). Pull the swing lever back to swing the upper to the right (away from the boom). Plug the swing lever to bring the upper to a stop, and then depress the swing brake pedal to hold the upper. Place the swing brake lock, with the swing pedal depressed, in the *ON* position to hold the upper stationary without the use of your foot.

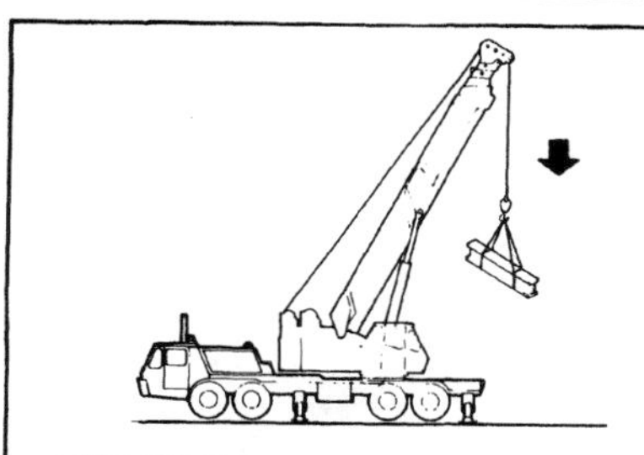

LOWERING THE LOAD. Push the main or auxiliary winch lever forward, to detent, to power down the boom. Push the winch lever past the detent to lower the load under controlled free-fall conditions.

CAUTION

Return the winch lever to the power down position slowly to avoid shock loading the winch and winch line.

Depress the button along side the winch lever to lower the load at high speed. The winch will lower at low speed when the button is not depressed.

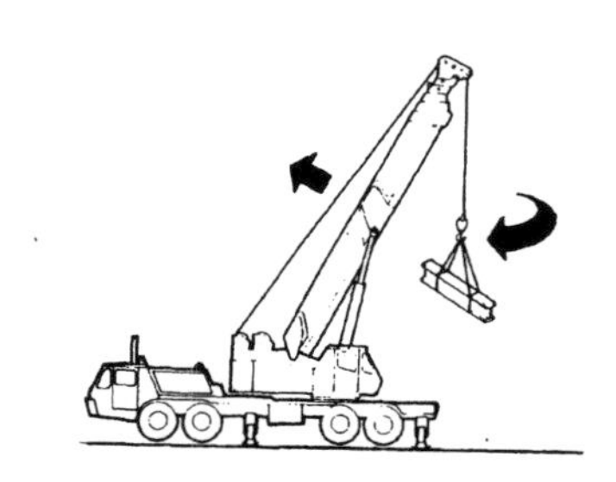

SPOTTING THE LOAD. Spotting the load requires accurate control of the boom and swing movements. It takes practice to locate the load at the exact spot without hunting or overshooting. Adjust the boom as required to accurately locate the load. Never extend the boom out so far that rating is exceeded. See the rating plate.

A655

Figure 2.21 *(Continued)*

high cost in relation to lift capacity giving telescopic cranes an ever-expanding share of the crane market. Even very large telescopic cranes enjoy this advantage, and some are very large indeed. Partial disassembly is usually required for transit, with latticed sections and counterweights hauled by truck for assembly at the site. Nonetheless shipping costs are moderate for the capacity of the crane and assembly times are reasonable. In comparison with other types of similar capacity, these cranes still get to work quickly after arrival at the site.

Telescopic cranes are often worked in tight quarters, even under conditions that are less than optimal. In roadwork, for example, lane restrictions can make full outrigger extension unfeasible. One solution

Figure 2.22 The latticed extension on this P&H CNT 280 crane is furnished with a slide-out pinned additional extension section, providing an appreciable increase in hook height. Extensions are used to increase hook height, not reach, but sometimes the added height enables the crane to reach past an elevated obstruction. (*Terex Corp.*)

Figure 2.23 Lorain LRT-15H fixed-cab rough terrain crane. A solid-section jib is mounted offset at the boom tip to increase lifting height. (*Koehring Company, Lorain Division.*)

is to utilize the "on tires" lift rating charts provided with most models. However, this leads to a substantial reduction in lift capacity.

Pressure from the construction industry has induced some crane manufacturers to offer partial outrigger extension as an enhancement over on tire ratings. Not all cranes can be so used, however. Only those cranes specially arranged and rated by the manufacturer are permitted these capabilities. Although cranes rated for partial outrigger extension have obvious advantages, they have drawbacks as well. These matters will be more fully discussed in Chap. 5.

2.4 Hammerhead and Luffing Jib Tower Cranes

Though tower cranes and mobile cranes evolved on opposite shores of the Atlantic Ocean, in the modern marketplace each has its own niche. Where the work is at great heights, or space is sparse, tower cranes are likely to be the choicest lifting machinery. Other conditions may favor tower cranes too; long term assignments and the need to sweep a large area are among these. There are tower cranes to satisfy the needs of nearly all construction operations, and they are adaptable to many industrial tasks as well. Tower cranes were used to build the world's tallest buildings, the twin towers of the Kuala Lumpur City Center, for mucking operations at a subway tunneling site in New York City, and to construct the Ekofisk One oil-drilling platform in the North Sea off Norway. They are also used at papermill woodyards and bridge construction sites. Small, lightweight easily transported tower cranes are common throughout Europe.

Contingents of this army of small cranes can be seen just about anywhere, except in North America, where mobile cranes predominate on small to medium sized projects. For larger projects, tower cranes have become a familiar sight in the United States and Canada. There is no limitation to the height of a high-rise building that can be constructed with a tower crane. The very high line speeds, up to 1000 ft/min (5 m/s), available with some models yield good production rates at any height. Some machines can operate in winds up to 45 mi/h (70 km/h), which is far above mobile-crane wind limits. High capacities, hook heights, and reaches now available enable tower cranes to service a far wider range of work than ever before.

Costs associated with installing and removing a typical large tower crane are high, but so is its productivity. Economics thus would indicate the use of these machines only for longer term projects with many lifts. This economic relationship is not absolute, however, and large tower cranes are commonly installed on rooftops in New York City for the relatively short-term task of erecting stone facing panels on buildings under construction. Lighter machines, particularly models that can be folded into a road-suitable trailer configuration, require less formal planning and installation cost.

Most tower crane designs follow the "Erector set" concept; that is, they are comprised of multiple components of common design. Furthermore, components are often shared with both larger and smaller models.

Electric motors are used to power all functions in many tower crane models, but North American- and Australian-made machines usually feature hydraulic drives and diesel prime movers. The control systems vary but the electric-driven cranes often include high- and low-speed ranges with stepped increments in each range; some, however, are equipped with solid-state step-less control. Friction and/or eddy-current brakes are provided on electric machines, usually "normally engaged," and automatic acceleration and deceleration rate-control devices for all motions are typical on the larger cranes. Remote controls are sometimes offered.

Towers are said to be *fixed* if the swing circle is mounted near the tower top as in Figs. 2.24 and 2.25 and to be of the *slewing* type if the swing circle is near the base as in Fig. 2.26. Slewing is the preferred term for the swing motion in the tower crane industry.

Figure 2.24 Richier 1268 saddle jib tower crane during construction of an apartment structure at North Bergen, New Jersey. Note the hoist line dead-ended at the jib tip. This keeps the hook at constant elevation as the trolley traverses on this French-made crane. (*Tower Cranes of America, Inc.*)

Jib types

Tower crane people use the term *jib* for the member that we in North America would call a boom. There are three jib types in general use. Jibs mounted to the tower in a horizontal or slightly upward inclined position are called *saddle,* or *hammerhead, jibs* and are illustrated in Fig. 2.24. A *luffing jib* is mounted with a pivot and derricking system to a machinery deck on the tower, as shown in Fig. 2.25; it is similar in many respects to a mobile-crane superstructure mounted on a tower. Other luffing jib tower cranes, with slewing-type towers, feature arrangements that are similar to a mobile-crane tower front end attachment (Fig. 2.26). Finally, there is an *articulated,* or *goose-neck, jib* (Fig. 2.27).

Saddle jibs. Saddle jibs feature a trolley suspended from the bottom of the jib that is capable of traveling the length of the jib to change the hook radius. The trolley includes the upper block and supports the

Figure 2.25 A Favco model 750 luffing boom tower crane at work as an external climber braced to the structure at this midtown New York construction site. The larger-sized tower segment just below the crane upperstructure is the climbing frame which is used to raise the crane for insertion of another tower section. (*A. J. McNulty & Co., Inc., photo by Robert Weiss.*)

load block suspended from it. The load hoist system is reeved with the dead end at the jib tip, which causes the load block to maintain constant elevation as the trolley changes radius. Trolley travel (hook radius) is controlled by an independent winch and rope system.

A second jib projects from the tower opposite the saddle jib. This is the *counterweight jib,* or simply the *counter jib,* which supports the load winch, power plant, and control panels as well as the counterweights. A small tower projects above the jibs, called the *top tower,* serving as an anchorage for the pendants supporting both jibs. Pendants may be wire ropes or, as on the larger cranes, steel bars. The top tower also houses the lead sheaves, giving sufficient distance to

Figure 2.26 KB-160 luffing jib tower crane with 143-ft (43.6-m) tower and 82-ft (25-m) reach. Capacity is 8.8 tons (8 t) for this rail-mounted crane, which is typical of construction cranes built in the former Soviet Union. The slewing-type tower telescopes from the bottom for self-erection and dismantling. (*VNIISTROIDORMASH.*)

the winch to maintain proper fleet angle. The operator's cabin is typically mounted just below the jib but above the slewing circle. With this arrangement, the operator has full view of the load.

With few exceptions, no obstruction can be permitted to prevent a saddle jib crane from slewing through a full 360°. The crane must swing freely with the wind, or *weathervane,* when out of service. It is not designed to accommodate storm winds taken from the side of the jib but is arranged to slew freely in order to present its smallest profile to the wind. Thus, at the completion of a work shift the operator must leave the crane with the slewing brakes disengaged. For hemmed-in sites, however, some cranes can be furnished with short jibs that can be locked in position when out of service and can endure storm-wind forces.

Tower cranes are usually designed for storm winds of about 80 mi/h (36 m/s) at jib elevation, but this may vary with the manufacturer or country of origin. The equivalent wind speed at the ground is about 65 mi/h (29 m/s). What that means is that special provisions must be considered at most coastal or hurricane-prone sites.

Many safety aids and operational limiting devices are provided with saddle jib tower cranes. A device that senses line pull and cuts winch

Figure 2.27 The upper portion of a Kroll K-103V articulated jib tower crane. On this Danish-made machine, a counterweight is used to aid in luffing, and the hook can be kept level as radius changes. (*F. B. Kroll A / S.*)

power when the permitted level is exceeded, together with another sensor mounted on the jib pendants, are intended to prevent overloads. Limit switches prevent overtravel of the trolley at either end. Another limit switch prevents the load block from striking the upper block, an occurrence known as *two blocking*. Limit switches can be set to prevent operation or to limit operating radius in any part of the slewing arc.

Because the tower is a cantilevered member subjected to high bending moments, each tower leg must be designed to resist alternating compressive and tensile forces. Given the usual stress levels and the expected number of lifetime loading cycles, leg design does not pose a difficult fatigue problem. The connections between tower sections are another matter, however, and are a critical design and maintenance problem. Very high strength bolts must be used, and it is essential that they be tightened or prestressed to carefully controlled levels. Failure to install tower connection bolts properly or to retighten them from time to time can result in bolt fatigue failure and tower collapse. This problem has led to the development of new tower connectors without bolts which are found on some cranes today.

Saddle jib cranes are available in a great many sizes, from very small, easily transportable devices to very large imposing machines developed for limited use at nuclear power plant construction sites. Although not the largest of the nuclear plant cranes, Fig. 2.28 illustrates what is certainly at the larger end of the spectrum.

Saddle jib cranes are by far the most popular of the tower cranes outside of North America, but luffing jib cranes are achieving greater popularity and domestically represent a high percentage of new orders. That trend is expected to continue together with the North American preference for diesel/hydraulic as opposed to electric-drive equipment.

Luffing jibs. Older-type tower cranes with luffing jibs utilize slewing towers as can be seen in Fig. 2.26. The tower is mounted on a slewing platform, which also carries the power plant and the counterweights, while the jib is supported and luffed by fixed pendant ropes running from the jib tip to pivoted struts at the tower top and then down to a spreader. The spreader holds the sheaves to which the running luffing ropes are reeved. The vertical ropes in the luffing system are roughly parallel to the tower and act to relieve the tower structure of most of the bending load. Thus, the tower on this type of crane is essentially a compression member.

More modern luffing jib machines are of the fixed-tower type; that is, they comprise a rotating superstructure on a slewing bearing at the top of the tower as shown in Fig. 2.25. Many of these machines resemble mobile-crane uppers on a tower, but in any case whether electric or diesel/hydraulic, all of the machinery is at the top. Capacities run to about 230 tons (209 t) and booms of as much as 240 ft (73 m) can be mounted.

For a given tower height and jib length, a luffing jib is capable of greater work heights than a saddle jib at all but the longest radii. The saddle jib changes radius by trolleying, which is much faster than luffing.

Luffing jib cranes must be left free to weathervane when out of service, but often their booms may be positioned at high angles (which vary with manufacturer and configuration). This makes the luffing jib crane very attractive for use at urban sites where the crane can be positioned to weathervane clear of taller structures alongside. However, if the boom is set at too high an angle, the crane may fail to weathervane.

Figure 2.28 Kroll K-1800 saddle jib tower crane of 55 tons (50 t) capacity. The maximum reach is 246 ft (75 m), and the maximum height under hook is 302 ft (92 m), dwarfing the nearby mobile cranes. The tower is 16.4 ft (5 m) square. (*F. B. Kroll A / S.*)

Articulated jibs. Because of their specialized nature, cranes of this type (Fig. 2.27) are uncommon. Each is so unique that as a class they defy general description other than to say that the jib has a pivot point somewhere in its middle area. Some models are *level-luffing;* that is, the hook elevation remains constant as radius changes. It is possible to provide either a trolley or a fixed-location hook or even a concrete pump-discharge line.

Articulated jibs are mounted on towers identical to those used with saddle jibs. Some designs are not required to weathervane. In this case the crane is made to withstand storm winds with the jib drawn in to minimum radius; there is then no need for a 360° obstruction-free slewing path. One type of crane has a hinged jib arranged so that the outer portion remains horizontal. As the jib is folded, the outer portion rises, giving increased height and the ability to pass over obstructions.

Tower mountings

Project site conditions and needs together with equipment type and capability dictate the type of tower mounting to be used. There are three basic mountings: the static, or fixed, base; the traveling base; and the climbing base.

Static base. This mounting is used with fixed-type towers. The tower is set on or into a foundation block constructed of concrete, onto a ballasted frame, or tied down with rock anchors as in Fig. 2.29. Some manufacturers prefer to use anchor bolts for attachment, while others

Figure 2.29 Tower base section for a high-capacity luffing boom tower crane, the Cornell TG-1900. The tower is anchored to a small concrete footing block, which in turn is held by rock anchors. The steel plating on the lowermost tower section serves to accommodate wedging loads when the crane leaves its static base and becomes an internal climbing crane. (*Photo by Jay P. Shapiro.*)

prefer to cast part of an expendable tower section into the concrete footing. The result is the same in either case; the tower is rigidly fixed to the foundation. A common optional base making use of knee braces is shown in Fig. 2.30. The tower and braces attach to a frame that provides a wide anchorage base and reduces footing mass.

The tower is erected as a freestanding cantilevered member, and the crane delivers vertical and lateral forces as well as overturning moments to the foundation. The vertical force is the weight of the crane plus the load for in-service conditions but only crane weight when it is out of service. The lateral load, induced by wind, is minor, but effects from slewing torsion may need to be considered. The overturning moment derives from wind and the loaded crane jib for in-service conditions or storm wind less dead-load moment in the direction of the counterweights for out-of-service conditions. In-service loading governs for short towers and their foundations, while out-of-service loading controls for tall towers and their foundations. The governing condition for the design of the tower requires that the overturning

Figure 2.30 Knee-braced base for a freestanding tower crane. The tower and braces stand on a cross frame, which rests on the support surface. The ends of the frame transmit the compressive and uplift reactions. (*Tower Cranes of America, Inc.*)

moment be applied across the diagonal of the tower; the governing case for design of the foundation may be different, however.

Tower foundations must be capable of resisting the applied loads with adequate protection against overturning. This can generally be accomplished with somewhat high soil bearing values, inasmuch as settlement does not affect crane operation or safety. Differential settlement is another matter; the entire subject of crane foundations is treated in detail in Chap. 6.

Static-mounted cranes erected to their full permitted freestanding height can then be braced to the structure being built. With periodically spaced braces, there is virtually no limit to the crane height that can be erected. The process by which the crane is self-erected to increase height is called *telescoping, jumping,* or *top climbing*. There are several methods by which this is done, but in general an additional section of tower is put into place near the top of the tower. By means of a mechanism, usually employing hydraulic rams, the crane is raised onto the new section, or a new section is inserted into a gap created in a climbing frame. Figure 2.25 shows a crane with a climbing frame while Fig. 2.31 shows a crane that climbs by adding a section at the top of the outer tower.

Figure 2.31 Richier model 1296SE at a nuclear power plant site. Note the telescoping equipment just below the point where the tower narrows at midheight. New sections of tower are put in place on top of the outer tower (the wider one), and the crane telescopes upward by means of a hydraulic ram, cross-beams, and dogs. (*Tower Cranes of America, Inc.*)

Models differ considerably with respect to the labor and time needed to climb. Moreover, the cost of bracing to the host structure can vary. These factors should be assessed during planning as they can have a significant impact.

The author's firm had the pleasure of designing several interesting and unusual telescoping crane installations for construction of hyperbolic cooling towers up to 400 ft (122 m) in base diameter and 530 ft (162 m) high. In each installation, a tower crane was mounted at the center of the cooling tower and climbed as the work progressed. Because the thin concrete-shell sides of the cooling towers did not have sufficient strength for bracing these cranes, sets of eight tower guys were installed at each of three levels as crane height increased, and were anchored at the strong reinforced-concrete ring at the structure base. Storm winds proved to be the governing loading case for each level of guying. Because of the steep guy angles, the catenary shape of the guy ropes made the rope spring rate far from linear, and complex procedures were needed to solve for the guy loads. The guys, of course, would interact with the tower, so that tower elastic deflections also had to be determined. The problem was not only that of designing the guys; the structural competence of the tower and its connections had to be verified for the unusual installation height and local wind conditions as well.

Traveling base. A traveling base comprises a tower platform on which *ballast* weights are also stacked, four sets of wheels or bogies, and steel rails. The tower is rigidly fixed to the platform, and knee braces are generally used as well. The bogies have electric drives.

The rails must be well supported so that they will not bend or settle excessively under operating loads. The rails must also be installed level to keep the tower plumb. A section of track, called a *parking track,* is set up with anchorage arrangements and added strength to resist the loads produced by storm conditions. Curved track can be used in the travel path. Travel bases enable one or more cranes to be arranged to provide hook coverage throughout a large construction or industrial site.

Cranes on travel bases can be telescoped in the same way as static-mounted cranes, but of course bracing cannot be used for increasing height. The crane can be assembled to a height easily erected by whichever mobile crane is available or can be self-erected in some cases and then built to full height by telescoping.

Climbing base. When a crane is installed within the confines of a building under construction, the crane can raise itself by use of a climbing base as work progresses. When arranged this way, it is called an *internal climbing* crane. The initial installation usually freestands on a static base, and transfer to the climbing apparatus takes place after the structure has risen several stories. The crane can sometimes be located in an elevator shaft, but when this is not practical, floor openings can be made to accommodate the crane.

A climbing base is a steel frame that transmits the weight of the crane to the host structure. In some arrangements, the base remains attached to the crane as it climbs while in others it is a detached component. If it remains attached, the base typically has some form of retractable ends arranged to clear the floor openings as the crane climbs. A detached climbing base is usually made to disassemble easily; the pieces would then be brought to the edge of the building for the crane to raise for its next use.

A climbing base works in tandem with a *climbing frame*. In traditional systems, the climbing frame supports a pair of *climbing ladders*. Climbing is carried out by hydraulic rams and *dogs* or pawls that enable the crane to advance up the rungs. Some models can climb up to ten stories in one operation. At the completion of a climb, the tower is wedged horizontally to the building structure. Wedges are placed at each corner of the tower at two levels so as to transfer horizontal forces and overturning moment. There are a number of variations from the climbing ladder system. Some models climb steel bars or threaded rods, and others are picked up by rope and winch systems.

Bars, rods and winches all rely on raising the crane using the same general principal; the ascending crane is suspended from the climbing frame by tension members. Alternatively, tower cranes can be jumped using push-type climbing systems. Some arrangements utilize long hydraulic rams that are capable of raising the crane a full floor height in one stroke. Others push down on a central pole; in principal this is a climbing ladder in reverse.

Crane weights are substantial [they can weigh 200 tons (180 t)], so that in most installations the climbing-frame reactions will overload the structure of any one floor. It is then necessary to install temporary shoring to distribute the loads to additional floor levels.

The topmost floor, after a climb, is often a wedging floor. With a fast construction sequence or during cold-weather operations it is critically important to be certain that the structure can support the wedging loads. This matter is discussed more fully in Chap. 6.

It is necessary to study the host structure carefully in order to select a crane location best suited for operational needs and for installation economy. We have seen instances where it was possible to avoid shoring altogether and others where fourteen stories of heavy timber shores were needed. Shoring represents considerable cost, since it must be moved up as climbing takes place. Climbing must be planned in advance and each mounting level predetermined. Inasmuch as climbing is usually performed at overtime and is a wind-sensitive operation, it can have an important effect on job progress and cost.

When the structure is completed, there is a tower crane on the roof and no way for the crane to disassemble or lower itself (Fig. 2.32). If the crane is within reach of a mobile crane, this will usually be the most economical means of removal. Another method requires that a derrick, usually a stiffleg derrick, be installed on the roof for dismantling and lowering the crane components piece by piece. Helicopters are occasionally used to remove tower cranes. However, components of large cranes are generally too heavy for helicopter lifts. Aside from the rigging problems, a safe flight path and a close-by staging area are needed.

The size of a climbing tower crane in no way limits the height of structure which it can build. It provides fast lifts and reasonably fast slewing speeds. The practical questions of how to support the crane on the building structure are the only limiting factor affecting the size of crane that can be arranged as a climber.

Figure 2.32 The same crane and project shown in Fig. 2.24. The concrete frame has been topped out, and the crane is being used to raise materials for the building enclosure and interior work. The work involved in disassembling and removing the crane can easily be visualized from this photograph. (*Tower Cranes of America, Inc.*)

2.5 Pedestal-, Portal-, and Tower-Mounted Cranes

Any type of crane previously discussed can be portal- or tower-mounted and any of the revolving superstructure cranes can be pedestal-mounted. The purpose of these mountings is to raise the crane boom or jib so as to clear obstacles or to gain better access to the work. Portal and tower mounts can provide a travel function as well (Fig. 2.33).

The cranes mentioned earlier in this chapter are essentially building service cranes—cranes designed to operate with varying loadings, only a small percentage of which are full structural-rated loads but all of which are lifted intermittently in single-shift work. Many cranes, particularly those with ratings controlled by stability, can endure more severe service. Many tower cranes are capable of heavier service at reduced load ratings, such as the portal-mounted tower crane in Fig. 2.33.

As severity of service increases, the proportion of lifted loads that are at full rating increases, the number of daily loading cycles increases, and the speeds and accelerations applied to the various mo-

Figure 2.33 Traveling portal-mounted Kroll K-400 at a shipyard in Falkenberg, Sweden. The portal mount permits the crane to pass over materials stored between the tracks. Note the ballast at the tower base. (*F. B. Kroll A / S.*)

tions may also increase. Thus, the possibility of fatigue damage may be introduced and the statistical probability of overload is increased, as is the probability that minor defects will cause trouble; on the whole, the structure assumes a lower level of reliability when a building service crane is subjected to this heavier service. The new reliability level may be entirely satisfactory, or a reduction in ratings may be necessary to improve reliability.

For heavy service, such as *duty cycle* work, more rugged machines are needed. Duty cycle work comprises steady work at a fairly constant short cycle time with fairly constant loading levels for one or more daily shifts. Ship unloading with a bucket is typical of duty cycle work and is but one of many industrial tasks requiring sturdy lifting equipment.

Revolver cranes are a series of large latticed boom cranes that resemble mobile-crane uppers, available in capacities to 3000 tons (2700 t). They are often barge-mounted, but when portal-, pedestal-, or tower-mounted they are used at such places as dam construction sites and shipyards. Many can be seen in the Scottish fabricating yards preparing drilling platforms and other facilities for use in the North Sea.

There are a number of crane types, called *ship unloaders* or *dockside cranes,* with either latticed or plated booms. These are ordinarily pedestal-, portal-, or tower-mounted, and many of them are level-luffing.

A pedestal is a fixed mounting to which the crane's slewing ring is attached. Pedestals may be made out of large steel tubes, mass concrete or a concrete tube, steel frameworks, or even a concrete-walled core section of a building. The pedestal serves as a means for raising the crane to the required working height and for anchoring it structurally by resisting all loads induced by the crane and its operation.

Cranes are portal-mounted to increase working height while limiting support weight, to permit transport equipment to pass below the crane, to allow crane travel, or for a combination of these reasons. A portal is a legged steel support structure. Most portal cranes travel on rails with electric-powered drives, but the cranes themselves may be electric-, diesel-, or diesel-electric-powered (Fig. 2.34).

When still more working height is needed, a steel-framework tower is mounted on the portal frame (see Fig. 2.33). Capacities of portal- and tower-mounted cranes may be governed by resistance to overturning. In this case, backward stability criteria must also be satisfied.

2.6 Overhead and Gantry Cranes

An overhead crane is a traveling machine that rides on a runway structure or pair of tracks above the work floor. The crane includes a

Figure 2.34 Kroll K-3000L portal-mounted crane. This crane with an unusual configuration has a capacity of 138 tons (125 t) at 82 ft (25 m) and 38.6 tons (35 t) at a reach of 197 ft (60 m). The height to its slewing circle is 121 ft (37 m); the crane is Danish-made. (*F. B. Kroll A / S.*)

bridge that spans between the tracks and a fixed or trolley-mounted hoisting system. Overhead cranes, also called *bridge cranes,* can be manually operated but when powered are almost invariably electric. They are sometimes formally referred to as overhead electric traveling (OET) cranes. Cranes may be top-running on rails or underslung, in which case they are suspended from, and ride along, the bottom flanges of the runway beams. Bridge girders may be either single or double. Cranes of 250 tons (225 t) capacity with 100-ft (30-m) spans are not unusual.

OET cranes are mainly used in industrial service but are also found at stone or concrete precasting yards, steel fabricating shops, and storage facilities.

A gantry crane is an OET crane mounted on legs which in turn ride on rails. The bridge, which can be a trussed or plate structure, may be cantilevered on either or both ends. When one end of the bridge is on legs and the other on a runway beam, the machine is called a semigantry crane.

One of the authors was astonished when he first saw the gantry crane in use at the Kockums AB shipyard at Malmo, Sweden. The crane is immense; the bridge spans some 600 ft (180 m), is the same distance above the ground, and has a capacity of about 1650 tons (1500 t)!

2.7 Unconventional Lifting Devices

Lifting of heavy objects must often take place in situations where a crane or derrick is impractical. The space may be confined or the overhead room restricted, or perhaps the weight or movement path cannot be handled by conventional equipment. Not uncommonly, these problems have been solved by multi-stage operations, or by onerous, time-consuming "brute force" rigging schemes. Classical rigging entails moving heavy objects by means of jacks, cribbing, rollers, or other such basic devices which provide mechanical advantage or temporary means of support, but at the expense of much time and labor.

Alternate equipment have been developed over the years to provide more efficient solutions to unconventional lifting problems. Some of these have developed niche markets in places like oil refineries and electric power plants.

Jacking towers

Jacking towers are used primarily for vertical movement of large loads. The lifting equipment can be assembled with relative ease in tight quarters or out-of-the-way locations where a large crane may not be suitable. There are two types of systems used for jacking loads with towers: those that push the loads up, and others that pull from above. The first type can be called the push-jacking system, but the second is generally called the strand jacking system.

A typical jacking tower is a frame supported on a foundation and perhaps guyed for stability. Towers may be triangular or rectangular in cross section. A pair of towers is usual, but four towers can be utilized when needed. When taller than the load, towers can be braced to one another and guys may not be needed or fewer guys employed; when shorter than the load, a combination of braces and guys may be used.

The push-jacking system uses a cross beam from which the load is suspended, hence towers must always be taller than the load. The cross beam in turn is suspended from yokes which climb within guide slots built into the near legs of each tower; a pair of hydraulic jacks at each end pushes the beam up until a locking plate, or dog, can be inserted and thus support the load. With the cross beam safely seated, the jacks can be retracted in preparation for the next increment of climbing. The jacks are short and remain in place; climbing is accomplished by inserting solid steel lifting pistons each time the jacks are retracted. Previously placed lifting pistons rest on the dog between jack strokes. There are two jacks located at the base of each tower; the lifting pistons are inserted, and the lock plates are located, at the base of the towers. Thus, all climbing work is accomplished from the ground.

Push-jacking system advantages include:

1. Positive load guidance by means of tower leg guide slots.
2. Jacking and insertion of lifting pistons is done from the ground.
3. The load is supported on solid steel elements when not on the jacks.
4. Hydraulic jacks lock if they accidentally loose pressure, maintaining the position of the load.
5. Rudimentary system is easily understood.
6. Synchronization and adjustments are easily made.
7. Field repairs to towers are not difficult to make.

Disadvantages of push-jacking systems are that:

1. Few units of this type are available.
2. Easy lateral movement is possible in the direction of the cross beam. Other adjustments may be possible by pulling on the suspended load or may have to be done after the load is landed.
3. The operation is somewhat slow with maximum lift speed about 25 ft (8 m) per hour.
4. Guys are needed and these should be preloaded because the towers have fixed bases. Preloading needs to be controlled and checked, and the guys induce axial loads in the towers and on the foundations.
5. As many lift operations extend for more than one day, wind and weather are factors.

A strand jacking system requires suspending loads from framing at the tower tops; the framing can be concentrically supported by the towers or may be supported on the “close” tower legs. Slings or lifting bars are attached to the load and run up to a lift beam. The lift beam is fitted to receive the lifting tendons which are typically a bundle of 18 mm diameter high strength wire rope strands selected to satisfy load weight.

The lifting tendons go up to strand jacks mounted on the overhead cross beam (really a pair of beams or trusses). The center hole strand jacks pull up the tendons stroke by stroke, gripping them for jack retraction, and repeating the cycle as needed. At the start, as the jacks begin to extend the upper grippers grab the tendons by means of friction and wedging forces. At the end of the stroke, the jacks are reversed and the lower grippers similarly engage, after which the upper grippers release. After full retraction, the cycle is repeated. Withdrawn tendon material is spooled. The power pack powering the jacks is normally located at top as well.

A pair of towers can support one or two point picks. Four towers can be used for either two or four point lifts. In the four-tower arrangement, a cross beam is employed to support each pair of load supporting jacks, and header beams support the cross beam (or beams). As heavy loads are often supported near the center of the pair of headers, these are commonly robust members. A four tower set-up can be self bracing, eliminating the need for external guys. The savings in guys and deadmen, however, would probably be exceeded by the costs of shipping and erecting two extra towers, header beams, and rigid bracing members. Therefore, a four tower arrangement would be used when there are clearance limitations on tower positioning or where the load must be moved between pick up and placement or when a four point lift is necessary.

When two towers are used, they may have to be anchored to a substantial moment resisting footing or would need guys to restrain wind and side loading during the operation as shown in Fig. 2.35. If the towers are guyed, the guys should be preloaded because of the base fixity. Preloaded guys imply larger deadmen for anchorages than passive guys.

Strand jacking offers precise control over the load. The system can also be configured to permit transverse and longitudinal load movement, and even rotation, for precise placement, but these features come at added cost. Typically, for loads of common magnitude, chain falls or cum-a-longs can be used to pull the load as needed for final alignment. If sufficient drift height has been provided, pulling forces are not very large and can be easily accommodated by the system, but must be allowed for in advance by the installation designer.

Although jacking equipment is at the top of the towers, controls and instrumentation can be on the ground. Pressure and stroke data for each jack (including guy preload jacks) can be continuously metered and can show up in readouts at an instrumentation station. One control lever operates the jacks and a second lever operates the gripper hydraulics. Jacks can be operated singly or simultaneously; speed can be varied as well. However, with the mechanical action taking place

Figure 2.35 A strand-jacking system utilizing two 300 ton (272 t) jacks to lift the 450 ton (408 t) vessel at this Venezuelan facility. The 270 ft (83 m) towers are braced to one another and guyed for stability. (*PSC Heavy Life, Inc.*)

at the top, including spooling of excess strand, it would be prudent for someone to be at the top to verify that equipment is functioning properly. TV cameras can also be considered for that purpose (see Fig. 2.36).

Though good synchronization requires reliable instrumentation, the strand jack method has a track record of many difficult heavy lifts. During the lift, the load is always held in the grippers when static, and on the center hole jacks when in motion.

Strand jacking system advantages are:

1. Precise vertical load control and the ability to adjust lateral out of plumb by operating one jack.
2. Control and monitoring from the ground.
3. High load capacity from simple towers that break down into small sections.
4. Moderately fast operation; typical loads can be lifted within a normal work day.
5. Adjustable guys which can readily correct alignment and plumbness to within tolerance even as the system changes from the unloaded to loaded condition.

Strand jacking system disadvantages are:

1. Tower bracing design, which is site specific.
2. Alignment and plumbness are critical.
3. Erection of the towers and guys comprise a significant construction operation. To raise the towers and set the heavy tower top equipment, a large mobile crane is needed.
4. Dual tower set ups require guys for stability. Because the towers have fixed bases, proper design and preloading of the guys is critical. Preloading limits tower movements when lateral loads arise, limiting bending stresses and displacements, but preloads also induce axial loads in the towers (and on the foundations) which have to be considered.

Hydraulic telescoping gantries

Hydraulic gantries are typically used for heavy lifting where mobile cranes or derricks are not feasible, where classic "jack and slide" procedures are awkward, or where other conditions make use of the gantry the most economical means to perform the work required. But, the inherent characteristics of hydraulic gantries are different from those of other rigging equipment.

Figure 2.36 160 ton (145 t) chemical tanks being lifted with strand jacks. As the load rises, the strands rise out of the top of the jacks and must be spooled or otherwise controlled to avoid damage. Jacking permits precise control of movements for any weight up to jack capacity.

There are two basic types of hydraulic telescoping gantries. Both types include two or four jacking units. Each jacking unit is made up of a wheeled structural box on which one or two hydraulic cylinders are mounted. A steel header beam is fitted across each pair of jacking units. In a typical operation, loads to be lifted are suspended from these beams by slings attached to lifting links fitted over the header. The assembly can be used in place for lifting, or can be made to travel along steel runway beams or on steel plates over concrete.

High lift, high capacity gantries incorporating hydraulic cylinders and utilized for rigging purposes are a fairly recent development, the first commercial units having come on the market during the early

1980s. At present there is no trade association nor is there an industry standard covering hydraulic gantry design, manufacture, testing, or use. Therefore, all aspects of these machines are subject to the judgment of the manufacturers and riggers.

Cranes, for example, will fail without warning when structurally overloaded to the limit, but may or may not overturn when overloaded with respect to stability. Gantries, on the other hand, can be made to refuse excessive loads by control of hydraulic pressure, and may lean considerably before becoming unstable or structurally collapsing (in the case of one type of gantry). But gantries can be sensitive to inadequacies in the strength, levelness, stiffness or arrangement of their support structures, and require much more deliberate attention to support than do cranes. The need for close engineering attention to the support of mobile cranes can be considered the exception; for gantries it is the rule.

The original and most common type of telescopic gantry utilizes large diameter multi-stage cylinders with as many as five stages and fully extended height of as much as 45 ft (13.7 m). The capacity for a pair of units can be as high as 1000 tons (900 t), but the capacity reduces with cylinder extension. The stages sequentially extend with the smallest diameter, and hence the weakest stage at the top. The lifting ability, structural strength and structural stability of the gantry leg depends entirely on the cylinders. Because lateral strength is limited by cylinder strength (which can be compared with a cantilevered column), these units are somewhat sensitive to side loading. Side loading can be introduced by a swinging payload (quick travel starts or stops), from passing over a poorly aligned runway track joint, or during the pulling of the payload for alignment on anchor bolts.

The second type of hydraulic telescoping gantry was developed to reduce sensitivity to side loading. These units have a single cylinder at each jacking unit, but the cylinder is enclosed in and braced by a telescoping steel box structure called a *lift boom* (Fig. 2.37). This innovation all but eliminates the threat of structural failure from side loading, but all telescopic gantries are still subject to overturning when pulled sideways while the vertical load on the unit is not very great. This can occur when one pair of jacking units is used to turn a load from horizontal to vertical, an operation sometimes called *tripping*.

Although each gantry leg is an independent structural entity, the legs typically act in concert. In the direction perpendicular to travel, the header beams dictate that lateral movement at the top of one leg must be exactly matched by movement in its partner. Similarly, the weight and rigidity of the load can cause the tops of the legs in the second pair of jacking units to move in response to movement in the first pair.

Figure 2.37 The lift cylinders are enclosed in telescoping steel boxes that add structural rigidity in this four-leg hydraulic gantry system. Note how the load is suspended from yokes that can be positioned anywhere along the headers. (*J & R Engineering Co., Inc.*)

Telescopic gantry equipment can be used to perform a number of functions besides simple lifting, or straight up and down lifting with no lateral movement. Once a load is lifted, the gantry system can be used to travel laterally with the load before lowering; in hydraulic gantry practice travel along the track is referred to as a propel movement. Side shifting can also be accomplished. Side shifting is the lateral movement of the load by rolling or sliding the lift links along the header beam. When this is done by pulling while using the end of the header beam as an anchorage, there will be no sideloading. However, sideloading will occur if the pulling anchorage is outside of the gantry. When it is necessary to do this, the system must be balanced using opposing guys or other means.

For tripping a load, ordinarily two pair of jacking units are used, but must be carefully controlled to avoid or limit side loading. When a single pair is used, holdback lines are needed to prevent overturning as the load passes the balance point.

Two pair of telescopic gantries are sometimes used with the load placed atop the header beams. The higher the load CG is above its base, with respect to base width, the more care is needed to keep each header beam, and the pair of beams, level to avoid tipping the load or causing excessive redistribution of loading among the cylinders.

Clearance and wear at slider pads of boom type gantries is inelegantly referred to as slop. Aside from slop, there are several possible causes for lateral displacement at the tops of gantry jacking units of both types, but displacement from out-of-level supports tends to be the largest. The entire gantry structure will respond to the influence of the levelness condition at each of the jacking units. Supports for jacking systems serve the single purpose of maintaining the levelness of jacking units individually and with respect to one another while transmitting loads to the ground.

When the top of a jacking unit moves with respect to its dead plumb position, the wheels supporting that jacking unit are no longer uniformly loaded; there are reactions at the wheels which change to counterbalance the shift at the top. Movement at the tops of the jacking units indicate that the load also moves, and that each unit will no longer be subjected to its original share of load weight. The structure used to support the gantry, that is, the track beams, must be capable of accommodating this reallocation of forces. The gantry structure relies on the track beams for support of its individual wheels and to maintain plumbness of the entire structure in static position and as the gantry travels.

The base of each jacking unit is typically furnished with eight steel wheels, two pair on each side, and two track beams are typically used on each side of a gantry set up when travel or propelling is contemplated. As typically aligned, pairs of wheels are centered on the track beams; usually, a steel bar is welded to the beams, fitting in the space between wheels in a pair, to keep the wheels nominally centered. The track beams receive the weight of the load and the gantry through jacking unit wheels, and serve to carry that weight to the ground or other support medium.

The designer of track beams must consider the local effects of the force in each wheel, including an allowance for increased force due to out-of-level conditions within specified tolerances. In addition, the beams must be strong enough to carry the wheel loads to the beam supports, and stiff enough to do this without deflecting excessively. When a beam deflects, it causes a pair of wheels, or the entire side of the jacking unit, to move downward. This will result in lateral movement at the tops of all jacking units. Such movement causes increased forces at some wheels (the wheels under the displaced jacking unit tops) and decreased forces at the other wheels; lateral movement implies a shifting of the lifted load, therefore jacking unit loads change. An increase in jacking unit load causes deflection to increase further, and the process continues until equilibrium is reached.

Deflections in track beams can be local, as at an individual wheel, or global as for a pair of wheels or the entire side of a jacking unit. Local effects can only be controlled by designing appropriate track beam details. Global effects are a function of how the track beams themselves are supported and of the characteristics of the particular track beam used. Track beam supports deliver load and gantry weight from the track beams to the ground or other support medium.

Support conditions may well vary along the length of track beams or between track beams. Each particular site at which hydraulic gantries are used must be independently evaluated for support conditions along the entire length of each individual track beam. The goal of such an evaluation is to verify that gantry track beams are supported in a manner that will maintain the levelness of the track beams, and therefore the gantry, throughout the rigging operation. Wherever non-typical conditions are found, special means must be utilized to create adequate support.

Some support conditions can be referred to as "hard points". A hard point is a support point that is, for practical purposes, non-yielding, or incapable of meaningful displacement. Examples of hard points would be solid rock, a substantial concrete mass, or steel bolsters which are in turn supported on a hard point. Any other support material will displace either elastically or permanently, the amount of displacement varying with the material. Natural wood or plywood displaces elastically at lower loading levels, and permanently at high levels. Soil displacement has both elastic and permanent components, and those components can vary with the nature of the soil, degree of compaction, and/or the moisture content of the soil. Soil displacement, or settlement, is time dependent; part of the settlement takes place immediately and the remainder as time progresses.

Displacement at any point along a track beam is the algebraic sum of the support medium displacement, track global displacement, and local wheel displacement. The track beam design process must substantiate that the differential displacements that will inevitably take place along the track beam or pair of beams are within tolerances that will not be detrimental to gantry operation and safety.

At any one point, a track beam which has initially been set level can go out of level when loaded if the shims at one or more support points are looser, thicker, or of different material than adjacent shims, or if the supporting material is of a yielding nature with respect to the support of the adjacent track beam.

The centering of the load, with respect to the jacking units, affects the distribution of load among the units and thus can affect lean. Likewise, if one header beam is permitted to go out of level, the "table leg effect" would cause a redistribution of loading. This is similar to a four-legged table on a hard floor. The table will rock and will be, in essence, supported on two diagonally opposite legs. In like manner, when one jacking unit is extended more than its partner, it and the diagonally opposite unit would try to carry the entire load—the equivalent to a rocking table. In actuality, those jacking units will tend to carry less than the full load because of the limited hydraulic pressure available and elastic effects, including the stretching of wire rope slings.

Chapter

3

Loads and Forces

The word "load" means something different to the engineer who works with cranes than to most people who operate or use them. Users and operators usually take this word to mean the weight suspended from the hook. The engineer, on the other hand, considers the suspended object as but one of several loads acting on the crane. Additional loads might include effects imposed by wind and acceleration of the crane motions; they might also include effects of jarring movements, side pulls, or out-of-level setup.

Engineers quantify loads from these various sources to gage their effects on the crane and to remedy problems they might otherwise cause. Analytical methods of the engineer, such as those furnished in this chapter, can be forbidding to most people. Nonetheless, non-engineer folks often gain a practical understanding of loads. In particular, crane operators and riggers who have experienced load effects in the field may be especially capable of connecting the engineering concepts described in this chapter with their personal observations.

Comprehension of how loads are created and how they affect the equipment is a key to the mastery of crane engineering, crane operation, and the practice of rigging. The mathematics and physics presented in this chapter is for all readers who may enjoy it, but it is primarily intended for engineers. The descriptive text, however, should help all readers relate the concepts described to their actual experiences on the job, and help them foresee, understand, and allow for what they may encounter in the future.

3.1 Introduction

In Chap. 1, we learned how to distribute the effect of boom dead weight and the hook load to the members of an elemental derrick, and how to calculate the support reactions induced by those loads. Chapter 2 introduced more complicated lifting machines with variations in the arrangement of components resulting in entirely different concepts and capabilities in load handling. These more complex contraptions may frighten us into thinking that they are too difficult to understand how they function. But such fear is unfounded. All things in nature must obey the same set of physical laws, laws that hold true whether we are dealing with cranes or with billiard balls. When these elaborate machines are broken down into basic components and evaluated step by step, even the most intricate of them can be understood. How they work is nothing more than the application of familiar physical processes.

At this point it is useful to distinguish between loads and the more general category of "forces". In the jargon of engineering, loads are distinguished from other forces because they induce reactions at external points of support. As installation designers and field people are largely concerned with support of the equipment, loads are for them *the* important consideration. Other forces, those that act on or within crane components rather than at the supports, are mainly the province of crane designers. The discussion that follows is about forces in general. In keeping with the target audience of this book, however, the thrust of the chapter is mainly about loads.

While cranes and derricks operate and go through their motions, they simultaneously generate forces and react to them. Some of these forces and their reactions are steady-state and others are time-varying. Forces are introduced by the operating environment, by the work being done, and by the inertia of the masses that have been put into motion, as well as by the weights of the crane components and the suspended load.

The operating environment imposes loads associated with wind, snow, ice, cold and heat, and on occasion, earthquakes. But in addition, the working environment includes more subtle forces induced by imperfections in the machinery and its setup. Out of levelness, and misalignment of the mountings, tracks, or support surfaces are included among such effects. So is friction when dust, mud, sand or stones interfere with the smooth working of moving parts.

The work being done by the crane's movements generate forces, even when imperfections and environmental influences are absent. The discrete motions of hoisting, swinging, trolleying, luffing, and travel each create a system of forces within the structure and the machinery works. No motion occurs without friction, and thus friction is an element of each moving system.

There are also those forces which result from overcoming the inertia of the suspended load and component masses in acceleration and deceleration processes for all functional motions. Several motions also introduce pendulum or centrifugal action in the lifted load, which is accompanied by radius change and/or lateral force.

For any defined state or condition it is possible to calculate with reasonable accuracy the forces developed and their effect on each machine component. But, it is seldom possible to define operating states or conditions accurately. Many of the forces described are time-variable in magnitude and in direction, as are the agencies introducing those forces. The lifted load itself is a widely dispersed variable!

How then do the machine designer and the installation designer deliver adequate reliability while maintaining reasonable cost? What properly defines the force systems for a particular lifting device?

Design loading concepts

Two general methods can be used to establish design loadings. The more common, the deterministic approach, is built on reason. The less prevalent probabilistic method is built on a statistical base.

The deterministic scheme is the concept underlying most codes, standards, and engineering procedures. In applying the deterministic approach, the designers or code writers subjectively decide on the loadings that are to govern the design. Ideally, these load levels are rationalized by combining sound engineering reasoning with ample experience.

Using the deterministic method, loads should be applied in sensible combinations. Some combinations are inconsistent and can be ruled out; a load lifted in the midst of a gale force wind is one example. Other loads are so highly improbable to act in concert that their combination should not be considered except in the most sensitive applications. Dynamic forces, particularly, are typically of very short duration and are unlikely to combine with other short-lived events. Simultaneous occurrence of peak values of lifted load, wind, and dynamic force would usually be considered highly improbable, for instance.

When true loadings are unpredictable and difficult to control, deterministic design may be somewhat empirical in the choice and use of the loads to be incorporated in the design. The intention is to produce a reasonable approximation to real life while creating a design with low probability of failure.

The probabilistic concept requires mathematical manipulation, based on statistical evidence, of the joint probability of occurrence of all forces that could act together during an interval of time and the evaluation of the time-varying magnitudes of the forces. The goal of the procedure is a design with a computed probability of failure that does not exceed acceptable levels. As a prerequisite for application of this concept, a body of statistics is needed that can completely describe each random variable characteristic of lifting-device operation as well as those environmental factors, material properties, and fabrication practices which are random variables. Evaluation and acceptance of permissible levels for probability of failure require economic analysis of the consequences of failure, including placing money values on loss of life and injury.

With an adequate statistical base, there is no question that the probabilistic approach toward design leads to a more economical and certain end product, but the cost of this design method is far greater than for deterministic design, especially when the cost of compiling the data base is included. As a general case, there is too much variety of hoisting equipment types and too great a range of applications to make the collection of adequate statistical data for crane design practical. It is not reasonable to expect that the full scope of needed data can be made available, and so compromises become necessary. If compromises are made to reduce the statistical data base, conservative deterministic parameters must be inserted to replace the missing data. Then, if too many trade-offs are made, the method can deteriorate into an illusion of probabilistic science and may indeed produce designs less economical and no safer than the deterministic methods so widely used.

Established deterministic design codes actually have an underlying probabilistic foundation. Certain aspects are explicitly derived from statistical data, and others because empirical decisions have been based on accumulated experience, an analogy for statistical data. Limit state design standards, which are increasingly prevalent in structural engineering practice, are derived from actual statistical data and probabilistic analysis. However, their final format is deterministic. Eventually the practice in crane engineering might be expected to follow a similar hybrid approach.

Classification of loads

Loads that affect cranes can be organized into three categories†: regular, occasional, and exceptional. Regular loads derive from gravity effects and from acceleration and deceleration induced by the crane drives and brakes. These are the loads that repeatedly occur during normal crane operation. The peak values of these loads do not usually occur simultaneously.

---

†From International Standard ISO 8686/1, Cranes and Lifting Appliances—Design Principles for Loads and Load Combinations—Part 1: General, clause 7, International Organization for Standardization.

The occasional load category includes loads that will affect the crane but are less frequent in occurrence. In-service wind, snow, ice, temperature, and for bridge, gantry, and portal mounted cranes, skewing, induce the loads in this group. Although in-service wind is almost always present, it infrequently reaches velocities as high as that used for design.

Lastly, exceptional loads are those which are infrequent; they are introduced by events such as out-of-service wind, erection and dismantling, testing, and earthquakes. The category of a load should not be taken as an indication of its importance. Exceptional loads, for example out-of-service wind, often govern the design of crane supports and some crane components. Each of the three categories includes loads that can be taken as static and those that are clearly dynamic.

3.2 Static Loads

The static loads associated with crane and derrick operations occur during a state of rest but also underlie dynamic loadings. They derive from the lifted load, from the dead weight of the machine, and from snow, ice, or other materials that can accumulate on the equipment parts.

As the crane or derrick goes through its motions and changes attitude, the distribution of the static loads changes, affecting the supports and various components. Dynamic effects come into play mainly during acceleration and deceleration and are superimposed on the static effects; they often will not be significant for mobile-crane installation designs but can be important for the supports of high-speed cranes in general and for some tower cranes.

Temperature changes produce effects that can be taken as static loading. Parts or components that are constrained in position by other portions of the structure built of a different material undergo loading with temperature change because of dissimilar coefficients of expansion.

For several crane types, wind loading is of very great significance. Since it includes both static and dynamic considerations and is an extensive subject, wind has been treated on its own in Sec. 3.4.

Lifted loads

The definition of lifted load is, quite simply, that which is given by the manufacturer of the particular machine in question. The definition is not the same for all equipment. For mobile cranes and many other types of lifting devices, it includes the hook block and overhaul weights as well as the live load and all accessories used to attach and hold the load. For hammerhead tower cranes and some large-capacity lifting machines where the hoisting-system reeving is left constant, the hook block and overhaul weights are taken as machine dead weight and not as part of the load. The rating chart or documentation for the device will state which items are to be included in the load.

When swing or slewing motion takes place, centrifugal force causes the load lines to go out of plumb and the load radius to increase. Likewise, inertia will cause the load to lag behind. Both effects introduce horizontal forces that act at the sheaves of the upper load block and at the load itself (for each effect, the two loads introduced are equal but opposite in direction). The magnitude and direction of the forces are such that the moments are balanced; that is, the vertical load on the hoist system multiplied by the horizontal displacement of the load must be equal to the horizontal force multiplied by the vertical distance from the load CG to the center of the upper-block sheave shaft (see Fig. 3.1). Although these forces accompany motion, their magnitude remains constant under constant velocity, so that they can be applied as static. Determination of the forces will be left to the following section on dynamic loads.

Whenever a load is lifted on a single hook, the center of gravity of the load will place itself vertically below the hook, regardless of the arrangement of the slings, lifting beam, or other attachments. This is

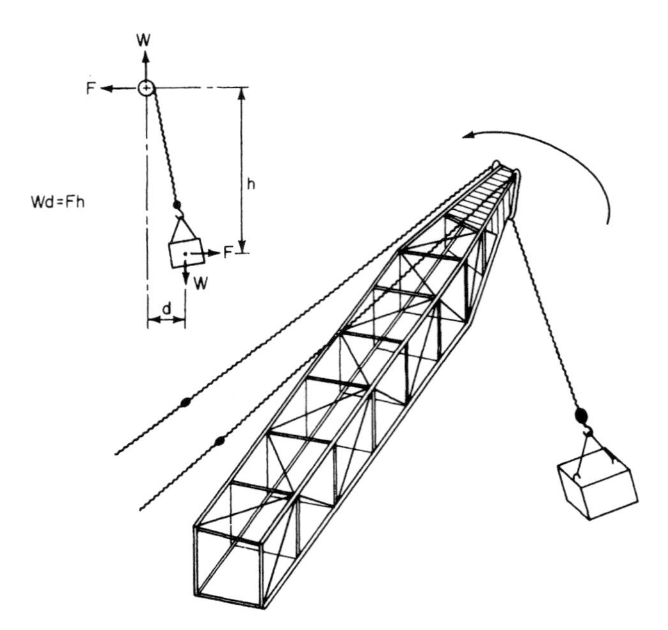

Figure 3.1 When a boom swings, the load lags behind and sweeps out to a greater radius as well.

dictated by the need for the sum of all forces and moments to be zero for a body in equilibrium. Failure to understand that fact is a source of jobsite accidents where the load slips or shifts while aligning itself below the hook. Needless to say, slings or lifting attachments must be arranged to place the hook rather close to a vertical line through the CG of the load before the lift begins.

Dead loads

When dead weights are calculated using handbook values of the unit weights of standard structural sections or from the nominal dimensions of plate stock, it may be advisable for particularly critical calculations to increase or decrease the weight determined. Normal permitted dimensional variations in mill products can produce weight differences of a few percent. If the case in point involves a number of different section sizes distributed roughly evenly in proportion to weight, statistically the variations should balance. But if the design includes little variation in sections, overweight or underweight can result. For custom built equipment a factor of 1.07 should be considered to allow for dimensional variations, fabrication tolerances, weld deposit etc. Where weight induces loading or stress, the appropriate factor should be used as a multiplier, but where weight acts to add stability, the factor becomes a divisor.

For any particular attitude of the lifting device, dead loads must be taken in their true positions, as many components move in relation to each other during machine functions. The location of the CG of a boom changes as the boom luffs, as does the CG of a boom foot mast and boom-hoist spreader, for example. In addition, several machine types are equipped with movable counterweights that shift as load radius changes.

Snow and ice loadings must be considered only for machines destined for use in areas known for considerable accumulations. Even then, only machines that are stability- or wind-sensitive need be investigated in the out-of-service condition. Strength margins are sufficient to permit the additional initial weight, but significant incrustations need to be broken loose before operations are started.

Machine types that are wind- or stability-sensitive suffer a double penalty when snow or ice buildup occurs. The added weight on booms or jibs diminishes the machine's stability moment or adds to structural moment, while the enlarged members offer increased wind resistance.

Effects of load distribution

When a machine superstructure swings, there is a redistribution of forces at the machine supports as mentioned earlier. For a fixed-base machine, compressive support reactions may change to uplift reactions that require tie-downs and vice versa. The reactions are readily found, however, by applying the laws of statics.

Machines supported on wheels, outrigger floats, or tracks are not to be tied down, and as a consequence the support points cannot accommodate uplift reactions. The reaction instead goes to zero, any additional uplift being distributed among the remaining supports. Procedures for determining support-point reactions for mobile cranes are fully derived in Chap. 5, but the same method is applicable to other machines that are not tied down.

The occurrence of zero load at a support point, such as an outrigger float on a mobile truck crane, does not necessarily signal an unsafe or critical condition. A mobile crane could lift an outrigger float clear of the ground within permitted load and operating conditions. This may be due partly to carrier-frame twisting deformation and partly to the result of outrigger beam deflection. Excessive lift-off under permitted conditions implies the possibility of inadvertent and uncontrolled radius increase and indicates poor carrier design.

When a second outrigger float shows signs of lifting, there can be no doubt that excessive load is being lifted for the operating environment and that the machine has reached the tipping point.

For wheeled machines that ride on rails, zero load at a supporting wheel is another matter. Should a wheel or bogie lift clear of the track, there is a good chance that horizontal forces or perturbations from vibrations or operation will cause derailment. Wheel flanges should not be permitted to rise above the crown of the track under any condition of operation.

Friction

Earlier discussions have shown how friction is an important factor in computing wire rope loads, lead line forces and overhauling weight requirements. Friction must, of course, be considered in the design of every mechanism. A characteristic of friction that cannot be overlooked is the decrease in frictional resistance from the static to the moving state; coefficients of friction for static parts in contact are higher than those for the same parts once motion has begun. This affects such things as the starting torque required of the drives for each functional motion and the strength of the drive anchorages.

Travel-system storm anchorages or braking systems that rely on friction must be designed with that characteristic in mind. Should a storm-wind peak gust overcome the static frictional resistance, lower-level winds or gusts could keep the machine in motion with disastrous results.

Mobile cranes, whether operating on tires, tracks, or outriggers, are not attached to the ground. Yet during operation swing acceleration and deceleration produce low-magnitude horizontal dynamic forces at the supports. These forces, however small, must be passed into tbe supporting surface through frictional resistance, which also prevents crane lateral movement. For that reason, many crane operators like to avoid metal-to-metal bearing contact at tracks or outrigger floats. They insert wood between metal bearing surfaces to increase the coefficient of friction.

Out-of-level supports

All cranes and derricks must be mounted level to within close tolerances. For mobile equipment, the usual specification permits a maximum of 1% out of level between supports. For tall, limber machines, such as tower cranes, an initial deviation from a level base will be amplified by beam-column action so that even more strict tolerances must be kept (see Chap. 6).

For machines with high CGs, such as tower cranes, a relatively small difference in levelness of the supports can result in a significant horizontal displacement of the CG. This, of course, produces a change in support reactions and component loadings and reduces resistance to overturning (see Chap. 4).

Out of levelness may induce difficulties in swinging as the swing path develops an "uphill" and "downhill" aspect. Responding to this phenomenon is a distraction to the operator, or worse, may cause a loss of control. Out-of-level mounting also produces a side pull on the boom over some of the swing arc and a change in the true, as opposed to the apparent, operating radius. That may be detrimental to structural capacity or stability.

Misalignment and skewness

The wheels of track-mounted equipment are flanged on one or both edges or have lateral guide rollers to keep the machine on the rails. In either case, permitted rail alignment and spacing tolerance is a function of the displacement the wheels can accommodate. Should displacement be excessive, side forces will develop on the wheels and tracks. At the very least, this will induce added load on the travel-train components and drive motors and cause accelerated wear of the parts in contact; it may also cause thermal cutouts in the motors to interrupt travel. At the other extreme, track fastenings could be caused to shear or become loosened with subsequent derailment and the possibility of overturning.

Similar problems can occur when a rail-mounted travel base goes askew. This is a possibility with machines such as overhead traveling cranes where the wheelbase is short in relation to the rail gage. On relatively lightweight cranes, the consequence of jamming due to skewness is usually limited to drive-motor stall or triggering of the overload protection.

Earthquake

An earthquake induces dynamic loading, but several of the most commonly used deterministic design codes have procedures for developing a static-load equivalent to represent the effect of earthquake forces. Many designers ignore earthquakes for any location other than parts of the American Southwest. The fact remains, however, that many other areas in the United States have experienced significant earthquake shocks within recorded history. The question is one of probability. There are earthquake maps that divide the country into earthquake zones according to potential risk.

For most cranes and derricks, probability favors the premise that the device will be unloaded should an earthquake occur. Further, for most earthquake zones, an unloaded machine will be able to withstand the shocks without damage. For permanent, highly sensitive, or long-term temporary installations of machines with high CGs, however, verification may be in order.

3.3 Dynamic Loads

In addition to the centrifugal force previously mentioned, dynamic loads are for the most part those associated with masses undergoing changes in motion as described by Newton's second law, $F = ma$. Dynamic loads are always time-varying in magnitude, position in space, or both.

All crane and derrick motions produce dynamic forces as the motions begin and end. This includes hoisting, trolleying, luffing, slewing, and travel motions, as well as counterweight movements for machines so arranged. The forces accrue to, and act on, the CG of all masses in the system that are experiencing acceleration or deceleration. Pendulum action is induced in the load by trolleying, slewing, and travel and is another dynamic phenomenon that must be taken into account. Finally, there are the rotating masses in the various drive systems producing local dynamic effects, but if the rotating mass is significant in relation to the mass of the entire crane, this will affect the translatory powering up or braking forces associated with the rotation.

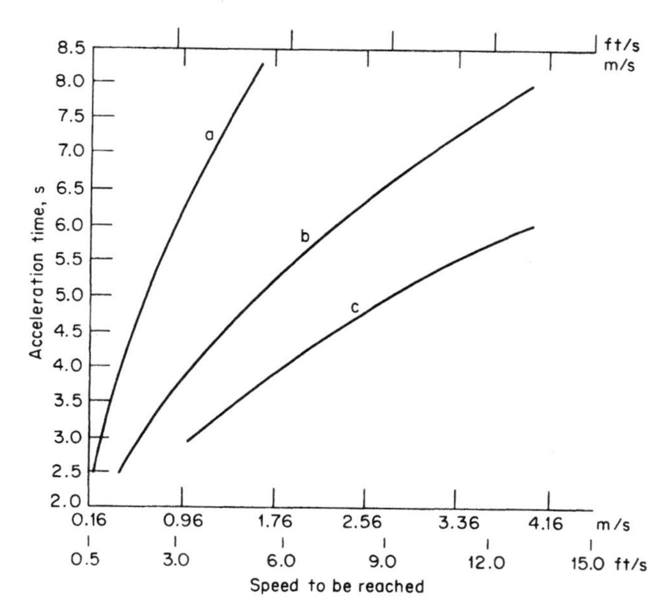

Figure 3.2 Nominal acceleration times suggested by the FEM for heavy lift equipment. *Curve a:* low and moderate speed with long travel, *Curve b:* moderate and high speed (normal applications). *Curve c:* high speed with high accelerations.

Linear motion

The force required to accelerate or decelerate a mass in linear motion can be found simply by applying Newton's law if the acceleration rate is known. (Except for machines with controlled acceleration rate, it is not.) In many instances, drive and braking systems are capable of far greater force than needed, and the operator rarely applies the full force. The extra capacity is required to overcome wind and other non-constant effects. The appropriate acceleration rate to be used in design is a matter for judgment, but the FEM† offers some guidance for heavy lifting equipment. Curves representing acceleration times suggested by the FEM are shown in Fig. 3.2. With acceleration time and final (initial) velocity given, average acceleration is the velocity divided by the time, of course. Many machines have drive systems that do not provide linear acceleration, but for most work a linear assumption is satisfactory since system inertia tends to linearize the acceleration.

†Fédération Europèene de la Manutention, "Rules for the Design of Hoisting Appliances," 2d ed., Paris, 1970, sec. I, Heavy Lifting Equipment.

For machines with brakes that will always fully engage automatically, the brake will be rated for braking force or for torque. When these values are mathematically referred to the point of contact of a wheel with a rail or a rope to its winch drum, the force inducing the acceleration becomes known and acceleration rates and times can be calculated.

Three conditions can create significant dynamic loading in a hoisting system: picking a load suddenly from a condition of rest, suddenly stopping a load being lowered, and sudden release of load, such as emptying a clamshell bucket or dropping a magnet load of scrap.

When a load is dropped in free fall,† the acceleration is retarded by friction at the sheaves and the inertia of the winch drum. Considering only gravity and friction, resulting acceleration is given by

$$a = g\,\frac{M_a}{r} \qquad (3.1)$$

where g = acceleration due to gravity
M_a = nominal mechanical advantage of system
r = lowering factor given in Eq. (1.1)

and the final velocity is found from

$$\nu = (2ah)^{1/2} \qquad (3.2)$$

which uses h as the height of fall. Should the brakes fully and suddenly engage, so that deceleration is virtually instantaneous, the load will not stop immediately. The hoist ropes will stretch, like springs, providing some additional movement. Taking the kinetic energy at the time of application of the brakes and adding the potential energy of the load W and dynamic rope stretch Δ, we can equate this with the potential energy in the ropes after stretching is complete. If F is the final total force in the ropes,

$$\frac{W\nu^2}{2g} + W\Delta = \frac{F\Delta}{2}$$

but the spring rate of the ropes is $k = F/\Delta$. Substituting this and rearranging, we get

$$F^2 - 2WF - \frac{Wk\nu^2}{g} = 0$$

$$F = W + \left(W^2 + \frac{Wk\nu^2}{g}\right)^{1/2}$$

†Standard free fall = 32.174 ft/s^2 = 9.8067 m/s^2.

$$\frac{F}{W} = 1 + \left(1 + \frac{k\nu^2}{Wg}\right)^{1/2} \tag{3.3}$$

Equation (3.3) shows that if the initial velocity is zero (that is, if the load is not dropped but is suddenly applied to a slack rope system), the effect on the ropes is to double the load. Theoretically this is the lower bound of dynamic amplification because any initial velocity will result in an effect more than twice the load. But the model used in the derivation of Eq. (3.3) is oversimplified. Winch drum and sheave inertia will retard acceleration more than indicated by Eq. (3.1), and, of course, neither automatic nor operator controlled brakes will cause an instantaneous stop.

In actuality, Eq. (3.3) overstates dynamic amplification by a factor of two or more depending upon the rate of application of the brakes and the elastic properties of the system. In general, the amplification effect will be between 1.0 and 2.0.†

Although the term *impact* is generally, and more correctly, applied to the striking of a body by another body, it is also used to describe the increase in load effect due to dynamic causes. The impact force F_i is then given by $F - W$, and impact can be expressed as

$$i = \frac{F - W}{W} = \frac{F_i}{W} \tag{3.4}$$

Operators do not often allow other than marginal loads to free-fall. Substitution of even small fall heights into Eq. (3.2) followed by solution of Eq. (3.3) will show that impact quickly rises to high multiples of the load. On machines that permit free fall, it is usual for an operator to ride the brake in order to modulate velocity and be in a position to maintain acceptable limits of deceleration for the final stop; this is done by judgment alone.

The FEM suggests values for impact for heavy lift equipment that are given by

$$i = \begin{cases} \dfrac{3\xi\nu}{10} & \text{U.S. customary units} \\ \xi\nu & \text{SI units} \end{cases} \tag{3.5}$$

The velocity ν in feet per second (meters per second) and

†Ibid, clause 7.1.4; as pointed out by Prof. Dr.-Ing. Rudolf Neugebauer, Technische Hochschule Darmstadt, Germany, in private correspondence. Although the mathematics describing this action is quite complex, the final result depends directly on the brake force actually applied, which is a difficult parameter to determine.

$$\xi = \begin{cases} 0.6 & \text{for overhead traveling and bridge cranes} \\ 0.3 & \text{for boom or jib cranes} \end{cases}$$

When Eq. (3.5) is used, a minimum value for i of 0.15 is contemplated and i is to remain constant for velocities exceeding $3\frac{1}{3}$ ft/s (1 m/s).

American mobile-crane manufacturers claim that impact loading is not an important condition for their type of crane and do not use it in design. They prefer their own deterministic loading conditions, which will be dealt with later. Their position can be defended with a strong logical argument. Few loads lifted by a mobile crane approach strength-governed ratings (most ratings are stability-governed) so that impact would rarely pose a structural threat. Loads that are in the structural range tend to be very heavy loads, which are understandably treated with respect by operators. As a short-duration dynamic condition, hoisting system impact has been found to impart too little energy to overturn a crane except in extreme cases (see Chap. 4).

A task force of the American Institute of Steel Construction (AISC) conducted a series of tests to determine how best to account for impact in derrick design. In their report† they state that "energy applied by motion at the hook is absorbed immediately and simultaneously by all elements of the derrick setup and cannot be separately identified in the ensuing elastic vibrations of the masses of the individual elements. . . ." The task force found that simply increasing the live load by an impact factor did not yield good correlation with test results. Instead, they recommend increasing axial as well as dead- and live-load bending stresses (but not lateral or side bending stresses) by an impact factor. This procedure produced values that closely match measured stresses. For lifting full structurally based rated loads, a factor of 20% is suggested, but the tests revealed that greater impact should be expected as loads decrease in relation to rating. The tests were deliberately carried out to produce extreme, or upper-bound, impact compared with normal, proper production operations.

The AISC task force makes another interesting and subtle observation that is particularly pertinent for derricks. Winches are manufactured so that there is a direct relationship between winch-line pull capacity and brake capacity. A winch properly matched to a derrick will be incapable of stopping rated loads instantaneously; the brakes will be sized so that a reasonable stopping distance can be expected. However, when an overcapacity winch is matched to a derrick, brake overcapacity will present a potential for excessive impact loading.

---

†"Guide for the analysis of Guy and Stiffleg Derricks," American Institute of Steel Construction, Inc., New York, 1974, p. 6.

Example 3.1 (*a*) An operator allows a 5000-lb (2268-kg) load to free-fall for 10 ft (3.05 m) before realizing that it is going too fast. In panic, he applies the brake fully, causing a virtually instantaneous stop. What peak force will develop in the rope if three parts of line were in use, friction loss can be taken as 2%, and the rope spring constant is 2300 lb/in per part (402.8 N/mm per part)?

$$1 - \mu = 1 - 0.02 = 0.98$$

From Eq. (1.1) for lowering load with three parts of line and assuming that $m = 0$, we have

$$r = \frac{1 - 0.98^3}{0.02 \times 0.98^3} = 3.12$$

The nominal mechanical advantage MA = 3 for three parts of line, so that from Eq. (3.1)

$$a = 32.2 \text{ ft/s}^2 \, \frac{3}{3.12} = 30.96 \text{ ft/s}^2 \ (9.44 \text{ m/s}^2)$$

and from Eq. (3.2)

$$\nu = [2(30.96 \text{ ft/s}^2)(10 \text{ ft})]^{1/2} = 24.88 \text{ ft/s} \ (7.58 \text{ m/s})$$

Equation (3.3) then gives the ratio of rope load to lifted load

$$\frac{F}{W} = 1 + \left[1 + \frac{3(2300 \text{ lb/in})(12 \text{ in/ft})(24.88 \text{ ft/s})^2}{(5000 \text{ lb})(32.2 \text{ ft/s}^2)}\right]^{1/2}$$

$$= 18.87$$

$$F = (5000 \text{ lb})(18.87) = 94{,}350 \text{ lb} \ (419.7 \text{ kN}) \qquad \textit{Ans.}$$

Impact has caused a theoretical increase in load of some 1787%!

(*b*) What maximum velocity would have to be maintained to keep impact from exceeding 30% if the stopping distance is not to exceed 5 ft (1.52 m)?

The stopping-distance requirement indicates deceleration through braking as the load comes to rest. Applying Newton's law gives

$$F_i = ma = 0.30W \qquad a = \frac{0.30W}{m}$$

but $m = W/g$; therefore

$$a = 0.30g = 9.66 \text{ ft/s}^2 \ (2.94 \text{ m/s}^2)$$

Inserting values into Eq. (3.2) gives

$$\nu = [2(9.66 \text{ ft/s}^2)(5 \text{ ft})]^{1/2} = 9.83 \text{ ft/s} \ (3.00 \text{ m/s})$$

Had the velocity been held to 9.83 ft/s (3.00 m/s), with the stop taking place over 5 ft (1.52 m) of vertical travel, the impact would have been limited to 30%. *Ans.*

(*c*) For the velocity developed in part (*a*), 24.88 ft/s (7.58 m/s), what stopping distance would be needed to hold impact to 30%?

Rearranging Eq. (3.2) and noting that for 30% impact we have found that deceleration will be 9.66 ft/s^2 (2.94 m/s^2), we get

$$h = \frac{v^2}{2a} = \frac{(24.88\ \text{ft/s})^2}{2 \times 9.66\ \text{ft/s}^2}$$

$$= 32.04\ \text{ft}\ (9.77\ \text{m}) \qquad \textit{Ans.}$$

(*d*) With the same initial velocity used above, what is the impact for a stopping distance of 5 ft (1.52 m)?

Rearranging Eq. (3.2) again, we have

$$a = \frac{v^2}{2h} = \frac{(24.88\ \text{ft/s})^2}{2 \times 5\ \text{ft}}$$

$$= 61.90\ \text{ft/s}^2\ (18.87\ \text{m/s}^2)$$

From Newton's law, $F = ma = Wa/g$,

$$F_i = \frac{W}{g}a = \frac{5000\ \text{lb}}{32.2\ \text{ft/s}^2}(61.90\ \text{ft/s}^2) = 9610\ \text{lb}\ (42.76\ \text{kN})$$

From Eq. (3.4)

$$i = \frac{9610\ \text{lb}}{5000\ \text{lb}} = 1.92 = \text{or } 192\%\ \text{impact} \qquad \textit{Ans.}$$

(*e*) What impact values does the FEM suggest should apply for heavy-lift boom-type cranes for the velocities above, 24.88 and 9.83 ft/s (7.58 and 3.00 m/s)?

The note under Eq. (3.5) indicates that there should be no increase for velocities above 3⅓ ft/s (1 m/s); therefore

$$i = \frac{3(0.3)(3\tfrac{1}{3}\ \text{ft/s})}{10} = 0.30 = 30\% \qquad \textit{Ans.}$$

The equation of motion. In applying the deterministic approach to dynamic problems, the time variation in loading must be fully known although it may be oscillatory or irregular. When the methods of structural dynamics† or vibration

†See R. W. Clough and J. Penzien, "Dynamics of Structures," McGraw-Hill, New York, 1975, and J. E. Shigley, "Mechanical Engineering Design," 2d ed., McGraw-Hill, New York, 1972, chap. 16.

theory are used, structural response is given in terms of displacements, and the time-varying moments and stresses can then be obtained.

A static problem has a single solution, but a dynamic problem has a series of solutions represented by an equation with time as an independent variable or by values at each of the times of interest. A more fundamental difference exists between static and dynamic situations, however. Under static loading, the external loads on a structure are in equilibrium with the moments and elastic forces developed within the structure. Under dynamic loading, the time-varying external loads induce accelerations and displacements in the structure and the ensuing time-varying inertial forces also enter into the equilibrium equation.

Cranes and derricks can generally be modeled as single-degree-of-freedom (SDOF) systems; i.e., a sufficiently accurate deflection time history of the structure can be formulated in terms of a single displacement or by means of an expression based on a single displacement. SDOF structures can be classified as rigid-body assemblages having springs to represent permitted elastic displacements at discrete points or continuously elastic structures deforming throughout their length. For both idealizations, SDOF behavior is forced on the system by assuming a displacement form or configuration that can be described in terms of a single displacement.

Newton's second law of motion, that the rate of change in momentum of a mass is equal to the force acting to produce that change, has been given earlier as $F = ma$. In differential equation form, with mass held as a constant, this can be expressed as

$$f(t) = m\ddot{z}(t)$$

where $f(t)$ = applied time-varying force vector
m = mass
$z(t)$ = position or displacement vector

The dot superscript represents differentiation with respect to time. When this is rearranged and stated as

$$f(t) - m\ddot{z}(t) = 0 \tag{3.6}$$

the second term can be called the inertia force resisting the acceleration, thus reflection d'Alembert's principle, which states that the inertia force associated with a mass is proportional to the acceleration of the mass. This principle permits dynamic situations to be expressed in terms of equations of equilibrium.

Equation (3.6) is an *equation of motion* that describes the relationship between displacement and the initiating dynamic load. When the differential equation is solved, the resultant expression will give displacements at any point in time. Since equations of motion reflect only dynamic effects, static loading effects must be added.

But Eq. (3.6) is a simple form of the equation of motion. A fully generalized form would be†

†Clough and Penzien, op. cit., pp. 34–36.

$$m^*\ddot{Z}(t) + c^*\dot{Z}(t) + k^*Z(t) = f^*(t) \tag{3.7}$$

$$m^* = \int_0^L m(x)[\psi(x)]^2\,dx + \Sigma m_i\psi_i^2 + \Sigma I_{0i}(\psi_i')^2 \tag{3.8}$$

$$c^* = \int_0^L c(x)[\psi(x)]^2\,dx + \Sigma c_i\psi_i^2 \tag{3.9}$$

$$k^* = \int_0^L k(x)[\psi(x)]^2\,dx + \int_0^L EI(x)[\psi''(x)]^2\,dx + \Sigma k_i\psi_i^2$$

$$- \int_0^L N(x)[\psi'(x)]^2\,dx \quad (3.10) \tag{3.10}$$

$$f^*(t) = \int_0^L f(x, t)\psi(x)\,dx + \Sigma f_i\psi_i \tag{3.11}$$

where the terms are generally defined in Fig. 3.3, but in addition, the displacement at any point x along the length of the structure is given in terms of the deflection $Z(t)$ at one selected point through use of the nondimensional *shape function* $\psi(x)$, so that $z(x, t) = Z(t)\psi(x)$. For displacement that is collinear with applied load, the shape function has the value of unity. The prime superscript represents differentiation with respect to x. When the generalized properties that vary along the length, $m(x)$, $c(x)$, $k(x)$, and $EI(x)$ and the axial loading $N(x)$, are in fact uniform and continuous or constant rather than varying with x, the constant values can be inserted into the equations. E refers to the modulus of elasticity of the member material, and I represents the moment of inertia of the cross-sectional area of the member; these values are used only when bending members are present. For mass moment of inertia I_0 see Fig. 3.4; note that this parameter applies only when a mass undergoes angular diplacement.

The m_i are lumped masses, the c_i are discrete dampers, the k_i are discrete springs, the f_i are discrete time-varying loads, and the ψ_i are the values of the shape function at those discrete points. In the mathematical model used for analysis, *lumped masses* are aggregates of mass concentrated at the point where the inertia forces associated with those masses would act.

The shape function provides a means for describing the displacement at any point in the system in terms of the primary displacement sought; since all displacements are then in terms of one displacement, the SDOF criterion is satisfied. In Fig. 3.5*a* the rigid bar AC has two lumped masses, two springs, and a hinge at A. If $Z(t)$ is the displacement at C, the displacement at B will be $\psi(x)Z(t) = (a/L)Z(t)$.

The shape function given in Fig. 3.5*b* for a continuously elastic side-loaded cantilevered column is a fairly accurate representation of the deflection curve. Although the relationship is not at all direct, a small error in the deflection curve assumed will yield a smaller error in calculation results. An approximation to the

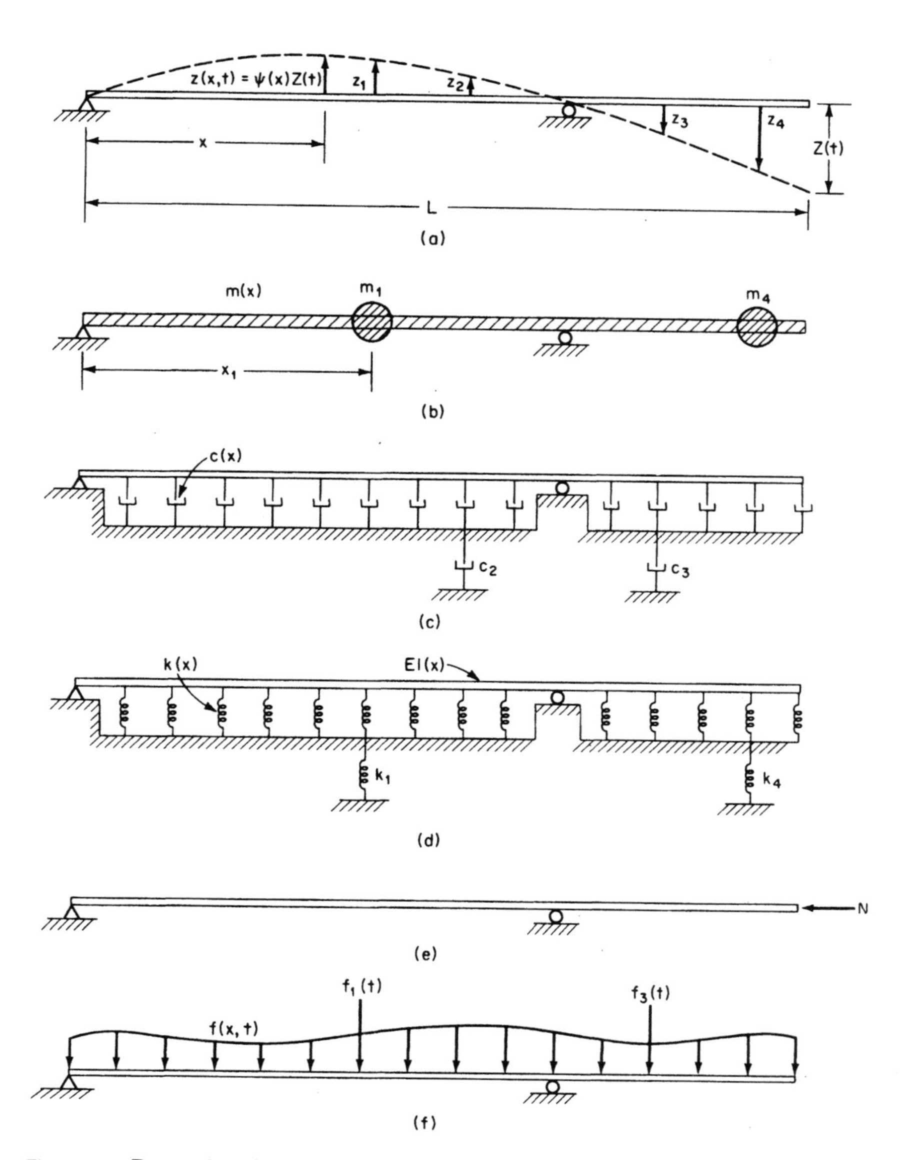

Figure 3.3 Properties of a generalized SDOF system: (*a*) assumed shape; (*b*) mass properties; (*c*) damping properties; (*d*) elastic properties; (*e*) axial loading; (*f*) applied loading. (*From R. W. Clough and J. Penzien,* Dynamics of Structures, *McGraw-Hill, New York, 1975, p. 35; used by permission.*)

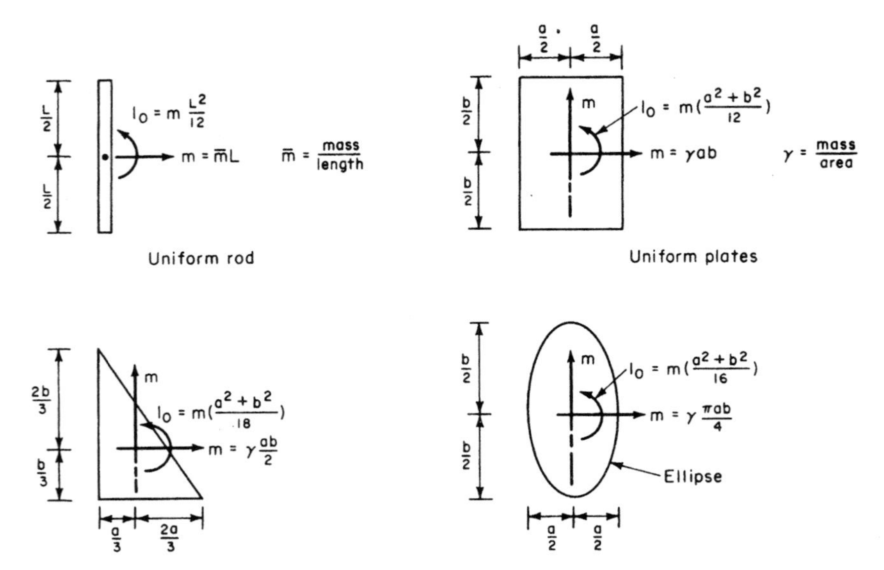

Figure 3.4 Rigid-body mass and mass moment of inertia. (*From R. W. Clough and J. Penzien,* Dynamics of Structures, *McGraw-Hill, New York, 1975, p. 24; used by permission.*)

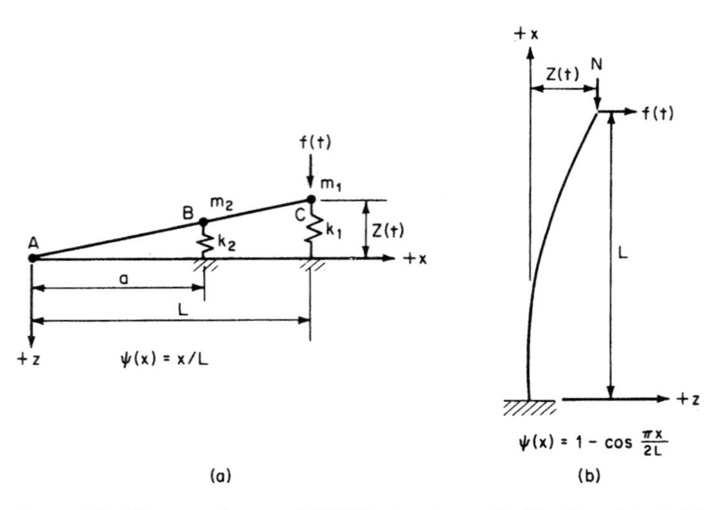

Figure 3.5 The two forms of SDOF structure idealization: (*a*) rigid-body assemblage and (*b*) continuously elastic structure. The shape function $\psi(x)$ is used to force SDOF behavior on the system by relating deflection $Z(t)$ to displacement at any point x.

true deflection curve, which is often quite complex, will therefore be satisfactory for most work.

Undamped vibrations. Many practical dynamic situations involve loadings of very short duration and, in any case, peak values of dynamic response are often the only values that are pertinent. Damping tends to moderate response with time, but immediate response is virtually free of damping effects. For determination of peak values, damping can generally be ignored. With the damping term dropped from Eq. (3.7), after dividing through by mass the expression becomes

$$\ddot{Z}(t) + \omega^2 Z(t) = 0 \qquad \text{where } \omega^2 = \frac{k^*}{m^*} \tag{3.12}$$

for undamped free response (i.e., response occurring after application of the dynamic loading has stopped). This common form of differential equation has the solution

$$Z(t) = C_1 \cos \omega t + C_2 \sin \omega t$$

The constants of integration C_1 and C_2 are evaluated by inserting the initial conditions, conditions at $t = 0$. For an initial displacement z_0, $C_1 = z_0$, and for an initial velocity $\dot{z}_0$, $C_2 = \dot{z}_0/\omega$. The equation then becomes

$$Z(t) = z_0 \cos \omega t + \frac{\dot{z}_0}{\omega} \sin \omega t \tag{3.13}$$

This equation clearly shows that displacement is composed of two components, one depending on initial displacement and the other on initial velocity. Figure 3.6 graphs each of the components and the full displacement plotted against time.

Further examination of Eq. (3.13) shows that the two terms are orthogonal vectors, so that the resultant, which is the maximum response amplitude, is

$$z_{\max} = \left[z_0^2 + \left(\frac{\dot{z}_0}{\omega} \right)^2 \right]^{1/2} \tag{3.14}$$

The argument ωt of the periodic cosine and sine functions causes the functions to repeat at angular intervals of 2π or at time intervals of $2\pi/\omega$. The time interval is called the *period of free vibration* and after substitution for ω becomes

$$T = 2\pi \left(\frac{m^*}{k^*} \right)^{1/2} \tag{3.15}$$

The *frequency of free vibrations,* the number of vibration cycles per unit of time, is then $\omega/2\pi$ or

$$f = \frac{1}{2\pi} \left(\frac{k^*}{m^*} \right)^{1/2} \tag{3.16}$$

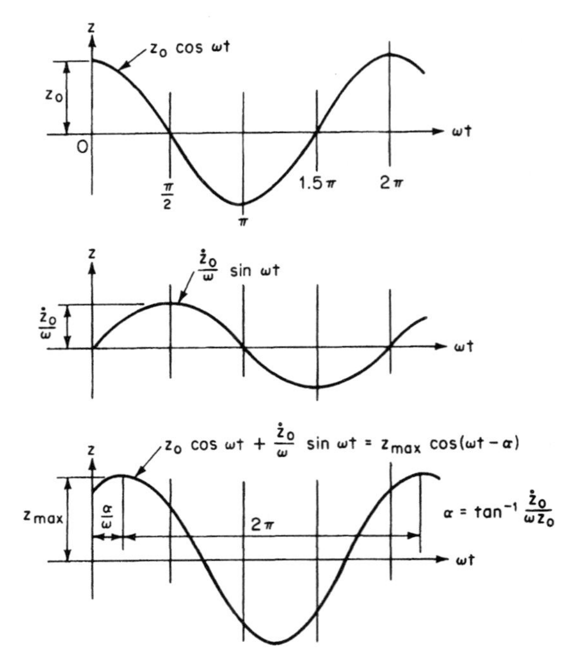

Figure 3.6 Dynamic displacement includes two components, one of which may be zero. An increment related to initial displacement is contributed by $z_0 \cos \omega t$, and $\dot{z}_0/\omega \sin \omega t$ is the component due to initial velocity. Adding the components yields system displacement, as shown in the third curve.

The parameter ω is often called the *circular frequency* because it has units of radians per unit of time.

The response of a system during loading or under continuing loading becomes more involved and is beyond the scope of our work at this point. For further information, the reader is referred to texts on structural dynamics or vibration theory.

The material above can be applied to Example 3.1. In the first part, initial displacement is zero and initial velocity has been calculated as 24.88 ft/s (7.58 m/s). The applied load is not time-varying, so that $f(t) = 0$; the system can be thought of as being loaded by the initial velocity. Generalized mass, from Eq. (3.8), is simply W/g because the shape function has a value of unity when the displacement is collinear with the applied load. The generalized spring rate, given in Eq. (3.10), is the summation of the spring rates in each of the three parts of line, or

$$k^* = (3 \times 2300 \text{ lb/in})(12 \text{ in/ft}) = 82{,}800 \text{ lb/ft } (1208 \text{ kN/m})$$

From Eq. (3.12)

$$\omega = \left[\frac{82{,}800 \text{ lb/ft}}{(5000 \text{ lb})/32.2}\right]^{1/2} = 23.09 \text{ rad/s}$$

Peak displacement is then given by Eq. (3.14)

$$z_{\max} = \left[0^2 + \left(\frac{24.88}{23.09}\right)^2\right]^{1/2} = 1.08 \text{ ft (328 mm)}$$

This entire displacement reflects elastic rope stretch, which is proportional to the initiating force divided by the spring rate. Therefore

$$F_i = (82{,}800 \text{ lb/ft})(1.08 \text{ ft}) = 89{,}400 \text{ lb (398 kN)}$$

but this does not include the static load of 5000 lb (2268 kg), so that the total force on the ropes is 94,400 lb (420 kN), which is essentially the value previously calculated.

With zero initial displacement, only the initial-velocity term contributes to displacement and this becomes a maximum when $\omega t = \pi/2$, as can be seen from Eq. (3.13). The total time lapse during which rope stretch takes place and impact force rises to full value is then $t = \pi/2\omega = 0.068$ s, a very short time indeed. Equation (3.13) gives us displacement values and therefore rope dynamic force at any time *t*. For intervals of 0.017 seconds we have

	Displacement		Dynamic rope force	
Time, *s*	ft	mm	lb	kN
0.017	0.41	125	33,950	151
0.034	0.76	232	62,900	280
0.051	1.00	305	82,800	368
0.068	1.08	328	89,400	398
0.085	1.00	305	82,800	368
0.102	0.76	232	62,900	280
0.119	0.41	125	33,950	151
0.136	0.00	0	0	0

Thus, in about ⅛ s the rope has become free of dynamic load and will remain so for about another ⅛ s, at which time loading will start again. Although the magnitude of the rope load is rather high, the very short duration of impulse loadings is such that it is unlikely that the crane will overturn unless the static load constitutes a substantial portion of the tipping load. For the data given, however, the rope would have failed.

In part (*b*) of the problem, impact was limited to 30%; therefore, rope stretch must be limited to

$$Z_{\max} = \frac{0.30 \times 5000 \text{ lb}}{82{,}800 \text{ lb/ft}} = 0.0181 \text{ ft (5.52 mm)}$$

The value of ω remains unchanged, so that using Eq. (3.12) gives

$$\ddot{Z}_{max} + 23.09^2(0.0181) = 0$$

$$\ddot{Z}_{max} = -9.65 \text{ ft/s}^2 \ (-2.94 \text{ m/s}^2)$$

where the minus sign indicates deceleration. This is the value previously determined that gave the velocity of 9.83 ft/s (3.0 m/s) asked for.

In part (*d*), acceleration was found to be -61.90 ft/s^2 (-18.87 m/s^2). From Eq. (3.12)

$$-61.90 + 23.09^2 Z(t) = 0$$

$$Z_{max} = 0.116 \text{ ft (35.4 mm)}$$

$F_i = 0.116$ ft (82,800 lb/ft) $= 9610$ lb (42.76 kN) which is the previous result that led to 192% impact.

Rotational motion

The dynamic effects accompanying rotational motion include inertial torque, centrifugal force, and load pendulum action. For many cranes, centrifugal force produces a twofold action. On the one hand, it creates a tensile force on a boom or jib that relieves the axial compressive forces to a minor extent. On the other hand, this same force acts more importantly to increase the overturning moment on the crane.

In the overall scheme of things, centrifugal force on the dead-load masses of cranes can usually be neglected, since swing speeds rarely exceed 1 r/min. However, centrifugal force throws the load out to an increased radius, enlarging the overturning moment (see Fig. 3.1). If n is the number of slewing revolutions per minute, we have

$$F_c = \frac{WR}{g}\left(\frac{\pi n}{30}\right)^2 \tag{3.17}$$

where F_c = centrifugal force
W = weight of load
R = operating radius
g = acceleration due to gravity

F_c will act horizontally at the upper load sheave shaft away from the axis of rotation and parallel to the horizontal projection of the boom centerline (see Chap. 4).

Load pendulum action is another matter. This dynamic motion is induced by the inertia of the load as the crane powers up or brakes during swinging, trolleying, or travel. The inertial force is horizontal, but it is tangential to the slewing arc (i.e., perpendicular to the centrifugal force). Consider a slewing boom; when braking occurs and the boom slows, the load will continue and pull forward on the boom. After the boom stops, the load will continue to swing, pendulum fashion.

The period of swing will vary with hoist line length L, the distance from load CG to the suspension point on the boom, with the relationship

$$T_p = 2\pi \left(\frac{L}{g}\right)^{1/2} \tag{3.18}$$

where T_p is the swing period. Line length, of course, is a random variable, which makes T_p a random variable as well, but T_p will increase with L. Should L be such that T_p corresponds to the natural period of vibration of the boom, resonance will occur. In a pure theoretical undamped system this will lead to a steady increase in vibration amplitude and eventual failure. In a real-life system with damping always present, amplitude will reach a peak value of some 5 to 10 times the effect of the same force statically applied. Fortunately, the structure period of vibration will always be shorter than load pendulum period and resonance will not take place. The FEM therefore suggests that the problem be simply resolved by applying load and dead-weight inertial forces at twice the values obtained when using mean acceleration. The factor 2 is taken to account for the elasticity in the system. For some very rigid devices, smaller values may be appropriate.*

During acceleration, the load will lag behind and the angle θ that the load hoist rope makes with the vertical is given by

$$\theta = \tan^{-1} \frac{a}{g} \tag{3.19}$$

where a is the tangential swing acceleration. Following this geometric representation, if W is the load weight, the inertial force F causing the lag can be expressed as

$$F = W \tan \theta = \frac{Wa}{g} = ma \tag{3.20}$$

which is the familiar basic inertial-force relationship. It should be noted that inertial force will vary inversely with stopping or powering-up time; i.e., if the time doubles, the acceleration rate and inertial force will halve.

---

*On the other hand, most crane operators soon learn to manage acceleration and deceleration rates so as to minimize or eliminate load pendulum action.

In terms of rotational motion the basic inertial equation can be stated as

$$T = J\alpha \tag{3.21}$$

where T = inertial torque
J = polar moment of inertia, moment of inertia of all rotating masses about axis of rotation = Σmr^2
α = angular acceleration

For any point on the rotating structure, the tangential acceleration $a = \alpha r$, where r is the radius to the point in question. The customary units for α are radians per second squared.

In calculating rotational inertial forces a convenient concept is that of *equivalent mass:* the single concentrated mass located at a particular radius and having the same rotational characteristics as the distributed masses it represents. It is a lumped mass. The concept can be expressed by the equation

$$m_e = \frac{md^2}{D^2}$$

where m_e at radius D is the equivalent of mass m whose CG is at radius d. This can be generalized as

$$m_e = \Sigma \frac{m_i d_i^2}{D^2} \tag{3.22}$$

to find the mass equivalent of a series of individual masses.

The inertial force F produced by angular acceleration α at a point at radius D from the center of rotation is therefore

$$F = m_e \alpha D = \frac{\alpha \Sigma m_i d_i^2}{D}$$

and the inertial torque is $T = m_e D^2 \alpha = \alpha \Sigma m_i d_i^2$.

To calculate bending moments induced in a swinging boom or jib by inertial forces, the equivalent mass of only those portions of the boom at greater radius than the moment plane are used. The inertial force acting at the CG of this mass is then found. The moment arm is the difference in radius between the moment plane, or cross section, and the CG of the mass.

Example 3.2 (*a*) What is the magnitude of the centrifugal force when an 11,000-lb (4990-kg) load (including the block) swings at 0.7 r/min at a radius of 100 ft (30.5 m)?

From Eq. (3.17)

$$F_c = \frac{(11{,}000 \text{ lb})(100 \text{ ft})}{32.2 \text{ ft/s}^2} \left[\frac{(0.7 \text{ r/min})(\pi)}{30}\right]^2 = 184 \text{ lb (817 N)} \qquad \textit{Ans.}$$

which is only about 1½% of the load.

(*b*) What load-suspension rope length will permit resonance to occur during slewing speed changes if the natural period of vibration of the boom is 1.2 s?

From Eq. (3.18)

$$1.2 \text{ s} = 2\pi \left(\frac{L}{32.2 \text{ ft/s}^2}\right)^{1/2}$$

$$L = 32.2 \left(\frac{1.2}{2\pi}\right)^2 = 1.17 \text{ ft (358 mm)} \qquad \textit{Ans.}$$

(*c*) If swing brakes are applied, causing the boom to come to a stop in 3.6 s from 0.7 r/min, what will be the inertial force induced by the load in part (*a*), and what angle will the load rope make with the vertical?

There are 2π rad in 360° (or 1 revolution), so that the angular velocity ϕ is

$$\phi = 0.7 \text{ r/min} \frac{2\pi}{60 \text{ s/min}} = 0.073 \text{ rad/s}$$

Then $\phi = \alpha t$, or

$$\alpha = \frac{0.073}{3.6 \text{ s}} = 0.020 \text{ rad/s}^2$$

Since tangential acceleration $a = \alpha r$,

$$a = 0.020(100 \text{ ft}) = 2.0 \text{ ft/s}^2 \text{ (0.61 m/s}^2\text{)}$$

From Eq. (3.20)

$$F = \frac{Wa}{g} = (11{,}000 \text{ lb}) \frac{2.0}{32.2} = 683 \text{ lb (3.04 kN)} \qquad \textit{Ans.}$$

which is the inertial force called for.

From Eq. (3.19)

$$\theta = \tan^{-1} \frac{2.0}{32.2} = 3.6° \text{ from the vertical} \qquad \textit{Ans.}$$

Example 3.3 A tower crane slewing motor can deliver 100,000 ft · lb (135.6 kN · m) of torque and is set for 1.0 r/min maximum slewing speed. If the entire slewing portion of the crane weighs 150 kips† (68.04 t) and the CG is located 30 ft (9.14 m) from the axis of rotation, how many seconds will

†1 kip = 1000 lb.

be required to power up to full swing speed in calm air? How many degrees of swing arc will be needed?

The mass moment of inertia (polar) of the system is

$$J = \frac{150 \text{ kips}}{32.2 \text{ ft/s}^2} (30 \text{ ft})^2 = 4192.5 \text{ kip} \cdot \text{ft} \cdot \text{s}^2 \ (579.6 \text{ t} \cdot \text{m} \cdot \text{s}^2)$$

From Eq. (3.21)

$$\alpha = \frac{T}{J} = \frac{100 \text{ kip} \cdot \text{ft}}{4192.5} = 0.024 \text{ rad/s}^2$$

$$t = \frac{\phi}{\alpha} = \frac{(1.0 \text{ r/min})(2\pi)}{0.024 \ (60 \text{ s/min})} = 4.36 \text{ s to full speed} \qquad \textit{Ans.}$$

$$\theta = \frac{1}{2} \alpha t^2 = 0.23 \qquad 0.23 \frac{360°}{2\pi} = 13.2° \text{ of arc traversed} \qquad \textit{Ans.}$$

Equations (3.7) and (3.12) have rotational-motion counterparts where $Z(t)$ is replaced by angular displacement $\phi(t)$. Carrying this substitution to the solution equation (3.13) gives

$$\phi(t) = \phi_0 \cos \omega t + \frac{\dot{\phi}_0}{\omega} \sin \omega t \tag{3.23}$$

and

$$\phi_{\max} = \left[\phi_0^2 + \left(\frac{\dot{\phi}_0}{\omega} \right)^2 \right]^{1/2} \tag{3.24}$$

where ϕ_0 is the initial angular displacement and $\dot{\phi}_0$ is the initial angular velocity. ω has the same definition as before.

3.4 Wind Loads

It should be understood at the outset that all practical calculations involving wind are approximate. This is dictated by the nature of wind, but it is not advisable to treat wind effects loosely; doing that will only lead to a coarser approximation and uncertainty in the end. More accurate and useful results are obtainable when the mathematical treatment of wind is carefully performed. This section describes several approaches to evaluating wind loads on crane structures. It concludes with a summary of the steps that must be taken, and the values that must be determined, in making a proper wind load analysis.

Wind is nothing more than a movement of air. Contact of the moving air with the earth's surface exerts a drag effect similar to friction. One can readily visualize that the drag induced by open prairie will be less than that caused by woodlands, which in turn will be less than that produced by a dense urban area. This drag causes wind speed to vary with height and with the coarseness of the landscape. Furthermore, topographical features, such as hills and valleys, channel wind to create lateral variations. At any particular location, actual wind measurements will show continuous random variation in both velocity and direction. Within the randomly varying pattern, a steady trend or average can be discerned for any time interval.† Velocity averaged over a defined period is reported rather than peak gust values.

National weather services furnish historical data as the basis for selecting wind speed values for design. Until recently, in the United States, the standard method for recording maximum values of wind speed was based on the time it takes one mile (1.6 km) of air to pass the monitoring point at standard elevation of 10 m (33 ft). This *fastest-mile* wind is an average of the gusts and lulls that occur during a maximum period of about 1 min. Gusts, which are random variations from that average, induce dynamic excitation as well as additional pressure. All cranes must resist both the static and dynamic components of the wind, but some cranes are particularly sensitive to the dynamic excitation.

A new wind standard was adopted in the United States during 1995.‡ This standard uses storm-wind data reflecting average velocity over a 3-s period, a period short enough to include the most powerful gusts. However, many local design codes no doubt still use the fastest mile wind for their design criteria. The same wind sample when measured using a 3-s average would be described as having higher velocity than it would as a fastest-mile wind. For example, in New York City, the fastest-mile wind speed (50 year recurrence) is 90 mi/h (40.2 m/s) per the old standard, while the 3-s average speed is 115 mi/h (51.4 m/s) under the new. Of course, calculation procedures must be different for the different averaging periods, because in the end design results should be similar if not identical. Therefore, the designer has to know which kind of wind data is being used, or calculations will produce results drastically above or below the proper values.

Wind-velocity pressure

Air at rest at sea level exerts a uniform pressure of about 14.7 lb/in^2 (101.4 kN/m^2) absolute. When air is in motion, this normal pressure

---

†For example, see Typical Wind Speed Recording in J. Kogan, "Crane Design: Theory and Calculations of Reliability," Israel Universities Press, Jerusalem, 1976, p. 212.

‡ASCE 7-95, Minimum Design Loads for Buildings and Other Structures, American Society of Civil Engineers.

is modified by a small amount at and near obstructions. The force wind exerts on a surface in its path and the negative force, or suction, on the leeward side of the object are due to these local low-level pressure changes. To put the phenomenon in scale, a change of only 1% in normal pressure is equivalent to 21 lb/ft^2 (1014 N/m^2).

The equivalent static pressure induced by wind is a function of the density of air, which varies with temperature, elevation, and barometric pressure. These variations are small, however, and are usually ignored in making practical wind calculations. The static-pressure relationship is given by

$$q = \tfrac{1}{2}\rho\nu^2 \tag{3.25}$$

where ρ is the density of air. For U.S. customary units, when velocity ν is given in miles per hour, the resultant pressure q is in pounds per square foot and Eq. (3.25) takes the form

$$q = \frac{\nu^2}{391} \approx \frac{\nu^2}{400} \tag{3.25a}$$

For velocity in meters per second and pressure in Newtons per square meter, the expression is

$$q = 0.613\nu^2 \approx \frac{5\nu^2}{8} \tag{3.25b}$$

With increase in elevation above ground surface, the drag effect of surface features becomes less pronounced, until an elevation is reached at which the wind is free of drag. This drag-free elevation is higher than heights that would be germane to most crane installations (700 to 1500 ft or 213 to 460 m); conversely, cranes can always be assumed to be installed in the drag-effect zone.

Various sources give different data on wind-velocity changes with height. A 1954 U.S. Navy publication† gives a power-law expression using an exponent of 1/7. But the Navy is concerned mainly with coastal areas, and the publication is offered as guidance for storm-wind design. For most engineering designs, storm winds are the only winds of interest.

Some velocity-variation data are given in Table 3.1, which differentiates between inland and coastal areas. This table seems to use a variable power-law exponent; the data presented appear to be rather

---

†*Tech. Publ.* NAVDOCKS TP-te-3, U.S. Navy, Bureau of Yards and Docks, Washington, May 15, 1954, p. 9.

TABLE 3.1 Fastest Mile of Wind for Various Height Zones above Ground†

Height zone, ft	Basic wind velocity, mi/h									
	60	67	75	80	85	90	95	100	115	130
FOR INLAND AREAS										
0–50	60	70	75	80	85	90	95	100		
50–150	70	80	90	95	100	105	110	120		
150–400	80	90	100	110	115	125	130	140		
400–700	90	100	115	120	130	135	145	150		
700–1000	100	110	125	130	140	145	155	160		
1000–1500	105	115	130	135	145	150	160	165		
FOR COASTAL AREAS										
0–50	60	70	75	80	85	90	95	100	115	130
50–150	85	95	100	105	110	115	120	125	140	150
150–400	115	125	130	135	140	145	150	155	170	180
400–600	140	150	160	165	170	175	180	185	190	195
600–1500	150	160	165	170	175	180	185	190	195	200

†1 ft = 0.30480 m; 1 mi/h = 0.44704 m/s = 1.609344 km/h.
SOURCE: William McGuire, "Steel Structures," Prentice-Hall, Englewood Cliffs, N.J., 1968; used by permission.

conservative, but they may have been modified to include gusting effects.

A Canadian researcher† has proposed power-law exponents of ⅐ for open country and coastal areas, 1/4.5 for wooded areas, towns, outskirts of cities, and rough coastal areas, and ⅓ for the centers of large cities. Davenport's values are accepted as authoritative and are used in an ANSI standard.‡ A set of curves using these values is given in Fig. 3.7.

The variation of velocity with height is given by

$$\nu = \nu_0 \left(\frac{h}{h_0}\right)^p \tag{3.26}$$

where ν = velocity at height h above adjacent ground
ν_0 = reference velocity at standard height h_0, 30 ft or 10 m
p = power-law exponent

For specific design applications, either Davenport's values or the data in Table 3.1 may be used, but Davenport's values are considered more representative of true conditions; gust effects must be added, however. The new wind standard

†A. G. Davenport, A Rationale for the Determination of Design Wind Velocities, proc. pap. 2476, *J. Struct. Div. ASCE*, vol. 86, no. ST5, May 1960, pp. 39–68.

‡ANSI A58.1–1982, Building Code Requirements for Minimum Design Loads in Buildings and Other Structures, American National Standards Institute.

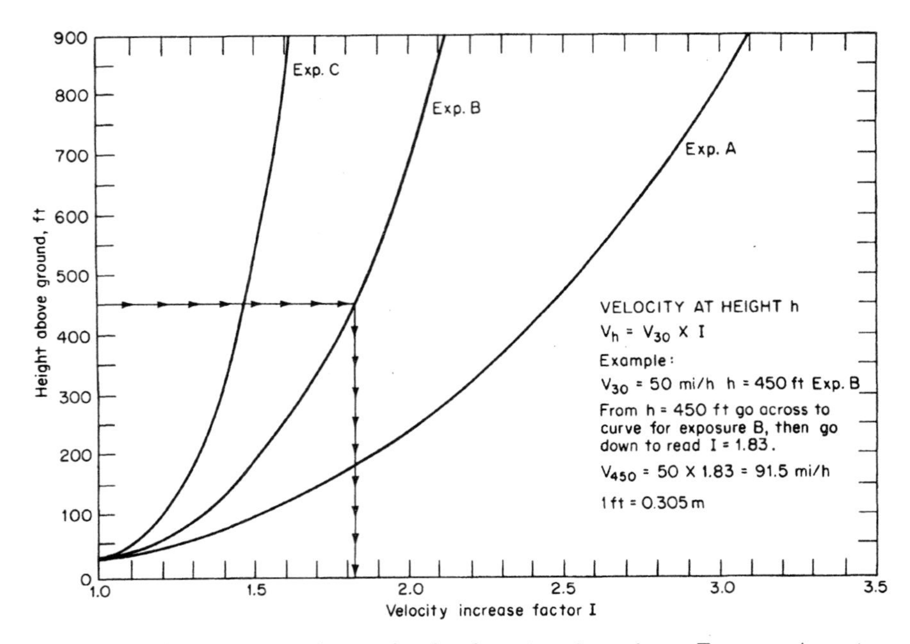

Figure 3.7 Velocity-increase factors for the three terrain regimes. *Exposure A:* centers of large cities and very rough hilly terrain. *Exposure B:* suburban areas, towns, city outskirts, wooded areas, and hilly terrain. *Exposure C:* flat, open country, coastal belts, and grassland.

given in ASCE 7-95 uses different power-law exponents and a different equation for pressure variation with height. This new method is explained later in the Chapter.

Wind pressure on objects

Equations (3.25*a*) and (3.25*b*) give the basic static pressure induced by wind on an object of undefined shape. Obviously a knife-edged object will cause little disturbance or pressure change in a windstream, while a large flat surface will have quite the opposite effect. Through research and tests, data have been gathered relating the shape of objects to the resistance they induce. Table 3.2 gives values for these force coefficients for a number of shapes common to crane designs.

When one object is in front of another identical object, the "shadow" effect must be considered. It is a parameter concerned with both shape and distance. Figure 3.8 gives values for the shielding coefficient η, which represents that part of the wind on the first of two bodies which acts on the second body. The coefficient can then be applied successively to additional bodies.

Finally, wind-pressure effects diminish if the wind is not head on to the surface under consideration. The effective pressure in the direction

TABLE 3.2 Wind-Pressure Coefficients, c_f

		f/b	C_f
Profiles, angles, box sections (small)		50	1.90
		40	1.70
		30	1.65
		20	1.60
		10	1.35
		5	1.30
		f/d	
Tubes	$dV < 50$ ft²/s where V = wind velocity, ft/s, and d is in feet	50	1.10
		40	1.00
		30	0.95
		20	0.90
		10	0.80
		5	0.75
	$dV \geqslant 50$ ft²/s	50	0.80
		40	0.75
		30	0.70
		20	0.70
		10	0.65
		5	0.60
		f/b	
Large box sections, over 14 in square and 10 by 18 in rectangular	$\frac{b}{c} \geq 2$	40	2.20
		30	2.10
		20	1.95
		10	1.75
		5	1.55
	$\frac{b}{c} = 1$	40	1.90
		30	1.85
		20	1.75
		10	1.55
		5	1.40
	$\frac{b}{c} = {}^1/_2$	40	1.40
		30	1.35
		20	1.30
		10	1.20
		5	1.00
	$\frac{b}{c} = {}^1/_4$	40	1.00
		30	1.00
		20	0.90
		10	0.90
		5	0.80
		f/b	C_f
Flat plates above ground level		$\geqslant 80$	2.00
		60	1.85
		40	1.75
		20	1.50
		15	1.40
		10	1.30
		$\leqslant 5$	1.20

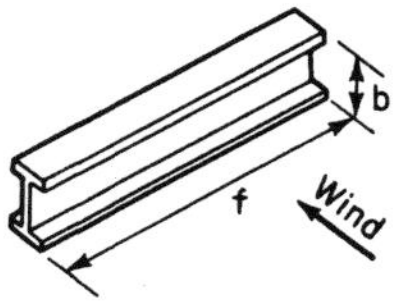

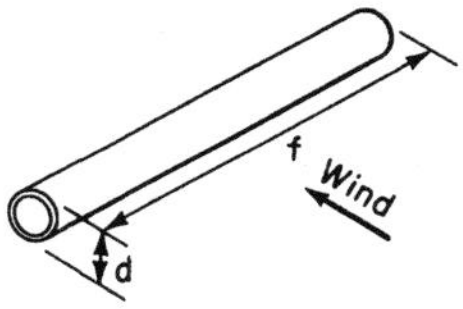

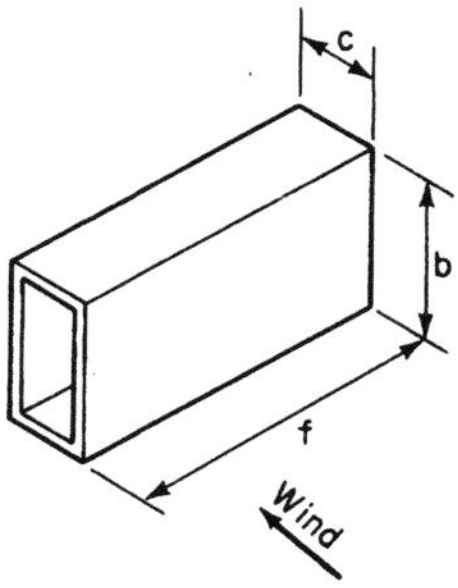

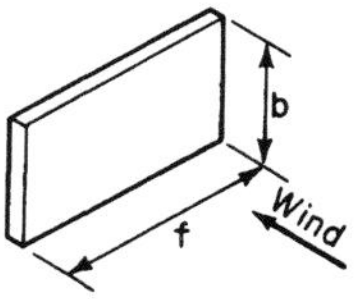

TABLE 3.2 (*Continued*)

Wire rope	Wind normal to rope				1.20
				φ†	
Latticed frames	Single frames, wind normal to face (use shielding factor η_m for multiple frames)	Profiles, angles, box sections, plates		≤0.05	1.95
				0.10	1.90
				0.20	1.75
				0.30	1.60
				0.40	1.45
				0.50	1.45
		Tubular members	$dV < 50$ ft²/s	≤0.05	1.30
				0.10	1.25
				0.20	1.20
				0.30	1.10
				0.40	1.05
				0.50	1.05
			$dV \geq 50$ ft²/s		0.80
	Assembled frames, square or triangular, wind normal to face	Square frames of profiles, angles, box sections, and plates		<0.025	4.0
				0.025–0.45	$4.13 - 5.18\varphi$
				0.45–0.7	1.8
				0.7–1.0	$1.33 + 0.67\varphi$
		Wind on diagonal, multiply normal by $1.0 + 0.75\sigma$			
				φ†	
		Triangular frames of profiles, angles, box sections, and plates		<0.025	3.6
				0.025–0.45	$3.71 - 4.47\varphi$
				0.47–0.7	1.7
				0.7–1.0	$1.0 + \varphi$
		Square and triangular frames with tubular members, multipliers for values above		<0.3	$^2/_3$
				0.3–0.8	$0.66\varphi + 0.47$
				0.8–1.0	1.0

†$\varphi = A_f/A_g$ where A_f = sum of face areas of members in frame and A_g = gross area enclosed by borders of frame.

of the wind becomes the head-on value multiplied by the square of the sine of the angle between the wind and the surface.

The force the wind exerts on an object can now be expressed as

$$F = qAC \tag{3.27}$$

where F = force on surface which acts in direction of wind
q = static pressure at height h of object
A = face area of object on which wind acts
C = configuration coefficient

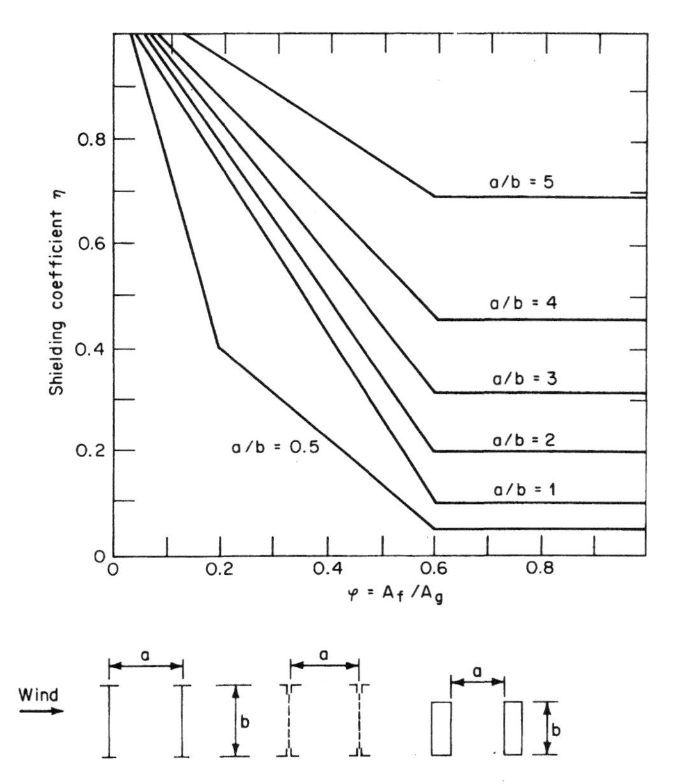

Figure 3.8 Shielding coefficient η. *(Reprinted by permission from Fédération Européene de la Manutention, "Rules for the Design of Hoisting Appliances," 2d ed., sec I, p. 23, Paris, 1970.)*

$= C_f$ for wind acting normal to object

$= C_f\eta_m$ for consecutive identical equidistant objects

$= C_f \sin^2 \theta$ for wind at angle to object

$C_f =$ force coefficient as given in Table 3.2

$\eta_m =$ cumulative effect of shielding on m identical equidistant objects

$= 1 + \eta + \eta^2 + \eta^3 + \cdots + \eta^{m-1} \leqslant (1 - \eta)^{-1}$

$\eta =$ shielding coefficient as given in Fig. 3.8

$\theta =$ acute angle between wind and surface

When a surface is at an angle to the wind and the effect normal to the surface is required, the force can be resolved in the ordinary manner. For wind blowing on the diagonal to a square tower, however, the value for C may be taken as $1.0 + 0.75\varphi$ times the head-on value for a single tower face with shielding of the second face taken into account. φ is the ratio of the solid area to the gross area of the face. The force will, of course, be taken to act on the diagonal.

A working value for q can be found by using Eq. (3.25*a*) or (3.25*b*) and adjusting for height by using Eq. (3.26), but these equations can be combined to form

$$q = \begin{cases} \dfrac{\nu_0^2}{391}\left(\dfrac{h}{h_0}\right)^{2p} & \text{U.S. customary units} \quad (3.28a) \\ 0.613\nu_0^2\left(\dfrac{h}{h_0}\right)^{2p} & \text{SI units} \quad (3.28b) \end{cases}$$

As will be more fully explained under the heading "Storm Winds and Gusts," the appropriate value to use for reference velocity ν_0 will depend on the characteristics of the local terrain.

When dealing with tall vertical objects, such as towers, the question arises: What height should be used? For a body whose top is at height h_2 and lower end at h_1, where $h_1 \geqslant h_0$, and with width b and shape factor C, the wind moment about ground level will be

$$M = C\int_{h_1}^{h_2} q(y)y\,dA$$

where $q(y) = q_0(y/h_0)^{2p} = q_0 h_0^{-2p} y^{2p}$, $y = 0$ at ground level, and the differential area is $dA = b\,dy$. Then

$$M = q_0 h_0^{-2p} Cb \int_{h_1}^{h_2} y^{2p+1}\,dy$$

$$= q_0 h_0^{-2p} Cb\,\frac{h_2^{2p+2} - h_1^{2p+2}}{2p + 2} \qquad (3.29)$$

When $h_1 = 0$,

$$M = \frac{q_0 Cb h_0^2}{2} + q_0 Cb h_0^{-2p}\,\frac{h_2^{2p+2} - h_0^{2p+2}}{2p + 2}$$

$$= q_0 Cb\left(\frac{h_0^2}{2} + h_0^{-2p}\,\frac{h_2^{2p+2} - h_0^{2p+2}}{2p + 2}\right) \qquad (3.30)$$

When it is necessary to take moments about a point at height h_3 ($h_3 \geqslant h_0$), the moment arm y above is replaced by $y - h_3$, which yields

$$M = q_0 h_0^{-2p} Cb\left(\frac{h_2^{2p+2} - h_3^{2p+2}}{2p + 2} - h_3\,\frac{h_2^{2p+1} - h_3^{2p+1}}{2p + 1}\right) \qquad (3.31)$$

The shear expressions are similarly found. When $h_1 \geqslant h_0$,

$$V = q_0 h_0^{-2p} Cb\,\frac{h_2^{2p+1} - h_1^{2p+1}}{2p + 1} \qquad (3.32a)$$

and when $h_1 = 0$,

$$V = q_0 C b \left(h_0 + h_0^{-2p} \frac{h_2^{2p+1} - h_0^{2p+1}}{2p + 1} \right) \tag{3.32b}$$

Equations (3.29) to (3.32) intrinsically reflect the continuous change in wind pressure with height and are therefore more accurate and convenient to use than other alternatives. One alternative is to divide the height into bands and to calculate a pressure approximate to each band. Another is to use the pressure at the top for the full height.

Example 3.4 (*a*) Calculate the static wind force per unit of tower height for the square tower shown in Fig. 3.9 with wind at 60 mi/h (26.8 m/s) at standard height. Take the velocity at midheight as representative of the entire tower and use Table 3.1 (coastal areas) for velocity variation with height.

First, the force coefficients are found for each of the members using Table 3.2. For the main-chord square tubes, the length-to-width ratio is

$$\frac{f}{b} = 21 \text{ ft} \, \frac{12 \text{ in/ft}}{3.5 \text{ in}} = 72 > 50$$

and therefore

$$C_f = 1.9$$

For the round battens, the horizontal members, the ratio is

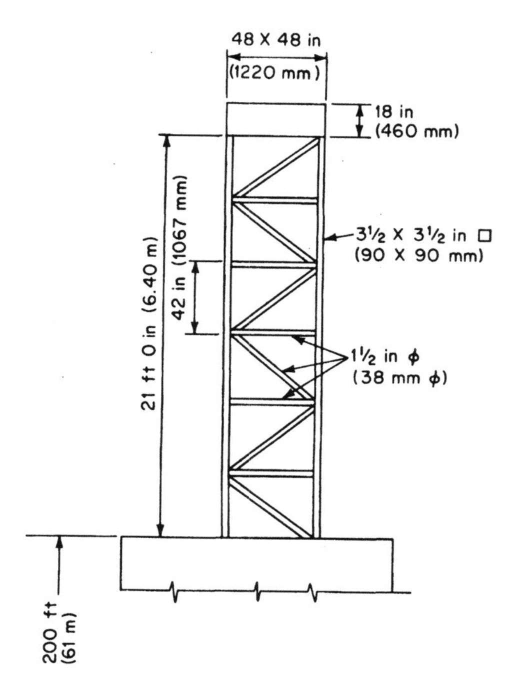

Figure 3.9 Tower dimensions for use in Example 3.4.

$$\frac{f}{b} = \frac{48 \text{ in} - 2(3.5 \text{ in})}{1.5 \text{ in}} = 27.33$$

The drag on cylindrical objects depends upon whether flow is smooth or turbulent, which in turn is a function of wind speed and tube diameter. From Table 3.1 at tower midheight wind velocity is to be taken at 115 mi/h (51.4 m/s) when base-height velocity is 60 mi/h (26.8 m/s):

$$115 \text{ mi/h} = 168.7 \text{ ft/s } (51.4 \text{ m/s})$$

$$dV = \frac{1.5 \text{ in}}{12 \text{ in/ft}} 168.7 = 21.1 < 50$$

Therefore by interpolation,

$$C_f = 0.94$$

The length of the round diagonals is

$$f = [(42 - 1.5)^2 + (48 - 2 \times 3.5)^2]^{1/2} = 57.6 \text{ in } (1.46 \text{ m})$$

$$\frac{f}{b} = \frac{57.6}{1.5} = 38.4 \qquad dV < 50$$

From Table 3.2

$$C_f = 0.99$$

by interpolation.

The shielding effect on the leeward face can be found from Fig. 3.8. The gross area of any one panel section of the tower is

$$A_g = (42 \text{ in})(48 \text{ in}) = 2016 \text{ in}^2 \ (1.30 \text{ m}^2)$$

and the sum of the face areas for all members in a panel is

$$\begin{aligned} A_f &= 2(3.5 \text{ in})(42 \text{ in}) + [48 \text{ in} - 2(3.5 \text{ in})](1.5 \text{ in}) \\ &\quad + (57.6 \text{ in})(1.5 \text{ in}) \\ &= 441.9 \text{ in}^2 \ (0.29 \text{ m}^2) \end{aligned}$$

The ratio $A_f/A_g = \varphi = 441.9/2016 = 0.22$, and the depth-to-width ratio a/b is unity for a square tower. From Fig. 3.8 the shielding coefficient η is found to be 0.72. This means that only 72% of the wind force on the windward frame will fall on the leeward frame. For $m = 2$, $\eta_m = 1 + \eta = 1.72$.

The products of the individual surface areas and their force coefficients can now be summed for a panel and the cumulative effect of shielding taken into account; then dividing by the panel length of 42 in = 3.5 ft (1.07 m) gives the unit effective wind area AC

$$AC = \frac{1}{3.5\ \text{ft}}\left[(3.5\ \text{ft})(2)(3.5\ \text{in})\frac{1.9}{12\ \text{in/ft}}\right.$$

$$+ (48\ \text{in} - 7\ \text{in})(1.5\ \text{in})\frac{0.94}{144\ \text{in}^2/\text{ft}^2}$$

$$\left.+ (57.6\ \text{in})(1.5\ \text{in})\frac{0.99}{144\ \text{in}^2/\text{ft}^2}\right](1.72)$$

$$= 2.40\ \text{ft}^2/\text{ft}\ (0.73\ \text{m}^2/\text{m})$$

From Eq. (3.25a), the velocity pressure for winds at 115 mi/h (51.4 m/s) is

$$q = \frac{115^2}{391} = 33.82\ \text{lb/ft}^2\ (1.62\ \text{kN/m}^2)$$

and then from Eq. (3.27)

$$F = 33.82(2.40) = 81.17\ \text{lb/ft}\ (1.18\ \text{kN/m}) \qquad \textit{Ans.}$$

Had Fig. 3.7 been used for velocity variation with height, for exposure C at 210.5 ft we would have had $I = 1.32$. Then $V = 1.32(60\ \text{mi/h}) = 79.2$ mi/h (35.4 m/s) and Eq. (3.25a) gives $q = 79.2^2/391 = 16.04\ \text{lb/ft}^2$ (768 N/m^2). As an alternative, Eq. (3.28) makes the adjustment directly

$$q = \frac{60^2}{391}\left(\frac{210.5}{30}\right)^{2/7} = 16.06\ \text{lb/ft}^2\ (769\ \text{N/m}^2)$$

The Davenport values produce pressures significantly lower than Table 3.1 because the table apparently includes an allowance for gusts while we have been using mean wind values.

If we use another interpretation of Table 3.2, applying the values under single latticed frames and then the shielding factor, the effective wind area comes to 2.32 ft^2/ft (0.72 m^2/m) instead of the previous value of 2.40. Table 3.2 distinguishes between frames with tubular members and frames of other shapes. To use these data for the structure in Fig. 3.9, a separate value must be found for each type of frame. The final value is determined by ratioing in proportion as the face area of each type of member relates to the face area of the entire frame. There are 294 in^2 (0.19 m^2) of square tubes in a panel and 147.9 in^2 (0.10 m^2) of round tubes out of the full panel face area of 441.9 in^2 (0.29 m^2). From Table 3.2 with $\varphi = 0.22$, for a square-tubed frame $C_f = 1.72$ and for a round-tubed frame $C_f = 1.18$. For the combined frame

$$C_f = \frac{(294\ \text{in}^2)(1.72) + (147.9\ \text{in}^2)(1.18)}{441.9\ \text{in}^2} = 1.54$$

$$AC = \frac{441.9\ \text{in}^2}{144\ \text{in}^2/\text{ft}^2}(1.54)\frac{1.72}{3.5} = 2.32$$

We can use Table 3.2 in yet another way and evaluate Fig. 3.9 using the data for frames assembled into a square. For an assembled frame of square tubes with $\varphi = 0.22$

$$C_f = 4.13 - 5.18\varphi = 2.99$$

and for a round-tubed frame $C_f = 2.99(2/3) = 1.99$. For the actual frame

$$C_f = \frac{294(2.99) + 147.9(1.99)}{441.9} = 2.66$$

Noting that shielding effects are already included in the assembled frame values, we get

$$AC = \frac{441.9}{144}\frac{2.66}{3.5} = 2.33$$

(*b*) What will be the force on the rectangular plated section at the top of the tower using Davenport's exponent?

We use values for large box sections in Table 3.2 to get

$$\frac{f}{b} = \frac{18 \text{ in}}{48 \text{ in}} = 0.38 \qquad \text{and} \qquad \frac{b}{c} = 1$$

so that $C_f = 1.4$

$$AC = 48\,\frac{18}{144 \text{ in}^2/\text{ft}^2}\,1.4 = 8.4 \text{ ft}^2 \ (0.78 \text{ m}^2)$$

$$h = 221.75 \text{ ft} \ (67.59 \text{ m})$$

$$q = \frac{60^2}{391}\left(\frac{221.75}{30}\right)^{2/7} = 16.31 \text{ lb/ft}^2 \ (781 \text{ N/m}^2)$$

$$F = 16.31(8.4) = 137 \text{ lb} \ (609 \text{ N}) \qquad \textit{Ans.}$$

(*c*) With the base wind at 100 mi/h (44.7 m/s), what force will develop on this part?

$$q = \frac{100^2}{391}\left(\frac{221.75}{30}\right)^{2/7} = 45.29 \text{ lb/ft}^2 \ (2.17 \text{ kN/m}^2)$$

$$F = 45.29(8.4) = 380 \text{ lb} \ (1.69 \text{ kN}) \qquad \textit{Ans.}$$

Storm winds and gusts

Most construction people are aware that some locales experience higher winds than others. Beyond location, however, maximum expected wind intensity varies over time intervals as well, such as the intensity accompanying the worst storm with a 25-, 50-, or 100-year average *recurrence period.* For example, the highest wind speed expected to occur in a particular location in any 5-year period may be 60 mi/h (26.8 m/s), while the 100 year maximum may be 110 mi/h (49.2 m/s). Wind maps are published giving lines of common maximum wind velocity, called *isotach lines*, for various return or recurrence periods. Figures 3.10 to 3.12 are 25-, 50-, and 100-year-recurrence-interval maps for extreme fastest mile winds in the United States. These maps can be used for the design of equipment or installations within any given zone when local conditions are taken into account and with wind gusts and their effects also considered.

Some crane types are more prone than others to excitation from wind gusts. A limber structure with its mass concentrated at the top would be especially prone. The tendency may be exacerbated if the base support is limber and elastic too. As many tower cranes fit this description, gust effects are an important consideration in the analysis of tower cranes.

Wind gusts cause a crane to oscillate unevenly because the gusts are random. Should a gust coincide in direction with crane displacement velocity, the gust will add energy to the system and will increase displacement. If the system energy carries displacement beyond a limiting value, the crane will no longer have a stabilizing moment and will overturn. Wind oscillation can also magnify stress and cause structural damage or failure.

Cranes with low natural vibration frequencies are more sensitive to gust action. Their longer oscillation periods mean that the magnitude of resonant gusts will be greater, and in any case the elastic response to all gusts will be amplified more than for a stiffer structure.

For a gust to be fully effective in loading a structure, say a crane tower, it must completely envelop the tower. Portions of the gust mass must extend from a position in advance of the windward side to beyond the leeward side. The gust depth must then be from 4 to 8 times tower depth, according to various researchers. For a 10 ft-square (3.05-m) tower and wind at 60 mi/h (26.8 m/s) this requires a gust of at least ½ to 1 s duration for sufficient depth, a part of the gust spectrum that includes the lower-energy gusts. But gusts that exceed the natural vibration period of the structure will act to produce a quasi-static response similar in effect to the mean wind. Subgusts within those longer-lasting gusts will then add a dynamically amplified component to response.

Following from the work of Davenport, Vellozzi and Cohen† have used a probabilistic derivation to produce a deterministic, generalized

---

†J. W. Vellozzi and E. Cohen, Gust Response Factors, proc. pap. 5980, *J. Struct. Div. ASCE,* vol. 94, no. ST6, June 1968, pp. 1295–1313.

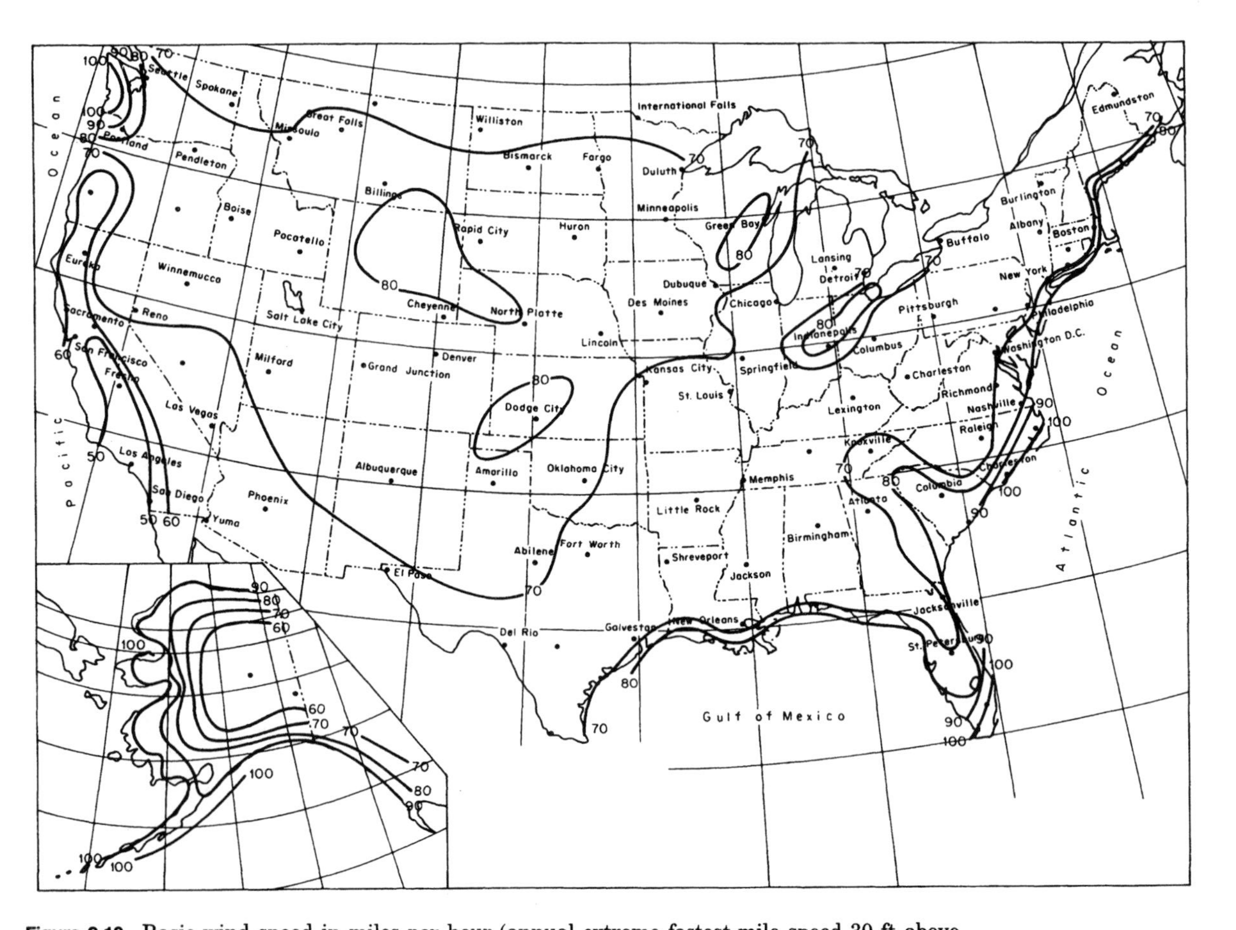

Figure 3.10 Basic wind speed in miles per hour (annual extreme fastest-mile speed 30 ft above ground): 25-year mean recurrence interval. *(From ANSI A58.1-1972; reprinted by permission of the American National Standards Institute, New York.)*

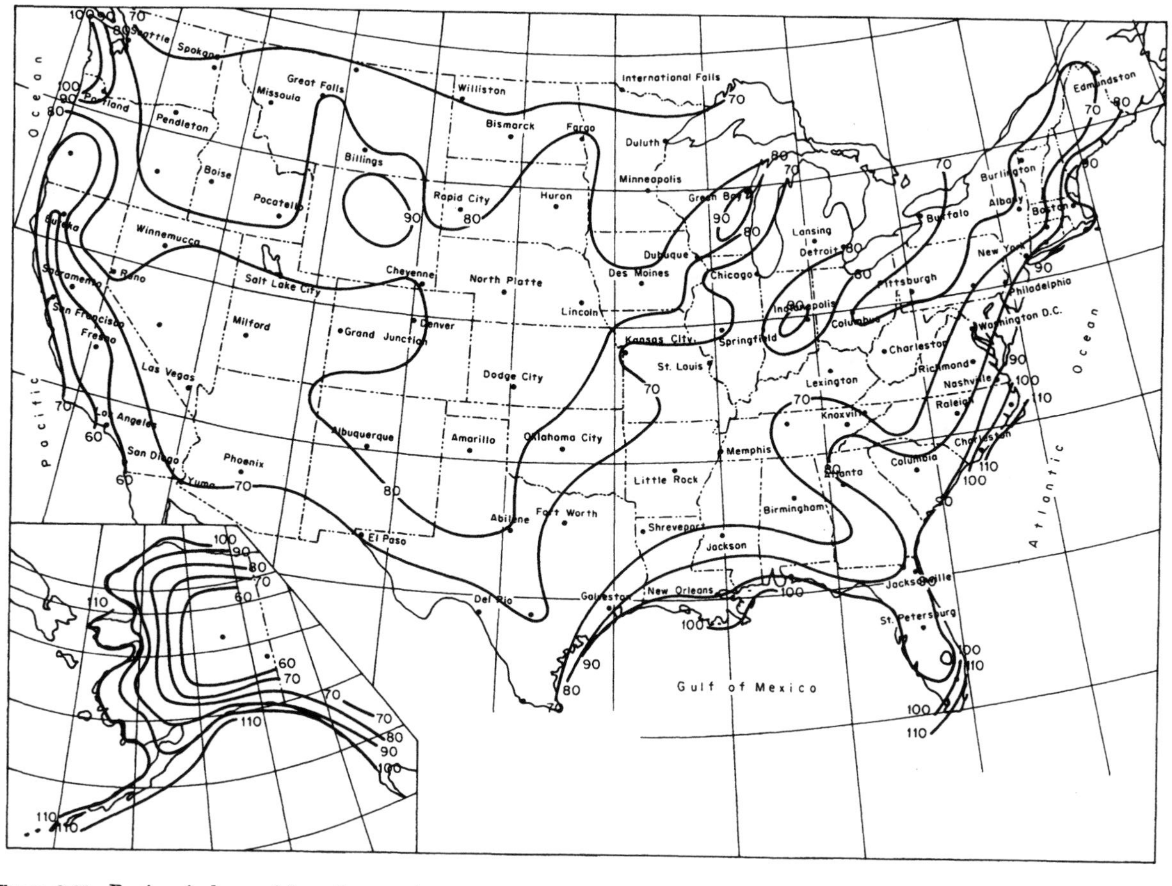

Figure 3.11 Basic wind speed in miles per hour (annual extreme fastest-mile speed 30 ft above ground): 50-year mean recurrence interval. (*From ANSI A58.1-1972; reprinted by permission of the American National Standards Institute, New York.*)

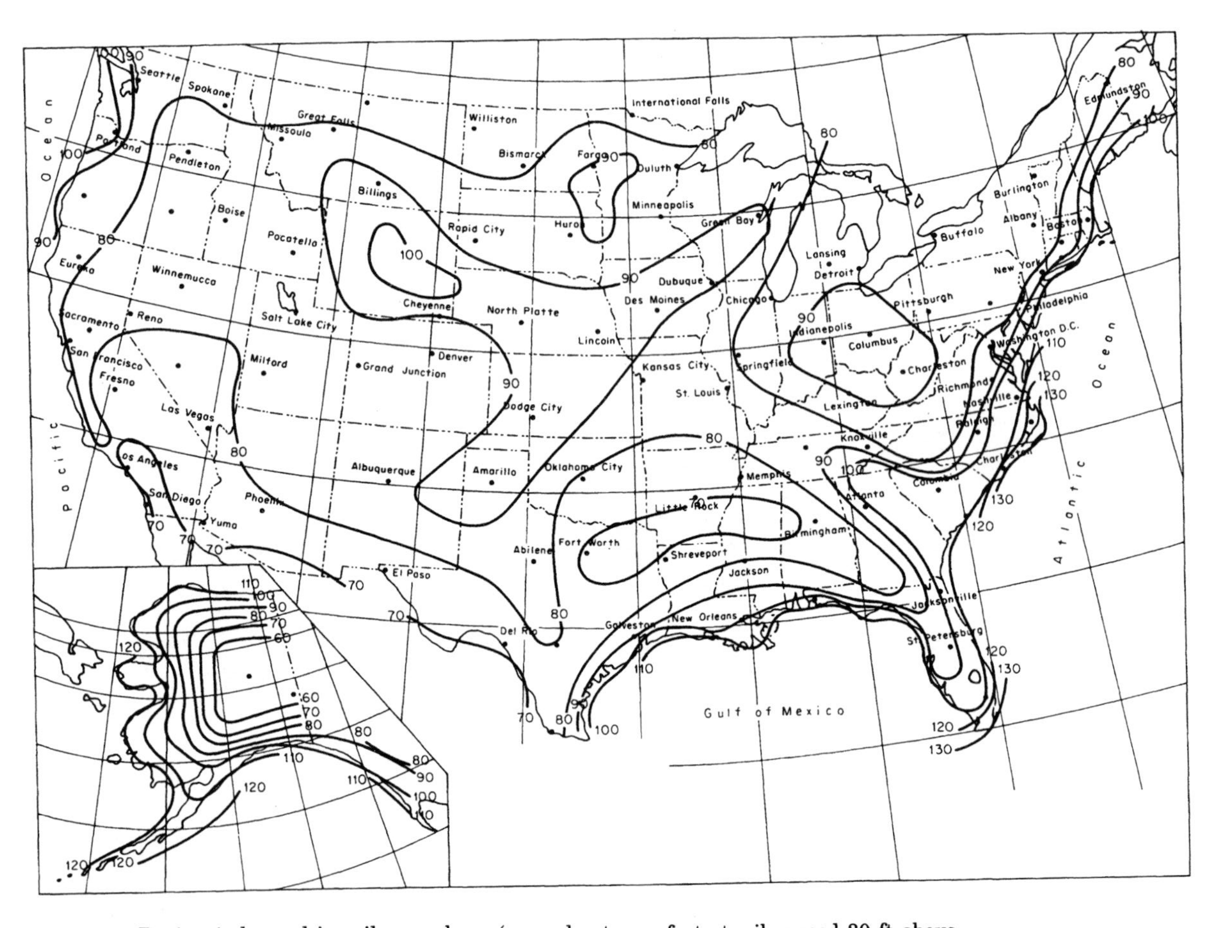

Figure 3.12 Basic wind speed in miles per hour (annual extreme fastest-mile speed 30 ft above ground): 100-year mean recurrence interval. (*From ANSI A58.1-1972; reprinted by permission of the American National Standards Institute, New York.*)

gust-effect procedure applicable to openwork or latticed structures, but as they themselves state, the procedure is approximate. Their method does, however, account for the wind-exposure environment and for the specific dynamic properties of the structure. But the designer is not given the opportunity to select a level of reliability, or conversely, a probability of failure, for the particular structure other than through selection of a storm-return period. Not all structures are of equal importance, nor do all structures pose an equal threat to life or property.

Vickery† claims, on the basis of his own theoretical derivation and of wind-tunnel and full-size test data, that the gust factors given by the Vellozzi and Cohen method are unconservative. The authors of the method responded that insufficient test data were available to support such a conclusion. We feel that the method is somewhat conservative when applied to cranes. Although that feeling is essentially intuitive, it is partially based on their use of averaged parameters more appropriate to buildings than to cranes.

The Vellozzi-Cohen method is reflected in the previously mentioned standard ANSI A58.1, but in the earlier 1972 edition,‡ where a procedure is given for determining a gust factor, a multiplier for the static wind loading to account for gusts.

Two distinct philosophies can be applied to the task of designing cranes or crane installations to resist gusts. In the first, one chooses a level of gusting that most probably will occur during the life of the crane and designs with a suitable margin of strength and stability to resist that gust. Implicit in this approach is that the probability of occurrence of gusts stronger than the design gust will be kept suitably low, say on the order of 10%. This concept is implied in the ANSI standard. If the overload gust, of low probability, occurs, it will have to be resisted by overstress or by a reduced margin of stability.

The second conceptual approach requires design such that the threshold of failure occurs on application of a gust with such a long mean recurrence period that an acceptable probability of failure is maintained. This approach requires wind statistics to be mathematically projected to reflect time periods much greater than the length of time records have been kept.

Another proposal for wind-gust levels is offered by the Factory Mutual Engineering Corporation,§ an agency supported by a group of

---

†B. J. Vickery, On the Reliability of Gust Loading Factors, *NBS Build. Sci. Ser.* 30, 1970, p. 102.

‡The 1982 edition presents another method for determining gust factors; for cranes, the results are close to the 1972 method. The gust factor provisions of ASCE 7-95, a replacement for ANSI A58.1, would apply for short, stiff crane structures.

§Wind Forces on Buildings and Other Structures, sec. 1–7 in "Loss Prevention Data," Factory Mutual Engineering Corp., Norwood, Mass., June, 1974.

insurance companies. They suggest gust factors of 1.59 for hilly or wooded terrain and 1.32 for flat open areas. The factors apply at an elevation of 30 ft (9.14 m) and are intended to be decreased to unity at 1500 ft (460 m). The factors apply to structures in general and do not include dynamic considerations directly.

The effect of a wind gust is such that it either can or cannot cause a crane to fail. Gust effects on a crane can therefore be taken as a Poisson process, inasmuch as each gust is statistically independent of any other gust. The probability that a critical gust will not occur in time T_w is then given by

$$P(T_r > T_w) = \exp(-\gamma T_w)$$

where γ = mean recurrence rate of occasions when the crane may reach incipient failure due to gusting

T_r = time during which the crane can be expected to remain in a wind-gust environment until the critical gust arrives

T_w = time during which the crane will remain installed at worksites

T_r, being dependent upon a random wind gust, is a random variable. The cumulative distribution function is then

$$P(T_r \leqslant T_w) = 1 - \exp(-\gamma T_w) \tag{3.33}$$

If a probability of failure due to gusting of 0.001 is selected for any single year of operation, the recurrence period $1/\gamma$ derived from Eq. (3.33) will be 999.5 years. This same equation will then show a probability of failure of 0.010 for 10 years of operation and 0.020 for 20 years. It will also show that the probability of an event occurring within its return period is 0.632.

But crane bases generally approximate a square. Wind typically poses potential danger for overturning when normal to a side; for structural failure wind must be normal for some crane types and on the diagonal for others. If "normal" and "on the diagonal" are extended to include a zone 12.9° to either side of these positions, the probability that a gust will be oriented to pose a threat can be taken as $25.8°/90° = 0.29$. The extended zones thus defined comprise those arcs within which wind effects are at least 95% of the dead-center values.

When the orientation probability is considered, design gusts can occur more frequently than 999.5-year mean intervals while still maintaining 0.001 probability of failure, because only 29% of all gusts would be expected to be effective.

Rational peak design wind speed selection can be achieved by studying actual design relationships. The nominal strength margin ordinarily used for structural design is 1.67. Then the relationship between design and failure wind speed should be

$$V(failure) = (1.67)^{1/2}\, V(design)$$
$$= 1.29\, V(design)$$

which implies a failure wind recurrence period of about 500 years if the recurrence period of the design wind is 50 years. But design codes in the United States usually permit allowable stresses to be increased by ⅓ when considering wind loading. Then

$$V(failure) = (1.67/1.33)^{1/2}\ V(design)$$
$$= 1.12\ V(design)$$

Assuming that dead load stress is 20% of allowable stress, and a ⅓ increase is applicable, the wind induced stress would be 1.13 times allowable, and

$$V(failure) = (1.67/1.13)^{1/2}\ V(design)$$
$$= 1.22\ V(design)$$

Stability based ratings for tower cranes and heavy duty type cranes are typically set at ⅔ of the static load that will induce overturning. It is interesting to note that this rating basis produces the same relationship between $V(failure)$ and $V(design)$ as the structural situation above. Thus, for the general case, one design wind speed can be chosen for both structural strength and stability evaluations.

Using information from ASCE 7-95†, the relationship between wind return periods for hurricane and other areas can be expressed in equation form

$$V_R = V_{50}C \tag{3.34}$$

where V_R is the wind speed associated with recurrence interval R, V_{50} is the pertinent 50 yr recurrence wind speed, and the factor C is

for hurricane areas $\quad C_h = 0.33 \log R + 0.44$

or other areas $\quad C_o = 0.23 \log R + 0.61$

With the directional probability 0.29 inserted into Eq. (3.33), the joint probability which is the combined probability that a strong enough gust will occur within the critical directional zone, the probability of structural or stability failure, becomes

$$P(T_r \leq T_w) = 0.29[1 - \exp(-\gamma T_w)] \tag{3.35}$$

Assuming a failure level wind recurrence interval R of 25 years, $\gamma = 1/25$. For each year, the crane remains on the site ($T_w = 1$), Eq. (3.35) reveals a probability of failure of 0.01, or 1%. For sites where it is unlikely that people would be present during high level storm winds, the risk would concern damage to property. For a lower probability of failure, say 0.001, the failure level wind recurrence period R would be about 250 years.

It is more comfortable to address probability of survival rather than failure; the probability of survival is simply one minus the probability of failure. The selection of an appropriate probability of survival is subjective and site specific. However,

†Ibid., Table C6-5.

when threat to life is minimal, another subjective determination, the decision would rest on the likely extent of property damage in the event of failure.

Example 3.5 (*a*) A crane is to be installed at a site for about 2 years in an area where the 50 year recurrence wind is 80 mi/h (35.8 m/s) and hurricanes do not occur. Dead load stress is about 20% of the allowable stress, and the allowable stress can be increased by ⅓ when wind induced stress is included. What design wind speed should be used if a survival probability of 0.999 is desired?

Using Eq. (3.35),

$$1 - 0.999 = 0.29[1 - \exp(-2\gamma)]$$

which by iteration gives $\gamma = 1/500$, or $R = 500$ years.

From Eq. (3.34),

$$V_R = 80(0.23 \log 500 + 0.61) = 80 \times 1.23$$

$$= 98 \text{ mi/h } (43.8 \text{ m/s})$$

$$V(design) = V(failure)/1.22 = 98/1.22$$

$$= 80 \text{ mi/h } (35.8 \text{ m/s}) \qquad \textit{Ans.}$$

(*b*) What design speed should be used if the crane remains at the site for 1 year with the same survival probability?

Using Eq. (3.35),

$$1 - 0.999 = 0.29[1 - \exp(-\gamma)]$$

which by iteration gives $\gamma = 1/250$, or $R = 250$ years.

From Eq. (3.34),

$$V_R = 80(0.23 \log 250 + 0.61) = 80 \times 1.16$$

$$= 93 \text{ mi/h } (41.6 \text{ m/s})$$

$$V(design) = V(failure)/1.22 = 93/1.22$$

$$= 76 \text{ mi/h } (34.0 \text{ m/s}) \qquad \textit{Ans.}$$

(*c*) Suppose the survival probability is reduced to 0.995 for a two year installation. How will this affect design wind speed?

$$1 - 0.995 = 0.29[1 - \exp(-2\gamma)]$$

which by iteration gives $\gamma = 1/115$, or $R = 115$ years.

$$V_R = 80(0.23 \log 115 + 0.61) = 80 \times 1.08$$
$$= 86 \text{ mi/h } (38.4 \text{ m/s})$$
$$V(design) = V(failure)/1.22 = 93/1.22$$
$$= 70 \text{ mi/h } (31.3 \text{ m/s}) \qquad Ans.$$

(*d*) A general-purpose construction tower crane has a useful life of perhaps 25 years, but installed duration at a typical construction site is short, say 2 years at most. As a general-purpose machine it may be installed at any geographical location. What value should be used for the design wind speed if the survival probability is 0.99?

$$1 - 0.99 = 0.29[1 - \exp(-2\gamma)]$$

which by iteration gives $\gamma = 1/60$, or $R = 60$ years.

$$C_h = 0.33 \log 60 + 0.44 = 1.027$$
$$C_o = 0.23 \log 60 + 0.61 = 1.019$$

$$V(design) = CV_{50}/1.22$$
$$= 0.84\ V_{50} \quad \text{for hurricane areas}$$
$$= 0.84\ V_{50} \quad \text{for other areas}$$

V_{50}, the 50-year recurrence peak wind velocity, is a function of the installation geographical location, but a serially-manufactured crane design has fixed properties regardless of installation location. The manufacturer, then, could prepare tables giving mast-height and jib-length limitations for a number of 50-year recurrence wind values and survival probabilities. With this information included in the crane documentation, installation designers could apply the proper crane configuration at any work site location with due consideration for local conditions.

Statistical wind verses real wind

When a crane cannot be removed from harm's way in the face of a major storm, it had better be able to withstand the worst that storm has to offer. As there is no possibility for the engineer to know in advance what intensity of wind a crane will be subjected to under such circumstances, a fictitious design wind must be substituted. The magnitude of the fictitious wind may be selected arbitrarily, but preferably it will be derived using statistical methodology. Given that the crane would not be working during this hypothetical storm, this is commonly referred to as the *out-of-service* condition.

Naturally, then, the maximum wind speed that a crane would be exposed to while working is called the *in-service* condition. Its intensity is not a statistical entity because it is possible to specify a limiting velocity for the crane during operation. When that limiting wind speed is exceeded, the crane must be taken out of operation. The in-service limit should typically be measured at a relevant elevation. Two thirds of the way up the boom might be selected as a suitable reference location, for example.

Mobile cranes are a special case. Their low out-of-service design wind levels make it necessary to lower the boom or take other protective measures when wind will exceed the design value, a value with a rather short return period. This requires reliance on local wind measurements, either jobsite anemometers or rough estimates which would include gusting levels (see Chap. 8, Table 8.4). the installation of a mobile crane is therefore not related to wind-map values, and gust-effect analysis is not needed.

When all modes of possible wind-induced failure are examined, it is entirely plausible for different design criteria to be required for various parts of the crane. For example, storm brakes that rely on friction will suffer a reduction in holding ability should a gust induce sliding, so that subsequent lower-level gusts could keep the crane in motion. To reduce the probability of sliding, the failure level wind could be used for design of this component, with a small margin of safety provided above it. Merely increasing the margin against sliding at the design wind speed does not provide a known increase in protection; using the failure wind gives a more certain measure of assurance.

Gust factors

For a given basic wind velocity at standard height, a gust factor at any other height can be calculated. The three methods given below are applicable to fastest-mile winds. The gust factor G_f is inserted into Eq. (3.27), making

$$F = G_f qAC \tag{3.36}$$

with G_f calculated at the top of the structure or at any height at which significant wind area is located. The parameter q must reflect height per Eq. (3.28), but also terrain exposure as discussed below. If we use the previously mentioned Factory Mutual parameters, height adjustment is provided by using the following (authors') equations

$$\text{Exposure } B\text{: } G_f = 1.59 - 0.59 \frac{h}{h_g} \tag{3.37a}$$

$$\text{Exposure } C\text{: } G_f = 1.32 - 0.32 \frac{h}{h_g} \tag{3.37b}$$

where exposure B represents suburban areas, towns, city outskirts, wooded areas, and rolling terrain, while exposure C is for flat, open country, coastal belts, and grassland. They apply for heights h greater than standard height (30 ft or 10 m) but less than or equal to h_g, which is 1500 ft (460 m). Note, however, that these values are appropriate only for relatively stiff structures such as portal and gantry cranes. They are unconservative for tower cranes.

When we use ANSI A58.1–1972,† the gust factor is given by

$$G_f = 0.65 + 1.95 \frac{\sigma}{\bar{P}} \tag{3.38}$$

in which $\sigma/\bar{P}$ is the ratio of the standard deviation of the wind loading to the mean wind loading. This parameter is in turn given by the expression (for latticed structures)

$$\frac{\sigma}{\bar{P}} = 1.7 \left[T\left(\frac{2h}{3}\right)\right]\left(\frac{PF_g}{\beta} + \frac{S}{1 + 0.001b}\right)^{1/2} \tag{3.39a}$$

and for solid structures

$$\frac{\sigma}{\bar{P}} = 1.7 \left[T\left(\frac{2h}{3}\right)\right]\left(\frac{0.785PF_g}{\beta} + \frac{S}{1 + 0.002b}\right)^{1/2} \tag{3.39b}$$

where $T(y)$ = exposure factor at height y, given by Fig. 3.13
$T(2h/3)$ = value of exposure factor at $y = 2h/3$
P = gust power factor given by Fig. 3.14
K_{30} = velocity-pressure coefficient at y = 30 ft (10 m) for exposure under consideration as given by Fig. 3.15
K_y = velocity-pressure coefficient at height y
V_{30} = wind velocity at standard height (map value), mi/h

†A different method for determining gust factors is given in the 1982 edition as "one such procedure . . . ," but the standard accepts any gust factor determined ". . . by a rational analysis that incorporates the dynamic properties of the main wind-force resisting system." ASCE 7-95 includes the same wording in reference to flexible structures such as most construction tower cranes.

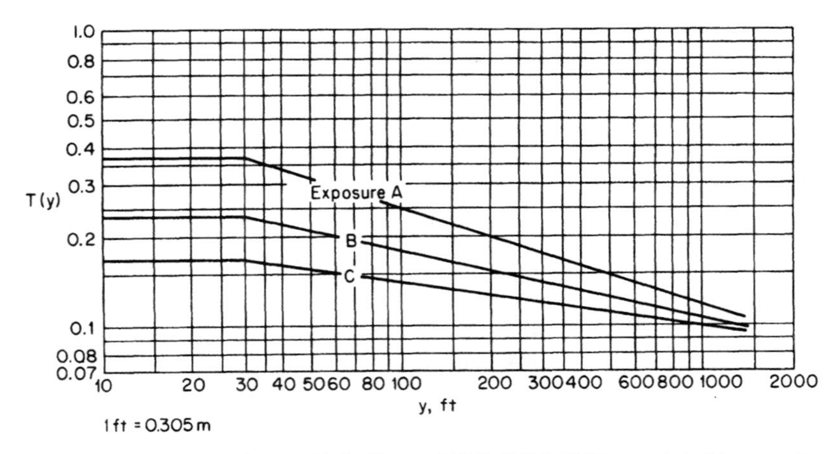

Figure 3.13 Exposure factor $T(y)$. (*From ANSI A58.1-1972; reprinted by permission of the American National Standards Institute, New York.*)

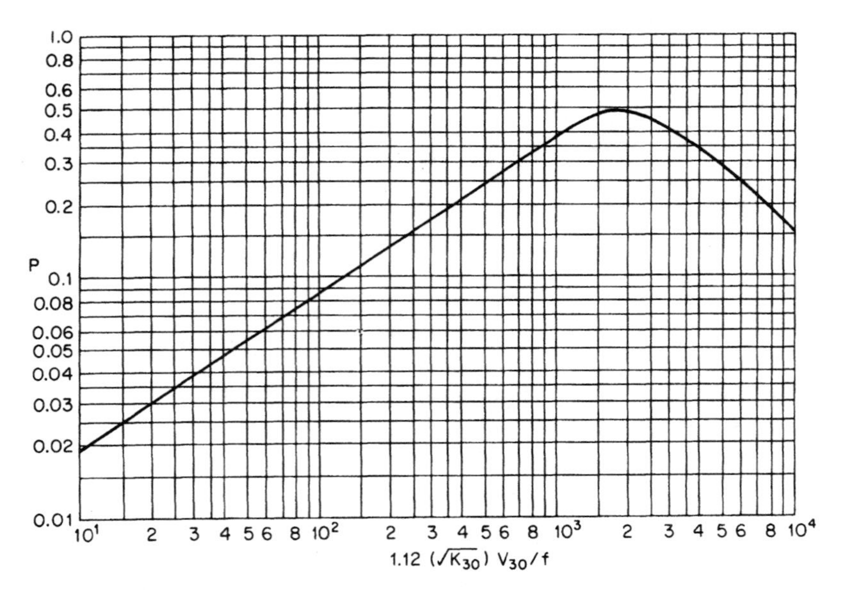

Figure 3.14 Gust power factor P. (*From ANSI A58.1-1972; reprinted by permission of the American National Standards Institute, New York.*)

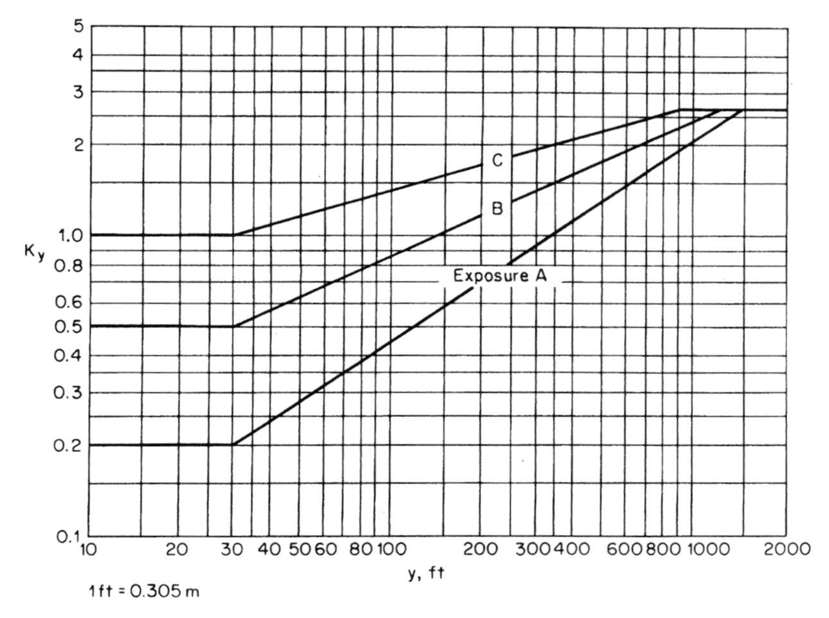

Figure 3.15 Velocity-pressure coefficients K_y. (*From ANSI A58.1-1972; reprinted by permission of the American National Standards Institute, New York.*)

f = natural frequency of vibration of structure in direction parallel to the wind, Hz†
F_g = gust correlation factor as given by Fig. 3.16
β = structure damping coefficient (percent of critical), can be taken as 0.01 for latticed steel structures
S = structure size factor as given by Fig. 3.17
b = average horizontal dimension of structure in direction normal to the wind, ft
h = height of structure measured from ground level, ft
h_e = portion of structure height exposed to wind, ft, for use in determining F_g and S

The exposures referred to are terrain regimes defined as follows:

Exposure A. Centers of large cities and very rough hilly terrain (corresponding to $p = 1/3$)‡

Exposure B. Suburban areas, towns, city outskirts, wooded areas, and rolling terrain (corresponding to $p = 1/4.5$)

†Hertz is the name of the SI unit for cycles per second (1 Hz = 1 c/s).

‡This exposure applies in cities when the upwind area for at least ½ mi (0.8 km) consists of buildings at least 50% of which are over six stories in height and when the next 1 mi (1.6 km) upwind is of exposure *B*.

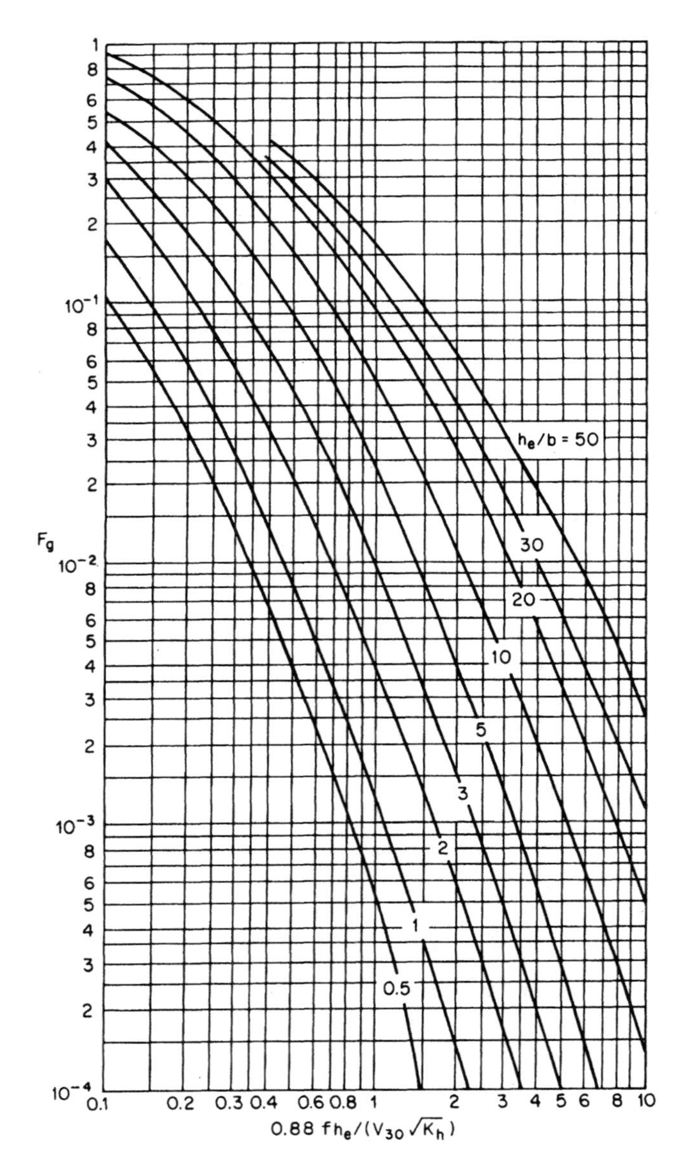

Figure 3.16 Gust correlation factor F_g. *(Adapted from ANSI A58.1-1972; by permission of the American National Standards Institute, New York.)*

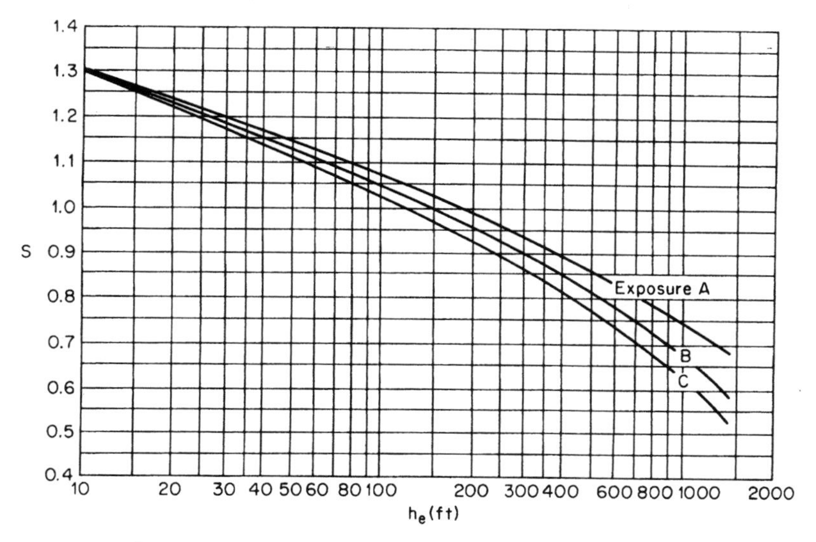

Figure 3.17 Structure size factor S. *(Adapted from ANSI A58.1-1972; by permission of the American National Standards Institute, New York.)*

Exposure C. Flat open country, coastal belts, and grassland (corresponding to $p = 1/7$)

Figure 3.15 deserves some study as it reflects interesting and practical wind characteristics. First, the velocity-pressure coefficient for exposure C remains constant at heights of 900 ft (275 m) and above, for exposure B at 1200 ft (365 m) and above, and for exposure A at 1500 ft (460 m) and above. These are the gradient heights for each of the terrain regimes, the heights above which local surface conditions do not influence wind speed. But wind maps and wind-speed records reflect exposure C conditions and are therefore not directly usable at exposure A or B sites. Design storm-wind levels must be adjusted from the map values to site values. The process used derives from the fact that wind speed at the gradient heights will be identical in each of the regimes.

To adjust wind speed from the design map value at base elevation, the speed at elevation 900 ft (275 m) is found by using Eq. (3.26) with $p = 1/7$. This speed will correspond to the speed at the gradient heights for exposures A and B. Equation (3.26) can be used again to solve for v_0 with h taken as the appropriate gradient height, p as the appropriate exponent, and v as the velocity at gradient height.

But Fig. 3.15 already includes the terrain-regime adjustment and carries the adjustment directly to velocity pressure. When we use values of K_y from Fig. 3.15, velocity pressure at any height is simply

$$q = K_y q_0$$

q_0 being the velocity pressure induced by V_{30}. With gust effects included

$$q = K_y G_f q_0 \tag{3.40}$$

However, when adjusting velocity rather than pressure, it is necessary to use $(K_y)^{1/2}$.

The coefficient K_y cannot be used in the same way with Eqs. (3.32) to (3.32*b*), as these equations already include height variation. To account for differences in exposure regime, the equations can be multiplied by the appropriate K_{30} value.

Some special suggestions are needed for the use of Eqs. (3.39*a*) and (3.39*b*). For cranes of nominally uniform construction (all latticed or all solidly enclosed) having a nominally uniform vertical distribution of wind-exposure area, one gust-response factor should be determined using the height to the top of the crane. This factor is to be used for all parts of the crane exposed to wind.

For cranes of mixed construction, where both solid and latticed portions contribute to wind-induced elastic deflections, two gust-response factors should be calculated, one for the latticed portion and one for the solid. The height of the highest latticed part should be used in determining the factor, and the height of the top of the solid portion should be used for its factor. A combined value is then used for the design. It is a weighted average of the two factors proportioned according to the respective wind-exposure areas.

For cranes with a nominally uniform vertical distribution of wind-exposure area but with a significant concentration of wind-exposure area at one height, two gust-response factors are calculated. The greater value should be used for design.

To determine the natural frequency of a crane structure, Eq. (3.16) may be used. However, there is a less burdensome method that yields an approximate result satisfactory for use in calculating gust-response factors. This method derives from consideration of basic theory and the fact that natural frequency is a physical property of a system and is therefore independent of system orientation. Taking a horizontal beam of stiffness k loaded with vertical load W [see Eq. (3.12)],

$$\omega = (k/m)^{1/2} = [g/(W/k)]^{1/2} = (g/y)^{1/2}$$

$$f = \frac{1}{2\pi}(g/y)^{1/2} \tag{3.41}$$

where y is the deflection induced by W and g is the acceleration caused by gravity.

For a tower crane subjected to wind forces, for example, we can approximate the natural frequency for use in calculating gust-response factors by using the static vertical loads as though they were acting horizontally. Then

$$y = \frac{h^3}{EI}\left(\frac{Q}{3} + \frac{wh}{8}\right) \tag{3.42}$$

where y = lateral displacement at the slewing ring
Q = weight of that part of the crane above the slewing ring
w = unit weight of the tower structure

TABLE 3.3 Values of Wind-Gust Parameters for Eq. (3.43)

Pulsation coefficient m_g			Frequency coefficient ξ	
Height				
ft	m	m_g	f, Hz	ξ
<65	<20	0.35	<0.17	3.3
65–130	20–40	0.32	0.17–0.20	3.2
130–200	40–60	0.28	0.20–0.25	3.15
200–260	60–80	0.25	0.25–0.33	2.95
260–330	80–100	0.23	0.33–0.50	2.65
330–650	100–200	0.21	0.50–1.0	2.25
650–1000	200–300	0.18	1.0–4.0	1.75
			>4.0	1

I = tower moment of inertia
E = modulus of elasticity
h = height to the slewing ring

A simplified gust-factor approach. Finally, there is a gust-factor equation suggested by Kogan.† His method, based on a probabilistic analysis using wind data from the former Soviet Union, results in a deterministic expression similar to Eq. (3.38). Kogan offers

$$G_f = 1 + m_g \xi \tag{3.43}$$

where m_g is a pulsation coefficient that varies with height and ξ is a coefficient that varies with crane natural frequency. Values for both parameters are given in Table 3.3. Because wind records from the former Soviet Union were based on 2-min averaging, Eq. (3.43) will yeild overstated gust factors when used with fastest-mile winds. The differences should be small, however, on the order of 3%.

Adjusting ANSI gust factors for cranes. Theoretical studies of gust effects on elastic structures are best performed using maximum hourly averaged winds. In this way, the gusting portion of the wind is clearly segregated, and the gusts will contain the full spectrum of frequencies. Equation (3.38) includes an adjustment to correct for the use of fastest-mile winds in the calculation procedure. The adjustment is in fact based on a 60 mi/h (26.8 m/s) fastest-mile wind. Restating Eq. (3.38) to show this factor leads to

$$G_f = \frac{1.0 + 3.0\, \sigma/\bar{P}}{1.54}$$

The restated expression now tells us two things: (1) a slightly more "finely

†J. Kogan, "Crane Design: Theory and Calculations of Reliability," Israel Universities Press, Jerusalem, 1976, p. 219.

tuned" gust factor is available by using an adjustment factor in the denominator that represents the actual design (map) wind level and (2) the second term in the numerator now shows that three standard deviations of wind loading are taken in the gust-factor value. This is an arbitrary or averaged value used by Vellozzi and Cohen.† Vickery‡ uses 3.5 but states that although the value will lie between 3.0 and 3.7 for a wide range of structures, the equation yielding those values gives overestimates for lightly damped low-natural-frequency structures, a category that includes many cranes. The number itself is actually the expected value of the ratio of the maximum to the root mean square value of dynamic displacement, which ultimately is related to the effective number of cycles of crane vibration during a wind-speed averaging period. The greater the number of cycles, the greater the probability that a response of a very large order will occur. Structure natural period of vibration therefore has a strong influence on this value.

There is a logical statistical basis for this number as well. Any period of wind activity will be inhabited by a family of gusts, including a full spectrum of frequencies. The high-frequency gusts, which individually have the least energy, are the most numerous members of the gust population. As frequency decreases, representation in the population decreases until the very lowest frequency gusts, which individually possess great energy, may appear very infrequently in any single time sample. Gusts at or near the natural frequency of the structure causes resonant effects, which greatly amplify the gust action beyond the level of an equivalent static force. Higher-frequency gusts produce progressively less effect. Lower-frequency gusts tend more toward a static-like response as frequency decreases.

It could be inferred, then, that if a structure's natural frequency is in the range of that of the most numerous portion of the gust population, the probability of response maximizing becomes increased. Stated differently, a structure with a high natural frequency is likely to be excited by resonant gusts fairly often, so the probability that excitation will occur in conjunction with a powerful low-frequency gust is greater than for a structure of low natural frequency. This should not obscure the fact that the low-frequency structure will experience the more extreme response amplitude, all other things being equal.

Restating Eq. (3.38) once more, we have

$$G_f = \frac{1.0 + \lambda \, \sigma/\bar{P}}{d^2} \tag{3.44}$$

After taking all aspects of the problem into account, including the statistical-probability implications of λ as representing the number of standard deviations of gust-induced displacement, the authors give in Table 3.4 their suggestions for values to be assigned to λ for several values of natural frequency.

The variable d in Eq. (3.44) converts fastest-mile speeds into average hourly speeds. Taking t as the averaging period, $t = 3600/v$, where v is the map value of design wind speed in miles per hour. Values for d are given in Table 3.5 for various averaging periods.

†Op. cit., p. 1296.

‡Op. cit., p. 97.

TABLE 3.4 Suggested Values for λ

f, Hz	≥1	0.5	0.33	0.25	0.20	0.167	0.143
T, s	≤1	2	3	4	5	6	7
λ	3.00	2.90	2.80	2.70	2.65	2.55	2.50

TABLE 3.5 Values for Conversion Factor *d*

t, s	2	5	10	30	60	100	200	500	1000	3600
d	1.53	1.47	1.42	1.28	1.24	1.18	1.13	1.07	1.03	1.00

Straight-line interpolation can be used in Tables 3.4 and 3.5.

Example 3.5 (*a*) A tower crane is to be installed at an exposure B site, and the map value for 25-year-recurrence storms is 70 mi/h (31.3 m/s). The tower is 7.5 ft (2.29 m) square and stands 150 ft (45.72 m) tall. If the natural frequency of vibration for this crane is 0.30 Hz, what will be the gust factor under ANSI provisions?

From Fig. 3.15

$$K_{30} = 0.5 \qquad \text{and} \qquad K_y = 1.02 \text{ for 150 ft (45.72 m)}$$

From Fig. 3.13

$$T(2h/3) = T(100) = 0.18$$

From Fig. 3.14

$$1.12\sqrt{K_{30}}\,\frac{V_{30}}{f} = 1.12\sqrt{0.5}\,\frac{70}{0.30} = 185$$

Therefore

$$\text{Gust power factor } P = 0.12$$

From Fig. 3.16

$$\frac{0.88fh_e}{V_{30}\sqrt{K_y}} = 0.88(0.30)\,\frac{150}{70\sqrt{1.02}} = 0.56$$

$$\frac{h_e}{b} = \frac{150}{7.5} = 20$$

Therefore,

$$\text{Gust correlation factor } F_g = 0.20$$

From Fig. 3.17

$$\text{Structure size factor } S = 1.0$$

Putting these values into Eq. (3.39a) and taking $\beta = 0.01$ gives

$$\frac{\sigma}{\bar{P}} = 1.7(0.18)\left(0.12\,\frac{0.20}{0.01} + \frac{1.0}{1 + 0.001 \times 7.5}\right)^{1/2}$$

$$= 0.564$$

The gust factor is then given by Eq. (3.38)

$$G_f = 0.65 + 1.95(0.564) = 1.75 \qquad \textit{Ans.}$$

(b) What would the gust factor be using Eq. (3.44)?

To find the conversion parameter d use $t = 3600/70 = 51.4$, which gives d the value of 1.25 by interpolation in Table 3.5. By interpolation, $\lambda = 2.76$ from Table 3.4, and

$$G_f = \frac{1.0 + 2.76(0.564)}{1.25^2} = 1.64 \qquad \textit{Ans.}$$

(c) What would be the gust factor under Factory Mutual recommendations?

Using Eq. (3.37a) for exposure B, we have

$$G_f = 1.59 - 0.59\,\frac{150}{1500} = 1.53 \qquad \textit{Ans.}$$

(d) What would the gust factor be using Kogan's equation (3.43)?

From Table 3.3, $m_g = 0.28$ for a height of 150 ft (45.72 m) and $\xi = 2.95$ for $f = 0.30$ Hz; therefore

$$G_f = 1 + 0.28(2.95) = 1.83 \qquad \textit{Ans.}$$

The ASCE 7-95† wind load provisions

The most important difference from earlier wind codes is the introduction of new criteria for measuring wind speed. Heretofore, fastest-mile wind speeds had been used on wind speed maps and in calculations. Fastest-mile winds are maximum wind speeds averaged over the time it takes for one mile of wind to pass the measuring point. For example, a 60 mi/h (26.8 m/s) wind will be averaged over one minute of time, a rather long period that averages out most gusts.

The new ASCE wind code introduces 3-second averaging, a rather short period that includes the most powerful gusts, but not the entire gust spectrum. Therefore, the very same wind sample, when expressed in terms of 3-s average speed, will be higher than the fastest-mile velocity. From the ASCE wind map of 50-year recurrence winds, the wind speed at New York City is 115 mi/h (51.4

†Ibid.

m/s), whereas thefastest-mile velocity is 80 mi/h (35.8 m/s), a very significant difference.

There are two additional differences of note in the ASCE code. Because wind speed is measured differently, the equation and the power-law exponents that adjust wind speed with height are different. The ASCE equation adjusts for both terrain regime and height with the following parameter, which is a multiplier for the basic wind pressure

$$K_z = 2.01(z/z_g)^{2/\alpha}$$

where α is a power-law exponent and z_g the gradient elevation for the terrain regimes. Applicable values of α and z_g are given in Table 3.6. Gust factors listed for stiff structures take values less than one; for flexible structures this code advises using gust factors derived from "rational analysis that incorporates the dynamic properties of the main wind-force resisting system."

It stands to reason that design wind forces for any one location and recurrence period should be the same, regardless of how wind speed is measured. But for hurricane prone areas of the United States, the ASCE wind map values seem to be based, at least in part, on simulation techniques rather than on recorded data, because of the paucity of such data. Using established procedures for converting fastest-mile hurricane winds into 3-s average winds implies that the new ASCE wind map is rather conservative. Again using New York City data, the 80 mi/h (35.8 m/s) fastest-mile wind converts to 98 mi/h (43.8 m/s), not 115 mi/h (51.4 m/s).

In order to utilize the methods for gust factor determination given in Eqs. (3.37), (3.38), (3.43), and (3.44), it is first necessary to convert the 3-s average design wind speed to fastest mile, but only for gust factor calculation purposes. Equation (3.44) and ASCE Fig. C6-1 furnish the means for doing this; the multiplier is 0.74 for hurricane areas and 0.8 elsewhere. To be applicable for use with 3-s average winds, the resulting gust factors must be multiplied by the square of those factors. This may produce gust factors with values less than unity, but that would be appropriate for use with 3-s average wind speeds.

Summary of procedures for calculating out-of-service wind forces

Because of the many complicated steps and procedures outlined in the preceding pages, the following guide is furnished to keep the reader on the straight and narrow path to proper and usable results.

TABLE 3.6 Height and Terrain Regime Adjustment Parameters for ASCE 7-95

Exposure category	α	z_g [ft(m)]
A	5.0	1,500 (457)
B	7.0	1,200 (366)
C	9.5	900 (274)
D	11.5	700 (213)

1. Determine design wind speed: this may come from project specifications, from a wind speed map, or from the rational analysis methodology given herein for short-term installations.
2. Determine applicable base level wind force: this may be given in project specifications, or in local design codes, but for cranes usually must be calculated. When calculated, values such as given by Eq. (3.25*a* or *b*) must be adjusted for terrain regime; the K_{30} values taken from Fig. 3.15 may be used for this purpose.
3. Determine shape factors: Table 3.2 and Fig. 3.8 provide the needed information. The direction of the wind with respect to the surface must also be considered.
4. Determine wind force at the applicable height: Eq. (3.28) is used for any one particular height, and Eqs. (3.29) through (3.32) are used for moments and shears on tall, slim vertical structures.
5. Determine the Gust Factor: Three methods are given in this Chapter, and a fourth mentioned. In each case, the method chosen should be appropriate for the particular structure. Some codes or data compilations include gust effects in wind pressure values given, but rarely would such data be appropriate for limber structures such as tower cranes. For stiff structures, any of the simplified methods should prove adequate.
6. The wind effect on a crane is the product of the wind force adjusted for height and terrain regime, the shape factor, the direction of the wind with respect to the surface, the exposed area, and the gust factor.

To close this discussion on wind, readers are again cautioned not to allow the extensive theoretical presentation and generous supply of mathematical equations, curves, and tables of data to distort their perspective. The study of wind and its effects may be based on scientific principles, but the wind at a particular site at any instant in time has studied neither science nor engineering; it is random and beyond precise prediction. At best, wind calculation results will be approximate. The goal in calculating wind effects should be an ample, not an absolute, protection against failure. That goal is often not easy to attain with assurance while avoiding unnecessary cost. It requires the engineer to perform an analytical study appropriate for the conditions, and to exercise sound judgment in the process of doing so.

Chapter

4

Stability Against Overturning

For cranes that are not anchored in place, stability is the primary factor controlling load ratings and the ability of the machine to perform useful work. Therefore elements that contribute to the stability of a crane and actions that diminish stability are things that are useful to know. This knowledge, both in the general and in the mathematical sense, will enhance our ability to evaluate and choose among crane configurations and types for particular jobsite tasks. It will also help us to establish jobsite policy and procedures to improve crane productivity and safety. In this chapter we shall review the essential elements of crane stability.

4.1 Introduction

One of the authors was once called upon to conduct a stability test of a telescopic truck crane at a contractor's operations yard. The crane operator had never done test work before but was experienced with this particular crane. For the test, a carefully weighed load was assembled, lifted just clear of the ground at close radius over the side of the crane, and slowly boomed out keeping the load just clear of the ground. When one outrigger float lifted free of the support surface, booming out was continued at a still slower rate. As the second outrigger float broke free, booming out was stopped, the load was eased to the ground, and the load radius was measured as the tipping radius for that load.

The test team took a lunch break at this point. Because of the slow, gentle procedures used, the operator had not felt the crane approach tipping; without that seat-of-the-pants indication of tipping familiar to many mobile-crane operators, he was convinced that we had not yet reached the tipping radius. At the end of the lunch break, the operator decided to check his intuition by trying to take the load out a little further. When the test engineer got to the scene, he started frantically waving his arms and shouting "Stop!" The load was being hoisted, but the load wasn't lifting; instead the counterweight-side outrigger floats had risen to about 2 ft (600 mm) clear of the ground!

That anecdote underscores a point pressed by mobile-crane manufacturers. *Know the load and the radius before making the lift—do not guess.*

Coupled with improved controls that permit very smooth starts and stops to motions, telescopic booms and long latticed booms, which are heavy and take up a large part of the tipping moment, can go from a stable to an unstable equilibrium state with no marked change in the operator's perception of machine condition. Moreover, these mammoth booms, once in motion, may develop inertia beyond the operator's ability to stop in time to prevent overturning. In times past, with the cranes of old, seat-of-the pants operation was not as risky as with modern equipment. Today, knowing the load and radius is a prerequisite for safe operation.

General concept of stability

Toward the end of Chap. 1, the main ingredients of the tipping equation were presented. It was shown how the moment of the machine superstructure and carrier about the tipping fulcrum resists overturning and hence is the stabilizing moment, also called the machine resisting moment or simply the machine moment. On the other hand, the weight of the boom and the lifted load were shown to induce overturning; they produce the tipping moment.

This simplistic representation correctly models the general static tipping problem as a matter concerning the weights of the load and the crane components as well as the location of the various CGs and the tipping fulcrum. The real-life tipping problem also includes the action of wind and to some extent dynamic forces, the elastic deformations of the components under load, the reliability of the supporting surface, and the levelness of the crane.

4.2 Mobile Cranes

Mobile cranes are stability-sensitive. Not only are most individual mobile-crane ratings governed by stability but the stability margin, the ratio of the tipping to the rated load, can be as low as 1.18 for a truck crane, which is less than the design factors used in the design of the structural components of the crane.

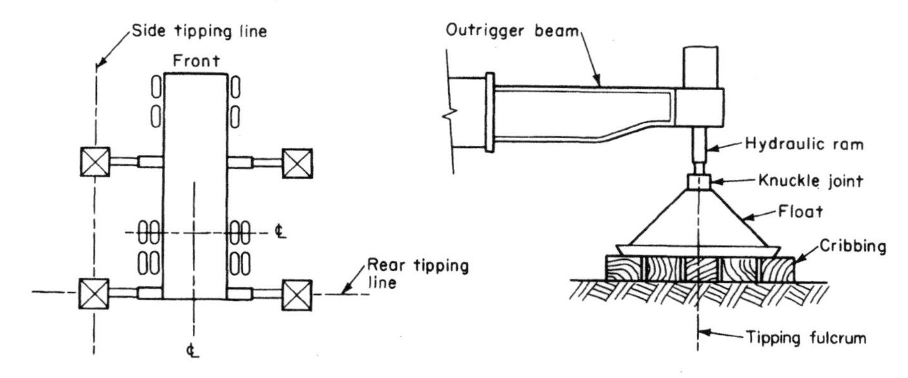

Figure 4.1 Tipping lines for a crane raised onto outriggers.

Stability-governed rated loads are established for a level machine on firm supports in calm air in the absence of dynamic effects. In the presence of any one or more of these detrimental conditions, the rated, or lifted, load must he reduced in order to maintain the stability margin. Jobsite tipping accidents can result from a failure to make the needed reductions. Some operators do not read the fine print on their rating charts and so assume that the ratings hold for all conditions of operation. On the contrary, the ratings are based on ideal conditions, and appropriate reductions, based on judgment, must be made for any variation from the ideal.

The weight and CG location of the various mobile-crane components are determined by calculation or by weighing. From these data stability based ratings are calculated, but the ratings and the accuracy of the weight and CG data are then confirmed by test.† For calculations, the location of the tipping fulcrum can easily be identified for each type or physical arrangement of crane.

Location of tipping fulcrum on outriggers

The tipping line is quite clearly defined for cranes on outriggers. Operating instructions usually require that the crane be raised clear of the ground so that the tires are free of crane weight. When this is done, the tipping lines shown in Fig. 4.1 apply.

When a portion of crane weight remains on the tires, the crane cannot be considered well stabilized. In the past, when screw jacks were used at the outrigger floats, the crane had to be set basically level while still on its tires. Although it was then possible to raise the crane

†SAE J765, Crane Load Stability Test Code, Society of Automotive Engineers, Warrendale, Pa., 1990.

off the tires by turning the superstructure first to one side and then to the other, in practice this was rarely done. More often, the screw jacks were tightened manually, using long steel bars for mechanical advantage, to load the floats without raising the machine off the tires. With the relatively short-boomed cranes of that time, there was little stability difficulty, as the floats controlled the levelness of the crane fairly well. With uneven ground surfaces, however, it was often difficult to level the crane properly in the first place.

With the larger, high-capacity, longer booms of today, all cranes are equipped with hydraulic jacks at the outrigger beams or floats. This is essential, as today's machines must be raised completely off the tires. If the tires are left in hard contact with the ground when the crane is initially set up, both the tires and the outrigger beams share the load. The pressure of the tires against the ground sets up a moment that acts against the resisting moment and effectively reduces it, thereby reducing crane capacity. The crane cannot develop its proper resisting moment without first tilting to unload the tires, but this in turn increases the radius and the overturning moment.

A crane is raised on its outriggers by means of hydraulic rams on the ends of the outrigger beams (Fig. 4.2) or by rams which push the beams down scissors fashion (Fig. 4.3). In both cases, the beams are

Figure 4.2 A large truck crane with hydraulic rams at the outboard ends of its outrigger beams. After the outrigger beams are extended by means of concealed rams, the end rams are activated to raise and level the crane. The worker in the picture is holding a remote control box from which he is controlling the individual outriggers. (*FMC, Crane and Excavator Division.*)

Figure 4.3 The outrigger beams on this crane are extended by concealed hydraulic rams, but to raise and level the crane, a scissors mechanism is used. The mechanism is powered by rams fixed to the sides of the truck chassis. After the machine is raised and leveled, eccentric cam safety locks positively hold the crane in place and prevent the machine from going out of level because of hydraulic bleeding. (*Harnischfeger Corporation.*)

extended horizontally by rams and are controlled individually, while the floats are mounted to a knuckle that permits pivoting in any direction. With the rams on the ends of the beams, in case of hydraulic-line rupture or loss of pressure, automatic valves prevent retraction and lowering of the crane. For the scissors arrangement, additional safety is possible. Rotating a handle mechanically locks the scissors beams in the raised position by means of a steel cam on an eccentric shaft.

Extension of outriggers

Outriggers must be fully extended as stipulated on the rating chart; otherwise, the ratings are not applicable. This can be illustrated by referring to Fig. 1.21, which allows us to formulate the stability relationship as

$$W_m d_m = W_b d_b + Wd$$

noting that W_m includes the weight of both the superstructure and the undercarriage. If we take, for example, a machine with

$$W_m = 319{,}000 \text{ lb } (144.7 \text{ t}) \quad d_m = 18.95 \text{ ft } (5.78 \text{ m})$$

$$W_b = 45{,}200 \text{ lb } (20.5 \text{ t}) \quad d_b = 115.5 \text{ ft } (35.2 \text{ m})$$

$$d = 250 \text{ ft } (76.2 \text{ m})$$

after rearranging, the tipping load is

$$W = \frac{W_m d_m - W_b d_b}{d} = 3300 \text{ lb } (1500 \text{ kg})$$

If the outriggers are 1 ft (300 mm) short of proper extension, or 92% of specified extension for the particular crane which the data represent, d_m decreases and d_b and d increase by that amount. When those values are substituted, W becomes 1830 lb (840 kg), or only 56% of the proper tipping load. With the outriggers short of proper extension by only 3 in (76 mm), the tipping load is nevertheless reduced by 11%! Those figures make it quite clear why operating instructions for truck cranes specify that the crane is to be operated only with the outriggers at specified extension. They also dramatically illustrate the importance of the true fulcrum location in defining calculated stability values.

Full outrigger beam extension often proves to be a problem for operation planners and crane operators at urban construction and highway reconstruction sites. In both situations, it is commonly found that limited space is available for positioning cranes because of the need to keep traffic lanes open. What is often done is that the outrigger beams on the traffic side are not extended at all.

There is a loss of capacity when the boom is opposite unextended outriggers because those floats, jacks, and outrigger beams are closer to the tipping fulcrum than they are supposed to be. If the operator relies on "on outriggers" ratings, the actual margin of stability may be seriously reduced. If "on rubber" ratings are used instead (which are always substantially less than those on outriggers), the crane may not be able to do its intended job; it will be severely limited in the work it is permitted to do. But, federal law (OSHA) requires that cranes be operated in accordance with the manufacturer's instructions; all manufacturers stipulate outrigger beam extension requirements on the rating chart. Therefore crane users in the United States are constrained by law to follow those requirements or to use "on rubber" ratings; undoubtedly, laws in other nations are similar.

There are no engineering reasons that prevent cranes from being rated for operation when the outrigger beams on one side are not extended and the boom is working on the opposite side or over the end.

If and when manufacturers provide such ratings, and the instructions and limitations necessary for working with them, working that way will become practical and legal. Until then, outriggers on both sides of the crane must be extended equally as stipulated on the rating chart.

Some mobile crane manufacturers now furnish rating charts for partially extended outriggers, but the authors are aware of none for outriggers extended on one side only. Only cranes specifically rated for partially extended outriggers may be so used, because aspects of the crane must be specially designed to accommodate new requirements. For example, outrigger beams or boxes are loaded differently when outriggers are partially extended; some outrigger designs could conceivably collapse under partial extension operating conditions. Moreover, there is a dramatic reduction in stability when a load is swung from over the end to over the side.

Partly extended outriggers are stiffer than fully extended ones. Because the operator will feel less "give" as loads are lifted and may therefore feel unduly secure, there may be a greater inclination for a seat-of-the-pants operator to lift excessive loads.

The stability equation can actually be a bit more complicated than previously given, especially for cranes finished with a boom foot (live) mast or a luffing jib. The weight of towers, masts and booms may act on either side of the tipping fulcrum—in some instances adding to stability and in other instances diminishing it. A live luffing component might cross from stabilizing to destabilizing as it goes through its range of motion.

The weight of the load ropes suspended from the boom tip is usually considered as crane dead weight, but when lowering a load below grade the rope below grade is taken as part of the lifted load. However, some manufacturers take all of the rope weight as part of load weight; when that is the case, a note to that effect will appear on the rating chart. Using the notations of Fig. 4.4, we can express the machine resisting moment as

$$M_r = W_m d_m + W_f(d_o - d_f) - W_b d_b - W_r d$$

where W_r is the weight of the suspended hoist ropes when taken as part of crane dead weight. As the crane booms out, d_f will increase until it exceeds d_o, and that term will be self correcting. The static overturning moment $M_o = Wd$. At the limit of stability, $M_r = M_o$; solving the tipping load

$$W = \frac{W_m d_m + W_f(d_o - d_f) - W_b d_b - W_r d}{d} \tag{4.1}$$

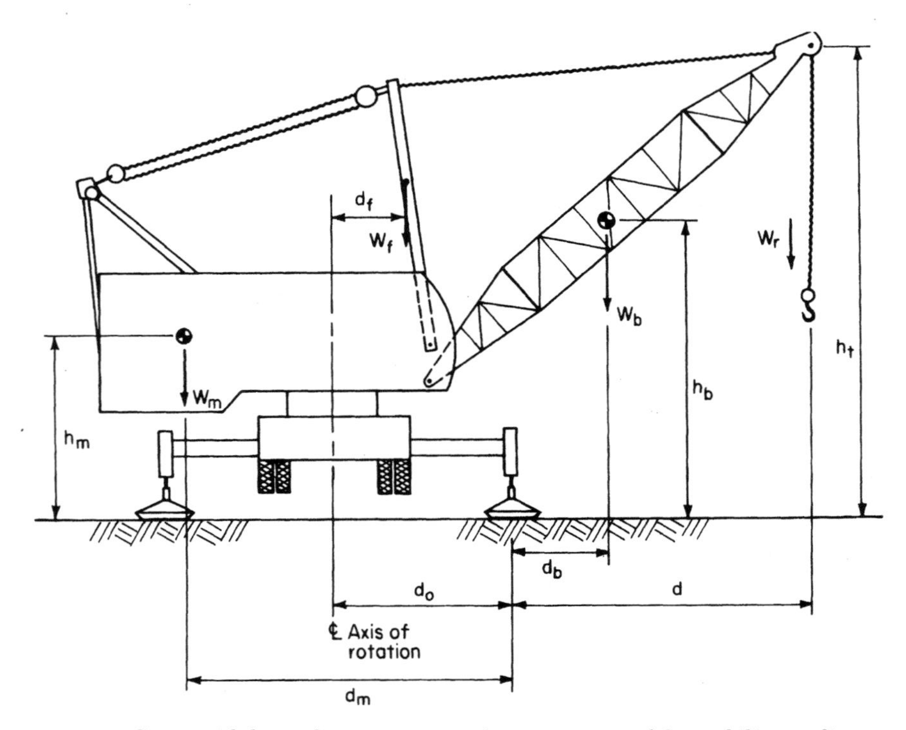

Figure 4.4 Crane with boom foot mast on outriggers, annotated for stability studies.

Figure 2.14 shows the typical condition wherein the front end of a truck crane carrier overhangs beyond the front outriggers. The weight of that portion of the frame, the driver's cab, and the carrier engine will then materially reduce the machine moment for operation over the front. The reduction is so great that the operating instructions for most machines of that type prohibit working in that sector. For those machines for which operation is permitted, a jack or outrigger generally is used at the very front of the crane.

The same restrictions do not apply to rough terrain cranes and two- to four-axle all terrain cranes having outriggers positioned at the four corners of the carrier. Those machines can safely pick loads over a full horizontal circle. Larger multiaxle all terrain machines have similar limitations as truck cranes.

Location of crawler crane tipping fulcrum

Crawler cranes must operate on a level surface or on a run of timber or other material laid to level the crane path. The crane itself has no leveling mechanisms. The bulldozer-like tracks or treads the crane uses for traveling are loose cast-steel segmented bands that rest on the ground for the purpose of spreading crane weight and reactions to a large bearing area. If the tracks were unfastened and stretched out on the ground, the crane could still operate and run back and forth on the tracks.

Thus the tracks do not control the location of the tipping fulcrum. The track rollers, seen in Figs. 4.5 and 4.6, ride on the tracks and define the position of the side fulcrum line.

Because the tracks are loosely pinned together and rest on the ground, the track opposite the tipping fulcrum is not fully effective in resisting side overturning. The weight of that part in contact with the ground cannot be included in the machine moment.

It would appear that when operating over the end of the machine, the tipping fulcrum is located vertically below the centerline of the shaft of the drive or idler tumblers (the large steel wheels at each end of the side frames that engage the track lugs). Shown in Fig. 4-6*a*. But typically crawler machines touch ground farther back at a track roller (Fig. 4.6*b*). This enables the machine to turn and to travel onto ramps and over soft ground more easily. With this configuration, the fulcrum is under the shaft of that first track roller. When the crane is operating over the end, part of the weight of each track cannot be included in the machine moment.

On many jobsites, to add a little more stability margin for operating over the ends, timber blocks are placed against the first track segment

Figure 4.5 Crawler side frame assembly with the drive sprocket at the far end and the idler sprocket in the foreground. The series of small wheels making contact with the track are the track rollers; they transmit load from the side frames to the track.

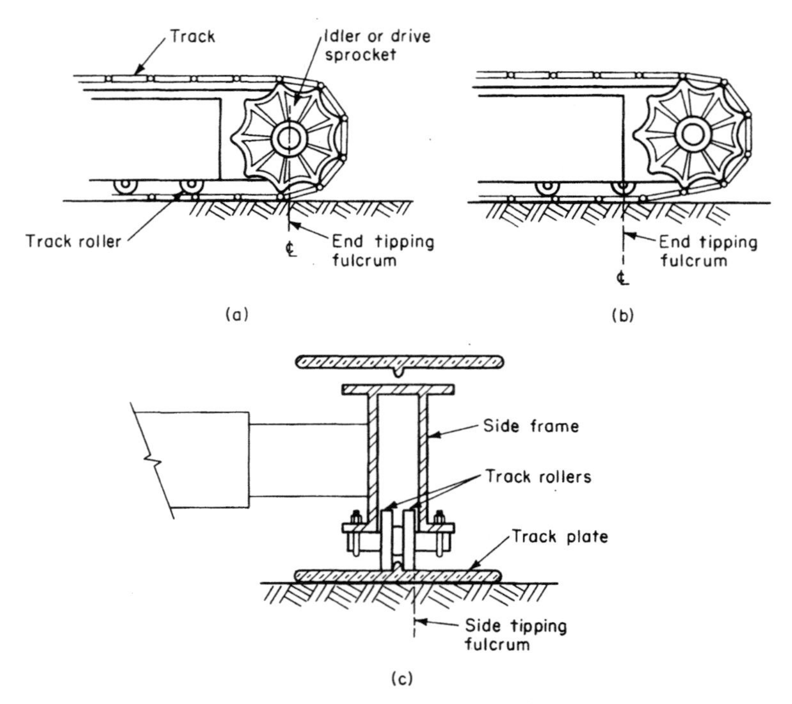

Figure 4.6 Crawler crane tipping lines showing (*a*) apparent over-end tipping fulcrum at sprockets, (*b*) actual over-end tipping fulcrum at first track roller, and (*c*) over-side tipping fulcrum. Any number of configurations of side frame, sprocket, track roller, and track are possible; the ones shown are merely illustrative.

that angles up from the ground. There can be no harm in using this procedure, provided the operator does not assume that rated loads can then be exceeded. In fact, some machines require blocking in order to raise and lower maximum boom lengths unassisted.

The stability of a crawler crane can be seriously diminished by improper blocking under the tracks. The problem can be subtle and go unnoticed during a casual inspection. In essence, this problem may occur when one or more of the track ends is not firmly supported, in effect shortening the tipping fulcrum. For example, when a few timbers are placed under a track to "firm up" a soft spot, those timbers may create a new tipping fulcrum short of the normal fulcrum.

Under such circumstances, a crane can tip over at substantially less than rated load. The blame is often mistakenly placed, with soft ground usually attributed as the cause. However, the real cause is that the tracks are not uniformly and firmly supported along their entire length. Another common instance of this inadequate support situation can occur when the track is partly supported on concrete and partly on the ground. The edge of the concrete may then become the effective tipping fulcrum, and insufficient stability may result. A crawler track may be able to accommodate imperfect support conditions near its center, but unless both ends are fully and firmly supported trouble is likely to ensue.

Location of tipping fulcrum for mobile cranes on tires

When cranes operate on tires, the location of the tipping fulcrum depends on the physical arrangement of the suspension system or on special provisions employed to create the fulcrum position desired. If an axle is spring-mounted, the spring position dictates the location of the fulcrum line. For unsprung axles pivoted to oscillate about the longitudinal centerline of the crane, the pivot controls the position of the fulcrum line.

When two axles are mounted on beams parallel to the crane centerline (often called *walking beams*) and each beam is pivoted to the crane frame (the whole assembly is called a *bogie*), the frame pivots then control the fulcrum location.

Other cranes are mounted on axles solidly affixed to the frame against rubber shock pads. For these machines, the center of the tires or the center of a pair of dual tires marks the location of the tipping line. Short of raising the crane onto its outriggers, this arrangement provides the greatest relative stability and hence the highest rated loads of all the "on-rubber" fulcrum positions discussed.

To increase stability for machines provided with sprung or oscillating axles or with bogies, mechanical blocking can be used. When the blocking is engaged, which is typically done using a control in the cab, the axles become locked to the frame and the tires then define the fulcrum line. The blocking arrangement is called an axle *lockout*. Fulcrum lines for each of the arrangements mentioned above, with and without blocking, are shown in Fig. 4.7.

Tires deflect when loaded, a fact that should not be overlooked. Moreover, when the crane is loaded with a moment, deflection will not be uniform and the crane will tilt somewhat. This action increases the load radius and also shifts the CG of the crane closer to the fulcrum line. Both effects reduce stability and introduce an element of risk to operation on tires.

Finally, stability is not the only limitation for operations on tires. Some ratings may be controlled by the ability of the tires or rims to support load.

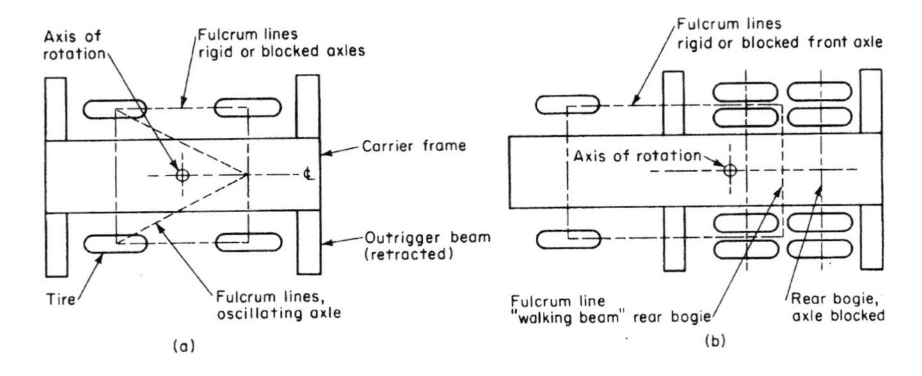

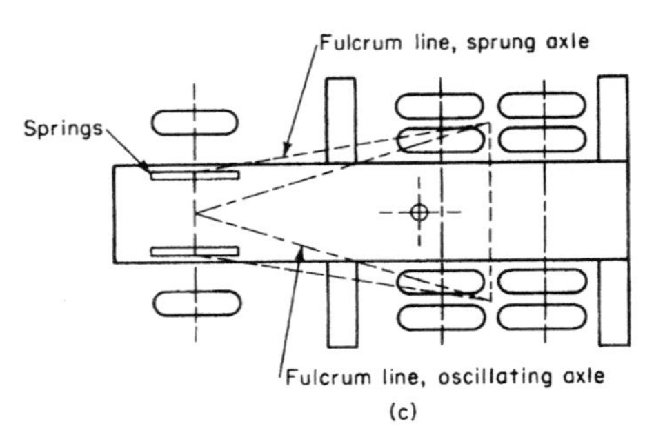

Figure 4.7 Tipping lines for cranes on tires with and without axle blocking: (*a*) rough terrain crane; (*b*) truck crane with rigid front axle; (*c*) truck crane with oscillating or sprung front axle.

Operating sectors

Looking down at a crane from above, if we mark the longitudinal centerline 0° when the boom is over the rear, then operation square over the side will be at 90°. For any angle less than 90°, the distance to the CGs of the rotating elements becomes shortened with respect to the side tipping fulcrum.

The parameters we used earlier in calculating the resisting moment were for the undercarriage and superstructure combined. When horizontal angles of operation are introduced, those two components must be segregated and Eq. (*a*) becomes

$$M_r = W_u d_u + W_c d_c + W_f(d_0 - d_f) \tag{4.2}$$

where W_u and d_u refer to the entire superstructure including the boom and hoist ropes and W_c refers to the undercarriage. d_c is the general tipping-fulcrum distance for the undercarriage CG, while d_{cs} specifically refers to the side distance and d_{ce} to the end (rear) distance. When the boom is at a horizontal angle θ to the longitudinal centerline, the expression for resisting moment must reflect this angle; Eq. (a) then becomes

$$M_r = W_c d_{cs} + W_u d_u \sin\theta + W_f(d_0 - d_f \sin\theta)$$

To clarify this point we now consider a crane without a boom foot mast so that the term in parentheses can be dropped. Equating the overturning moment M_0 with the resisting moment gives

$$M_o \sin\theta = W_c d_{cs} + W_u d_u \sin\theta$$

$$M_o = \frac{W_c d_{cs}}{\sin\theta} + W_u d_u \qquad \theta >> 0°$$

For any angle other than 90°, sin θ will be less than unity and the first term on the right becomes amplified. With otherwise constant parameters, as θ changes from 90°, the equality is maintained only if M_0 increases (i.e., only if the tipping load increases). This equation shows that the least value for M_o and the tipping load occurs at θ = 90°.

As θ decreases, a point is reached at which the end (rear) fulcrum line becomes effective rather than the side. This will occur as the boom passes the intersection of the end and side fulcrum lines. With the boom over the intersection, M_o achieves its maximum value (i.e., the tipping load is at its greatest).

When stability is contemplated in this way, it becomes quite clear that the tipping load varies with θ. But to take advantage of this fact would require a very complex form of rating chart indeed. Furthermore, practical study of crane use and field requirements would show that little is likely to be gained from such a chart whereas safety could be severely impaired.

Rating charts published by mobile-crane manufacturers present stability data in a limited way. If one set of ratings is given, the least value for the entire permitted operating arc (horizontal) is supplied. If two sets of ratings are listed, they will be for θ = 0° and θ = 90°. At any other horizontal angles, the stability margin will actually be somewhat greater than the specified value.

Considerations discussed above together with the need for simple, readily understood charts have given rise to the definitions of operating arcs shown in Fig. 4.8.† The sectors indicated are somewhat

†See also SAE J1028, Mobile Crane Working Area Definitions, Society of Automotive Engineers, Warrendale, Pa., 1989.

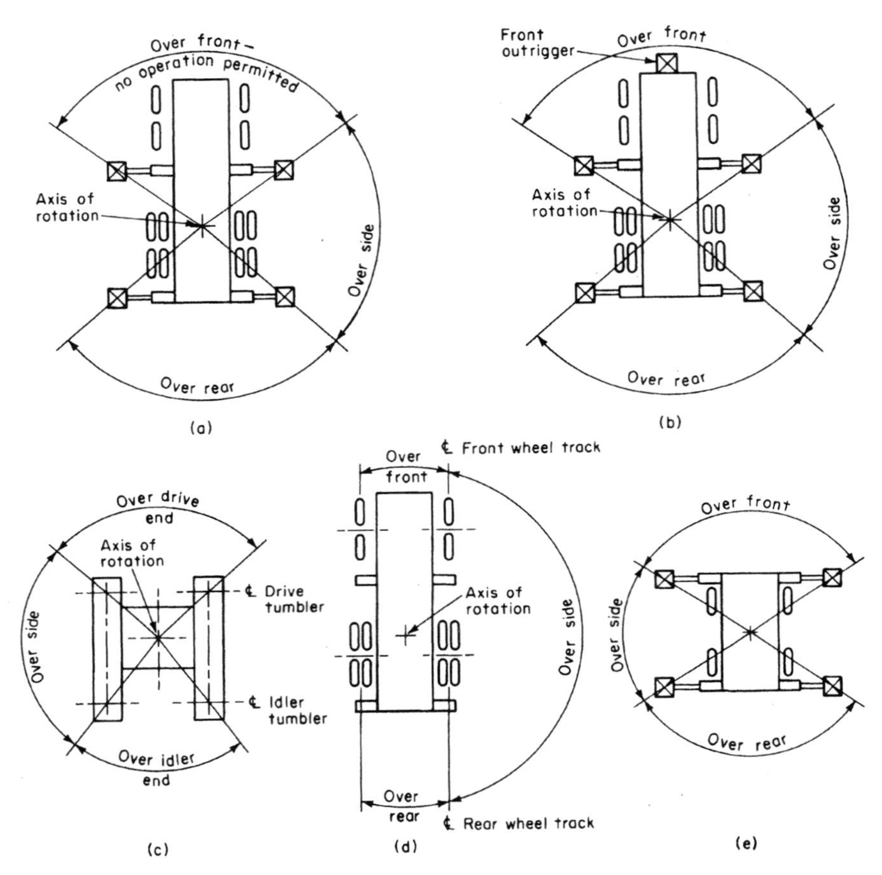

Figure 4.8 Mobile-crane operating sectors for various crane configurations: (*a*) truck crane on outriggers, normal configuration; (*b*) truck crane on outriggers with front outrigger; (*c*) crawler-mounted crane; (*d*) truck crane on tires; (*e*) rough terrain crane on outriggers.

standardized, but a particular manufacturer may still specify different arcs or different arcs may be assigned for a particular crane model if design parameters justify it. For example, some truck cranes without front outriggers are nonetheless rated for operation in that quadrant.

Maximum true lifting capacity, and therefore the greatest stability margin, is obtained when a truck crane boom is operated directly over a rear outrigger float. For a crawler crane, it is obtained when the boom is over the intersection of the side and idler end tipping lines. This should not be construed to mean, however, that it is permissible to exceed rated loads when the boom is in those positions. More than stability considerations are contemplated when rated loads are set, and exceeding rated loads, aside from the legal implications, can cause damage or failure in one or more of several components.

Effect of out-of-level operation

When a machine is mounted out of level and operated with the boom over the low side, the resisting moment will be less than the design value because the tilt raises the machine CG and moves it closer to the tipping line.

For a statically positioned crane, the operating radius and boom position are not affected because the location of the load does not change. However, for a crane that picks up a load while level and then swings 90° or so to a low side or end, and for a crane traveling with load, variations from level will alter the operating radius and boom angle. Therefore, two cases must be examined: the first where the radius remains constant, and the second where the radius increases due to out of level.

For the first case, taking machine CG height as h_m and horizontal distance from the tipping line as d_m (see Fig. 4.4), the radial distance of the CG from the tipping line r_1 will be

$$r_1 = (h_m^2 + d_m^2)^{1/2}$$

and the angle γ which the radial line makes with the horizontal is then

$$\gamma = \tan^{-1} \frac{h_m}{d_m}$$

Should the crane be mounted out of level at an angle α to the true horizontal with the boom on the low side, the radial line to the CG will be at angle $\gamma + \alpha$ to the horizontal and the distance to the tipping line will be reduced to

$$d'_m = r_1 \cos (\gamma + \alpha)$$

$$= r_1 (\cos \gamma \cos \alpha - \sin \gamma \sin \alpha)$$

but α is a small angle, therefore using small angle geometry (and angles measured in radians), $\cos \alpha \approx 1$, $\sin \alpha \approx \alpha$ and

$$d'_m = r_1 \cos \gamma - \alpha r_1 \sin \gamma$$

$$= d_m - \alpha h_m$$

For the simple case of a static out-of-level crane without a boom foot mast, the tipping load of Eq. (4.1) can be restated as

$$W = \frac{W_m(d_m - \alpha h_m) - W_b d_b - W_r d}{d} \tag{4.3a}$$

or in other words, the reduction in tipping load is brought about solely by the term containing the out-of-level angle α, which diminishes the distance to the tipping line. In percentage terms, the loss in stability is then

$$L_o = \frac{W_m \alpha h_m}{Wd} \times 100$$

Earlier we used a large crane to demonstrate the effects of outrigger beams not properly extended. Taking the data for this same crane (see Extension of Outriggers earlier in this section) and adding rope weight W_r = 500 lb (227 kg) and CG height h_m = 6.1 ft (1.86 m), we can directly find the tipping load for a level crane. For $\alpha = 0$, Eq. (4.3*a*) will give the tipping load

$$W = \frac{319{,}000(18.95 - 0 \times 6.1) - 45{,}200 \times 115.5 - 500 \times 250}{250}$$

$$= 2800 \text{ lb (1270 kg)}$$

The usual installation tolerance for levelness is given† as 1% maximum out of level; that is, the crane can be mounted to a slope not exceeding 1:100, which is an angle of 0.01 radians. The loss of capacity is then

$$L_o = \frac{319{,}000 \times 0.01 \times 6.1}{2800 \times 250} \times 100 = 2.8\%$$

For a static truck crane, the out of level slope causing the crane to be at the brink of tipping while lifting rated load (85% of the tipping load) is

$$15\% = \frac{319{,}000 \times \alpha \times 6.1}{2800 \times 250} \times 100$$

$$\alpha = 0.054 \text{ rad} = 5.4\% = 3.1°$$

which is equivalent to a difference of 1.3 ft (395 mm) across the 24 ft (7.3 m) outrigger spread, an extreme condition indeed!

For the second case, where the radius increases, the radial distance from the tipping line to the boom tip r_2 will be

†Clause 5-3.4.6 of ASME B30.5, *Safety Code for Mobile and Locomotive Cranes,* American Society of Mechanical Engineers, New York, 1994.

$$r_2 = (h_t^2 + d^2)^{1/2}$$

and the angle β which the radial line makes with the horizontal is then

$$\beta = \tan^{-1}\left(\frac{h_t}{d}\right)$$

With the crane mounted out of level at angle α to the true horizontal with the boom on the low side, the radial line to the boom tip will be at angle $\beta - \alpha$ to the horizontal and the distance to the tipping line will increase to

$$\begin{aligned} d' &= r_2 \cos(\beta - \alpha) \\ &= r_2(\cos\beta \cos\alpha + \sin\beta \sin\alpha) \\ &= r_2(\cos\beta + \alpha \sin\beta) \\ &= d + \alpha h_t \end{aligned}$$

The tipping equation now becomes

$$W = \frac{W_m(d_m - \alpha h_m) - W_b(d_b + \alpha h_b)}{d + \alpha h_t} - W_r \tag{4.3b}$$

Using the same example as for the first case, with $\alpha = 0.01$, $h_t = 191.7$ ft (58.4 m), and $h_b = 99.9$ ft (30.4 m), the loss in capacity is

$$\begin{aligned} \text{Loss} &= 2800 - \{[319{,}000(18.95 - 0.01 \times 6.1) \\ &\quad - 45{,}200(115.5 + 0.01 \times 99.9)]/ \\ &\quad (250 + 0.01 \times 191.7)\} - 500 \\ &= 284 \text{ lb } (129 \text{ kg}) \end{aligned}$$

which represents a 10% loss in capacity, much greater than the 2.8% loss of the first case.

Such calculations unequivocally demonstrate the out-of-level sensitivity of the second case, where the radius increases due to the out-of-level condition. They also give insight into the loss of capacity that can be expected as a result of frame twist and outrigger beam deflection, carrier characteristics that can also shift the CG toward the tipping line while causing radius to increase in non-static situations.

Furthermore, cranes poorly supported on soft ground will go out of level over time. Even for well supported machines, unless the level-

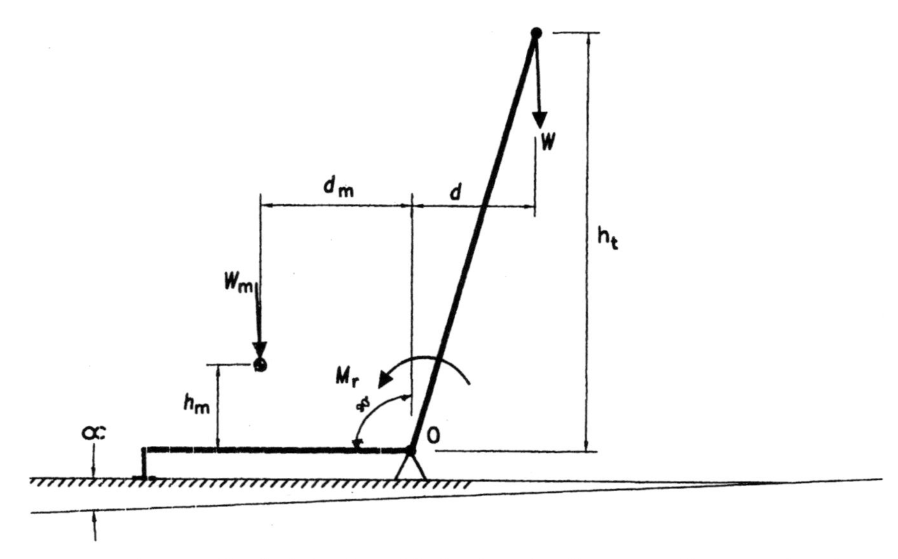

Figure 4.9 A generalized abstract model of a free-standing crane. The crane is not precisely level, and has load W suspended from the boom tip. M_r is the static moment resisting overturning, including the effect of the out-of-level base, and W_m is the weight of the entire crane including the boom.

ness of the crane is monitored and corrected from time to time, the crane may go out of level and lose capacity.

On most mobile cranes the CG is low, and the loss in capacity is small for a static crane kept level within manufacturer's specifications, as has been demonstrated. But there are crane attachments which raise CG height, such as tower attachments for latticed boom cranes and luffing jibs for telescopic cranes. Those machines are more sensitive to out-of-level conditions.

Crane stability can be examined from another perspective†, a view that considers the work required to rotate the crane to the point of tipping rather than only static balance. Using the parameters given in Fig. 4.9, the point O is the tipping fulcrum of that generalized abstract model of a crane, and the crane is loaded by lifted load W together with horizontal static forces not shown on the model. The crane is supported on a surface at angle α to the horizontal; it is out of level. A moment function $F(\alpha)$ is induced by all forces acting on the crane and includes out-of-level effects. The coordinates d_m, h_m, h_t and d in Fig. 4.9 are determined for $\alpha = 0$; they are properties of the crane and remain constant

†This concept of stability was presented in earlier editions of this work, but has been further developed by Prof. Anatoly A. Zaretsky as part of his method for evaluating crane stability under dynamic loading, covered later in this chapter. Professor Zaretsky is the Head of Laboratory of the Scientific-production Association of Construction and Road Building Machinery, Moscow, Russia.

when the crane tilts. As in the derivation of Eq. (4.3), take the radial distance from O to the point of action of W_m as r_1 at angle γ to the horizontal; likewise the radial distance to the boom tip is taken as r_2 at angle β to the horizontal. Note however, that W_m is now taken as the weight of the entire crane including the boom. Then the dead weight moment $M_D = W_m r_1 \cos \gamma$, the lifted load moment $M_W = Wr_2 \cos \beta$, and M_{hor} will be taken as the moment of horizontal static forces about O including steady-state wind, centrifugal force, etc.

The moment on the crane now becomes ($\alpha = 0$)

$$F(0) = M_D - M_W - M_{\text{hor}} = M_{\text{net}} \tag{4.4}$$

but, as the crane is rotated clockwise through the out-of-level angle $\alpha > 0$

$$\begin{aligned} F(\alpha) &= W_m r_1 \cos(\gamma + \alpha) - Wr_2 \cos(\beta - \alpha) - M_{\text{hor}} \\ &= W_m r_1(\cos \gamma \cos \alpha - \sin \gamma \sin \alpha) \\ &\quad - Wr_2(\cos \beta \cos \alpha + \sin \beta \sin \alpha) - M_{\text{hor}} \end{aligned}$$

Reintroducing the coordinates given in Fig. 4.9 by using their equivalents $d_m = r_1 \cos \gamma$, $h_m = r_1 \sin \gamma$, $d = r_2 \cos \beta$, $h_t = r_2 \sin \beta$, introducing a new term, $S_h = W_m h_m + Wh_t$, and restating the moment terms $M_D = W_m d_m$, $M_W = Wd$, after simplifying becomes

$$F(\alpha) = M_D - M_W - M_{\text{hor}}/\cos \alpha - S_h \tan \alpha$$

But, α is a small angle. Therefore, using small angle geometry (and the angle measured in radians), $\cos \alpha \approx 1$, $\tan \alpha \approx \alpha$, and

$$F(\alpha) = M_D - M_W - M_{\text{hor}} - S_h\alpha = M_{\text{net}} - S_h\alpha \tag{4.4a}$$

But, two stability cases must be considered. First, the situation where radius increases due to out-of-level effects; tilting of the entire crane makes Eq. (4.4*a*) applicable, and

$$M_r = M_{\text{net}} - S_h\alpha \tag{4.5}$$

where M_r is the static moment resisting overturning including the effect of an out-of-level base.

Second, a static crane where the radius is not changed by the out-of-level condition; distance d of Fig. 4.9 holds constant because neither the load nor boom positions are affected by the out-of-level angle α, therefore,

$$M_r = M_D - M_W - M_{\text{hor}} - W^*h^*\alpha \tag{4.6}$$

where W^* is crane weight less boom weight, and h^* is the height of the point of action of W^*.

$S_h = W_m h_m + Wh_t$ is a new parameter for use in stability calculations, a parameter that recognizes the heights of the crane CG and lifted load as playing a direct role in evaluating the stability of a crane when out-of-level conditions cause a change in radius. In like manner, the parameter W^*h^* is used for a crane not subject to radius change.

Using the same data and crane examined under the heading Extension of Outriggers in Sec. 4.2, the data will now take another form

$$W_m = 364{,}700 \text{ lb } (165.4 \text{ t}) \quad d_m = 1.92 \text{ ft } (585 \text{ mm})$$

$$h_m = 17.98 \text{ ft } (5.48 \text{ m})$$

$$W^* = 319{,}000 \text{ lb } (144.7 \text{ t}) \quad h^* = 6.1 \text{ ft } (1.86 \text{ m})$$

$$d = 250 \text{ ft } (76.2 \text{ m}) \quad h_t = 191.7 \text{ ft } (58.43 \text{ m})$$

For a dead-level crane ($\alpha = 0$) and $M_{\text{hor}} = 0$, Eq. (4.4) gives

$$M_{\text{net}} = 364{,}700 \times 1.92 - 250W$$

At the limit of stability, $M_{\text{net}} = 0$, therefore the tipping load

$$W = 2800 \text{ lb } (1270 \text{ kg}) \text{ as before.}$$

For a crane that is 1% out of level ($\alpha = 0.01$), using Eq. (4.6), that is, for the case where radius does not change

$$M_r = 0 = 364{,}700 \times 1.92 - 250W - 319{,}000 \times 6.1 \times 0.01$$

$$W = 2723 \text{ lb } (1235 \text{ kg})$$

for a loss of 2.8% of capacity, the same as calculated before.

Solving for the out-of-level angle that would cause loss of stability while lifting a rated load (85% of the tipping load)

$$M_r = 0 = 364{,}700 \times 1.92 - 250 \times 2800 \times 0.85 \times 319{,}000 \times 6.1\alpha$$

$$\alpha = 0.054 \text{ rad} = 5.4\% = 3.1° \text{ as previously found.}$$

Considering a crane where conditions induce radius change due to out of level, from Eq. (4.5) for 1% out of level

$$S_h = 364{,}700 \times 17.98 + 191.7W$$

$$0 = 364{,}700 \times 1.92 - 250W - (364{,}700 \times 17.98 + 191.7W)\, 0.01$$

$$W = 2519 \text{ lb } (1143 \text{ kg})$$

for a loss of 10% of capacity, as before.

Though an out-of-level crane operating at constant radius does not necessarily lose much stability, there are other concerns created by this condition. As an unlevel crane swings, it must go "uphill and downhill", with the swing drive first struggling to overcome added resistance and then freewheeling downhill. Together with wind, this could make for uncomfortable and less controlled crane operation. With the superstructure stopped in a cross-slope position, the crane may not stay in place without use of the swing brake because the imbalance introduced by the out-of-level condition may overcome frictional resistance in the swing bearing. That situation can make it difficult for an operator to hold the load in position for precise load placement, and present danger to landing crew personnel. For the operator, this is similar to working against a cross wind, but whereas the wind cannot be controlled (except to stop when it is excessive), the levelness of the crane certainly can be. With a properly leveled crane, load handling difficulties can be minimized, and real crane capacity maximized.

As a crane traveling with load traverses an out-of-level portion of the travel path, on a down slope the crane will tilt and the load radius will increase; both the load and the boom weight will move further from the tipping line. Thus, the tipping load will decrease more severely than in the static case. More capacity will be lost. In addition, the load will swing like a pendulum with additional deleterious effect.

The actual loss of capacity caused by the crane being out of level will vary with crane model, configuration, and radius. A representative comparison between figures for the particular crane and configuration in the earlier calculations for static (no radius change) and non-static (radius change) conditions follows

Percent loss of capacity (for a particular crane)		
Percent out of level	Static crane	Non-static crane
0.5	1.4	5.0
1	2.8	10.0
2	5.5	19.9
5	13.9	48.8
10	27.8	94.1

The figures are for one example and will vary markedly for different cranes, configurations, and operating radii. Moreover, out-of-level operation can induce side loading of the boom and jib. In extreme cases, those added side load stresses may lead to failure.

Special considerations

With a crane installed to true level on adequate supports with the unloaded hook set precisely to radius in calm air, immediately on lifting the rated load for that radius the manufacturer's rating chart will be exceeded and the lift will be improper! First, the outrigger beams will deflect and the carrier frame will twist causing the radius to increase. Second, the boom pendant and boom-hoist lines' ropes will stretch under load, further increasing the radius. The result will be a load at a greater radius than the rating was selected for. The operating radii given in the load rating charts are *loaded radii.* A lift at full rated load must be made at a shorter radius to allow for the deflections and stretch. The load can then be boomed out to the required radius, although experienced operators can usually predict the radius change between the unloaded and loaded states quite accurately and allow for it in advance.

Wind affects stability in two ways; by acting on the load and by acting on the crane itself. Stability ratings for mobile cranes do not take either into account. It thus behooves the crane user (operator) to beware that crane ratings may need to be reduced for wind effects. This unfortunately must usually be done in accordance with the user's (operator's) judgment rather than published information.

When designing cranes, full structural rated loads are taken together with a side wind of 20 mi/h (8.9 m/s) and a side loading at the boom tip of 2% of the rated load. Those loads have been found to be a good representation of normal operating conditions including a small but reasonable dynamic allowance. But rated loads are infrequently lifted and although winds at 30 mi/h (13.4 m/s) apply 2.25 times the force of a 20 mi/h wind, for many lifts the crane could safely operate at the higher wind loading. Unfortunately, the operator has no way of being sure of when those conditions are present.

To make matters worse, the wind force on the load is not known. For most loads lifted in winds up to 20 mi/h the force will be negligible, but for loads with larger sail areas and in higher winds, the operator is again left without guidance. The unsatisfactory possibilities here are twofold: (1) unsafe lifts may be attempted, and (2) safe lifts may be canceled out of uncertainty and fear.

As far as boom strength is concerned, wind from the side of the boom is important. For stability, however, wind blowing from behind the operator is more critical. When the wind blows from this direction, it applies force to the boom and to the load that adds to the overturning moment; it has the same effect as adding load at the hook. But when the wind blows into the operator's face, the opposite effect occurs. Wind in this direction will reduce the overturning moment. However, a boom at a high angle can be blown backwards over the cab.

In Chap. 8, an approximate field method is given that can be used to check both strength and stability under wind conditions. It enables jobsite personnel to determine whether or not a lift can be made.

Stability-based ratings

In many countries it is accepted that stability-based ratings should be set as a percentage of the tipping load. Just what percentage to use is arbitrary, and judgments vary on this point. In the United States and Canada we have used 85% for truck cranes and 75% for crawler cranes and tower front end attachments. Elsewhere 75%, 66⅔%, and other figures have been selected.

When load ratings are established this way, a crane carrying a rated load can, under some circumstances, have a slender margin against tipping. That is especially so when the margin is measured against the *tipping moment,* a quantity that is composed of contributions from both the boom and the load.

A long latticed boom or a fully extended telescopic boom at a large radius can induce a moment approaching tipping without ever picking a load. Indeed many telescopic boom range diagrams include blanked-out areas, radii at which it is not permitted to reach with an unloaded boom because of stability concerns. When a rated load which is either 75% or 85% of the tipping load is then added at the hook, the margin measured against the tipping moment can be razor thin.

An alternate approach then, is to use moments instead of loads as the basis for stability ratings. The logic is that the whole machine tips, therefore the ratings should be based on the stability of the entire crane. On the other hand, the logic of load based ratings is that ratings are established as a protection against lifting excessive loads, therefore it is the load that matters and should be the basis for ratings.

Most crane engineers follow the load approach, but nevertheless we must all face the fact that for long booms at long radii the tipping load may be considerably less than 1% of machine weight. Under such conditions, the crane is very sensitive to both wind and dynamic effects and can easily be overturned. But it is also a fact that cranes behave very well at short radii with present rated loads.

Wrestling with this problem and the conflicting judgments and theories on the subject, a committee of the International Organization for Standardization (ISO) has agreed on a simple and practical rating system that almost marries the two approaches. It is expressed by the equation

$$\text{Rated load} = \frac{T - 0.1f}{1.25}$$

where T is the tipping load and f is the weight of the boom referred to the boom tip or boom tip weight.

Initially, it was proposed that 10% of boom tip weight be deducted from the tipping load to account for the dynamic effect of braking after booming out. But although later tests showed this to be a variable rarely exceeding 5%, the probability of braking to a sharp stop with rated load exactly at rated radius is rather remote.

The authors' view (and many other engineers have come to feel the same way) is that the 10% figure, however arbitrary, makes a good practical adjustment to the rating curve. It has little or no effect at short radii, where adjustment is not needed, but as radius increases it becomes more prominent. At maximum radius, rated load may be reduced to two-thirds of the tipping load for most cranes, but less for very long booms. But the adjustment is a function of boom weight and therefore more pronounced in heavy booms, such as telescopic types, than in lightweight tubular booms. Thus, after a fashion, it is a self-adjusting rating modifier that strikes a good balance between safety and machine utilization.

There is another side to this coin, however. Looking over the data used for stability calculations earlier in this Section, the ISO formula would reduce the rated load to just over 500 lb (227 kg) or about 18% of the tipping load at that very long radius. As a practical matter, this rating method will eliminate some lifting at long radii.

The ISO rating formula is now utilized for mobile-crane ratings in Italy, Germany, the United Kingdom, Finland, and Poland.

4.3 Tower Cranes

Tower crane CGs are very high in relation to the usual base widths. This is particularly true for crane types which have their counter-weights mounted at the very top of the structure, but less so for those types which are counter-weighted at the base. The implication to be drawn from this characteristic is that tower cranes should be more conservatively rated than mobile cranes, and they are. Rated loads are generally taken at two-thirds of the tipping load, but other stability checks are necessary as well.

An important feature of most types of tower cranes, affecting both strength and stability considerations, is the flexibility of the mast, or tower, structure. When the mast is a cantilever, it deflects under the action of unbalanced dead, live, wind, and dynamic loads. When a substantial portion of crane mass is located at the top of the mast, that mass acts on the primary elastic deflections, causing increased deflection and an amplification of the bending moments. These tower crane masts are beam-columns.

The action is not quite the same for all tower crane masts. With base counterweighting, less mass is at the mast top, but in addition, when pendant lines run parallel to the mast, they relieve the mast of most of the applied moment but add axial compression. These masts, too, are beam-columns.

For a crane in service, wind can be taken from either behind the operator or broadside on the jib, the latter case often governing. With side wind, wind and primary load produce moments acting at 90° to each other. The overturning moment is then the resultant of these moments. The resultant can have any orientation, but prudence dictates that for stability calculations it should be considered when in the least favorable position.

It is usual, especially for tall cranes, for the critical stability condition to develop with the crane out of service and under the influence of storm winds. Weathervaning jibs expose a smaller area to the wind and produce wind moments that are opposite in direction to the dead-load moment, which is controlled by the counterweight. Nonetheless, storm-wind moment usually governs for these cranes.

Stability ratings of tower cranes are determined by calculation, as it is too dangerous to verify tipping loads by test. Three primary stability conditions must be verified:

Basic stability. Crane under static loading in calm air with the rated load not to exceed 62.5% of the static tipping load

Dynamic stability. Crane in service with wind and other dynamic loads as appropriate and with rated load taken as not more than 74% of the dynamic tipping load

Stability under extreme loading. Crane out of service and subjected to 120% of storm wind effects

Under the first two conditions, the tipping load, i.e., that load which brings the crane to the point of incipient overturning, is found. The lesser of the calculated loads is then used as the load rating. For the third case, the crane must not overturn under the stipulated loads.†

An additional criterion that must be satisfied concerns backward stability, that is, stability against overturning toward the counterweights. In mobile-crane practice counterweighting is limited by controlling the location of the CG of the unloaded crane with respect to the backward tipping fulcrum when the shortest boom is at its highest boom angle. In tower crane practice, the greater sensitivity to stability

†The figures given are from ISO 12485, Stability Requirements for Tower Cranes, International Organization for Standardization, Geneva, Switzerland, 1997.

due to the high CG location requires that in addition to static backward stability, the crane must remain stable under dynamic conditions such as the spring-back effect that accompanies release of load when operating with a clamshell bucket or a magnet, or landing a load too quickly.

Elastic deflections of tower cranes

Mast elastic deflections are caused by load eccentricity and by wind and are amplified by beam-column action, also called the P-Δ effect. If M_c is the net moment about the crane centerline (Fig. 4.10) in the absence of wind, the displacement of the mast top from the centerline δ_c is given by

$$\delta_c = \frac{M_c}{Q} \frac{1 - \cos kh}{\cos kh} \tag{4.7}$$

where Q is the weight of the slewing portion of the crane (at the mast top) plus the load and one-third of the mast weight,

$$k = \left(\frac{Q}{EI}\right)^{1/2}$$

E is the modulus of elasticity of the mast materials, and I is the moment of inertia of the mast cross section. The trigonometric arguments are in radians, and Eq. (4.7) intrinsically includes beam-column effects.

With wind introduced, taking W_w as the wind force on the concentration of exposure area above the slewing circle and w as the wind force per unit of length on the mast, we get

$$\delta_w = \frac{1}{Qk} \left[W_w(\tan kh - kh) + wh \left(\tan kh - \frac{kh}{2} \right) - \frac{w}{k} \frac{1 - \cos kh}{\cos kh} \right] \tag{4.8}$$

Taking moments about the crane base gives

$$M_c' = M_c + Q\delta_c = \frac{M_c}{\cos kh} \tag{4.9}$$

and for the wind moment M_w (noting that $M_w = W_w h + wh^2/2$)

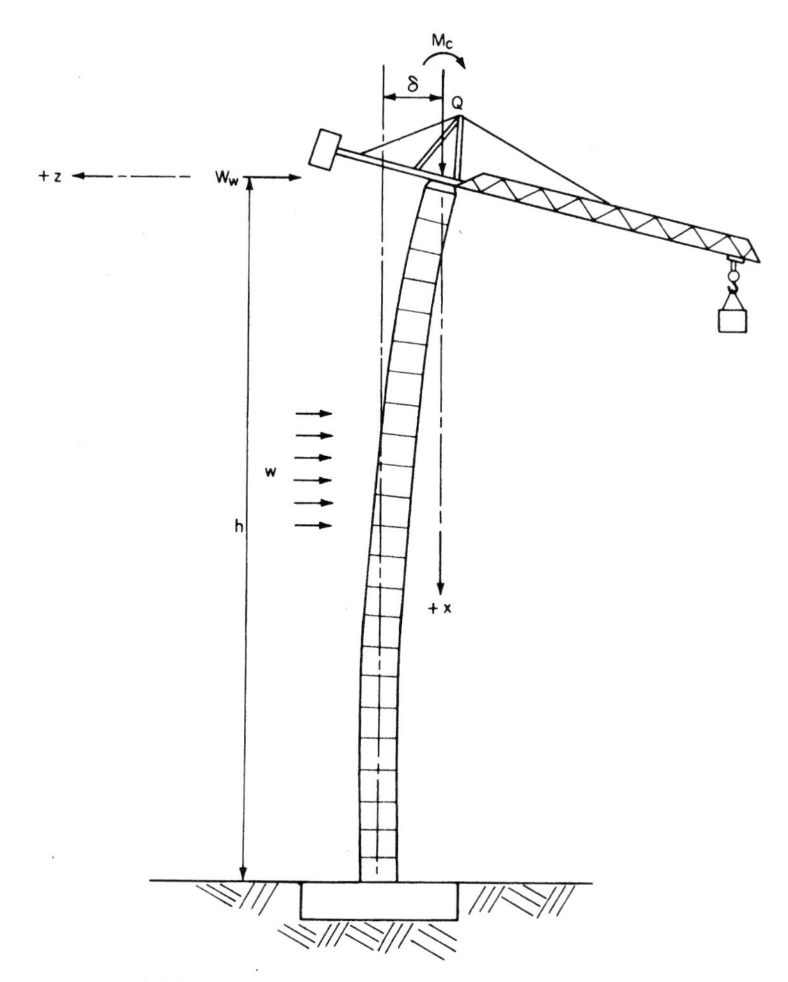

Figure 4.10 Model of tower crane annotated for calculation of elastic deflection.

$$M'_w = M_w + Q\delta_w = (W_w + wh)\,\frac{\tan kh}{k} - \frac{w}{k^2}\,\frac{1 - \cos kh}{\cos kh} \qquad (4.10)$$

The augmented moments of Eqs. (4.9) and (4.10) are used in both stability and strength calculations. For total deflection, δ_c and δ_w can be added vectorially. The moments can be similarly added.

Example 4.1 (*a*) A tower crane has an unbalance moment of 1500 kip · ft (2034 kN · m) about the axis of rotation while operating in calm air. The weight of that portion of the crane above the slewing ring plus the load and one-third of the mast weight is 100 kips (45.36 t), and mast height is 150 ft (45.72 m). For $E = 30{,}000$ kips/in^2 (200 GN/m^2) and $I = 45{,}000$ in^4 (1,873,000 cm^4), what will be the mast top deflection and the base moment?

From Eq. (4.7)

$$k = \left(\frac{100 \text{ kips}}{30{,}000 \text{ kips/in}^2)\ (45{,}000 \text{ in}^4)}\right)^{1/2}$$

$$= 2.722 \times 10^{-4} \text{ rad/in } (1.072 \times 10^{-2} \text{ rad/m})$$

$$\cos kh = \cos\,[2.722 \times 10^{-4})(150 \text{ ft})(12 \text{ in/ft})]$$

$$= 0.8824$$

$$\delta_c = \frac{1500 \text{ kip} \cdot \text{ft}}{100 \text{ kips}} \frac{1 - 0.8824}{0.8824} = 2.00 \text{ ft (610 mm)} \qquad \textit{Ans.}$$

and from Eq. (4.9)

$$M_c' = \frac{1500}{0.8824} = 1699.9 \text{ kip} \cdot \text{ft } (2305 \text{ kN} \cdot \text{m}) \qquad \textit{Ans.}$$

(*b*) If the wind force on the side of the jib is 1200 lb (5.34 kN) and the unit wind force on the mast is 7.5 lb/ft (109.5 N/m), what will be the wind deflection at the mast top and the base moment due to wind?

From Eq. (4.8)

$$\tan kh = \tan\,[(2.722 \times 10^{-4})(150 \text{ ft})(12 \text{ in/ft})]$$

$$= 0.5333$$

$$\delta_w = \Bigg\{(1200 \text{ lb})[0.5333 - (2.722 \times 10^{-4})(150)(12)] + (7.5 \text{ lb/ft})(150 \text{ ft})\left[0.5333 - (2.722 \times 10^{-4})(150)\frac{12}{2}\right] - \frac{7.5(1 - 0.8824)}{(12)(2.722 \times 10^{-4})(0.8824)}\Bigg\} \Bigg/ (100{,}000 \text{ lb})(2.722 \times 10^{-4})$$

$$= 2.59 \text{ in (66 mm)} = 0.22 \text{ ft} \qquad \textit{Ans.}$$

and from Eq. (4.10)

$$M_w' = [1200 \text{ lb} \times 150 \text{ ft} + (7.5 \text{ ft/lb})(150 \text{ ft})^2/2 + 100{,}000 \text{ lb} \times 0.22 \text{ ft}]/1000 \text{ lb/kip}$$

$$= 286.4 \text{ kip} \cdot \text{ft } (388.3 \text{ kN} \cdot \text{m}) \qquad \textit{Ans.}$$

(*c*) How much moment amplification has been introduced by the beam-column effect?

For the load moment:

$$\left(\frac{1699.9}{1500} - 1\right)(100) = 13.3\% \qquad Ans.$$

For the wind moment the unamplified wind moment is

$$W_w h + \frac{wh^2}{2} = \frac{1200(150) + 7.5(150^2/2)}{1000 \text{ kip/ft}}$$

$$= 264.4 \text{ kip} \cdot \text{ft } (358.4 \text{ kN} \cdot \text{m})$$

Then

$$\left(\frac{286.4}{264.4} - 1\right)(100 = 8.3\% \qquad Ans.$$

(*d*) The moments for load and wind are at right angles to each other. What are the resultant moment and deflection?

$$M = (1699.9^2 + 285.9^2)^{1/2} = 1723.8 \text{ kip} \cdot \text{ft } (2338 \text{ kN} \cdot \text{m})$$

$$\delta = (2.00^2 + 0.22^2)^{1/2} = 2.01 \text{ ft } (613 \text{ mm})$$

To derive Eqs. (4.7) and (4.8) it is necessary to make use of the concepts of elastic stability.† For a beam-column, the basic expression is the differential equation

$$EI \frac{d^2z}{dx^2} = -M_x$$

for coordinate axes, as shown in Fig. 4.10. When a cantilevered beam-column is loaded with only an axial force and a moment at its upper end,

$$M_x = M_c + Qz$$

$$EIz'' = -M_c - Qz$$

$$z'' + \frac{Q}{EI} z = -\frac{M_c}{EI} \qquad (a)$$

where the prime notation refers to differentiation with respect to x. Substituting $k^2 = Q/EI$, the general solution to (*a*) becomes

$$z = A \cos kx + B \sin kx - \frac{M_c}{Q} \qquad (b)$$

†See S. P. Timoshenko and J. M. Gere, "Theory of Elastic Stability," McGraw-Hill, New York, 1961.

but the constants of integration A and B are unknown. They can be evaluated by inserting boundary conditions that define the requirements this beam-column must meet. The first, $z(0) = 0$, states that deflection must be zero at the top (with respect to the chosen origin for the coordinate axes). Substituting this into Eq. (*b*), we have

$$0 = A \cos 0 + B \sin 0 - \frac{M_c}{Q} \qquad A = \frac{M_c}{Q}$$

Second, the slope at the fixed end must be 0, or $z'(h) = 0$

$$z' = -\frac{M_c}{Q} k \sin kx + Bk \cos kx$$

$$0 = -\frac{M_c}{Q} \sin kh + B \cos kh \qquad B = \frac{M_c}{Q} \tan kh$$

and with the constants evaluated Eq. (*b*), the expression for elastic deflection at any point in the beam-column, simplifies to

$$z = \frac{M_c}{Q} (\cos kx + \tan kh \sin kx - 1)$$

Maximum displacement will occur at the base, or fixed end, $z(h) = \delta_c$

$$\delta_c = \frac{M_c}{Q} \left(\cos kh + \frac{\sin^2 kh}{\cos kh} - 1 \right)$$

$$= \frac{M_c}{Q} \frac{1 - \cos kh}{\cos kh}$$

which is Eq. (4.7).

Under wind loading only, but with the axial force also acting

$$M_x = W_w x + \frac{wx^2}{2} + Qz$$

$$z'' + k^2 z = -\frac{W_w x + wx^2/2}{EI}$$

$$z = C \cos kx + D \sin kx - \frac{W_w x + wx^2/2}{Q} + \frac{w}{Qk^2}$$

Inserting the boundary condition $z(0) = 0$ gives

$$0 = C + \frac{w}{Qk^2} \qquad C = -\frac{w}{Qk^2}$$

and from the slope condition $z'(h) = 0$

$$z' = -Ck \sin kx + Dk \cos kx - \frac{W_w + wx}{Q}$$

$$0 = \frac{w}{Qk} \sin kh + Dk \cos kh - \frac{W_w + wh}{Q}$$

$$D = \frac{W_w + wh}{Qk \cos kh} - \frac{w}{Qk^2} \tan kh$$

The maximum wind-induced deflection δ_w, measured at the fixed end, is then given by $z(h) = \delta_w$

$$\delta_w = -\frac{w}{Qk^2} \cos kh + \frac{W_w + wh}{Qk} \tan kh - \frac{w}{Qk^2} \tan kh \sin kh - \frac{W_w h + wh^2/2}{Q} + \frac{w}{Qk^2}$$

$$= \frac{1}{Qk}\left[W_w(\tan kh - kh) + wh\left(\tan kh - \frac{kh}{2}\right) - \frac{w}{k}\frac{1 - \cos kh}{\cos kh}\right]$$

as given in Eq. (4.8). These deflection expressions intrinsically include the secondary effects of the axial load acting on the primary displacement, the P-Δ effect.

Static-mounted cranes

A crane affixed to one or more foundation blocks can rely on the added mass to resist overturning. The added mass may enhance stability both by providing ballast and by shifting the tipping fulcrum. Thus, the more spread out the footing, the less massive it needs to be to impact stability. This benefit must be balanced against the strength of the footing and interference from nearby obstructions. Further, the footing must be stiff enough to limit crane displacement.

When a single mass is used as a crane foundation, its edges are taken as the tipping lines. However, for this assumption to be valid, the underlying subgrade must have adequate bearing capacity. It is insufficient, thus, to consider the problem of stability in isolation; the bearing pressure under the crane foundation must be evaluated too. In some special cases, the tipping fulcrum should not be taken at the edge of the footing. For example, the tipping line for a pile foundation is routinely taken through the centers of the edge piles.

Most crane footings are rectangular and will most readily tip towards the fulcrums that are closest to the crane. The situation is more complicated when the footing is asymmetrical or when the potential overturning moment is not uniform in all directions. Oftentimes, the loading on the foundation must be evaluated for both working conditions and for storm exposure. This is treated more fully in Chap. 6 which covers tower cranes.

Foundation settlement is of minor importance insofar as stability is concerned, but differential settlement can cause a problematic loss of plumbness. The result may be difficulty in swinging and a diminution in load control, or the crane may be prevented from weathervaning properly. For a climbing tower crane that will later be braced to the building, the loss of plumbness may spoil the alignment of the crane to the intended attachment points on the building. Differential settlement may be caused by irregular soil conditions or by a prevailing lopsidedness in soil bearing pressure.

Cranes are usually more tolerant of footing settlement than buildings; they may indeed experience greater settlement because of their dynamic back-and-forth loading. Joining a crane footing to building footings should therefore be done with some caution due to the potential undesirable effects on the building structure.

Traveling bases

The ballast required for traveling tower cranes is mounted right on the travel base itself, generally in the form of blocks of precast concrete. In one direction, the rail centers define the tipping line. In the other direction, the wheel-shaft centers mark the tipping fulcrum for cranes with four wheels. When four bogies with eight or more wheels are used, the bogie pivots control tipping. Bogies are used to equalize the load on a set of wheels.

Rail lines must be constructed and maintained in such manner that the crane will operate under conditions assumed in the design. A dip in the track will reduce stability by causing the crane to go out of plumb. The dip can be introduced by initial inaccuracy in placing the track, by a short stretch of poor supporting soil, or by local settlement over a period of time. If supports are placed too far apart, the track may deflect between supports, causing the crane to lean. Mechanical rail splices introduce the possibility of rail-top misalignment at the joint; this produces a horizontal dynamic force when hit by a wheel.

Travel bases must be constructed to offer a high degree of resistance to rotation; knee braces are often installed for this purpose. A base that lacks sufficient rigidity will permit the mast to lean, with a subsequent loss of stability. Should the mast bottom rotate through an angle φ, the top of the mast will be displaced φh if φ is taken in radians. This displacement will introduce a new moment $Q\varphi h$, which in turn will cause additional displacement. The final displacement δ_b due to base rotation is then

$$\delta_b = \varphi \frac{\tan kh}{k} \tag{4.11}$$

and the added moment is

$$M_b' = \frac{Q\varphi}{k} \tan kh \tag{4.12}$$

where the rotation angle φ is given by

$$\varphi = \frac{\Sigma M}{2a^2 k_0} \tag{4.13}$$

and a is the distance between the tipping lines in the direction of the moment. k_0 is the rotational spring rate of the travel base. Only moments that act in the same direction are included in the summation.

One stability condition peculiar to traveling cranes concerns application of travel brakes. Braking action decelerates the crane, inducing inertia forces at the crane masses so that overturning moments result. Travel-brake stopping rates must therefore be controlled to prevent the crane from overturning.

Buffers, or stops, are placed at the ends of sections of travel trackage, but the crane is not supposed to strike them. If the crane should suffer the high deceleration accompanying a buffer strike, it would surely overturn. Instead, automatic brake trippers are placed at a distance from the buffers. The buffers serve only in the event that the brakes fail to stop the crane fully in the allotted distance.

4.4 Barge Mounted Cranes

Floating cranes continually go out of level. Cranes, by their nature, are never in balance because the counterweight effect must prevail for stability governed cranes while for pedestal mounted cranes the lifted load often prevails. This means that a crane always imposes a moment on the barge in addition to crane and load weight. Since most barges are much longer than they are wide, only operations over the side of the barge usually need to be evaluated for stability.

Consider first a crane positioned dead center on a barge. With the boom pointed along the length of the barge, the barge will be dead level in the side to side direction because there will be no moment applied, provided no other items of equipment or materials are stored on the barge. When a mobile crane swings over the side of a barge, it will impose a moment about its tipping line that is in the direction of the counterweight. However, with respect to the barge centerline, the maximum moment will be towards the load, and the barge will tilt, or *list*. A pedestal mounted crane will most likely produce its governing tilt condition at the maximum value of rated load times radius and will tilt the barge downward on the boom side as well.

A crane that moves about the barge is similar to a mobile crane on the centerline. As the crane travels from barge centerline towards the side of the barge, the weight of the crane causes the barge to tilt down towards that side. When the crane picks up a load, the list angle of the barge will increase.

Barges of usual sizes used for cranes would not be expected to list excessively, but potential listing must always be checked because mobile crane ratings will have to be adjusted if the tilt is more than 1%. Pedestal crane ratings may also require adjustment. For small angles of tilt, an approximate method may be used to determine tilt angles. First, taking a pedestal crane mounted on the barge centerline, consider a barge of width B, length L, and weight Q; the list angle α in radians is given by

$$\alpha = \frac{12M}{B^3 L w} \tag{4.14}$$

where M is the moment the crane imposes on the barge, w is the density of water [62.4 lb/ft^3 (1000 kg/m^3) for fresh water and 64 lb/ft^3 (1025 kg/m^3) for salt water] and α is in radians. To convert α to degrees, multiply by 180 and divide by π. Equation (4.14) is applicable providing the entire barge bottom remains below the surface of the water. This can be checked by calculating barge draft s. Taking W_t as the weight of the crane and load

$$s = \frac{Q + W_t}{BLw} \tag{4.15}$$

If $\alpha \leq 2s/B$, the entire bottom will remain below the surface. The problem becomes more complex if part of the bottom rises above the surface, and Eq. (4.14) is no longer applicable.

If barge weight Q is not known, and as a practical matter it seldom will be known, it can now be easily calculated using Eq. (4.15). The draft s can be measured with or without the crane, and either Q or $Q + W_t$ (less load weight) can then be calculated from barge dimensions. Draft might be measured at each of the four corners of the barge, and the average used in Eq. (4.15).

Equation (4.14) has been derived by balancing the moment applied to the barge with the righting effect of the triangle of water displaced by the tilt; it is sufficiently accurate for angles of tilt of a few degrees. Equation (4.15) equates barge plus crane weight with the weight of the water thereby displaced.

When a mobile crane is either centered on the barge or moves off the centerline towards the side of the barge, the moment M becomes

$$M = (W_m + W)d_t - M_r$$

$$= (W_m + W)d_t - M_{\text{net}} + S_h\alpha$$

$$\alpha = \frac{(W_m + W)d_t - M_{\text{net}}}{B^3Lw/12 - S_h} \qquad (4.16)$$

with M_r taken from Eq. (4.5) for a level crane and d_t is the offset distance of the crane from centerline of barge to the appropriate crane tipping line.

Barge tilt causes the crane to go out of level with a similar affect on crane lift capacity as discussed earlier for an out-of-level support surface. Once barge tilt angle has been determined, the effect on capacity is given by Eq. (4.5), but only for mobile cranes. Barge tilt angles less than or equal to 1% (100 times α) will be within the mobile crane manufacturer's level limits and will not affect rating chart capacities. Pedestal crane rating adjustment may require structural analysis.

The equations given above are only approximate. They are intended as a quick and simple checking means to assess barge suitability or to gauge relevance of standard rating charts. Further work is required before any crane can be utilized on a barge, such as given in the more refined equations in the technical section below. In addition to consideration of stability and basic crane ratings, however, side loading must be evaluated. When the list is 1% or more, it is likely that side loading effects will require rating reductions.

There is an important difference in how cranes and lifted loads behave for barge mounted machines as opposed to land machines. Whenever the crane lifts or places a load that is not on the barge, the barge list changes. Lifting loads causes the crane to lean towards the load, while landing a load produces the opposite effect. As the crane takes a strain on the load, the barge and crane commences tilting and the boom tip moves away from the barge centerline. Assuming the boom tip started out directly above the load CG, on lifting, the load will swing out. Operator's or their signal persons normally try to compensate for this by judgment, but it would be unusual for that compensation to produce precise results—the load will swing either away from the crane or towards it. This is virtually unavoidable and must be expected and allowed for.

When a mobile crane operates from a single position on a barge, it may be desirable to fix the crane in position. For a crawler crane, this could be accomplished by blocking the crawler tracks with wood blocking to prevent movement. To hold the blocking in place, steel stops can be welded to the barge deck. For further security, four chains can be used to hold the tracks down and prevent overturning. However, when this is done, the chains should be left sufficiently slack to permit the tracks to lift a few inches. This will minimize the possibility of a structural overload by letting the operator know that the crane has started to tip. When a truck crane is mounted on a barge in a fixed position, similar means can be used to prevent overturning. In this case, the chains would be placed at the outer ends of the outrigger beams and also left slack.

A far more complex situation arises should crane operations be conducted in rough water. Here, both the crane barge and the load barge will be moving vertically due to wave action, and may be moving in opposite directions. This leads to the possibility that the crane will suddenly snatch the load off the deck of the load barge, causing significant dynamic reactions, and then roughly land it on deck again. Those reactions pose the threat of damage or tipping for the crane and damage to the load. This is a special situation that is beyond the scope of this book, but some insights into the problem can be gained from an SAE Recommended Practice document.†

A more exact solution will have to account for movement of the CGs of the barge and crane that takes place as the barge lists. That movement adds to the moment acting on the system. Figure 4.11 shows a level barge and a barge tilted by a moment. The weight of the displaced water equals the weight of the barge, crane, and lifted load, Q_t. Therefore, the weight of the displaced water will remain constant if Q_t is constant, regardless of the magnitude of system moment or how much the barge lists. As the barge tilts to angle α, a triangular wedge of water is created. The offset of the center of mass of that wedge from the barge centerline creates a moment which resists the applied moment. But, for a pedestal crane mounted on the barge centerline, tilting of the barge causes the CG of Q_t to displace adding to the applied moment. The amplified applied moment acting to tilt the barge now becomes

$$M_a = M + Q_t h_q \alpha$$

†SAE J1238, *Rating Lift Cranes on Fixed Platforms in the Ocean Environment,* Society of Automotive Engineers, Warrendale, PA, 1978.

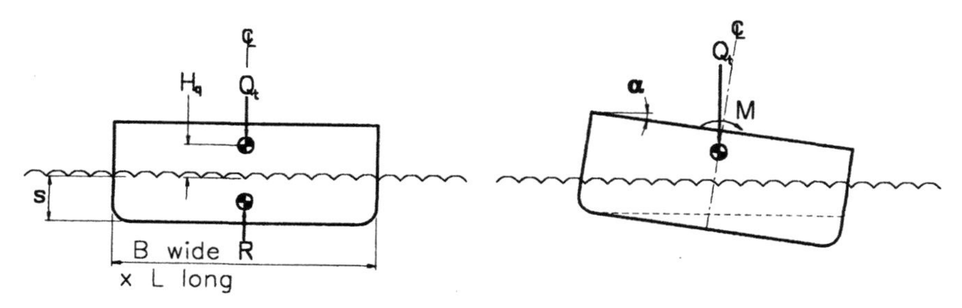

Figure 4.11 Diagrammatic view of a crane barge when level and when caused to list by applied moment M. Q_t is the combined weight of the barge, the crane, and the lifted load. R is the weight of the displaced water, which must be equal to Q_t.

where M is the applied moment introduced by the crane and load and h_q is as shown in Fig. 4.11. The buoyancy of the displaced water must support barge plus crane and load weight and balance the amplified applied moment.

The cross-sectional area of the displaced water will be sB whether the barge is level or listed. Using small angle geometry, the cross-sectional area of the displaced wedge of water will be $B^2\alpha/2$, and the centroid of that triangular area will be located at distance $B/6$ from the centerline of the barge. The centroid of the remaining portion of sB (the part that is not included in the wedge), will be centered with respect to the barge centerline. Therefore, the centroid of the entire mass of displaced water will be located

$$d_w = \frac{\frac{B}{6} \times \frac{B^2}{2}\alpha}{sB} = \frac{B^2\alpha}{12s}$$

from the centerline of the barge. d_w is the moment arm for the righting force generated by the displaced water to resist the amplified applied moment M_a. Dividing M_a by Q_t yields the eccentricity of the vertical loads, which must be equal to the eccentricity of the righting force because the forces themselves are equal and the entire system is in equilibrium.

$$\frac{B^2\alpha}{12s} = \frac{1}{Q_t} \times (M + Q_t h_q \alpha)$$

$$\alpha = \frac{12M}{B^3Lw - 12Q_t h_q} \tag{4.17}$$

which is more accurate than the angle calculated using Eq. (4.14) by virtue of the adjustment in the denominator.

Mobile cranes, whether or not positioned on the barge longitudinal centerline, will impose moment M_r on the barge, as given by Eq. (4.5), and

$$\frac{B^2\alpha}{12s} = \frac{1}{Q_t} \times [(W_m + W)d_t - M_{\text{net}} + S_h\alpha + Q_t h_q \alpha]$$

$$\alpha = \frac{(W_m + W)d_t - M_{\text{net}}}{B_3Lw/12 - S_h - Q_t h_q} \tag{4.17a}$$

which is similar to Eq. (4.16), but with an adjustment to account for the CG shift.

Once tilt angle α has been determined, Eq. (4.5) can be used to calculate mobile crane stability ratings that are consistent with barge list and the appropriate ratio of rated to tipping load, but this need only be done should list exceed 1%.

The preceding analysis has been concerned only with the stability aspects of crane ratings on a listing barge. Another important consideration is side loading, because crane booms are typically sensitive to side loads. For a mobile crane on the barge centerline, side loading will be maximum when the boom is oriented 45° to the centerline. Therefore, the moment M in Eq. 4.17 is calculated with respect to the side of the barge when the boom is at a 45° horizontal angle, or in other words 0.71 times its usual value. Now, if 71 times α calculated over the side is less than or equal to 1%, side loading will be within permitted limits. If greater than 1%, the rated load must be reduced accordingly, and the crane manufacturer should be consulted for this purpose. However, clearly conservative reduced ratings may be obtained by dividing the rated load by $(1 + 35.5\alpha)$.

Mobile cranes that travel about the barge will experience side loading during most lifts, regardless of boom horizontal angle. For those installations, the crane manufacturer should be consulted.

Example 4.2 A mobile crane is mounted on a barge away from the centerline so that it imposes a moment of 10,000,000 lb · ft (13,558 kNm). The barge is 75 ft (22.9 m) wide, 200 ft (61.0 m) long with gross weight of 2300 kips (10,230 kN) including crane weight. The barge is in fresh water. How much will the barge tilt? Will the mobile crane rating chart be valid?

Without using a full set of mobile crane data, a rough approximation is available by using Eq. (4.14)

$$\alpha = \frac{12x10{,}000{,}000}{75^3x200x62.4} = 0.023 \text{ rad} = 1.31°$$

Check using Eq. (4.15)

$$s = \frac{2{,}300{,}000}{75x200x62.4} = 2.46 \text{ ft (749 mm)}$$

$\alpha < 2 \times 2.46/75 = 0.066$ which means that the entire bottom remains submerged and the solution is valid. The barge will tilt about 2.3% and the mobile crane rating chart will not be valid because tilt is more than 1%. Therefore, a more refined study using Eq. (4.17*a*) will be needed. *Ans.*

4.5 Other Cranes

Our discussion in Chap. 2 showed that the range of crane types and configurations is rather extensive. A detailed treatment of stability considerations for each is not practical, nor would much be gained by it. The concepts dealt with under mobile, tower and barge mounted cranes apply equally to all cranes, and the rating criteria given for tower cranes generally apply for cranes other than mobile and floating cranes. The configuration, operating functions, site conditions, and performance requirements for any particular crane will then indicate the relative importance of the several factors that either detract from stability or enhance it.

Few other crane types are built with masts of the height or flexibility of tower cranes. Therefore, the effects of deflections, the P-Δ effect, can generally be ignored when studying stability of other crane types. Some special considerations may need to be evaluated, however. When cranes are intended for operation at high speed, dynamic effects take on greater importance. Some cranes must operate on inclined tracks or on curves, and these requirements introduce both static and dynamic considerations. Some revolving cranes have little or no physical attachment at the slewing ring, so that the ring can be taken as the tipping line, and many cranes must be mounted on unsymmetrical bases. None of those conditions vary the basic aspect of the stability problem, however, nor do they change the criteria for stability.

The limits on inaccuracies permitted in the placement of tracks are not dealt with in the ANSI standards for the various crane types. For tolerance parameters, we must turn to the FEM† for guidance. The limits specified are intended by FEM for permanently installed cranes; for temporary (construction) installations, the criteria may be considered too restrictive.

FEM specifies that the distance between rails must be kept within ±3 mm (⅛ in), while the difference in elevation between any two points perpendicularly across the rails may not exceed 0.15% with a maximum of 10 mm (⅜ in). Any two points along the track within one wheelbase length must be kept to within ±3 mm (⅛ in) in elevation for gages up to 3 m (10 ft), or within ±0.1% of the wheelbase for wider gages. Rail straightness is controlled by the requirement permitting no more than ±1 mm (1/25 in) deviation in any 2 m (6.5 ft) of length. Misalignment of the rail joints is not permitted at all.

†Fédération Européene de la Manutention, "Rules for the Design of Hoisting Appliances," 2d ed., Paris, 1970. See also ISO 8306, Cranes—Overhead travelling cranes and portal bridge cranes—Tolerances for cranes and tracks, International Organization for Standardization, Geneva, 1985.

For centrifugal force, FEM suggests that the effect be applied to the load only and ignored for the crane components; our own work indicates that this may on occasion be inappropriate when evaluating stability, particularly for long-boomed cranes.

Think stability. An awareness of the factors affecting stability for any particular machine will lead to clear insights into aspects of installation quality control which are essential to safe and productive operations. It may also lead to more economical installations as knowledge replaces uncertainty in the planning process.

4.6 Dynamic Stability

Dynamic loads were discussed in Chap. 3. Those affecting stability include centrifugal force, inertial forces associated with travel, inertial effects from braking while lowering a load, and to some extent the harmonic loads from hoisting system acceleration and deceleration. The latter is not likely to cause overturning except under extraordinary circumstances. Complex situations do develop, however, for cranes with heavy loads (with respect to rated load) undergoing rapid speed changes, for cranes with high CGs, when misalignments are present at track joints, and for other travel path irregularities. For some of those situations, the parameters can be taken from FEM tolerance criteria, which will not be treated in detail here.

Dynamic loads are often considered in the stability equation by statically applying the maximum values they achieve and requiring that the crane remain stable when so loaded. That approach has the advantage of ease of calculation, and usually proves conservative, but not always, as will be demonstrated. The error produced by this simple method can be quite large, however, when compared with more accurate, but more elaborate, techniques.

In this section, a procedure will be presented for evaluating the effect on stability of short duration dynamic forces resulting from acceleration or deceleration of travel and hoist system drives. This method offers a powerful tool for crane operation planners to avoid overturning and reduce risk in crane operations where dynamic events may pose a threat to stability.

Centrifugal force

When a crane swings, centrifugal force, as given in Eq. (3.17), will cause the load to move to an increased radius (Fig. 4.12). The new radius will be a function of the load-line length, an entirely random parameter of no practical use. Both the load and the centrifugal force can be taken as acting on the crane at the boom tip, however, as the

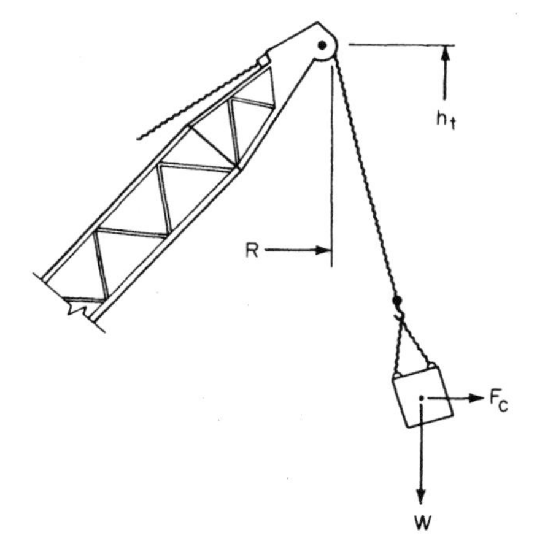

Figure 4.12 When a crane swings centrifugal force F_c causes the load to move to an increased radius.

ropes pass the loads to the crane at this point. Taking moments, we then find

$$W' = \frac{WR + F_c h_t}{R} = W + F_c \frac{h_t}{R} \tag{4.18}$$

where W' is the effective vertical load acting at radius R that is equivalent in effect to both W and F_c.

Many mobile cranes are capable of a rapid swing that can induce a strong centrifugal force. When the crane is also lifting a rated load, the addition of centrifugal force can bring the crane to the point of tipping. The following example illustrates.

Using the crane data supplied earlier in treating out-of-level operation, which for that particular crane describes $R = 262$ ft (79.9 m) and $h_t = 191.7$ ft (58.4 m), and taking a load that is 85% of the tipping load while the crane swings at 1 r/min, from Eq. (3.17) we get

$$F_c = \frac{0.85(2800)(262)}{32.2}\left(\frac{\pi}{30}\right)^2 = 212.4 \text{ lb } (994.8 \text{ N})$$

Equation (4.18) yields the effective load at radius R

$$W' = 0.85(2800) + 212.4\,\frac{191.7}{262} = 2535 \text{ lb } (11.3 \text{ kN})$$

which shows that centrifugal force has taken up almost 40% of the difference between rated and tipping load. Of course, centrifugal force acts on the boom as well; it can be treated as though acting at the boom CG. At the load radius R, the effective load is then found to be about 3300 lb (1497 kg), which is greater than the tipping load. The indication is that the crane will overturn.

However, other factors may mitigate the opportunity for the crane to tip. For example, the high centrifugal force might not occur when the crane is in its least stable orientation, or the crane may swing past that least stable point before the slow tipping process takes hold. Moreover, most crane operations are carried out over too short a swing arc for a high rotational speed to develop.

Centrifugal force is less problematic for tower cranes. Slewing speed is modulated by the drive system and the ratings make allowance for load dynamics. The high CGs of tower cranes make these precautions necessary.

Inertial forces affecting stability

When establishing lift ratings based on crane stability, it has been traditional to simulate the short-lived inertial forces acting on a crane by adding peak dynamic values to static loads, by using arbitrary dynamic factors, or by establishing ratings by taking a percentage of the tipping load. Over time, those approaches have proven to be suitable for rating purposes, but they do not offer a realistic approximation to actual dynamic effects. In some instances such as in the investigation of failures, these methods may be valid only as early stage approximations.

The traditional methods are in effect force-based systems in which static and dynamic components are added algebraically. But, in order to tilt a crane and cause it to overturn, work is required. Dynamic events introduce energy into the crane, such as when a crane traveling with a load makes a quick stop, or when a load being lowered is abruptly stopped before reaching the ground. When that energy is sufficient to do the work needed to overturn the crane, it will overturn. If the energy is not sufficient to carry the crane to the critical tilt angle, it may tilt but not overturn. The static loads acting on the crane define the tilt angle at which the crane will overturn and therefore the work required to tilt the crane to the limit. The greater the static forces with respect to the tipping load, the less energy required to do the work and cause failure. Among the static loads, wind from an unfavorable direction produces an effect similar to adding load on the hook—gusting adds further loading as well as causing a dynamic "bounce."

Dynamic (inertial) forces resulting from acceleration or deceleration furnish energy input. The closer the crane is to tipping because of its static loads, the more sensitive it will be to dynamic disturbances; intuitively, an experienced crane operator knows that motions should be less abrupt when operating close to capacity. An analysis of the physics of crane dynamics affirms this intuition.

The ability to determine how dynamic actions affect crane stability can be a useful tool in the hands of crane installation planners, particularly for situations pushing the norms of established practice. As this section will show, each element in the dynamic stability equation offers planners insights into the means available to ameliorate or control dynamic effects and thereby reduce risk and thus improve the probability of success. On the other hand, dynamic stability analysis can also show when unexpected dynamic episodes will offer no stability threat, permitting risk control assets to be more favorably employed.

Figure 4.9 can be considered a generalized dynamic model of a crane.† It will be assumed that the load is suspended close to the boom tip for one or a few parts of hoist line, or further from the tip for multi-part lines. This assumption is made to minimize the effect of harmonic spring effects on the hoist system so that the entire crane can be considered as a single degree of freedom (SDOF) system.‡ The overturning static force from load weight can then be taken to act at the boom tip. Horizontal static forces as from steady state wind or centrifugal force are not shown on Fig. 4.9, but will be included in the moment of static forces acting on the crane, Eq. (4.4), and the equations that follow. Wind gusts are not considered.

Newton's second law offers an equation of motion describing the response of the loaded crane of Fig. 4.9 to the action of very short duration dynamic forces, or after the cessation of mechanism acceleration or deceleration. The differential equation is

$$J\ddot{\varphi} = F(\varphi) \tag{4.19}$$

where J is the moment of inertia of crane and load masses about point O of Fig. 4.9, the tipping line, φ is the crane angle of tilt (positive clockwise) and $F(\varphi)$ is the non-linear function reflecting the moment about O of all forces acting on the crane. Angles are in radians and the dot superscript reflects differentiation with respect to time.

After integration, Eq. (4.19) becomes

$$J\dot{\varphi}^2/2 = \int_{\varphi_o}^{\varphi} F(\varphi)d\varphi \tag{4.19a}$$

where φ_o represents a stable equilibrium condition under static loading; it is the

†The method for evaluating dynamic stability that follows has been taken from SAE Technical Paper 972721, *Overturning Stability of a Free Standing Crane Under Dynamic Loading,* Anatoly A. Zaretsky and Howard I. Shapiro, Society of Automotive Engineers, Warrendale, PA, Sept. 1997.

‡Further work by Prof. Zaretsky has demonstrated that these results are valid for multi-degree-of-freedom systems. Therefore, the close-to-the-boom-tip assumption need not be imposed. The mathematical proofs were offered in private correspondence received just before publication of this book.

tilt angle of the crane at the instant the dynamic event commences. But noting the general expression for kinetic energy in a rotating system

$$T = J\dot{\varphi}^2/2$$

Equation (4.19*a*) can be interpreted in a different way. It can be thought of as expressing a basic relationship where the left side portrays system kinetic energy and the integral on the right side reflects the work that will be done as the crane tilts from φ_o to φ. This interpretation reveals that there is a direct link between input energy and resulting crane tilt. At its upper limit, φ will be the angle of unstable equilibrium φ_{un}. Therefore, the minimum amount of energy required to tilt the crane to φ_{un} is the energy that will produce overturning failure.

Also, this crane mathematical model can be viewed as being comparable to a nonlinear conservative pendulum, since dissipating forces have not been taken into account in the crane model. It is reasonable to ignore damping, because only part of the first oscillating cycle of the crane is of interest and very little dissipation occurs during the first cycle. Pendulum oscillations go on without end once they begin, but as a pendulum passes the point of stable equilibrium, both T_{max} and $\dot{\varphi}_{\max}$ will occur. Using the pendulum analogy motivates a criterion for crane dynamic stability

$$T_{\max} = J\dot{\varphi}^2_{\max}/2 \leq \int_{\varphi_o}^{\varphi_a} F(\varphi)d\varphi \qquad (4.20)$$

If the input energy is less than or equal to the work required, the crane will not rotate beyond φ_a, or overturn if φ_{un} is the limit. A value for φ_a, the upper limit of the integral, may be selected that is less than φ_{un}. This may be done to provide a margin of protection against overturning, or in the case of rail mounted cranes, to prevent the wheel flanges from rising above the track and in that way avoid possible derailment when the crane returns to the static state.

From an equilibrium expression including all static forces acting on the crane, φ_{un} can be found; the procedure is similar to that used in deriving Eqs. (4.5) and (4.6), but substituting the tilt angle φ for the out-of-level angle α. This will lead to

$$F(\varphi) = M_r - S_h\varphi \qquad (4.20a)$$

where the terms M_r and S_h are as defined for Eqs. (4.5) and (4.6); M_r intrinsically includes out-of-level effects. When the crane has tilted to the point of unstable equilibrium, $F(\varphi) = 0$, the static forces that had induced moments are balanced, and φ is at its limit φ_{un}

$$0 = M_r - S_h\varphi_{un}$$

$$\varphi_{un} = M_r/S_h$$

We will now assume a spring at the left crane support in Fig. 4.9 to reflect elastic effects at that support. The actual support spring response is often so stiff that it can be ignored in practical calculations, but we must use it to develop the theoretical behavior of the crane. The static loading on the crane will therefore cause some elastic subsidence at the spring support point inducing a negative tilt angle

$$\Delta = -M/cb$$

where c is the support spring constant and b is the distance between the spring support and fulcrum point O. The crane will now be at rest at the initial tilt angle φ_o

$$(\varphi_o < 0) \qquad \varphi_o = \Delta/b = -M/c_o \qquad M = -\varphi_o c_o$$

where $c_o = cb^2$. At any time that $\varphi < 0$, the moment of the spring reaction will balance the moment of crane forces [as reflected in Eq. (4.20*a*)] that induced the spring reaction. For the initial state, $\varphi = \varphi_o$, the entire moment must be balanced because the crane is in static equilibrium; therefore

$$M = F(\varphi_o)$$

$$-\varphi_o c_o = M_r - S_h \varphi_o$$

$$\varphi_o = -M_r/(c_o - S_h)$$

The defining points have now been developed for a plot relating moment $F(\varphi)$ and tilt angle φ; $F(\varphi_o) = 0$, $F(0) = M_r$, $F(\varphi_{un}) = 0$. Figure 4.13 shows that curve with straight lines connecting the points. As will be explained later, using straight lines when perhaps curved lines would be more exact does not affect the accuracy of this analysis. From Fig. 4.13, the sloped legs of the curve yield the following expressions

$$F(\varphi) = M_r + (c_o - S_h)\varphi \qquad \varphi < 0$$

$$F(\varphi) = M_r - S_h \varphi \qquad \varphi \geq 0 \qquad (4.20b)$$

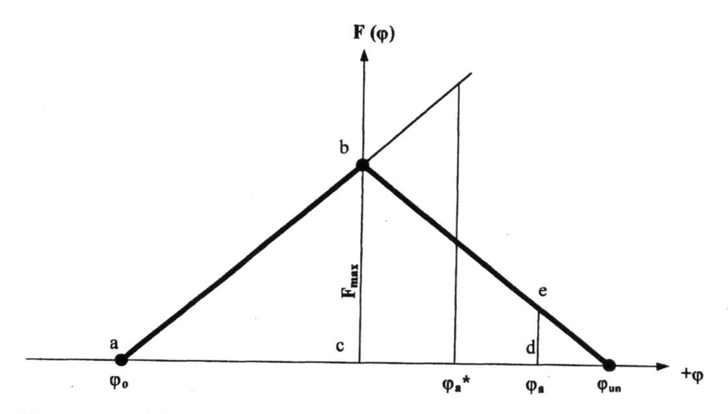

Figure 4.13 A functional plot of the moment $F(\varphi)$ acting on the crane as it tilts from its initial condition of static equilibrium at angle φ_0.

Equation 4.20*b* and Fig. 4.13 reveal that the maximum moment experienced by the crane occurs when the spring support becomes fully unloaded.

The area under the curve of Fig. 4.13 between φ_o and φ_a is equivalent to the value of the integral on the right side of Eq. (4.20); $I(\varphi)$ includes the areas of triangle abc and trapezoid bcde.

$$I(\varphi) = \mathrm{A_{abc}} + \mathrm{A_{bcde}} = 0.5\, M_r^2/(c_o - S_h) + M_r^2(2 - \mu)\mu/S_h$$

$$= 0.5\, M_r^2[S_h/(c_o - S_h) + (2 - \mu)\mu]/S_h \qquad (4.20c)$$

where $\varphi_a = \mu\varphi_{\mathrm{un}}$ and $0 < \mu \leq 1$.

With the exception of T_{max}, all of the values in the stability criterion of Eq. (4.20) have been developed. Note that the stability criterion depends on the area bounded by the points defining the curve of the integral between the limits φ_o and φ_a, but does not depend on the shape of the curve because the work done by the system is conservative. For a conservative system, the work done is a function of the end points, not the path taken. Whether the lines connecting the points in Fig. 4.13 are curved or straight does not affect the value of the integral. For that reason, we can use any function $F^*(\varphi)$ to replace the function $F(\varphi)$, in order to simplify the function, provided the equality of the integrals is maintained. Thus

$$\int_{\varphi_o}^{\varphi_a} F(\varphi)d\varphi = \int_{\varphi_o}^{\varphi_a^*} F^*(\varphi)d\varphi \qquad (4.20d)$$

provides the rule that can be used to simplify the system by linearization.

Figure 4.14 depicts a linear system of loading wherein the dynamic moment $P(t)$ attains full value instantaneously at $t = 0$, and maintains constant value

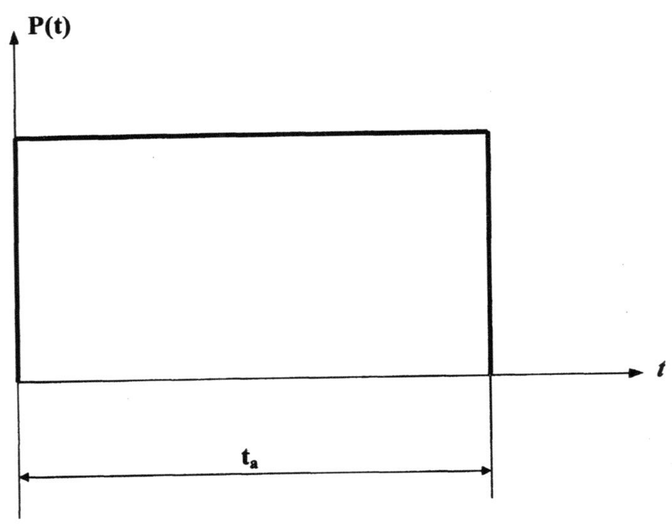

Figure 4.14 Hypothesized moment/time model for linearizing the effects of dynamic events on the crane of Fig. 4.9.

until $t = t_a$, at which time it diminishes instantaneously to zero. We will examine this linear system with the goal of manipulating it through use of the relationship of Eq. (4.20*d*) to produce a practical solution to Eq. (4.19). This linear system can be described by a form of the equation of motion for an undamped SDOF system common in the literature

$$J\ddot{z} + (c_o - S_h)z = P(t) \tag{4.20e}$$

where the new variable z, $z = (\varphi - \varphi_o)$, is used to make the solution to Eq. (4.20*e*) independent of the initial conditions since $\dot{z} = \dot{\varphi}$ and $z(0) = 0$. With the additional boundary condition $\dot{z}(0) = 0$, the general solution to Eq. (4.20*e*) is

$$z = [P/(c_o - S_h)](1 - \cos \omega t)$$

where $\omega^2 = (c_o - S_h)/J$ and ω is the circular frequency of loaded crane free oscillations. The solution applies for $0 < t \le t_a$, but we will also need displacement z during the interval $t > t_a$. After using a two-step solution and mathematical manipulation, this yields

$$\dot{z} = 2P[\omega/(c_o - S_h)](\sin 0.5\ \omega t_a) \cos \omega\ (t - 0.5t_a) \tag{4.20f}$$

Maximum velocity $\dot{z}_{\max}$ will be given by Eq. (4.20*f*) when the cosine term equals one. But, after multiplying the right side of Eq. (4.20*f*) by $0.5\ \omega t_a/0.5\ \omega t_a$, Eq. (4.20*f*) can be further simplified and then equated to $T_{\max}$ in consideration of the energy criteria, Eq. (4.20), to yield

$$T_{\max} = 0.5\ J[(Imp/J)(\sin 0.5\ \omega t_a)/0.5\ \omega t_a]^2 \tag{4.20g}$$

where $Imp = Pt_a$, an impulse of the moment P. Designating a new variable

$$k_t = (\sin 0.5\ \omega t_a)/0.5\ \omega t_a$$

one can note that the limit of $k_t = 1$ as $t_a \to 0$. By substituting $t_o = 2\pi/\omega$, the natural period of loaded crane oscillations

$$k_t = (\sin \pi t_a/t_o)/\pi t_a/t_o \tag{4.21}$$

The remaining portion of Eq. (4.20*g*) can be interpreted as the magnitude of $T_{\max}$ produced by instantaneous acceleration T_o. The k_t term, where $k_t \le 1$, reflects that the event takes place during an actual time interval t_a. For the general case

$$T_{\max} = T_o k_t^2 \tag{4.22}$$

The natural form of T_o is $0.5\ J(Imp/J)^2$ as in Eq. (4.20*g*), but from the law of conservation of energy we can obtain a more practical expression

$$T_o = 0.5 \sum m_i v_j^2 \tag{4.23}$$

where m_i is a mass (or part of the crane) moving at linear velocity v_j when the *j-th* mechanism is applied.

If M_d is taken to represent the moment induced by all dynamic forces causing tilting, the crane will remain stable provided $M_d \leq M_r$. Therefore the stability limit has been reached should $M_d = M_r$.

In order to make the kinetic energy of the non-linear system of Eq. (4.19) equal to the kinetic energy of the linear system of Eq. (4.20*e*), the linearization rule of Eq. (4.20*d*) must be satisfied. This can be accomplished by setting the value of the integral given by Eq. (4.20*c*) equal to T_{max} of Eq. (4.22) and noting that at the limit $M_d = M_r$. Manipulation then yields the following criterion for dynamic stability

$$M_d = \{2T_o k_t^2 S_h/[S_h/(c_o - S_h) + (2 - \mu)\mu]\}^{1/2} \leq M_r \qquad (4.24)$$

Equation (4.24) defines the maximum value that M_d can take without causing the crane to tilt beyond the limit φ_a, and represents but another way to express the criterion of Eq. (4.20). M_d is an equivalent moment representing the effects of the time-varying dynamic forces that impose kinetic energy T_{max} on the crane; it is conceptual rather than physical. M_d cannot be used for structural calculations because it will not discover peak values for stress; it is a parameter for use only in stability determinations.

Because T_{max} has been derived using extreme value considerations [the cosine term of Eq. (4.20*f*) has been set equal to 1], and some energy may in fact be dissipated by spring action in the hoist ropes, M_d represents the maximum value that this static equivalent moment can take. Therefore, if $M_d \leq M_r$ the crane tilt angle will be less than or equal to, but will not exceed, φ_a.

When two or more drive mechanisms simultaneously induce dynamic forces from acceleration or deceleration, the kinetic energy from each mechanism can be added in the following manner

$$T_C = \sum T_{oi} k_{ti}^2 \qquad (4.25)$$

where T_{oi} is the kinetic energy of the mass accelerated by the i-th mechanism and k_{ti} takes into account the time during which the acceleration or deceleration of that mechanism takes place. T_C is used in place of $T_o k_t^2$ in Eq. (4.24) when more than one mechanism is engaged simultaneously. Therefore, M_d is not the algebraic sum of the effects of multiple dynamic inputs. When mechanisms do not engage simultaneously, their effects will be out of phase with one another, and the sum of non-simultaneous dynamic actions will never produce more energy than reported by Eq. (4.25) for simultaneous events.

Plotting Eq. (4.21) will show that k_t will be zero when t_a/t_o takes full integer values, but magnitudes of the ratio t_a/t_o in the vicinity of 1 and greater produce such small energy values when inserted into Eq. (4.24) that the ensuing moment M_d will be of little consequence; therefore, only mechanisms where $t_a/t_o < 1$ need be considered.

Every loading event will cause a crane to tilt, however small the tilt angle will be. The rigidity term c_o tends to be very large for most actual support situations when compared with the moment S_h. That results in the first term in the denominator of Eq. (4.24) being small, often falling between 0.005 and 0.025; in those instances, that term can be eliminated to simplify the equation, which then becomes

$$M_d = \{2T_o k_t^2 S_h/[(2 - \mu)\mu]\}^{1/2} \tag{4.26}$$

Rearranging Eq. (4.24), and taking the limit value for M_d, produces an equation for the maximum energy from a single mechanism that can be tolerated

$$[T_o k_t^2]_{\max} = 0.5M_r^2[S_h/(c_o - S_h) + (2 - \mu)\mu]/S_h$$
$$= \varphi_{\text{un}} M_r[S_h/(c_o - S_h) + (2 - \mu)u] \tag{4.27}$$

or for several mechanisms

$$[T_c]_{\max} = \varphi_{\text{un}} M_r[S_h/(c_o - S_h) + (2 - \mu)u] \tag{4.27a}$$

The static equivalent moment M_d can be calculated and compared with the static resisting moment M_r. There is no mathematical correlation between those moments to inform designers when adequate protection from failure has been achieved, other than $M_d \leq M_r$. Designers and planners are therefore left with their experience and judgment.

The dynamic stability equations, Eq. (4.24), (4.26), (4.27), and (4.27*a*), include variables that present planners with opportunities to reduce and control overturning risk when dynamic forces are meaningful. Each parameter offers insights that can be useful. First, there are obvious measures for increasing the static moment M_r, such as minimizing the operating radius; disassembling a portion of the lifted load, when feasible (which will also reduce S_h); establishing more severe limits for levelness, especially for travel; limiting swing speed to reduce centrifugal force and hence M_{hor}; and reducing the ambient wind speed limit at which operations will be permitted to go forward to reduce M_{hor}. Second are the measures available that relate to the mechanisms. For each mechanism that could be engaged simultaneously, suppressing motion speeds will reduce the energy parameter T_o, and if measures can be taken to be certain that braking stop times can be increased, k_t values will be reduced accordingly. Increasing the number of parts of line and use of power load lowering are viable means for accomplishing reductions in both T_o and k_t for the hoist mechanism, but that mechanism does not often produce strong dynamic effects. The travel drive mechanism usually induces a strong influence. Lastly, using the shortest length of boom consistent with maintaining adequate clearances will increase M_r because the effective dead load moment will increase, and S_h will decrease as well.

Example 4.3 To demonstrate dynamic stability effects, we will consider a crane similar in size to the one used earlier in studying out-of-level effects, but for this evaluation the crane will be a crawler type machine with a shorter boom [250 ft (76.2 m) v. 320 ft (97.5 m)] at a shorter radius [70 ft (21.3 m) v. 250 ft (76.2 m)], and hence a heavier load. The parameters for this crane are

Lifted load

$W = 62{,}800$ lb (28.5 t) $\quad d = 58$ ft (17.7 m) $\quad h_t = 249.4$ ft (76.0 m)

Crane wt.

$$W_m = 351{,}300 \text{ lb } (159.3 \text{ t}) \; d_m = 13.83 \text{ ft } (4.2 \text{ m})$$

$$h_m = 19.37 \text{ ft } (5.9 \text{ m})$$

Less Boom

$$W^* = 319{,}000 \text{ lb } (144.7 \text{ t}) \qquad h^* = 6.1 \text{ ft } (1.9 \text{ m})$$

$M_{\text{hor}} = 147{,}500$ lb·ft (200 kNm) from wind

Tilt limit is overturning, φ_{un}

Part of the travel path is 1.5% out-of-level (which exceeds limits set by most crane manufacturers)

Travel speed $v_1 = 1.5$ ft/s (0.46 m/s) Stopping time $t_1 = 0.5$ s

Hoist lowering speed (two parts) $v_2 = 3.0$ ft/s (0.9 m/s) Stopping time $t_2 = 0.5$ s

Loaded crane natural period $t_o = 2.25$ s

First, we will determine if the crane will be stable following a sharp travel stop.

From Eq. (4.4) $M_{\text{net}} = 351{,}300 \times 13.83 - 62{,}800 \times 58 - 147{,}500$

$$= 1{,}068{,}579 \text{ lb·ft } (1449 \text{ kNm})$$

$$S_h = 351{,}300 \times 19.37 + 62{,}800 \times 249.4$$

$$= 22{,}467{,}000 \text{ lb·ft } (30{,}461 \text{ kNm})$$

Eq. (4.5) $M_r = 1{,}068{,}579 - 22{,}467{,}000 \times 0.015$

$$= 731{,}574 \text{ lb·ft } (992 \text{ kNm})$$

$$t_1/t_o = 0.5/2.25 = 0.22$$

Eq. (4.21) $k_t = \sin(0.22\pi)/0.22\pi = 0.92 \qquad k_t^2 = 0.85$

Eq. (4.23) $T_o = 0.5\,(351{,}300 + 62{,}800)\,1.5^2/32.17$

$$= 14{,}481 \text{ lb·ft } (19.6 \text{ kNm})$$

Eq. (4.26) with φ_{un} as the limit, $\mu = 1$ and the equation simplifies to

$$M_d = (2 \times 14{,}481 \times 0.85 \times 22{,}467{,}000)^{1/2}$$

$$= 743{,}697 \text{ lb·ft } (1008 \text{ kNm})$$

$M_d > M_r$. The crane is unstable.

Measures that can be taken to make the crane stable include:

(*a*) Tighten ambient wind limitations at the site; limit wind speed to 75% of the previously considered value. $M_{\text{hor}} = 0.75^2 \times 147{,}500 = 82{,}969$ lb·ft (112.5 kNm)

$$M_r = 731{,}574 + 147{,}500 - 82{,}969$$

$$= 796{,}105 \text{ lb·ft } (1079 \text{ kNm}) > M_d \quad \text{OK}$$

(*b*) Reduce travel speed to 1.0 ft/s (0.3 m/s)

$$T_o = 14{,}481 \times 1^2/1.5^2$$

$$= 6436 \text{ lb·ft } (8.7 \text{ kNm})$$

$$M_d = 743{,}697\,(6436/14{,}481)^{1/2}$$

$$= 495{,}798 \text{ lb·ft } (672 \text{ kNm}) < M_r \quad \text{OK}$$

(*c*) Level the travel path to 1% maximum grade

$$M_r = 731{,}574 + 0.005 \times 22{,}467{,}000$$

$$= 843{,}909 \text{ lb·ft } (1144 \text{ kNm}) > M_d \quad \text{OK}$$

Had this evaluation been performed by traditional methods, i.e., using Newton's second law and the well known equation $F = ma$:

$$M_d^* = (351{,}300 \times 19.37 + 62{,}800 \times 249.4)(1.5/0.5)/32.17$$

$$= 2{,}095{,}000 \text{ lb·ft } (2841 \text{ kNm})$$

which overstates travel stop dynamic effects by about 180% of the value determined by the method presented above.

For the hoisting function $t_2/t_o = 0.5/2.25 = 0.22$ which means that k_t^2 remains 0.85.

$$T_o = 0.5 \times 62{,}800 \times 3^2/32.17 = 8785 \text{ lb·ft } (11.9 \text{ kNm})$$

$$M_d = (2 \times 8785 \times 0.85 \times 22{,}467{,}000)^{1/2} = 579{,}252 \text{ lb·ft } (785 \text{ kNm})$$

which is less than M_r. The crane is stable. Even with an instantaneous stop ($k_t^2 = 1$), the crane will remain stable. Because crawler cranes are rated at

75% of the tipping load, hoist system dynamic effects rarely furnish a threat to stability. When evaluated using traditional methods

$$M_d^* = 62{,}800 \times 58 \times (3.0/0.5)/32.17 = 679{,}341 \text{ lb·ft } (921 \text{ kNm})$$

which is an overstatement of only about 17% compared with this method.

For illustrative purposes, assume that this crane is rail mounted and tilting will be limited to a rise of 1 in (25 mm) so that the wheel flange does not lift above the rail top. The crane base dimension B = 21 ft (6.4 m). With a 1% out-of-level limit

$$\varphi_{un} = 843{,}909/22{,}467{,}000 = 0.04 \text{ rad}$$

$$\varphi_a = 1/(21 \times 12) = 0.004 \qquad \mu = 0.004/0.04 = 0.1$$

and for travel dynamic effects

$$M_d = 743{,}697/[(2 - 0.1)0.1] = 3{,}914{,}195 \text{ lb·ft } (5307 \text{ kNm})$$

What this result reveals is that M_r must be very high indeed if a crane is to respond to dynamic excitation by tilting through a very small angle. It is interesting to note that when using the traditional method for dynamic analysis, this last problem cannot be solved because the traditional method does not address tilt angle limitation.

Chapter

5

Mobile-Crane Installations

Cranes and airplanes have more in common than the sky they share, contrary to first impressions. The early history of aviation belongs to the barnstormers and bush pilots who pushed their rudimentary craft to the limits with hardly a thought for risk. Likewise, crane operation was once a seat-of-the-pants skill governed largely by the notion that lifting equipment could be used at the very boundary of stability.

The cranes and aircraft of those early days were machines with modest capabilities that responded almost like extensions of the operators themselves. The high performance characteristics of modern equipment does not, however, come without tradeoffs. One is a loss of intuitiveness—in other words, now control of the machine requires more intellect and less gut reaction. Another concerns risk of loss in the event of mishap. The early machines were relatively small and inexpensive, and pilots frequently survived crashes while crane operators usually walked away. Not so today where property damage claims can be stratospheric, and the public is often in harm's way. Moreover, society today is far less tolerant of accidents.

These factors—the intellectual demands engendered by modern machines and the often terrible consequences of accidents—have fostered the discipline of risk management. An underlying rule of this practice is that a balance must be struck between safety and economy. Successful mobile crane management follows this rule.

A successful mobile crane operation is not simply one that has been carried out without mishap. True success has been achieved only when, in addition, the operation has been executed at the lowest possible cost consistent with tolerable risk. The measurement of risk management success is imprecise, however, and may be hard to appreciate by considering only an individual operation. But, over time, repetitive accident-free and productive crane use will show up favorably on the bottom line, and that will be the true gauge of success.

There are a number of proven risk management measures, such as worker training and rigorous equipment maintenance, to name but two. However, after those prerequisites, the key risk management measure for mobile crane use is preplanning. Planning needs will vary widely from operation to operation, to be sure. Sometimes a single telephone call to order the right crane is sufficient. At the other extreme, some operations demand months of preparation.

A typical planning process will entail, at the very least, some consideration of the crane's capacity and reach, of how it will fit into the site, and of how it will be supported.

This chapter is concerned with the planning process. It offers a detailed presentation of the steps to be taken and the considerations to be addressed when choosing the crane needed to do the work, evaluating site access, calculating operating clearances and loads imposed by the crane, designing or specifying support for the crane, reviewing working load considerations, and minimizing risks. Knowing these elements of planning can help decide which are necessary for any one situation, or how elaborate the planning process must be. Mathematical procedures are introduced and explained which may not always be required, but are clearly worth the effort when they are needed. However, planning a crane installation is an art requiring extensive field experience and knowledge of cranes. The material given here can be put to good use by those who have that experience and knowledge, and can be invaluable to those who wish to gain it.

5.1 Introduction

A large construction site in a New Jersey city was located just across a long eight-lane bridge over a railroad right of way. All traffic going to the job had to pass over that city-owned bridge. True to form for the New York metropolitan area, there were no load-limit signs at the bridge.

The project builders had agreed to protect the city's interests by requiring proof that all large cranes using the bridge for access would do so without causing overload.

We designed installations for several large cranes at this site, including a 250-ton (227-t) truck crane to be used for placing precast concrete facing units. Review of the bridge drawings quickly showed that this crane could not pass over in normal travel configuration; as a matter of fact, in order to avoid overload it was necessary to send the crane out without any part of the boom mounted, without the front bumper counterweight, and with a main counterweight section removed. Even then, with considerable weight reduction achieved, we still had to lay a path of timber pontoons and cross the bridge while all other traffic was stopped.

When the crane-rental firm was informed of the road configuration required, they expressed surprise—after all, they had been running over that bridge in full array for years!

Had we been overconservative in our restrictions? No, not at all; we had utilized every bit of posted capacity in that structure and kept stresses up to the specified limits. But the rental firm's reply helped to explain why the city had to make extensive structural repairs not long before.

In many instances, mobile crane planning should start with consideration of the transit route to the jobsite. Each individual crane model has its own unique characteristics that may affect the need to check bridge and culvert capacity, turning clearances, width, clearance under overpasses, and telephone lines. Power lines require close scrutiny and might need to be turned off, relocated, or shielded; higher voltages need more than the minimum 10 ft (3 m) clearance. Construction site gates and access roads are often too constricting for a large crane. Abrupt changes in grade can cause outriggers or the chassis of a truck crane to hang up. If site area is limited, space may need to be found to marshal trucks in the neighborhood until their loads are needed, not only for assembly of the crane, but also for the work the crane will be performing. Large, modern, heavy-capacity telescopic cranes with luffing jibs may require ten or more truckloads of components.

Site space is often at a premium, particularly at urban jobsites. A crane cannot be assembled without sufficient laydown space, unless elaborate alternate procedures are worked out and additional equipment provided. Latticed booms and luffing jibs require a generally level stretch of ground somewhat longer than the member to be assembled, and consideration must be given to the space required for the trucks delivering the sections. Often an assist crane is needed for boom and counterweight assembly. Should power lines, or similar dangers, be present at or adjacent to the site, proper distances must be allowed for both the assist and main cranes.

Most jobsites have ample space to permit most cranes to move about unhindered. But often a crane must be positioned in a constricted space, among obstacles, or confined by structures under the ground. Under such difficulties, the positioning of the crane to do work is the crux of making a proper and efficient installation. Here is the meeting point of job needs: positioning defines crane capabilities and limitations, site-configuration constraints on swing, reach, and load positioning for hookup, and, most important, site conditions as they affect or control support of the crane.

Reach and capacity have to be sufficient to pick up the loads, swing them, and place them where needed; the path of both load and boom must be clear of obstructions throughout their range of movement during the entire operation. The boom must be long enough to raise the load to the required height without danger of collision between the load and the boom, or between the boom or load and any structure or other object. On occasion it will be necessary to check counterweight and gantry swing clearances, and to verify that adequate space is available to maneuver the crane into place. And, after completing the work, the crane must be able to extricate itself and lower its boom for dismantling.

Constraints affect productivity. For example, if a crane hook cannot place material throughout the area where it is needed, there will be a cost incurred in moving the material by other means.

The following sections offer mathematical solutions to many installation planning problems, but mathematics alone will not solve the problems. A theoretical understanding of constraints should be tempered with an awareness of how these limitations play out at the site—how they ultimately affect safety and productivity. Site visits are important, but situations at construction sites can change quickly, particularly those concerning access, obstructions, and ground conditions.

Computer programs are now available to assist in planning; these are useful accessories to the planning process. They offer speed, because crane data is in computer memory, and graphic output that promotes three dimensional visualization of the operation. But all computer programs attempt to emulate real life situations by means of mathematical simulation, and therefore are not perfect. Planners using programs consequently must become alert to their limitations and take those limitations into account.

Whether planning is done using computer simulation or the mathematical solutions given in this chapter, planners must note that the pure world of mathematics does not fully reproduce the real world of the jobsite. Cranes will not be positioned accurately to the inch, indeed, often not to the foot. Buildings are not made in a machine shop; their tolerances may be measured in inches, and temporary items may be in place projecting beyond the clean outlines shown on the design drawings. When cranes swing and raise or lower loads, inertial forces cause parts of the crane to displace from their ideal static positions.

Although the computer or the calculator may yield clearance numbers to several decimal places, those numbers are theoretical and imply precision that seldom exists in the real world. Be pragmatic and do not let those presumably accurate numbers permit you to cut your installation parameters too close. Otherwise you will find it necessary to make last minute changes on the fly, adding risks whereas you had been intending to reduce them. Computer images and mathematical solutions are only inputs for the truly essential planning ingredient: an imaginative mind. A learner with imagination can develop viable proposals and options to consider for installation planning, but an experienced individual is needed for the final decisions.

5.2 Transit to the Site

Given that each crane model has unique dimensions and a specific gross weight and distribution of weight among axles, and given that each state, province, or other subdivision independently promulgates road weight and dimension regulations, it follows that each crane model must be individually examined for transit moves. In addition, each route must be examined for width and height limitations as well as bridge or culvert capacities. Beyond these seemingly formidable obstacles, limitations and characteristics of the individual crane types themselves must be considered.

Although rough terrain cranes are roadable, their low road speed and uncomfortable ride limit self-powered transit moves to distances of only a few miles. For longer runs, the crane must be carried on a trailer, although recent trends have produced growing numbers of all terrain cranes which can comfortably travel at highway speed.

The gross weights of rough terrain cranes are low enough to allow passage over most roads and bridges. Likewise, widths and heights do not often prove troublesome. Many models are provided with four-wheel steering, making them very maneuverable and enabling them to be operated in tight quarters—even within buildings if the floors are able to support the weight.

Crawler cranes are capable of self-powered travel about the jobsite, but for transit they must be trucked or carried by rail. Smaller cranes can be driven directly onto drop-bed trailers, often with the basic boom mounted, to make the move in one piece. As weight and width increase, it becomes necessary to remove more and more components in order to remain within road or rail limitations. For rail transit, railroad traffic people need to be consulted, as limitations vary with the routing. To move the largest crawler cranes, 15 or more truck trailers or as many as 11 rail cars may be needed. Many new crawler models have been designed for ease of assembly, and some for self assembly.

Truck cranes perform all but exceedingly long transit moves under their own power. Most cranes are designed to be extremely flexible in this regard, with as many as 15 or 20 individual components removable for axle-load reduction or balance.

Manufacturers of truck cranes publish tables of axle-load adjustment data (Fig. 5.1). Although format varies between manufacturers,

Use table below to determine weight adjustments to conform with local highway regulations. Item 1 or Item 1A is the base figure (total weight). From this item deduct "minus" figures or add "plus" figures shown. All figures indicate weight in pounds.

Item no.	Item	Total weight or adjustment†	Boom over front of carrier		Boom over rear of carrier	
			Front bogie	Rear bogie	Front bogie	Rear bogie
1	Complete standard machine with 50-ft boom lowered to travel position, with single sheave hook block (manual outriggers)	133,985	23,723	110,262	47,968	86,017
1A	Same as Item 1 except with hydraulic outriggers	137,065	24,178	112,887	48,432	88,642
EFFECT OF REMOVING						
2	Cast counterweight	−26,000	+11,200	−37,200	−20,600	−5,400
3	Single-sheave hook block	−745	−1,000	+255	+425	−1,170
4	50-ft boom with upper spreader, guy lines, hoist line, boom hoist line but without hook block	−7,140	−12,800	+5,660	+10,160	−17,300
5	25-ft boom upper section with guy lines	−3,275	−7,250	+3,975	+7,425	−10,700
6	25-ft boom lower section with spreader	−3,430	−3,470	+40	+2,640	−6,070
7	Front outrigger beams (manual or hydraulic)	−2,240	−1,340	−900	−1,340	−900
8	Rear outrigger beams (manual or hydraulic)	−2,240	+700	−2,940	+700	−2,940
9	Front outrigger housing	−2,000	−1,200	−800	−1,200	−800
10	Rear outrigger housing	−2,000	+625	−2,625	+625	−2,625
11	Front hydraulic extension cylinders	−180	−110	−70	−110	−70
12	Rear hydraulic extension cylinders	−180	+60	−240	+60	−240
13	Front hydraulic vertical cylinders	−960	−560	−400	−560	−400
14	Rear hydraulic vertical cylinders	−960	+290	−1250	+290	−1,250
15	Four aluminum outrigger floats	−440	−185	−255	−185	−255
16	Pin connected rear frame section	−2,080	+640	−2,720	+640	−2,720

EFFECT OF ADDING						
17	Boom backstops	+1,420	+1,075	+350	−550	+1,975
18	Two-sheave hook block instead of standard	+840	+2,520	−1,680	−2,760	+3,100
19	Three-sheave hook block instead of standard	+1,270	+3,280	−2,550	−2,800	+4,070
20	Four-sheave hook block instead of standard	+1,820	+5,470	−3,650	−4,800	+6,620
21	Fairlead	+850	+430	+420	−90	+940
22	Aluminum front fenders instead of standard	−350	−300	−50	−300	−50
23	Aluminum rear fenders instead of standard	−580	0	−580	0	−580
24	Tagline winder	+650	+980	−330	−750	+1,400
25	Front bumper counterweight	+14,000	+19,000	−5,000	+19,000	−5,000
26	Cummins NH220 engine	+600	+600	0	+600	0
27	Caterpillar 1673 engine	+300	+300	0	+300	0
28	Hydraulic outriggers instead of manual outriggers	+3,080	+445	+2,625	+455	+2,625

†Minus figures indicate weights to be deducted from adjusted Item 1 or Item 1A figure at head of its column. Plus figures indicate weights to be added to adjusted Item 1 or Item 1A figure at head of its column.

Boom Length	A	B	C	D	E	F	G	H	J	K	L
50′ 0″	16′ 10 3/16″	53′ 4″	12′ 11½″	71′ 11 3/16″	50″	6′ 11¼″	42″	19′ 2″	7′ 1¼″	76½″	32′ 6¼″

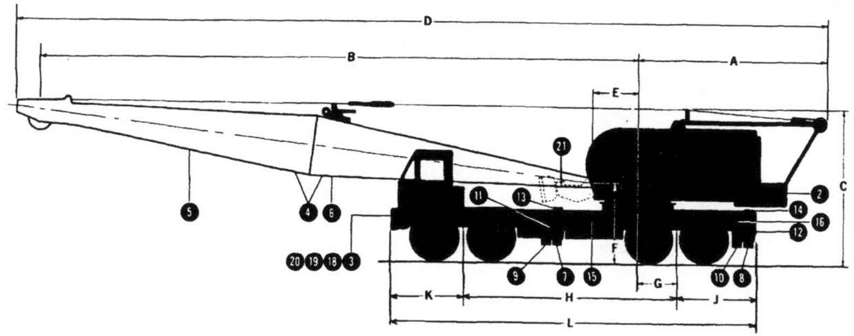

Figure 5.1 Typical truck crane axle-loading adjustment chart and general dimension diagram. (*Harnischfeger Corporation.*)

the general scheme of adjustments is the same; the tables are used to find just which, if any, components need to be removed to satisfy axle loading regulations. They also can be used, with fully or partially assembled cranes, to give the axle-load information necessary to check on-site travel moves or moves over critical bridges, ramps, or floors. For this work, additional data and some calculations beyond addition and subtraction may be required.

For the crane of Fig. 5.1, item 1A gives the gross weight as 137,065 lb (62,172 kg) for a standard machine with a 50-ft (15.2-m) boom mounted and with hydraulic outriggers. Under the heading "Effect of Adding," the weights of optional equipment are listed.

Assume that it is necessary to have nearly uniform loading on each axle. This can be accomplished by placing the boom over the rear of the carrier. Referring to Fig. 5.1, we can make the following adjustments:

	Item no.	Front bogie, lb	Rear bogie, lb
Basic machine	1A	48,423	88,642
Remove	4	+10,160	−17,300
	8	+700	−2,940
	10	+625	−2,625
	12	+60	−240
	14	+290	−1,250
	16	+640	−2,720
Final loads		60,898 (27,623 kg)	61,567 (27,926 kg)

The axle loads are balanced to within about 1% by removing the boom and the entire rear outrigger assembly. To produce minimum transit weight, removal of items 2, 7, 9, 11, 13, and 15 will reduce the front bogie load by 23,995 lb (10,884 kg) and the rear bogie load by 7,825 lb (3,549 kg). This will yield a transit weight of 90,645 lb (41,116 kg), or a reduction of about one-third from the original weight. Further reduction is possible by undecking the superstructure and moving that component separately. Some of the larger machines are specially designed for quick undecking.

Crane components carried on flat bed or lowboy trailers are usually wide loads, and may be high as well, often requiring a check of overhead clearance at underpasses. Truckers need to be aware that latticed booms, jibs, and extensions do not get along well with chains. Chain tightening can bend, nick, or wrinkle sections and seriously reduce their ability to support load. Even nylon strap tie downs can bend boom or jib chords. Therefore tie downs should be located only where the chords are supported by diagonal members, and padding should be inserted between chains and load supporting members. Some experienced riggers visit local carpet shops to get cut-offs and remnants, using them as padding.

5.3 Travel on Site

On-site roads are usually unpaved, and often poorly compacted, uneven, and badly graded. That difference implies a need for definite travel paths and oftentimes some corrective earthwork. Only rough terrain and all terrain cranes are capable of negotiating uneven and soft ground. Crawler cranes with short booms may also be able to travel over bad ground, but then the boom must be positioned at a low angle to avoid bouncing backward and into the boom stops. An alternative to compacting soft ground would be timber mats, steel plates, or a layer of crushed stone, gravel, or even brickbats to firm up the surface and spread wheel or track loads to a wider area. Availability and cost of the materials and equipment to place them are the only parameters that need to be considered, provided the result will be ample support for making the move without damaging the crane or bogging down.

Sometimes a crane must travel with a load on the hook, an operation that requires a higher standard of ground surface preparation. Avoidance of side slope is of particular importance. Travel under load will be covered later in this chapter.

The place of assembly for a large truck crane may need to be some distance from the operating location. Most of the rules for travelling under load apply to assembled long-boom cranes with an empty hook. In addition, the boom angle and orientation (over front or rear) may need to be considered with respect to axle loads. The goal is to optimize loadings without removing components, but equal bogie loadings often are not the optimum travel arrangement. Many truck cranes are built with more axles supporting the rear than the front, and the optimum loading will be proportioned to axle capacities. Generally, the front axle(s) should be favored so that the steering mechanism is not overburdened. When in doubt, check with the manufacturer before making the travel move.

By adjusting the boom angle, boom weight can be utilized to shift load between the front and rear bogies as needed to optimize axle loadings. The example below shows how to equalize axle loadings, but the same method can be used to attain desired unequal loadings as well.

By taking the axle loadings for the full crane less basic boom and adding the effect of the boom length actually mounted (by taking the moment of the boom weight about either bogie) it is possible to determine the optimum boom angle for travel. With very short booms it may not be possible to balance axle loads, but with long booms balance can be achieved when the boom is placed over the front of the carrier. Axle loadings may be quite high, but once they are known, the travel path can be prepared to accommodate them (Fig. 5.2).

Example 5.1 The crane of Fig. 5.1 is to be operated with 180 ft (54.9 m) of boom. It will be equipped with hydraulic outriggers, front bumper counterweight, and boom backstops. What boom angle must be maintained for the crane to travel with balanced axle loads while fully assembled? What will be the axle loading? The boom (including guy lines and spreader) weighs 17,100 lb (7756 kg), and the boom CG is located 85 ft (25.9 m) from the boom foot pin measured along the boom centerline.

For the boom over the front of the carrier:

	Item no.	Front bogie, lb	Rear bogie, lb
Basic crane	1A	24,178	112,887
Add boom backstop	17	+1,075	+350
Add bumper counterweight	25	+19,000	−5,000
Remove basic boom	4	−12,800	+5,660
Total crane less boom		31,453 (14,267 kg)	113,897 (51,663 kg)

For balanced axle loads, each bogie must carry (including boom weight)

$$(31{,}453 + 113{,}897 + 17{,}100)(\tfrac{1}{2}) = 81{,}225 \text{ lb } (36{,}843 \text{ kg})$$

Figure 5.2 The soft access road could not support the wheel loads imposed by this crane. An attempt to right the crane by using the outriggers also failed because the load was not spread to a large enough bearing area.

This requires that the front bogie loading be increased by

$$81{,}225 - 31{,}453 = 49{,}772 \text{ lb } (22{,}576)$$

as its share of the boom effect.

Crane dimensions are given in the diagram in Fig. 5.1. The distance from the axis of rotation to the centerline of the rear bogie is 42 in (106.7 cm) and from boom foot pin to axis of rotation is 50 in (127.0 cm); the wheelbase is 19 ft 2 in = 19.17 ft (5.84 m). Taking moments about the rear bogie, we have

$$49{,}772 \text{ lb} = 17{,}100 \text{ lb} \frac{(42 \text{ in} + 50 \text{ in})/(12 \text{ in/ft}) + (85 \text{ ft})(\cos\theta)}{19.17 \text{ ft}}$$

$$= 892.02(7.67 + 85 \cos\theta)$$

$$\cos\theta = 0.5662 \qquad \theta = 55.5° \text{ boom angle} \qquad \textit{Ans.}$$

For a two-axle bogie, axle loads will be

$$\tfrac{1}{2}(81{,}225) = 40{,}613 \text{ lb } (18{,}422 \text{ kg}) \qquad \textit{Ans.}$$

The result can be checked by taking moments about the front bogie

$$\frac{17{,}100(85 \cos 55.5° + 7.67 - 19.17)}{19.17} = 32{,}688 \text{ lb } (14{,}827 \text{ kg})$$

This should equal the required reduction in rear bogie loading of 113,897 − 81,225 = 32,672 lb (14,820 kg), which it very closely does.

When checking or designing structures that must support travel loadings, it is often prudent to increase the calculated axle loads by 10 or 15%. This will allow the actual boom angle to vary somewhat from the calculated exact value and will account for dynamic effects. Impact can be minimized by providing smooth travel-path surfaces and by maintaining very low travel speeds, but wind effects and braking action cannot be eliminated. Needless to say, moves of long-boom cranes may have to be postponed when the wind is up, not only for the effect on axle loads but more importantly because of concern for sideways stability.

Truck cranes have a relatively narrow travel base, and the tires bring about considerable "give"; these machines are therefore sensitive to transverse out-of-level conditions and wind. Because of that, during moves of long-boom truck cranes many supervisors add an element of security by extending the outriggers fully. The floats are lowered so that they skim about 2 in (50 mm) clear of the ground surface. For this procedure, the crane must travel at very slow speed, the travel path must be firm and very well graded, and watchers must be posted at each outrigger float. With these precautions the crane can be moved with reasonable assurance that a gust of wind or a soft spot in the travel path will not cause the crane to turn over, but forward motion must be stopped quickly when a float touches the ground.

When traveling long-boom crawler cranes, a balance check can be made using procedures given later in this chapter. However, the counterweight usually prevails so that ground pressure at the rear of the tracks will ordinarily be much greater than pressure at the front. This is a suitable situation for travel. When the leading edge of the tracks are lightly loaded, they can advance cleanly without the resistance offered by tracks depressed into the soil. When making turns, the tracks can slide over the ground surface rather than push the soil out of the way. Machines so powerful that they can manage travel up a 30% grade would not be expected to have difficulty when maneuvering heavily loaded tracks, but they do. Steering places greater strain and demand for power on crawler components than any other action. This can be appreciated while watching crawler cranes travel on jobsites.

Negotiating grades can be a particular problem for all mobile cranes unless addressed in the planning work. Whether going up or down a grade, the CG of the crane moves either to the rear or front and may affect stability fore and aft. But, stability can be controlled by adjusting boom angle and using the weight distribution methods given in this chapter. For some components it may be necessary to estimate CG height, as this information is not furnished in the crane data.

Axle weight data provided by the manufacturer (such as in Fig. 5.1) are for a level crane; they will be affected by grades. But from that data, the horizontal location of the CG of the crane (less the boom) can be determined and the CG height estimated component by component and then combined for the whole. Stability on grades can be determined with reasonable accuracy by applying the horizontal and vertical CG positions.

When a crawler crane enters a sharp downgrade (or leaves an upgrade), there is potential for sudden uncontrolled movement. As the machine approaches the transition, the leading edge of the crane's rigid base will start to project out over open space. As the crane CG passes the abrupt change, the leading edge will drop to close the gap.

To avoid damage, the crane should be stopped before the transition and be crept slowly forward over this critical point. Alternatively, after the crane is stopped just short of the break point, the operator can slowly lower the boom, which will gradually and smoothly bring the CG over the transition. However, a travel surface with a smooth gradual changeover is preferable to a break point, but not always feasible to arrange.

Example 5.2 The truck crane of Fig. 5.1, which was the subject of Ex. 5.1, will be traveling down a 20% grade with a 180 ft (54.9 m) boom over the front. Before entering the downgrade, what angle should the boom be positioned at to obtain equal axle loading while on the grade?

A 20% downgrade is a slope in which the drop is 20 units for each 100 horizontal units; in other words it is an angle whose tangent is 0.2, or 11.3°. Figure 5.1 furnishes data which can be used to find the horizontal position of the CG of the crane without its boom. Noting that the wheelbase, or distance between bogie centers, is 19.17 ft (5.84 m), measured from the rear bogie center, that horizontal distance C is

$$C = \frac{31{,}453 \text{ lb} \times 19.17 \text{ ft}}{31{,}453 + 113{,}897} = 4.15 \text{ ft } (1.26 \text{ m})$$

Figure 5.1 gives the distance from the rear bogie to the boom foot as E plus G, or 92 in which is 7.67 ft (2.34 m), and that figure must be added to the boom CG horizontal position given in Ex. 5.1.

The CG height B_h for the boom at angle θ to the horizontal (before entering the slope), taking account of the boom foot height of 6.94 ft (2.12 m) is

$$B_h = 6.94 + 85 \sin \theta$$

The height of the CG of the crane (less the boom) can be calculated with some accuracy by taking component by component and using an estimated CG height for each. However, let us assume that for this crane the figure is 5.0 ft (1.52 m), and note that these calculations are not very sensitive to CG height estimation errors. We now have sufficient information to balance axle loads by taking moments about the rear bogie center with the crane on an 11.3° downslope.

$$0.5 \times 19.17 \cos 11.3 = \{145{,}350(4.15 \cos 11.3 + 5.0 \sin 11.3) + 17{,}100[(7.67 + 85 \cos \theta)\cos 11.3 + (6.94 + 85 \sin \theta)\sin 11.3]\}/(145{,}350 + 17{,}100)$$

which simplifies to

$$1.0 = 2.22 \cos \theta + 0.44 \sin \theta$$

Solving by iteration gives 75° which is the boom angle before the crane enters the slope that will produce uniform axle loadings on the slope. While on the slope, the boom will be at 75 − 11.3 = 63.7° to the horizontal.

Had the height of the CG for crane less boom been estimated at 4 ft (1.22 m) instead of 5 ft (1.52 m), the boom angle would come to 74°. This result supports the assertion that the calculation is not very sensitive to the CG height estimation for the crane less boom.

5.4 Lift and Swing Clearances

It is embarrassing, to say the least, to send a crane to a job only to find that it is incapable of placing the loads where needed. There is seldom an excuse for this kind of failure which is usually the result of inadequate planning. Some accidents occur because field crews try to make do with equipment that is not adequate for the work at hand.

Lift clearances

Most crane rating charts provide boom-point height data as well as rated load for each combination of boom length and operating radius permitted. The maximum achievable hook height can be quite a bit less than this figure; however, the minimum distance from boom point to hook is often listed in the documentation. When those data are not given, a generous allowance should be made, 4 to 8 ft (1.2 to 2.4 m) for smaller cranes and 6 to 18 ft (1.8 to 4.6 m) for large machines. Not only does this figure vary with boom size and style but it is a function of load-block capacity and reeving as well. Some smaller cranes, however, require 16 ft (4.9 m) of clearance, which shows that it is advisable to use actual values whenever possible instead of relying on an arbitrary allowance.

For loads that must be placed above grade (on a roof, for example), achievable hook height needs to be great enough to accommodate the height of the load and of the slings or other lifting equipment and to permit handling clearance as well. When calculating the height needed, it may be necessary to check at an intermediate as well as at the final placement radius, such as while passing over a parapet wall. Needless to say, height clearance between the load and any obstruction must be allowed; this is a matter for judgment, but 6 ft (2 m) should be the minimum at longer boom lengths where a signalman is not in position to watch the load closely. For shorter boom lengths or when a signalman can check the clearance, 2 ft (0.6 m) may be acceptable.

If sufficient clearance is not provided, the hook block may strike the boom head because the operator's attention is focused either on the signalman or on the point where the load comes closest to the obstruction. Even if the hook block is visible to the operator, its proximity to the boom point is difficult to judge from the operator's cab.

There is great danger in overraising the hook block, called *two blocking*. Mobile-crane winches must be very powerful to satisfy the needs of both very short and very long booms. They may be capable of enough force to break the hoist line. If the operator is not able to sense the change in engine tone as it starts to lug with the increased load following two blocking, the line can break and the load may drop. With a light load on the hook, the winch can pull the boom back over the cab. While new cranes are furnished with devices to sense and prevent two blocking, proper planning will avoid the need for such protection. An anti-two-block device can prevent an accident, but not the cost and lost time of having to unexpectedly add boom sections.

Rating charts do not indicate jib-point elevations, but the crane documentation will include a *range diagram* (Fig. 5.3), a copy of which is often mounted in the crane cab as well. A range diagram is a side view of the crane marked off with horizontal height lines and vertical radius lines. In addition, there are arcs that trace the position of the boom tip (for each boom length) as radius changes; also shown are arcs for each jib length. Jibs may be mounted parallel to the boom or at an angle to it (usually not more than 45°, although some manufacturers limit *offset angles* to 30° maximum). Jib offset positions are also shown on some range diagrams. Finally, radial lines emanating from the boom foot indicate boom angles as a further help in evaluating boom position. For most work, the rough data obtainable with a range diagram are sufficient to verify that the job can be done and that boom and jib lengths and jib offset will offer satisfactory clearance.

When conditions appear to be too close to permit reliance on the range diagram, calculations or larger-scale drawings must be made. Boom and jib templates can prove to be very useful for this work. The reach of the crane in Fig. 5.4 is clearly limited by clearance between the boom and wall top.

Swing clearances

It is not difficult to study and check clearances when the boom is at right angles to a wall of a building or other obstruction. But more often than not either the crane must be placed too close to the wall to permit a swing to 90° or several loads must be placed at different locations. With the boom at a horizontal angle to the wall and a vertical angle to the ground, the clearance problem now becomes trigonometrically complicated. Although some of the following derivations are rather difficult, they are included in the general text because of the importance of the results.

Assume a crane with boom of length L, width B, and depth D. In the notation of Fig. 5.5, the operating radius will be

$$R = t + L \cos \theta \tag{5.1}$$

when boom angle θ is known. Conversely, when the radius is known

$$\theta = \cos^{-1} \frac{R - t}{L} \tag{5.2}$$

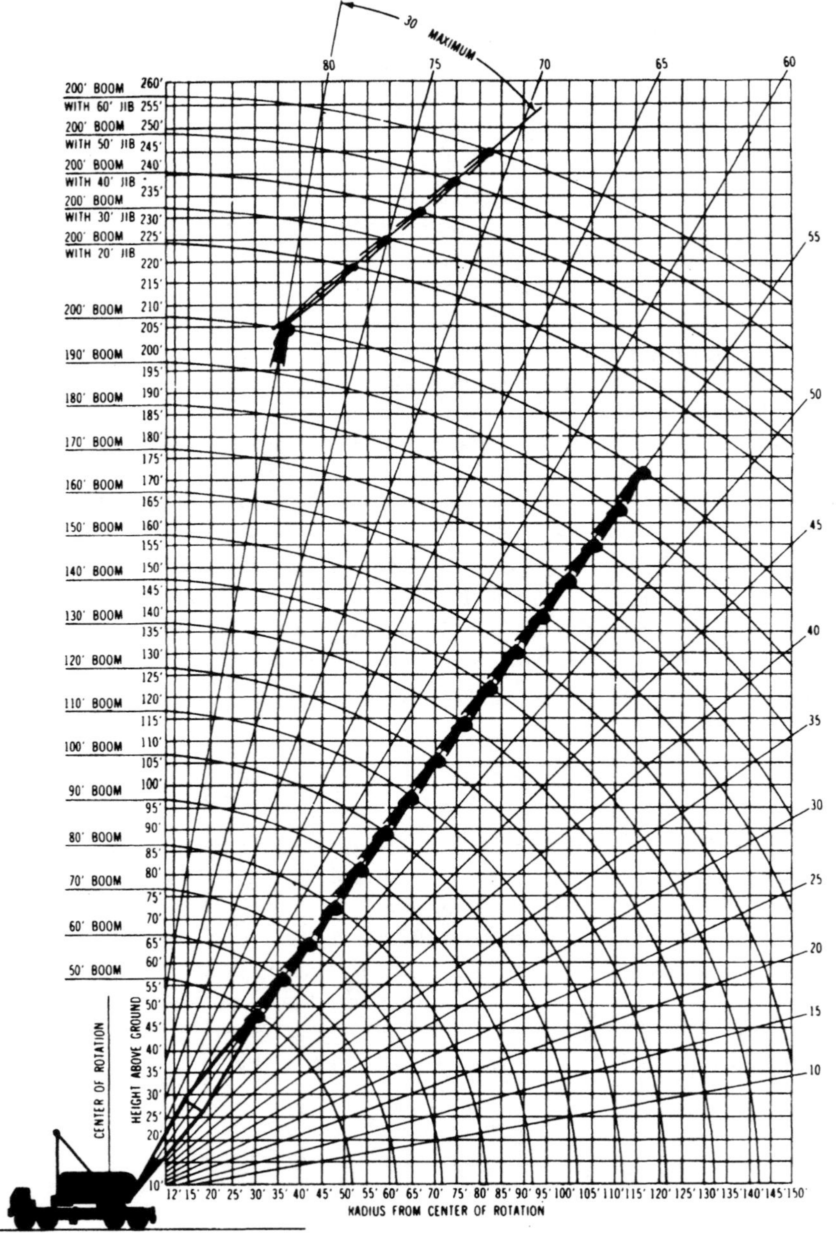

Figure 5.3 Typical range diagram for a latticed boom crane. Such diagrams are adequate for most height and swing clearance checks. (*Harnischfeger Corporation.*)

Figure 5.4 A GCI crane with 125-ft (38-m) latticed boom and 20-ft (6-m) jib reaches over the containment structure wall to place loads inside this Midland, Michigan, nuclear plant. The reach is clearly limited by clearance between the boom and the wall top. (*General Crane Industries, Ltd.*)

For telescopic cantilevered booms where the boom foot pin is behind the axis of rotation, t is taken as negative.

The horizontal angle φ the boom makes with the wall or other obstruction is given by

$$\varphi = \sin^{-1} \frac{g + e}{R} \tag{5.3}$$

The dimension e, it should be noted, represents the distance the load must be placed beyond the face of the wall.

The load-clearance problem, with $e = 0$, can be dealt with as in the previous section, as it is no more than a matter of checking heights. But as the boom passes over the edge of the wall, the one boom chord that is closest to the wall must be kept suitably clear. The main longitudinal boom members, or chords, support high-magnitude compressive forces, particularly when the crane is under load, so that even a small lateral thrust, such as from bumping the wall edge, could induce collapse.

The line that measures the shortest distance between the closest boom chord and the wall edge is perpendicular to both of those lines,

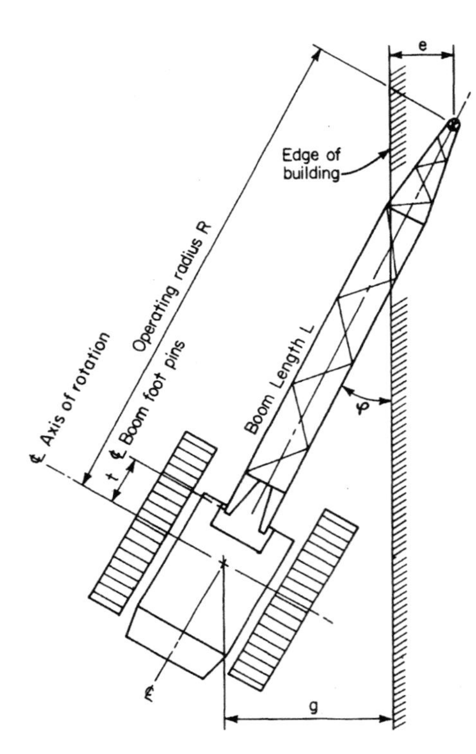

Figure 5.5 Plan view of a crawler crane showing the position parameters used for clearance calculations when the boom is at an angle to the wall.

but the three lines do not lie in one plane. Let the length of that clearance line be denoted as C. The direction cosines (Fig. 5.6) for the clearance line are

$$\alpha_c = \frac{x_c}{C} \qquad \beta_c = \frac{0}{C} = 0 \qquad \gamma_c = \frac{z}{C}$$

referring to the X, Y, and Z cartesian axes respectively. For the boom chord (see Fig. 5.6), let l be the length of the line scribed on the surface running from a point at the height of the wall edge to the point of intersection with the clearance line. The direction cosines of this line are

$$\alpha_b = \frac{-x_b}{l} \qquad \beta_b = \frac{y_b}{l} \qquad \gamma_b = \frac{z}{l}$$

Expressing the projections of the boom chord line onto the cartesian axes in spherical coordinate form,

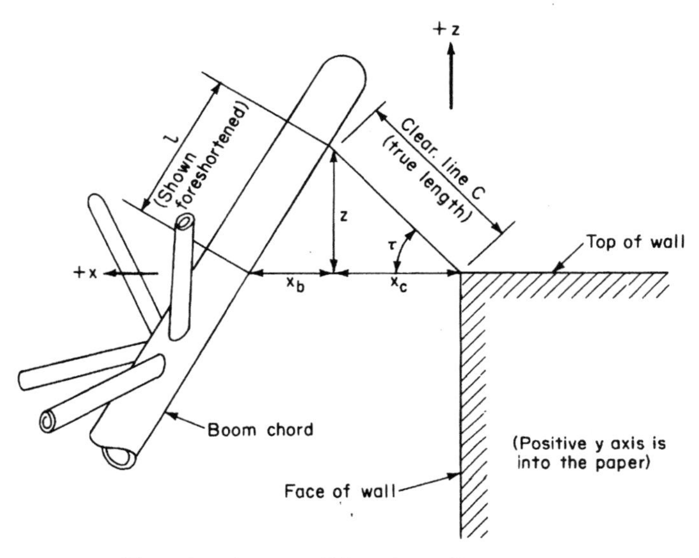

Figure 5.6 Elevation view parallel to the wall.

$$x_b = -l \sin \varphi \sin (90° - \theta) = -l \sin \varphi \cos \theta$$

$$y_b = l \cos \varphi \sin (90° - \theta) = l \cos \varphi \cos \theta$$

$$z = l \cos (90° - \theta) = l \sin \theta$$

permits the direction cosines to be restated as

$$\alpha_b = -\sin \varphi \cos \theta \qquad \beta_b = \cos \varphi \cos \theta \qquad \gamma_b = \sin \theta$$

But since lines l and C are normal, they must conform to the requirement that

$$\alpha_b \alpha_c + \beta_b \beta_c + \gamma_b \gamma_c = 0$$

$$-\frac{x_c}{C} \sin \varphi \cos \theta + \frac{z}{C} \sin \theta = 0$$

$$\frac{z}{x_c} = \frac{\sin \varphi}{\tan \theta} = \frac{g + e}{R \tan \theta}$$

Let the angle the clearance line makes with the horizontal (Fig. 5.6) be τ. Then

$$\tau = \tan^{-1}\frac{z}{x_c} = \tan^{-1}\frac{\sin\varphi}{\tan\theta} = \tan^{-1}\frac{g+e}{R\tan\theta} \tag{5.4}$$

The horizontal line between the top of the wall and the closest boom chord (lying in the same vertical plane as C) has length k (Fig. 5.7), comprising the segments (Fig. 5.6)

$$k = -x_b + x_c = l\sin\varphi\cos\theta + C\cos\tau$$

$$= \frac{l\sin\varphi\sin\theta}{\tan\theta} + C\cos\tau$$

but $l\sin\theta = z = (x_c\sin\varphi)/(\tan\theta) = (C\cos\tau\sin\varphi)/(\tan\theta)$. After substitution and manipulation this becomes

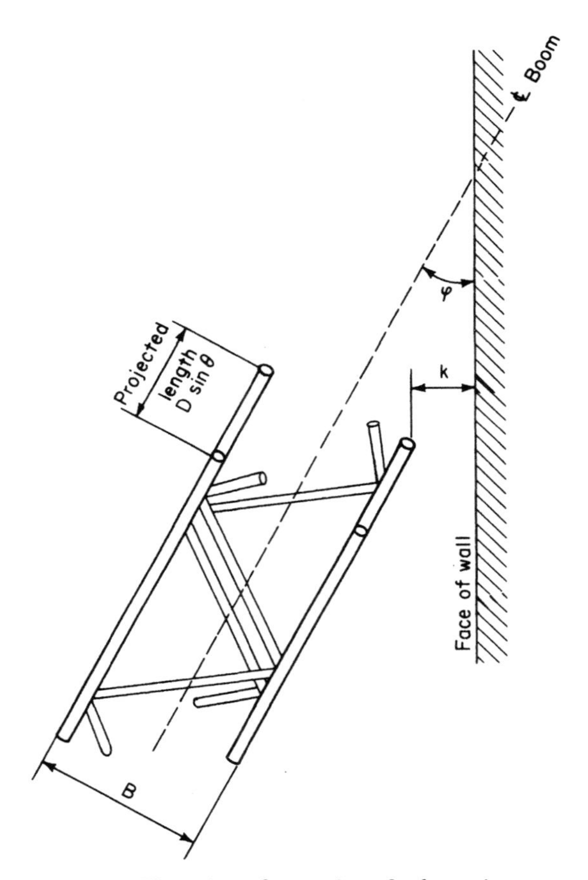

Figure 5.7 Plan view of a portion of a boom in proximity to a wall.

$$k = \frac{C}{\cos \tau} \tag{a}$$

From Figs. 5.5 to 5.8 it can be seen that the length k can also be described in terms of readily observable dimensions and angles as

$$\begin{aligned} k &= \left[R - t - \frac{H - h + C \sin \tau + (D/2) \cos \theta}{\tan \theta} - \frac{D}{2} \sin \theta\right] \sin \varphi \\ &\quad - \frac{B}{2} \cos \varphi - e \\ &= \left(R - t - \frac{H - h}{\tan \theta} - \frac{D}{2 \sin \theta}\right) \sin \varphi \\ &\quad - \frac{B}{2} \cos \varphi - e - \frac{C \sin \tau \sin \varphi}{\tan \theta} \end{aligned} \tag{b}$$

By taking k as given in Eq. (b), substituting into Eq. (a), and rearranging we find that the clearance C is given by

$$C = \frac{1}{A}\left[\left(R - t - \frac{H - h}{\tan \theta} - \frac{D}{2 \sin \theta}\right) \sin \varphi - \frac{B}{2} \cos \varphi - e\right] \tag{5.5}$$

where

$$A = \frac{1 + \sin^2 \tau}{\cos \tau} \tag{5.6}$$

For a crane with given dimensions L, D, B, t, and h and for known installation parameters R, e, g, and H, it is possible to solve for boom clearance C by using Eqs. (5.1) to (5.6). Once the clearance is calculated, however, its acceptability is a matter of judgment. The factors entering into the judgment equation include the accuracy of the field data, the length of the boom, visibility of the clearance point to the operator or signalman, wind speed, and operator experience. Other factors may have to be considered, such as when setting a load in a precise location which usually requires additional maneuvering room. We cannot call on mathematics to decide these issues because some of the constraints on clearance judgments are not quantifiable, and others are not known in advance. Therefore, a review of the constraints will be helpful.

As mentioned earlier in this chapter, positioning of mobile equipment is not precise, and obstruction dimensions rarely match exactly those shown on drawings. Dimensional questions alone make clearance calculations a theoretical exercise, and indicate that calculated results must be used with discretion. Also, the longer the boom, the

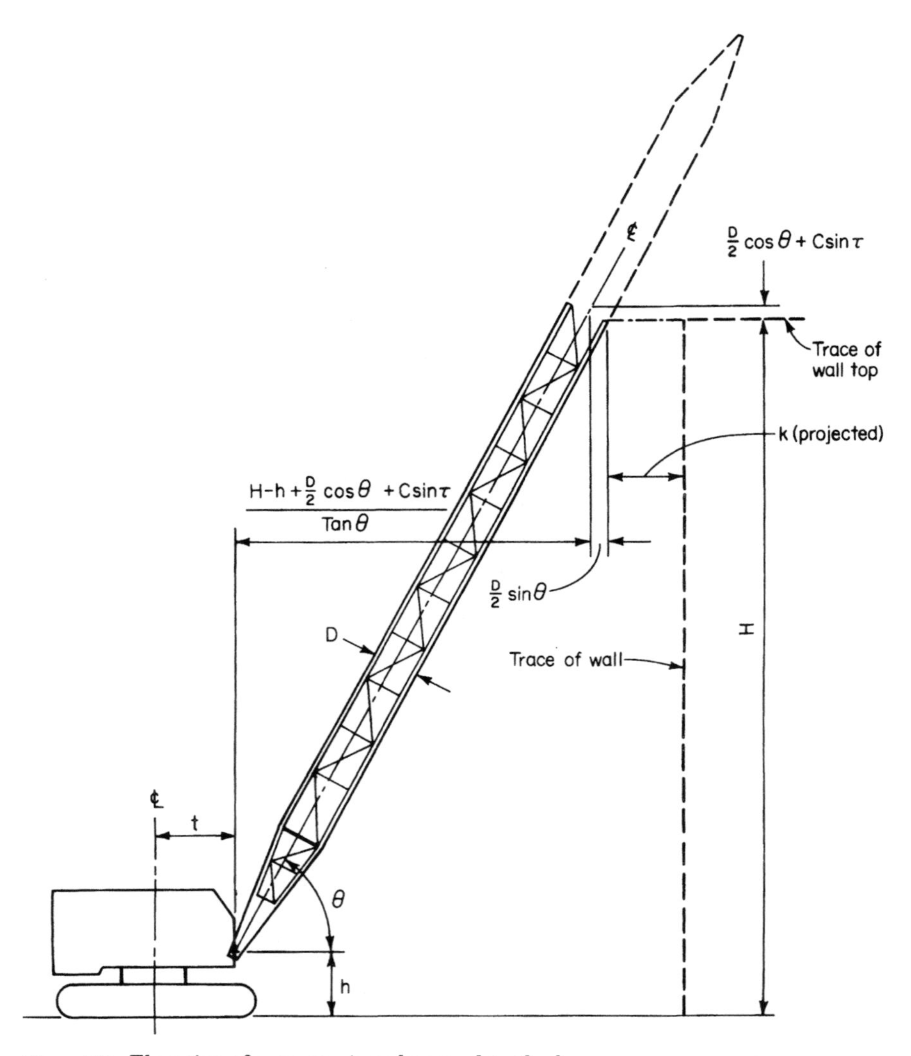

Figure 5.8 Elevation of a crane viewed normal to the boom.

greater its tendency to move about, varying from its theoretical position as a result of crane and load movements. Working crane booms are never still, even in calm air. Furthermore, the visibility of the clearance area is of critical importance for avoiding contact. Needless to say, if the signalman is close by the clearance point and in clear view of the operator or in continuous radio contact, control will be optimized and clearances can be held to relatively small values. The operation plan must stipulate the signalman's position when conditions dictate the use of small clearances. Radio signaling protocols need to be clear, positive, and understood by the participants. And lastly, wind speed will not be known in advance, but when the load has a large wind area it will be a strong factor in deciding clearance needed, or reason for postponing a particular lift. It may be a hard decision for field supervisors to make, but delaying the work is surely preferable to collapsing the boom and dropping the load.

For job planning, making clearance judgments is unavoidable. As a rule of thumb and as guidance for making these difficult decisions, 2 ft (0.6 m) of clearance should prove adequate for short booms and 4 ft (1.2 m) for long booms. With a signalman in close proximity for monitoring clearance and operating speeds kept low, these values can be cut in half, but only in conjunction with a strict limitations concerning wind speed.

With all parameters but H given and a value for C preselected, by manipulation of Eq. (5.5) the maximum permissible wall height is given by

$$H = h - \frac{D}{2 \cos \theta} + \left(R - t - \frac{CA + e}{\sin \varphi} - \frac{B}{2 \tan \varphi} \right) \tan \theta \tag{5.7}$$

where A is determined from Eq. (5.6). Equation (5.7) is useful for the construction planning of high-rise concrete structures. It determines the maximum floor height that can be serviced by the boom and in this way indicates when a jib or increased boom length will be needed. The changeover point is relevant not only for the timing of the installation work but because jib capacity often necessitates reduction in the size of the concrete bucket used. Therefore, the changeover can also mark the onset of reduced productivity in concrete placement.

When all parameters are known except the boom length, which needs to be chosen, the following equation can be used:

$$L = \frac{H - h}{\sin \theta} + \frac{D}{2 \sin \theta \cos \theta} + \frac{(CA + e)/\sin \varphi + B/(2 \tan \varphi)}{\cos \theta} \tag{5.8}$$

but, of course, the boom angle θ is unknown. Therefore, start with a guess of boom length and find θ from Eq. (5.2). Proceed through Eqs. (5.3) to (5.6) and (5.8) and calculate a value for L. Rounding out the calculated value to the next longer standard boom length, repeat the calculations until it is certain that the last standard boom length used is correct. If the trial value of boom length is the next standard increment above the calculated value, the shortest boom length giving the specified clearance has been found.

The attitude of the load during placement (with respect to the boom) must also be considered, as it may be the critical clearance factor.

When using a crane with a tower attachment, for close-in operations, a midboom hoist line offers increased load placement flexibility (Fig. 5.9) for cranes so fitted.

Swing clearance with jib mounted

If the boom point remains above the top of the obstruction, the problem is quite simple. Taking a jib of length J, width b, and depth d that is mounted at an offset angle μ to the boom gives

$$R = t + L \cos \theta + J \cos (\theta - \mu) \tag{5.9}$$

When the radius is given and the boom angle θ must be found, this can be done by iterating

$$\frac{R - t}{L} = \cos \theta + \frac{J}{L} \cos (\theta - \mu) \tag{5.10}$$

until θ has been determined to satisfactory accuracy (about 0.1° for long booms at high angles). Once θ is known, calculations can proceed as for a boom alone; it will be *boom* not *jib* clearance that governs.

When the boom point is below the obstruction height, the angle the jib clearance line C_j makes with the horizontal can be expressed as

$$\tau_j = \tan^{-1} \frac{\sin \varphi}{\tan (\theta - \mu)} \tag{5.11}$$

while the horizontal distance between the top of the wall and the closest jib chord is

$$k_j = \frac{C_j}{\cos \tau_j} \tag{c}$$

From Fig. 5.10, the length k_j can also be given by

$$\begin{aligned} k_j = \Bigg[& R - t - L \cos \theta \\ & - \frac{H - L \sin \theta - h + (d/2) \cos (\theta - \mu) + C \sin \tau_j}{\tan (\theta - \mu)} \\ & - \frac{d}{2} \sin (\theta - \mu) \Bigg] \sin \varphi - \frac{b}{2} \cos \varphi - e \\ = \Bigg[& R - t - L \cos \theta - \frac{H - L \sin \theta - h}{\tan (\theta - \mu)} - \frac{d}{2 \sin (\theta - \mu)} \Bigg] \sin \varphi \end{aligned}$$

Figure 5.9 At constricted urban locations, a tower attachment can provide hook access over the entire worksite until the structure reaches tower height. A midboom hoist line provides hook coverage at close reaches. The Manitowoc 4100W crane in this photograph has erected 14 stories and can continue with full access to the twentieth floor. Above that level, a hopper will have to be placed near the corner of the building and motorized buggies will distribute the concrete. Boom clearance considerations will control positioning of the hopper. (*S & A Concrete Corp.*)

$$- \frac{B}{2} \cos \varphi - e - \frac{C \sin \tau_j \sin \varphi}{\tan (\theta - \mu)} \qquad (d)$$

Using k_j as given by Eq. (d), substituting into Eq. (c), and rearranging, we find the clearance

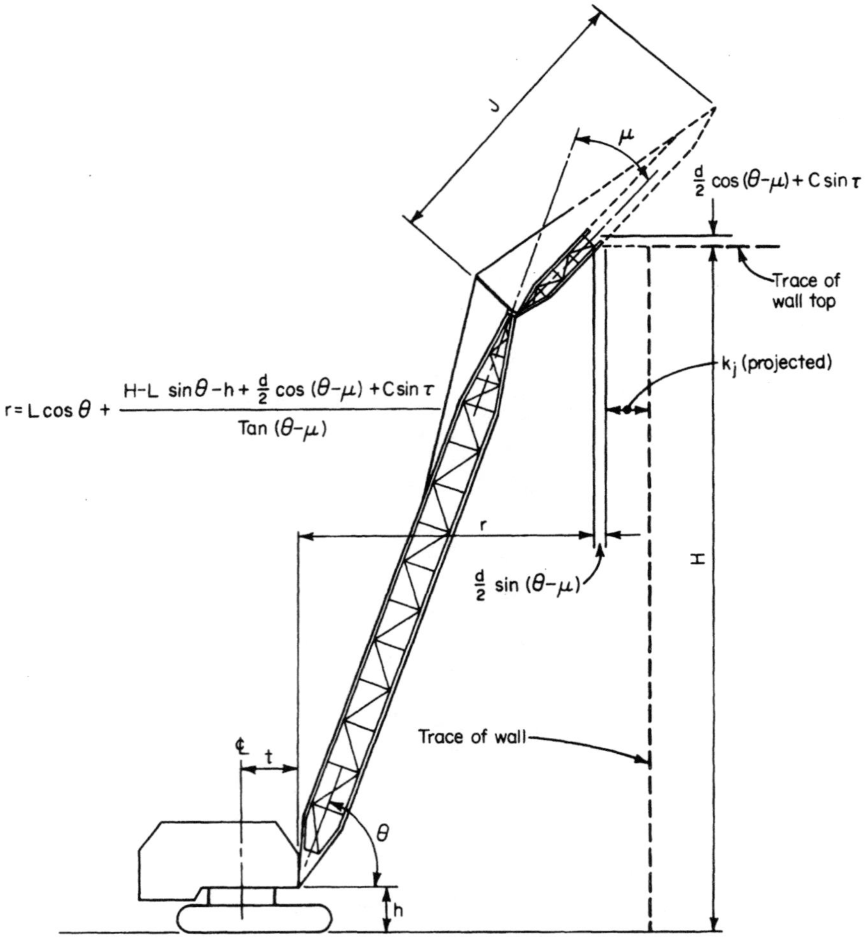

Figure 5.10 Elevation of a crane viewed normal to the boom.

$$C = \frac{1}{A_j}\left[\left(R - t - L\cos\theta - \frac{H - L\sin\theta - h}{\tan(\theta-\mu)} - \frac{d}{2\sin(\theta-\mu)}\right)\sin\varphi - \frac{b}{2}\cos\varphi - e\right] \qquad (5.12)$$

where
$$A_j = \frac{1 + \sin^2\tau_j}{\cos\tau_j} \qquad (5.13)$$

As in the case for the boom alone, given all crane dimensions and the installation parameters R, e, g, and H, it is possible to calculate jib clearance by using Eqs. (5.9) to (5.13), but the acceptability of the clearance value is again a matter of judgment, and further checks must be made, as described following Eq. (5.15) below.

Jibs are used for the single purpose of increasing lift height. In some instances, the increased height means that an obstacle can be cleared and the reach can thereby be increased, but it is the added height that makes this possible. Since with added height come added wind effects, greater elastic movement, and a reduction in operator perception, a generous clearance should be allowed, certainly not less than 4 ft (1.2 m) unless the signalperson is close by.

With a preselected value for C_j, the maximum obstruction height leaving this clearance is given by

$$H = h + L \sin \theta - \frac{d}{2 \cos (\theta - \mu)} + \left(R - t - L \cos \theta - \frac{CA_j + e}{\sin \varphi} - \frac{b}{2 \tan \varphi}\right) \tan (\theta - \mu) \quad (5.14)$$

which is convenient for use in planning high-rise construction work. It determines the floor height that can be serviced with any boom-and-jib combination, subject to further checks described below.

For jib-length selection when all other parameters are known, an iterative procedure can be used with the expression

$$J = \frac{H - h - L \sin \theta}{\sin (\theta - \mu)} + \frac{d}{2 \cos (\theta - \mu) \sin (\theta - \mu)} + \frac{(CA_j + e)/(\sin \varphi) + b/(2 \tan \varphi)}{\cos (\theta - \mu)} \quad (5.15)$$

With a trial jib length and offset angle μ, Eq. (5.10) can be iterated for θ. Using Eq. (5.15), compare the calculated value of J with the trial value. If the trial value is the next longer available jib length, the shortest acceptable jib has been found and the specified clearance will be maintained.

Misleading results can be derived from jib-clearance calculations unless clearance of the boom itself is also verified. To check boom clearance with the jib mounted (remember that the jib hook fixes boom position) new values are needed for some of the parameters. For the clearance line to the boom

$$\tau = \tan^{-1} \frac{\sin \varphi}{\tan \theta}$$

where both φ and θ are the values calculated for boom plus jib. The radius of the boom tip will be

$$R_t = t + L \cos \theta$$

and the position parameter e_t of the boom tip will be

$$e_t = e - [J \cos (\theta - \mu)] \sin \varphi \tag{5.16}$$

The parameter e_t locates the centerline of the boom, at the tip, with respect to the face of the structure. Therefore, the actual tip clearance C_t can be derived from the horizontal dimensions. Working from this point, we get

$$C_t = \frac{-e_t - (d_t/2) \sin \theta \sin \varphi - (b_t/2) \cos \varphi}{A} \tag{5.17}$$

where d_t is the boom tip dimension analogous to D, b_t is similarly defined, and A is given by Eq. (5.6). If this clearance value is not acceptable, jib clearance does not control, the calculated maximum service height is not correct, and a new height must be calculated for clearance with respect to the boom [using Eq. (5.7)].

A second point on the boom must also be checked, the point at which the tapered tip section begins (Fig. 5.11). The clearance at this point is similarly derived and is given by

$$C_{st} \frac{[L_t \cos \theta - (D/2) \sin \theta] \sin \varphi - e_t - (B/2) \cos \varphi}{A} \tag{5.18}$$

where L_t is the length of the tapered tip section. If both C_t and C_{st} have been found deficient, or if C_{st} alone is too small, the service height with adequate clearance will be given by Eq. (5.7) using e_t for e and R_t for R; R_t is the radius of the boom tip.

With both C_t and C_{st} adequate, the jib controls and the previously calculated service height is correct.

When the tip clearance C_t is not acceptable but the start of taper clearance C_{st} is, the maximum service height is controlled by clearance measured at some point between these positions on the boom. The boom dimensions at the critical height will fall between D and d_t for boom section height and B and b_t for width, say d' and b'. These values will be a function of maximum service height, which in turn will be affected by the values. The parameter d' is given by

$$d' = d_t + \frac{(D - d_t)[L - (H - h + c \sin \tau)/(\sin \theta)]}{L_t} \tag{5.19}$$

b' is given by a similar expression containing the unknown height H.

Figure 5.11 With 350 ft (107 m) of boom and 100 ft (30.5 m) of jib, the operator cannot even see the jib, let alone gauge clearances or load height. Radio or telephone communication between the operator and a signalman at the working level is essential for both productivity and safety. The building shown is being topped out at 46 floors. Note that the tapered tip section appears parallel to the building face.

Using an estimated value for H, we first find approximate values for d' and b' and insert them into the service-height equation

$$H = h - \frac{d'}{2 \cos \theta} + \left(R_t - t - \frac{b'}{2 \tan \varphi} - \frac{CA + e_t}{\sin \varphi} \right) \tan \theta \quad (5.20)$$

With this height put into Eq. (5.19) more refined values will result. Several trials may be necessary until H, d', and d' come into balance, as the convergence may be slow. The value for H thus determined, however, will be the maximum height of structure for which the specified clearance will be maintained.

Note that in the preceding analysis it is assumed that the crane boom will be crossing over a single hurdle which is the height of the wall or structure to be cleared. More complex situations are to be found in practice, such as buildings with multiple setbacks as in Fig. 5.12. The crane location on this actual installation drawing was chosen because swinging over the corner of the structure offered better reach onto the roof. Clearance at each setback was checked using the methods outlined in this section.

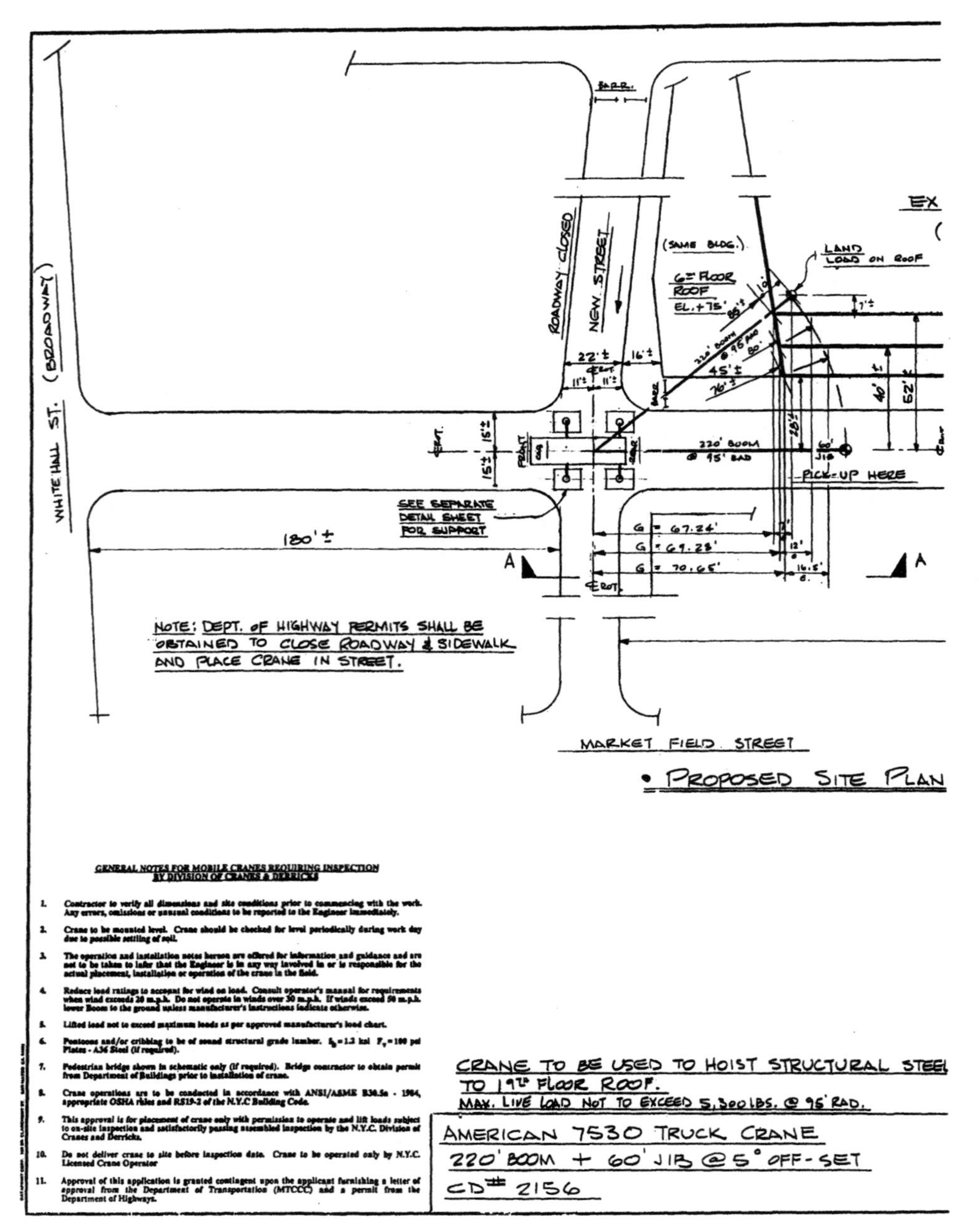

GENERAL NOTES FOR MOBILE CRANES REQUIRING INSPECTION BY DIVISION OF CRANES & DERRICKS

1. Contractor to verify all dimensions and site conditions prior to commencing with the work. Any errors, omissions or unusual conditions to be reported to the Engineer immediately.
2. Crane to be mounted level. Crane should be checked for level periodically during work day due to possible settling of soil.
3. The operation and installation notes hereon are offered for information and guidance and are not to be taken to infer that the Engineer is in any way involved in or is responsible for the actual placement, installation or operation of the crane in the field.
4. Reduce load ratings to account for wind on load. Consult operator's manual for requirements when wind exceeds 20 m.p.h. Do not operate in winds over 30 m.p.h. If winds exceed 50 m.p.h. lower Boom to the ground unless manufacturer's instructions indicate otherwise.
5. Lifted load not to exceed maximum loads as per approved manufacturer's load chart.
6. Pontoons and/or cribbing to be of sound structural grade lumber. $f_b = 1.2$ ksi $F_v = 100$ psi Plates - A36 Steel (if required).
7. Pedestrian bridge shown in schematic only (if required). Bridge contractor to obtain permit from Department of Buildings prior to installation of crane.
8. Crane operations are to be conducted in accordance with ANSI/ASME B30.5a - 1984, appropriate OSHA rules and RS19-2 of the N.Y.C Building Code.
9. This approval is for placement of crane only with permission to operate and lift loads subject to on-site inspection and satisfactorily passing assembled inspection by the N.Y.C. Division of Cranes and Derricks.
10. Do not deliver crane to site before inspection date. Crane to be operated only by N.Y.C. Licensed Crane Operator
11. Approval of this application is granted contingent upon the applicant furnishing a letter of approval from the Department of Transportation (MTCCC) and a permit from the Department of Highways.

Figure 5.12 Installation drawing for a mobile crane that was needed to hoist structural steel to the roof of an existing building. Narrow streets and multiple building setbacks made the planning for this project an exercise in boom clearance calculations.

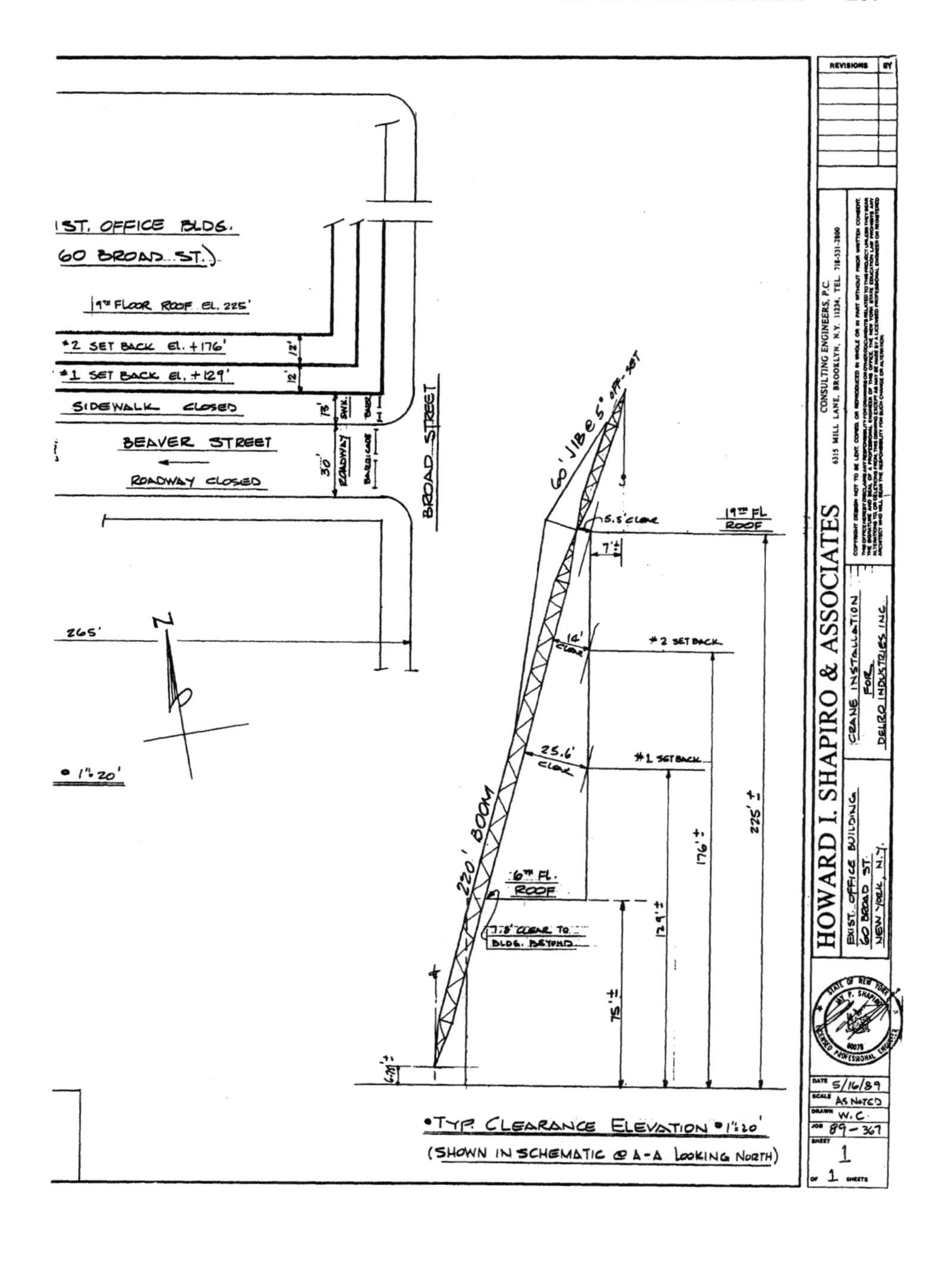
EXIST. OFFICE BLDG.
(60 BROAD ST.)
19TH FLOOR ROOF EL. 225'
#2 SET BACK EL. +176'
#1 SET BACK EL. +129'
SIDEWALK CLOSED
BEAVER STREET
ROADWAY CLOSED
BROAD STREET
265'
N
@ 1" = 20'
60' JIB @ 5° OFF-SET
220' BOOM
5.5' CLEAR
7'±
19TH FL ROOF
14' CLEAR
#2 SET BACK
25.6' CLEAR
#1 SET BACK
6TH FL. ROOF
7.8' CLEAR TO BLDG. BEYOND
225'±
176'±
129'±
75'±
•TYP. CLEARANCE ELEVATION @ 1" = 20'
(SHOWN IN SCHEMATIC @ A-A LOOKING NORTH)
REVISIONS
BY
HOWARD I. SHAPIRO & ASSOCIATES
CONSULTING ENGINEERS, P.C.
6315 MILL LANE, BROOKLYN, N.Y. 11234, TEL. 718-531-2800
EXIST. OFFICE BUILDING
60 BROAD ST.
NEW YORK, N.Y.
CRANE INSTALLATION
FOR
DELRO INDUSTRIES INC.
DATE 5/16/89
SCALE AS NOTED
DRAWN W.C.
JOB 89-367
SHEET 1
OF 1 SHEETS

With a little imagination, these methods can be applied to mobile cranes with tower attachments or to derricks, tower cranes, and other equipment. For instance, they were used to assure adequate clearance during the tower crane boom erection shown in Fig. 5.13.

The mobile crane in Fig. 5.13 is, of course, telescopic. Telescopic booms are in essence cantilevered beams; they bend under load. There is no quick but accurate way to estimate these deflections which can be quite large for the very long booms in current use. But, for tele-

Figure 5.13 Tower crane erection and dismantling are rigging projects where many of the most critical operations depend on maintaining clearance between the erection crane boom and the load. Placement of this Grove TM 1275 truck crane was controlled by adjacent buildings and by space needed on the street to assemble the tower crane boom at a reach within the crane's capacity. (*Photo by Jay P. Shapiro.*)

scopic cranes the radii on the rating chart are loaded radii, that is, they take into account deflection under rated load. Even though the clearance methods given here do not take deflection into account, they are valid and useful. That is because the equations given assume that the boom will follow a straight line from boom foot to head pin. But boom angles, when given on the rating chart, are measured at the boom base which does not deflect. (For the same radius, the actual boom angle will be greater than the calculated one.) For that reason, clearance calculations should use boom angles calculated from the radius rather than a chart angle. Then if the calculations show adequate clearance between boom and load, the actual clearance will be somewhat greater because of boom deflection. Using rating chart boom angles will cause clearances to be over estimated for extended telescopic booms; calculated angles are therefore more appropriate and conservative.

Example 5.3 (*a*) A crane is to be used for placing concrete on a high-rise structure. What maximum height H can be serviced while maintaining 4 ft (1.2 m) clearance between the boom and the edge of the structure? Crane dimensions are t = 5 ft (1.52 m), h = 6.75 ft (2.06 m), $D = B$ = 7.83 ft (2.39 m), and boom length L = 320 ft (97.53 m). Installation parameters are R = 90 ft (27.43 m) and g = 25 ft (7.62 m). The crane will be lifting a concrete bucket that will be discharged into a hopper. The point of discharge will be 3 ft (0.91 m) beyond the building edge.

From Eq. (5.2)

$$\theta = \cos^{-1} \frac{90 - 5}{320} = 74.60°$$

From Eq. (5.3)

$$\varphi = \sin^{-1} \frac{25 + 3}{90} = 18.13°$$

From Eq. (5.4)

$$\tau = \tan^{-1} \frac{\sin 18.13°}{\tan 74.60°} = 4.90°$$

The maximum height will then be given by Eq. (5.7) after finding A with Eq. (5.6):

$$A = \frac{1 + \sin^2 4.90°}{\cos 4.90°}$$

$$= 1.011$$

$$H = 6.75 - \frac{7.83}{2 \cos 74.60°} + \left[90 - 5 - \frac{4(1.011) + 3}{\sin 18.13°} - \frac{7.83}{2 \tan 18.13°}\right] \tan 74.6$$

$$= 175.0 \text{ ft } (53.34 \text{ m}) \qquad \textit{Ans.}$$

(*b*) If the crane is moved farther from the building face so that g becomes 40 ft (12.19 m), what will be the effect on H?

If we hold the radius constant, Eq. (5.3) gives the new value of φ as 28.54° and Eq. (5.4) shows that τ changes to 7.50° and A becomes 1.026. With these values inserted into Eq. (5.7), H = 220.5 ft (67.20 m). The 15 ft (4.57 m) increase in g has led to an increase in H of 45.5 ft (13.87 m). *Ans.*

(*c*) If g is increased to 87 ft (26.52 m), the boom will be perpendicular to the building face. What maximum work height can be reached before it becomes necessary to add a jib?

It is now given that φ = 90°, and from Eq. (5.4) τ has become 15.40° and A = 1.110. Using Eq. (5.7) once more, we get H = 273.6 ft (83.38 m) *Ans.*

(*d*) With the boom at minimum radius of 60 ft (18.29 m) the boom angle is 80.10°. Keeping the boom perpendicular to the building, what maximum height can now be reached?

There is no change in φ, which remains 90°, but τ is now 9.90° and A = 1.045. Inserting values into Eq. (5.7) yields H = 258.0 ft (78.63 m), or a reduction in height from the previous trial. *Ans.*

Example 5.3 shows that with radius held constant maximum work height increases with dimension g (Fig. 5.5) until the optimum point is reached. The optimum will occur when the boom is perpendicular to the structure. However, for any angle of boom to structure, there is an optimum radius that will give maximum height. Considering all conditions, greatest height is achievable with minimum e, φ = 90°, and radius optimized, but changing radius may not offer a significant increase in height. For the example, the optimum radius is 110 ft (33.53 m), but this gives a work height only 1.7 ft (0.52 m) greater than the 90-ft (27.43-m) radius used in parts (*a*) to (*c*).

Example 5.4 What boom length would be needed to place a load 26 ft (7.92 m) in from the edge of a roof 72 ft (21.94 m) high? The crane must be placed so that g = 36 ft (10.97 m) exactly. Crane dimensions are t = 4 ft (1.22 m), h = 6.5 ft (1.98 m), D = 5 ft (1.52 m), and B = 5.5 ft (1.68 m); a clearance of 4 ft (1.2 m) will be used.

From the data, H = 72 ft (21.94 m) and e = 26 ft (7.92 m). An educated guess can be made for the boom angle with the assumption that the boom will be perpendicular to the wall. The distance normal to the wall from boom foot to wall is 36 − 4 = 32 ft (9.75 m). If the vertical height from roof edge to boom centerline is estimated to be about 20 ft (6.1 m), the boom centerline above the roof edge will then be approximately 72 − 6.5 + 20 ≈ 85.5 ft (26.1 m) above the boom foot. The boom angle θ will then be

$$\theta \approx \tan^{-1} \frac{85.5}{32} \approx 70°$$

With the boom perpendicular to the wall, $R = 36 + 26 = 62$ ft (18.90 m). Then from Eq. (5.1)

$$62 = 4 + L \cos 70° \qquad L \approx 170 \text{ ft } (51.8 \text{ m})$$

Equation (5.4) will give the clearance-line angle

$$\tau = \tan^{-1} \frac{1}{\tan 70°} = 20.0°$$

and after A has been found with Eq. (5.6), Eq. (5.8) will give the first calculated value for boom length

$$A = \frac{1 + \sin^2 20.0°}{\cos 20.0°} = 1.189$$

$$\begin{aligned} L &= \frac{72 - 6.5}{\sin 70°} + \frac{5}{2 \sin 70° \cos 70°} \\ &\quad + \frac{(4 \times 1.189 + 26)(\sin 90°) + 5.5/(2 \tan 90°)}{\cos 70°} \\ &= 167.4 \text{ ft } (51.02 \text{ m}) \end{aligned}$$

Therefore, 170 ft (51.8 m) was a good guess and is the boom length needed. To check, find the angles needed:

$$\theta = \cos^{-1} \frac{62 - 4}{170} = 70.1° \qquad \tau = \tan^{-1} \frac{1}{\tan 70.1°} = 19.90°$$

and then from Eq. (5–6)

$$A = \frac{1 + \sin^2 19.90°}{\cos 19.90°} = 1.187$$

which is inserted into Eq. (5.5) to find the clearance

$$\begin{aligned} C &= \frac{\left(62 - 4 - \dfrac{72 - 6.5}{\tan 70.1°} - \dfrac{5}{2 \sin 70.1°}\right) \sin 90° - (5/2) \cos 90° - 26}{1.187} \\ &= 4.74 \text{ ft } (1.45 \text{ m}) \end{aligned}$$

The actual clearance is greater than the minimum specified. To be certain, let us try a 160-ft (48.8-m) boom. The boom angle becomes

$$\theta = \cos^{-1} \frac{62 - 4}{160} = 68.75° \qquad \varphi = 90° \qquad \tau = 21.25° \qquad A = 1.214$$

and $C = 3.17$ ft (0.97 m), which is less than allowed. The 170-ft (51.8-m) boom is the shortest that can perform the work and maintain the specified clearance. *Ans.*

Example 5.5 (*a*) Given the crane and installation dimensions of part (*a*) of Example 5.3 with an 80-ft (24.4-m) jib added at 10° offset angle, what maximum height can be reached? Jib dimensions are $d = b = 2.5$ ft (0.76 m), and C_j is to be taken as 4 ft (1.2 m). Boom tip dimensions $b_t = d_t = 2.5$ ft (0.75 m), and the tapered tip section is 40 ft (12.2 m) long.

The boom angle is found by iterating Eq. (5.10)

$$\frac{90 - 5}{320} = \cos\theta + \frac{80}{320}\cos(\theta - 10°)$$

$$0.266 = \cos\theta + 0.250\cos(\theta - 10°)$$

$$\theta = 79.7°$$

The horizontal angle of the boom to the building φ remains unchanged from part (*a*) of Example 5.3 at 18.13°, but the clearance-line angle to the horizontal changes and is given by Eq. (5.11)

$$\tau_j = \tan^{-1}\frac{\sin 18.13°}{\tan(79.7° - 10°)} = 6.57°$$

The clearance parameter A_j, from Eq. (5.13), is then

$$A_j = \frac{1 + \sin^2 6.57°}{\cos 6.57°} = 1.020$$

The maximum height achievable with jib mounted is then found by using Eq. (5.14)

$$H = 6.75 + 320\sin 79.7° - \frac{2.5}{2\cos 69.7°} + \left[90 - 5 - 320\cos 79.7° - \frac{4(1.020) + 3}{\sin 18.13°} - \frac{2.5}{2\tan 18.13°}\right]\tan 69.7°$$

$$= 321.3 \text{ ft } (97.92 \text{ m})$$

based on jib-clearance criteria. Boom tip height $= 6.75 + 320\sin 79.7° = 321.6$ ft (98.02 m), which is just greater than H. Checking boom clearance, from Eq. (5.4) we find

$$\tau = \tan^{-1}\frac{\sin 18.13°}{\tan 79.7°} = 3.24°$$

and the boom tip position parameter e_t is given by Eq. (5.16)

$$e_t = 3 - (80 \cos 69.7°) \sin 18.13° = -5.64 \text{ ft } (-1.72 \text{ m})$$

The clearance parameter A is given by Eq. (5.6)

$$A = \frac{1 + \sin^2 3.24°}{\cos 3.24°} = 1.005$$

Equation (5.17) then gives the boom tip clearance

$$C_t = \frac{5.64 - (2.5/2) \sin 79.7° \sin 18.13° - (2.5/2) \cos 18.13°}{1.005}$$

$$= 4.05 \text{ ft } (1.23 \text{ m}) > 4 \text{ ft } (1.2 \text{ m}) \qquad \text{OK}$$

The clearance at the start of the taper section is given by Eq. (5.18)

$$C_{st} = \frac{\left(40 \cos 79.7° - \dfrac{7.83}{2} \sin 79.7°\right) \sin 18.13° + 5.64 - \dfrac{7.83}{2} \cos 18.13°}{1.005}$$

$$= 2.93 \text{ ft } (0.89 \text{ m}) < 4 \text{ ft } (1.2 \text{ m})$$

which is not acceptable. Clearance height will then be given by Eq. (5.7). The boom radius is 5 + 320 cos 79.7° = 62.2 ft (18.96 m) and

$$H = 6.75 - \frac{7.83}{2 \cos 79.7°} + \left[62.2 - 5 - \frac{4(1.005) - 5.64}{\sin 18.13°} - \frac{7.83}{2 \tan 18.13°}\right] \tan 79.7°$$

$$= 262.5 \text{ ft } (80.01 \text{ m}) \qquad \textit{Ans.}$$

Boom clearance governs so as to maintain the specified 4 ft (1.2 m) value. Adding the 80-ft (24.4-m) jib increased H by 87.5 ft (26.7 m).

(*b*) Using the same radius, R = 90 ft (27.43 m), but with the boom set perpendicular to the building, what height can be achieved?

This position requires that g = 87 ft (26.52 m), and of course, φ = 90°. The clearance angle τ_j increases to

$$\tau_j = \tan^{-1} \frac{1}{\tan 69.7°} = 20.30°$$

$$A_j = \frac{1 + \sin^2 20.30°}{\cos 20.30°} = 1.195$$

$$H = 6.75 + 320 \sin 79.7° - \frac{2.5}{2 \cos 69.7°} + \left[90 - 5 - 320 \cos 79.7° - \frac{4(1.195) + 3}{1} - \frac{2.5}{2 \tan 20.30°}\right] \tan 69.7°$$

$$= = 362.9 \text{ ft } (110.61 \text{ m}) \qquad \textit{Ans.}$$

The calculated maximum service height H must be approaching the jib point height, making clear the need to check lift clearance. The jib point height will be

$$\text{Point height} = 6.75 + 320 \sin 79.7° + 80 \sin 69.7°$$
$$= 396.6 \text{ ft } (120.88 \text{ m})$$

which means that only 396.6 − 362.9 = 33.7 ft (10.27 m) is available above the roof to accommodate the load clearance, load, slings and attachments and the minimum distance from hook to jib point. This figure must be checked against the particular equipment to be used for the work in order to be certain that lifts to that height can actually be made. In a boom-clearance check R_t = 62.22 ft (18.96 m) but g = 87 ft (26.52 m); therefore boom clearance is obviously OK.

Clearance of other crane appurtenances

In tight quarters, the boom might not be the only part of the crane with a clearance problem. The tailswing radius may encroach over an active traffic lane or face interference from an obstruction. Some large cranes have pendants and tall boom foot masts that could run afoul of trees, buildings or electric power lines.

Front end projections sometimes interfere with obstructions too, when the crane is close to the work. Cabs, slack cables and transition pieces need scrutiny when some models operate in constricted spaces.

Dealing with these clearance questions is not so much a mathematical exercise but more one of awareness. The clearance dimensions of projecting or overhanging components are usually given in the crane documentation, most often in general dimension diagrams. Particularly in tight quarters installations, each movement of the crane should be reviewed as part of planning; it is during such a review that other appurtenance clearance situations should be revealed and rectified.

5.5 Crane Loads to the Supporting Surface

The full body of data needed to perform the calculations in this section are not ordinarily given in crane manufacturer's literature. However, the calculation methods that follow were originally presented in the first edition of this book which was published during 1980. Since then, an increasing number of crane users have been making use of these relatively easy procedures for designing adequate support systems tailored to their cranes and site situations.

Crane owners, users, and installation designers can request needed data from the crane manufacturers. You will find them mostly cooperative and forthcoming. Everyone benefits from intelligent installation practice.

With the data at hand, any crane-installation support situation can be dealt with and structural supports, when needed, can be designed as part of the planning process.

For our purposes, let us assume that we have the following body of data for each crane to be considered:

Carrier. Weight and CG horizontal distance from the axis of rotation

Superstructure. Weight, including counterweights, and CG horizontal distance from the axis of rotation

Boom. For each boom length (for both latticed and telescopic booms), weight and CG location coordinates, including the effects of guy lines, upper spreader, and boom foot mast

Jib. For each jib length, weight and CG location coordinates, including the effects of guy lines and jib mast

For boom and jib data, the CG locations should be given in terms of a distance along the centerline measured from the foot pin and an offset above and perpendicular to the centerline (Fig. 5.14).

It will be convenient to transform the boom and jib CG location data from cartesian to polar coordinate form

$$\theta_b = \tan^{-1}\frac{y_b}{x_b} \qquad L_b = (y_b^2 + x_b^2)^{1/2}$$

$$\mu_j = \tan^{-1}\frac{y_j}{x_j} \qquad J_j = (y_j^2 + x_j^2)^{1/2}$$

where θ_b and L_b and μ_j and J_j define the position of the boom and jib CG respectively. If data are provided in polar form, the conversion step can be eliminated, of course. Polar data will enable us to express the moment of the boom about the axis of rotation as

$$M_b = W_b[t + L_b \cos(\theta + \theta_b)] \tag{5.21}$$

and with a jib mounted, the moment becomes

$$M_{bj} = M_b + W_j[t + L\cos\theta + J_j\cos(\theta - \mu + \mu_j)] \tag{5.22}$$

The entire crane structure above the swing circle can be replaced mathematically by a moment and a vertical load. If the weight of the superstructure, less boom and jib weights W_b and W_j, is called W_u and its CG is located horizontally from the axis of rotation a distance d_u,

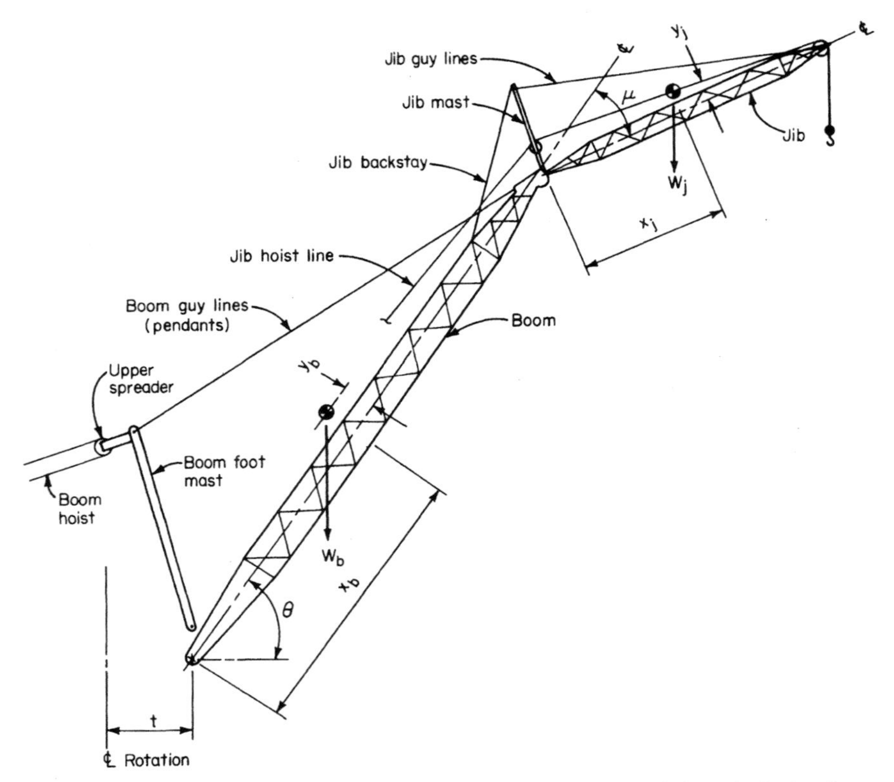

Figure 5.14 Elevation of a boom showing the parameters needed for calculating boom moment.

then the moment for operating radius R, including the lifted load W and the weight of the suspended hoist ropes W_r, is

$$M_u = M_b + (W + W_r)R - W_u d_u$$

or

$$M_u = M_{bj} + (W + W_{rj})R - W_u d_u \tag{5.23}$$

when a jib is being used and the vertical load is given by

$$V_u = W_b + W + W_r + W_u$$

or

$$V_u = W_b + W_j + W + W_r + W_u \tag{5.24}$$

when a jib is being used. Both moments and vertical loads act at the axis of rotation.

Crane manufacturers prefer to segregate the weight of the counterweights from the balance of the superstructure weight. In that event, the $W_u d_u$ term of Eq. (5.23) and the W_u term of Eq. (5.24) will each be replaced by two terms.

Truck cranes

The normal operating state for a truck crane is on outriggers. At this point, let us consider only cranes with four outriggers.

There is no reason to assume that the centroid of the outriggers will coincide with the projection of the axis of rotation onto the ground, although many cranes are arranged in just such a manner. Instead, let us take the outrigger centroid as being to the rear of the axis of rotation projection a distance x_0 along the longitudinal centerline of the crane (Fig. 5.15). If the outrigger centroid should be forward of the axis projection, x_0 is taken as a negative quantity in the equations that follow.

The carrier weight will be called W_c with its CG located a distance d_c forward of the outrigger centroid and along the longitudinal axis. The net moment of all loads about the outrigger centroid for operations over the rear of the crane will therefore be

$$M_{nr} = M_u - V_u x_0 - W_c d_c$$

while for operations over the side the net moment is

$$M_{ns} = M_u$$

In addition to the moments, the outriggers must support the vertical load

$$V = V_u + W_c \tag{5.25}$$

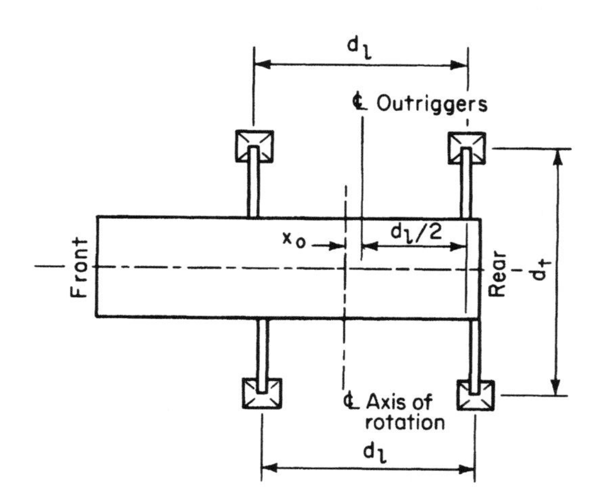

Figure 5.15 Plan view of a truck crane. Note how the outrigger beams are offset from one side to the other to permit retraction.

There are two distinct concepts that can be applied for distributing moments and vertical loads to the individual outrigger floats and hence to the ground. The first method considers the crane as a rigid body; when the points representing outrigger reactions are plotted on a three-dimensional graph, they will all lie in a common plane because the body delivering the loads is rigid.

The second method considers the crane as an elastic frame; for a typical crane, the outrigger reactions will not lie in a common plane when plotted. The results from both methods satisfy the requirements of equilibrium, but values will be different except when the outrigger pattern is a square.

Method one Using the first method, if we take the distance between outriggers (Fig. 5.15) in the longitudinal direction as d_l and in the transverse direction as d_t, for operations over the rear, each outrigger will carry

$$P = \frac{V}{4} \pm \frac{M_{nr}}{2d_l}$$

where the second term is added for the two rear and subtracted for the two front outrigger floats.

For operations over the side, the distribution is not uniform because of the eccentricity of both carrier and upper weights. For identification, let us label the front outrigger reaction on the boom side as P_{fb} and on the counterweight side as P_{fc}. Similarly, the rear outrigger loads will be called P_{rb} and P_{rc}. The loads are then given by

$$P_{fb} = \frac{V}{4} + \frac{M_{ns}}{2d_t} + \frac{W_c d_c + V_u x_0}{2d_l}$$

$$P_{fc} = \frac{V}{4} - \frac{M_{ns}}{2d_t} + \frac{W_c d_c + V_u x_0}{2d_l}$$

$$P_{rb} = \frac{V}{4} + \frac{M_{ns}}{2d_t} - \frac{W_c d_c + V_u x_0}{2d_l}$$

$$P_{rc} = \frac{V}{4} - \frac{M_{ns}}{2d_t} - \frac{W_c d_c + V_u x_0}{2d_l}$$

The pattern is quite clear; each outrigger float receives an equal share of vertical loading, an effect from the transverse moment and an effect from the longitudinal moment. The numerical values of each moment effect remain constant (for a particular load, operating radius, and boom position relative to the longitudinal axis); only the signs change. The same sort of distribution pattern occurs during the more complex case of operations with the boom at an angle to the crane longitudinal centerline.

Let α be the horizontal angle between the boom and the longitudinal axis of the crane measured from the rear. Then

$$M_{nr} = M_u \cos \alpha - V_u x_0 - W_c d_c$$

$$M_{ns} = M_u \sin \alpha \qquad (5.26)$$

where M_{nr} is the portion of the moment acting over the rear and M_{ns} is the portion over the side. When these moments are distributed to the individual outriggers together with vertical loads, the outrigger reactions become

$$P_{fb} = \frac{V}{4} + \frac{1}{2}\left(\frac{M_{ns}}{d_t} - \frac{M_{nr}}{d_l}\right)$$

$$P_{fc} = \frac{V}{4} - \frac{1}{2}\left(\frac{M_{ns}}{d_t} + \frac{M_{nr}}{d_l}\right)$$

$$P_{rb} = \frac{V}{4} + \frac{1}{2}\left(\frac{M_{ns}}{d_t} + \frac{M_{nr}}{d_l}\right)$$

$$P_{rc} = \frac{V}{4} - \frac{1}{2}\left(\frac{M_{ns}}{d_t} - \frac{M_{nr}}{d_l}\right) \qquad (5.27)$$

Equations (5.26) and (5.27) give the outrigger loads for any boom orientation with no need to refer to special equations for full over-the-rear or over-the-side positions.

When handling rated loads at long radii, it is not unusual for the calculated value for one outrigger reaction to be negative. This is analogous to the condition where the crane imparts no load on an outrigger beam or even lifts a float. (When two floats lift or two reactions have negative calculated values, the crane is in an overload condition and into the process of tipping.) The physical meaning of the minus sign is that the ground must pull down on the float, which, of course, it cannot do. The reactions must therefore be rebalanced to give zero reaction at the negative float, thus matching the mathematical with the physical conditions.

If we take the negative reaction as having absolute value a, it is necessary to adjust the reactions without changing the moment or vertical load equilibrium. To do this, the negative reaction is set to zero (a is added to it), a is added to the reaction diagonally across from the heretofore negative outrigger float, and a is subtracted from each of the remaining two outrigger reactions. If P_{rc} had a calculated value of $-a$, set $P_{rc} = 0$, add a to P_{fb}, and subtract a from both P_{rb} and P_{fc}. The result is that there has been no change in the total of the vertical reactions or in outrigger moments about the centroid.

The sum of all outrigger reactions must equal V. Likewise, the sum of all moments about the crane longitudinal centerline and about any transverse line must both be zero if the system is to be in equilibrium. These are the check conditions that must be satisfied by any set of outrigger reactions.

For several reasons actual outrigger reactions will not coincide with calculated values. No truck, rough terrain, or all terrain crane is truly rigid, input data is not precise, and neither wind nor dynamic effects have been considered in the calculations. But more accurate values are not needed because crane cribbing or other supports should never be designed with a thin margin against failure. Cribbing is not only furnished to prevent situations as shown in Fig. 5.16, but to keep the crane level and free of rocking motions. Ground support failure under a loaded crane is often catastrophic. If the small cost of placing adequate structural materials under the outrigger floats is thought of as

Figure 5.16 This photograph demonstrates that on poor ground it becomes essential for outrigger floats to be placed on timber or steel supports that will spread the load to a larger bearing area.

an insurance premium, the money will be well spent. There are few situations where bare outrigger floats are sufficient without some form of cribbing to spread the load. Means for designing cribbing using different kinds of materials are presented later in this chapter.

The reader may have noticed that the method presented does not account for the offset of the outrigger beams from one side to the other in the fore and aft direction. If dimensions used reflect the centers of the pairs of outrigger beams, the offsets will have no meaningful effect on the calculated values. However, when a crane is supported on a structure instead of the ground, those offsets can be very important and should be allowed for in support and site layouts.

Several loading cases usually need to be considered in order to discover governing values for design of supports for any given installation. Obviously, each load should be checked at its maximum radius and at one or more swing angles from picking to placing the load. When several loads are to be handled, the condition causing the maximum load moment, that is, load times radius, will be the situation to use for calculations. Other cases to check include minimum radius with no load on the hook, where counterweight effects may produce controlling outrigger reactions; several swing angles require checking. For long booms, the erection condition with the boom out flat over the rear could produce maximum loading on the rear outriggers.

Front outriggers On some crane models an outrigger is placed at the front (Fig. 5.17) to permit operation in that sector. Although a five-support pattern is statically indeterminate, an approximate value for the front outrigger reaction can be found. If it is assumed that the fifth outrigger will remain essentially unloaded when the crane is stable on four outriggers, it follows that the fifth outrigger will act only when the rear outriggers become unloaded. This will be a condition of three-point support that can readily be solved by taking moments. Because of frame elasticity, the assumption should yield reasonable results for a front outrigger initially set without excessive preload.

Example 5.6 A truck crane will be lifting 35 kips (15.9 t) (including hook block and slings) over the rear at 50 ft (15.24 m) radius, swinging full over the side and then booming out to 70 ft (21.33 m) to place the load. Find the maximum reaction at each outrigger during this operation. Crane data are as follows:

$t = 5.0$ ft (1.52 m)	$x_0 = 0.58$ ft (177 mm)
$d_t = 24$ ft (7.31 m)	$d_l = 22$ ft (6.71 m)
$\theta(50) = 72.5°$	$\theta(70) = 64.3°$
$W_r = 0.35$ kip (160 kg)	$W_b = 27$ kips (12.25 t)

Figure 5.17 The front outrigger is needed in order to set the next beam. When used, it extends the over-the-side operating zone for this machine toward the front to the line from the axis of rotation through the front outrigger, the line on which this photograph was taken.

Load block and sling weight = 3 kips (1.36 t)

Boom length = 150 ft (45.7 m)

L_b = 68.0 ft (20.73 m) θ_b = 5.0°

	Weight		CG location to axis of rotation		
Component	kips	t	ft	m	
Superstructure	90	40.8	6.0	1.83	toward counterweights
Counterweights	94	42.6	17.5	5.33	toward counterweights
Carrier	126	57.2	3.67	1.12	toward front or from
			4.25	1.30	outrigger centroid

Since the requirement is to find maximum loading at each outrigger, six combinations of position and condition must be checked:

	$W + W_r$		Radius		
Position	kips	t	ft	m	Horizontal angle x
Over rear	3.35	1.52	50	15.24	0°
Over rear	35.35	16.0	50	15.24	0°
Over corner	3.35	1.52	50	15.24	$\tan^{-1}(24/22) = 47.49°$
Over corner	35.35	16.0	50	15.24	$\tan^{-1}(24/22) = 47.49°$
Over side	3.35	1.52	50	15.24	90°
Over side	35.35	16.0	70	21.33	90°

For the unloaded case over side, 50 ft (15.24 m) radius was chosen because 70 ft (21.33 m) will control for the loaded case for the outrigger under the boom. At 50 ft it is possible for the outrigger under the counterweight to experience its maximum. Actually, there is no way to tell which radius the operator will choose for return, so that the more critical should be assumed unless the operator is to be given specific instructions.

The boom moment, given by Eq. (5.21), is needed for both operating radii

$$M_b(50) = 27[5 + 68 \cos (72.5° + 5.0°)]$$

$$= 532.4 \text{ kip} \cdot \text{ft } (721.8 \text{ kN} \cdot \text{m})$$

$$M_b(70) = 27[5 + 68 \cos (64.3° + 5.0°)]$$

$$= 784.0 \text{ kip} \cdot \text{ft } (1062.9 \text{ kN} \cdot \text{m})$$

The superstructure moment is needed for three combinations, both loaded and unloaded at 50-ft (15.24-m) radius and loaded at 70 ft (21.33 m). Using Eq. (5.23), we get

$$M_u(50, 3.35) = 532.4 + 3.35(50) - 90(6.0) - 94(17.5)$$

$$= -1485.1 \text{ kip} \cdot \text{ft } (-2013.5 \text{ kN} \cdot \text{m})$$

$$M_u(50, 35.35) = -1485.1 + (35 - 3)(50)$$

$$= 114.9 \text{ kip} \cdot \text{ft } (155.8 \text{ kN} \cdot \text{m})$$

$$M_u(70, 35.35) = 784.0 + 35.35(70) - 90(6.0) - 94(17.5)$$

$$= 1073.5 \text{ kip} \cdot \text{ft } (1455.5 \text{ kN} \cdot \text{m})$$

Equation (5.24) gives the superstructure vertical load which is needed for both the loaded and unloaded cases

$$V_u(3.35) = 27 + 3.0 + 0.35 + 90 + 94 = 214.35 \text{ kips } (953.5 \text{ kN})$$

$$V_u(35.35) = 214.35 + (35 - 3) = 246.35 \text{ kips } (1095.8 \text{ kN})$$

The total vertical loads include carrier weight, Eq. (5.25), and for the unloaded and loaded conditions

$$V(3.35) = 214.35 + 126 = 340.35 \text{ kips } (1513.9 \text{ kN})$$

$$V(35.35) = 246.35 + 126 = 372.35 \text{ kips } (1656.3 \text{ kN})$$

With these preliminary data prepared, we are ready to explore each combination of position and condition. For the over-the-rear position, using Eq. (5.26), $M_{ns} = 0$ by observation, while for the unloaded condition

$$M_{nr}(50,\ 3.35,\ \text{rear}) = -1485.1 - 214.35(0.58) - 126(4.25)$$

$$= -2144.9 \text{ kip} \cdot \text{ft } (2908.1 \text{ kN} \cdot \text{m})$$

The individual outrigger loads are then found by using Eq. (5.27)

$$\frac{V}{4} = \frac{340.35}{4} = 85.1 \text{ kips } (378.5 \text{ kN})$$

$$\frac{M_{nr}}{d_l} = \frac{-2144.9}{22} = -97.5 \text{ kips } (-433.7 \text{ kN})$$

$$P_{fb} = P_{fc} = 85.1 + \tfrac{1}{2}(97.5) = 133.8 \text{ kips } (595.2 \text{ kN})$$

$$P_{rb} = P_{rc} = 85.1 - \tfrac{1}{2}(97.5) = 36.3 \text{ kips } (161.5 \text{ kN})$$

In the same position, for the loaded condition

$$M_{nr}(50,\ 35.35,\ \text{rear}) = 114.9 - 246.35(0.58) - 126(4.25)$$

$$= -563.5 \text{ kip} \cdot \text{ft } (-764.0 \text{ kN} \cdot \text{m})$$

$$\frac{V}{4} = \frac{372.35}{4} = 93.1 \text{ kips } (414.1 \text{ kN})$$

$$\frac{M_{nr}}{d_l} = \frac{-563.5}{22} = -25.6 \text{ kips } (-113.9 \text{ kN})$$

$$P_{fb} = P_{fc} = 93.1 + \tfrac{1}{2}(25.6) = 105.9 \text{ kips } (471.1 \text{ kN})$$

$$P_{rb} = P_{rc} = 93.1 - \tfrac{1}{2}(25.6) = 80.3 \text{ kips } (357.2 \text{ kN})$$

For the over corner position, when unloaded, we find (Eq. 5.26)

$$M_{nr}(50,\ 3.35,\ \text{corner}) = -1485.1 \cos 47.49^\circ - 214.35(0.58) - 126(4.25)$$

$$= -1663.3 \text{ kip} \cdot \text{ft } (-2255.2 \text{ kN} \cdot \text{m})$$

$$M_{ns}(50,\ 3.35,\ \text{corner}) = -1485.1 \sin 47.49^\circ$$

$$= -1094.8 \text{ kip} \cdot \text{ft } (-1484.3 \text{ kN} \cdot \text{m})$$

$$\frac{V}{4} = \frac{340.35}{4} = 85.1 \text{ kips } (378.5 \text{ kN})$$

$$\frac{M_{nr}}{d_l} = \frac{-1663.3}{22} = -75.6 \text{ kips } (-336.3 \text{ kN})$$

$$\frac{M_{ns}}{d_t} = \frac{-1094.8}{24} = -45.6 \text{ kips } (-202.9 \text{ kN})$$

$$P_{fb} = 85.1 + \tfrac{1}{2}(-45.6 + 75.6) = 100.1 \text{ kips } (445.3 \text{ kN})$$

$$P_{fc} = 85.1 - \tfrac{1}{2}(-45.6 - 75.6) = 145.7 \text{ kips } (648.1 \text{ kN})$$

$$P_{rb} = 85.1 + \tfrac{1}{2}(-45.6 - 75.6) = 24.5 \text{ kips } (109.0 \text{ kN})$$

$$P_{rc} = 85.1 - \tfrac{1}{2}(-45.6 + 75.6) = 70.1 \text{ kips } (311.8 \text{ kN})$$

and when loaded in this position

$$M_{nr}(50, 35.35, \text{corner}) = 114.9 \cos 47.49° - 246.35(0.58) - 126(4.25)$$

$$= -600.7 \text{ kip} \cdot \text{ft } (-814.4 \text{ kN} \cdot \text{m})$$

$$M_{ns}(50, 35.35, \text{corner}) = 114.9 \sin 47.49°$$

$$= 84.7 \text{ kip} \cdot \text{ft } (114.8 \text{ kN} \cdot \text{m})$$

$$\frac{V}{4} = \frac{372.35}{4} = 93.1 \text{ kips } (414.1 \text{ kN})$$

$$\frac{M_{nr}}{d_l} = \frac{-600.7}{22} = -27.3 \text{ kips } (-121.4 \text{ kN})$$

$$\frac{M_{ns}}{d_t} = \frac{84.7}{24} = 3.5 \text{ kips } (15.6 \text{ kN})$$

$$P_{fb} = 93.1 + \tfrac{1}{2}(3.5 + 27.3) = 108.5 \text{ kips } (482.6 \text{ kN})$$

$$P_{fc} = 93.1 - \tfrac{1}{2}(3.5 - 27.3) = 105.0 \text{ kips } (467.1 \text{ kN})$$

$$P_{rb} = 93.1 + \tfrac{1}{2}(3.5 - 27.3) = 81.2 \text{ kips } (361.2 \text{ kN})$$

$$P_{rc} = 93.1 - \tfrac{1}{2}(3.5 + 27.3) = 77.7 \text{ kips } (345.6 \text{ kN})$$

When operating over the side $\alpha = 90°$ and for the unloaded condition at 50 ft (15.24 m) radius

$$M_{nr}(50, 3.35, \text{side}) = -214.35(0.58) - 126(4.25)$$

$$= -659.8 \text{ kip} \cdot \text{ft } (-894.6 \text{ kN} \cdot \text{m})$$

$$M_{ns}(50, 3.35, \text{side}) = -1485.1 \text{ kip} \cdot \text{ft } (-2013.5 \text{ kN} \cdot \text{m})$$

$$\frac{V}{4} = \frac{340.35}{4} = 85.1 \text{ kips } (378.5 \text{ kN})$$

$$\frac{M_{nr}}{d_l} = \frac{-659.8}{22} = -30.0 \text{ kips } (-133.4 \text{ kN})$$

$$\frac{M_{ns}}{d_t} = \frac{-1485.1}{24} = -61.9 \text{ kips } (-275.3 \text{ kN})$$

$$P_{fb} = 85.1 + \tfrac{1}{2}(-61.9 + 30.0) = 69.2 \text{ kips } (307.8 \text{ kN})$$

$$P_{fc} = 85.1 - \tfrac{1}{2}(-61.9 - 30.0) = 131.1 \text{ kips } (583.2 \text{ kN})$$

$$P_{rb} = 85.1 + \tfrac{1}{2}(-61.9 - 30.0) = 39.2 \text{ kips } (174.4 \text{ kN})$$

$$P_{rc} = 85.1 - \tfrac{1}{2}(-61.9 + 30.0) = 101.1 \text{ kips } (449.7 \text{ kN})$$

The final condition has the machine loaded at 70 ft (21.33 m) radius

$$M_{nr}(70, 35.35, \text{side}) = -246.35(0.58) - 126(4.25)$$

$$= -678.4 \text{ kip} \cdot \text{ft } (-919.8 \text{ kN} \cdot \text{m})$$

$$M_{ns}(70, 35.35, \text{side}) = 1073.5 \text{ kip} \cdot \text{ft } (1455.5 \text{ kN} \cdot \text{m})$$

$$\frac{V}{4} = \frac{372.35}{4} = 93.1 \text{ kips } (414.1 \text{ kN})$$

$$\frac{M_{nr}}{d_l} = \frac{-678.4}{22} = -30.8 \text{ kips } (-137.0 \text{ kN})$$

$$\frac{M_{ns}}{d_t} = \frac{1073.5}{24} = 44.7 \text{ kips } (198.8 \text{ kN})$$

$$P_{fb} = 93.1 + \tfrac{1}{2}(44.7 + 30.8) = 130.9 \text{ kips } (582.3 \text{ kN})$$

$$P_{fc} = 93.1 - \tfrac{1}{2}(44.7 - 30.8) = 86.2 \text{ kips } (383.4 \text{ kN})$$

$$P_{rb} = 93.1 + \tfrac{1}{2}(44.7 - 30.8) = 100.1 \text{ kips } (445.3 \text{ kN})$$

$$P_{rc} = 93.1 - \tfrac{1}{2}(44.7 + 30.8) = 55.4 \text{ kips } (246.4 \text{ kN})$$

Summarizing the outrigger loads, we have

Position	Condition, ft	P_{fb}	P_{fc}	P_{rb}	P_{rc}
Over rear	Unloaded, 50	133.8	133.8	36.3	36.3
Over rear	Loaded, 50	105.9	105.9	80.3	80.3
Over corner	Unloaded, 50	100.1	145.7	24.5	70.1
Over corner	Loaded, 50	108.5	105.0	81.2	77.7
Over side	Unloaded, 50	69.2	131.1	39.2	101.1
Over side	Loaded, 70	130.9	86.2	100.1	55.4
Maxima, kips		133.8	145.7	100.1	101.1

Method two When the second distribution method is used, the crane carrier frame and outriggers are treated as beams and the imposed concentrated vertical loads projected to the supports accordingly. Using parameters from the first method and Eq. 5.26, the net moment acting on the crane

$$M_{net} = (M_{nr}^2 + M_{ns}^2)^{1/2} \tag{5.28}$$

The resulting eccentricity of total vertical load V is

$$e = M_{net}/V$$

at angle β measured from the rear with respect to the longitudinal centerline.

$$\beta = \tan^{-1}\left(\frac{M_{ns}}{M_{nr}}\right) \qquad M_{nr} > 0$$

$$\beta = 180° + \tan^{-1}\left(\frac{M_{ns}}{M_{nr}}\right) \qquad M_{nr} < 0$$

$$\beta = \alpha \qquad M_{nr} = 0 \tag{5.29}$$

The position ratios for the vertical load resultant are then

$$e_l = \frac{e}{d_l}\cos\beta = \frac{M_{net}}{V \cdot d_l}\cos\beta$$

$$e_t = \frac{e}{d_t}\sin\beta = \frac{M_{net}}{V \cdot d_t}\sin\beta \tag{5.30}$$

A general expression can be formulated to reflect the portion of vertical load distributed to each outrigger float

$$P = V\left(\frac{d_l/2 \pm e \cdot \cos\beta}{d_l}\right)\left(\frac{d_t/2 \pm e \cdot \sin\beta}{d_t}\right)$$

$$= V(\tfrac{1}{2} \pm e_l)(\tfrac{1}{2} \pm e_t)$$

using the same outrigger float notation used for the first method

$$P_{rb} = V(1/2 + e_l)(1/2 + e_t)$$

$$P_{rc} = V(1/2 + e_l)(1/2 - e_t)$$

$$P_{fb} = V(1/2 - e_l)(1/2 + e_t)$$

$$P_{fc} = V(1/2 - e_l)(1/2 - e_t) \tag{5.31}$$

Example 5.7 (*a*) Using the crane and data of Ex. 5.6, find the outrigger reactions when the unloaded boom is over a rear corner at 50 ft. (15.2 m). Use Method Two.

Using values already found in Ex. 5.6

$$V(3.35) = 340.35 \text{ kips } (1513.9 \text{ kN})$$

$$M_{nr}(50,3.35, \text{corner}) = -1663.3 \text{ kip} \cdot \text{ft } (-2255.2 \text{ kN} \cdot \text{m})$$

$$M_{ns}(50,3.35, \text{corner}) = -1094.8 \text{ kip} \cdot \text{ft } (-1484.3 \text{ kN} \cdot \text{m})$$

$$d_t = 24 \text{ ft } (7.31 \text{ m}) \qquad d_1 = 22 \text{ ft } (6.71 \text{ m})$$

From Eq. (5.28)

$$M_{net} = (1663.3^2 + 1094.8^2)^{1/2}$$
$$= 1991.3 \text{ kip} \cdot \text{ft } (2699.8 \text{ kN} \cdot \text{m})$$

and from Eq. (5.29)

$$M_{nr} < 0 \quad \therefore \quad \beta = 180° + \tan^{-1}(-1094.8/-1663.3) = 213.35°$$

The vertical load resultant position ratios given by Eq. (5.30) are

$$e_1 = \frac{1991.3}{340.35 \times 22} \cos 213.35 = -0.222$$

$$e_t = \frac{1991.3}{340.35 \times 24} \sin 213.35 = -0.134$$

Now there is sufficient data to calculate outrigger reactions using Eq. (5.31)

$$P_{rb} = 340.35(1/2 - 0.222)(1/2 - 0.134) = 34.6 \text{ kips } (153.9 \text{ kN})$$

$$P_{rc} = 340.35(1/2 - 0.222)(1/2 + 0.134) = 60.0 \text{ kips } (266.9 \text{ kN})$$

$$P_{fb} = 340.35(1/2 + 0.222)(1/2 - 0.134) = 89.9 \text{ kips } (399.9 \text{ kN})$$

$$P_{fc} = 340.35(1/2 + 0.222)(1/2 + 0.134) = 155.8 \text{ kips } (693.0 \text{ kN})$$

Comparing these results with Method One:

	Method one	Method two
P_{rb}	24.5 kips (109.0 kN)	34.6 kips (153.9 kN)
P_{rc}	70.1 kips (311.8 kN)	60.0 kips (266.9 kN)
P_{fb}	100.1 kips (445.3 kN)	89.9 kips (399.9 kN)
P_{fc}	145.7 kips (648.1 kN)	155.8 kips (693.0 kN)

(*b*) Using Method Two, find outrigger reactions when the crane lifts a 35 kip (15.9 kg) load over the rear at 50 ft (15.2 m) radius.

Using some results from Ex. 5.6

$$M_{nr}(50{,}35.35,\ \text{rear}) = -563.5\ \text{kip} \cdot \text{ft}\ (-764.0\ \text{kN} \cdot \text{m})$$

$$M_{ns}(50{,}35.35,\ \text{rear}) = 0 \qquad V = 372.35\ \text{kips}\ (1656.3\ \text{kN})$$

Then $M_{net} = M_{nr}$, $\beta = 0$ and

$$e = -563.5/372.35 = -1.51 \quad e_1 = -1.51/22 = -0.069$$

$$P_{rb} = P_{rc} = 372.35(\tfrac{1}{2} - 0.069) \times \tfrac{1}{2} = 80.3\ \text{kips}\ (357.2\ \text{kN})$$

$$P_{fb} = P_{fc} = 372.35(\tfrac{1}{2} + 0.069) \times \tfrac{1}{2} = 105.9\ \text{kips}\ (471.1\ \text{kN})$$

exactly the same results found with Method One because there is no load eccentricity with respect to the rear.

(*c*) Using Method Two, find outrigger reactions for the same load as in part (*b*), but over the side at 70 ft (21.3 m) radius.

From Eq. 5.6:

$$V(35.35) = 372.35\ \text{kips}\ (1656.3\ \text{kN})$$

$$M_{nr}(70{,}35.35,\ \text{side}) = -678.4\ \text{kip} \cdot \text{ft}\ (-919.8\ \text{kN} \cdot \text{m})$$

$$M_{ns}(70{,}35.35,\ \text{side}) = 1073.5\ \text{kip} \cdot \text{ft}\ (1455.5\ \text{kN} \cdot \text{m})$$

Eq. (5.28) $\quad M_{net} = (1073.5^2 + 678.4^2)^{1/2} = 1269.9\ \text{kip} \cdot \text{ft}\ (1721.8\ \text{kN} \cdot \text{m})$

Eq. (5.29) $\quad M_{nr} < 0 \qquad \beta = 180 + \tan^{-1}(1073.5/-678.4) = 122.29°$

Eq. (5.30) $\quad e_l = (1269.9/372.35 \times 22) \cos 122.29 = -0.083$

$$e_t = (1269.9/372.35 \times 24) \sin 122.29 = 0.120$$

Eq. (5.31) $P_{rb} = 372.35(\frac{1}{2} - 0.083)(\frac{1}{2} + 0.120) = 96.3$ kips (428.4 kN)

$P_{rc} = 372.35(\frac{1}{2} - 0.083)(\frac{1}{2} - 0.120) = 59.0$ kips (262.5 kN)

$P_{fb} = 372.35(\frac{1}{2} + 0.083)(\frac{1}{2} + 0.120) = 134.6$ kips (598.7 kN)

$P_{fc} = 372.35(\frac{1}{2} + 0.083)(\frac{1}{2} - 0.120) = 82.5$ kips (367.0 kN)

Comparing these results with Method One:

	Method one	Method two
P_{rb}	100.1 kips (445.3 kN)	96.3 kips (428.4 kN)
P_{rc}	55.4 kips (246.4 kN)	59.0 kips (262.5 kN)
P_{fb}	130.9 kips (582.3 kN)	134.6 kips (598.7 kN)
P_{fc}	86.2 kips (383.4 kN)	82.5 kips (367.0 kN)

Which method furnishes consistently accurate results? The answer is probably neither. Actual results will depend on the stiffness characteristics of the individual crane model and the support surface. True results will probably lie between the values calculated by the two methods (except for dynamic and wind effects). But, if outrigger supports are properly designed, either method will prove satisfactory.

Crawler cranes

The axis of rotation of a crawler crane may or may not coincide with the centroid of the track bearing surfaces; on larger machines the axis is often to the rear of the centroid. (The rear is defined as the end containing the *drive sprockets*.) For generality, we will take the axis of rotation at distance x_0 (Fig. 5.18) to the rear of the bearing center. Let d_l be the effective bearing length of the tracks, w the width of the tracks, and d_t the center-to-center transverse distance between the tracks.

Between the drive sprocket and the front *idler sprocket* are a series of smaller diameter wheels called *track rollers*. Crane weight and moment effects pass from the crawler *side frames* to the track rollers, thence to the *track shoes,* and finally to the support surface or the ground. Track shoes are the individual segmented plates that make up the crawler tracks. Both front and rear sprockets are usually raised somewhat, although they may not appear to be, and therefore it is the track rollers that define the length of bearing. For some cranes, it is necessary to block the front of the tracks in order to raise long booms or boom and jib combinations. When that is the case, the bearing length extends at least to the sprocket center if not further.

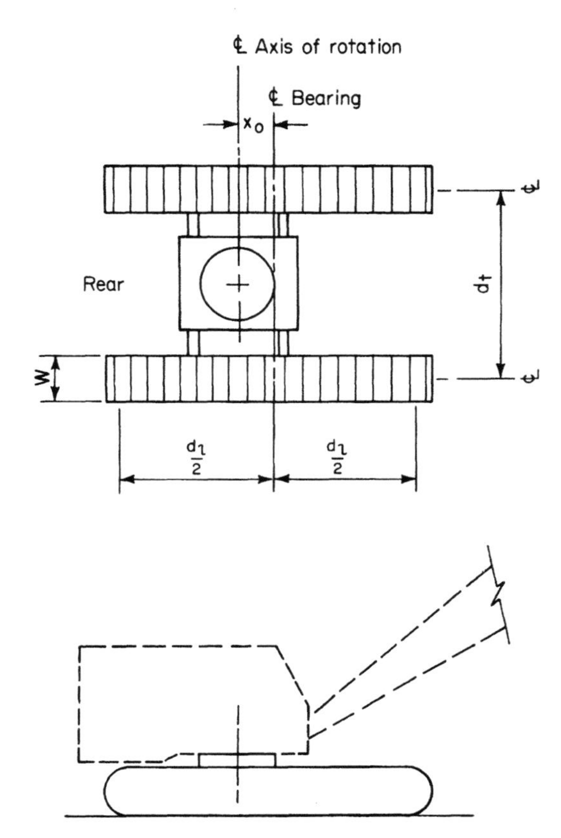

Figure 5.18 Annotated views of a crawler crane.

When cranes operate on yielding soils, the tracks may press into the ground creating the appearance of sprocket to sprocket bearing length. But in making track pressure calculations, it is not prudent to make use of that larger area. Supports designed using the area defined by the track rollers will be more certain to keep the crane level.

With the boom at a horizontal working angle α from the longitudinal centerline, measured from the front, the net moment applied at the center of bearing will be

$$M_{nf} = M_u \cos \alpha - V_u x_0 - W_c d_c \tag{5.32}$$

over the front, assuming that the undercarriage CG is at distance d_c behind the bearing centroid, and

$$M_{ns} = M_u \sin \alpha \tag{5.33}$$

over the side. The total vertical load is

$$V = V_u + W_c \tag{5.34}$$

If the crane were perfectly balanced with respect to the centroid of the track-bearing surfaces, $M_{nf} = M_{ns} = 0$, the load would be equally shared between the tracks and uniformly distributed along their length. Each track would carry $V/2$. But, if $M_{ns} \neq 0$, the distribution of load between the tracks cannot be equal. The difference in track loading must produce a ground reaction moment equal and opposite to M_{ns}.

Taking the reaction under the more heavily loaded track as R_h and under the more lightly loaded track as R_l,

$$V = R_h + R_l$$

The difference between R_h and R_l is caused solely by M_{ns}, which motivates the expressions

$$R_h = V/2 + \frac{M_{ns}}{d_t}$$
$$R_l = V/2 - \frac{M_{ns}}{d_t} \tag{5.35}$$

The reactions (resultants) of Eq. (5.35) satisfy the equilibrium requirements $\Sigma V = 0$ and $\Sigma M_{\text{side}} = 0$. What remains is to take into account the effects of the moment over the front, M_{nf}.

The front moment controls the longitudinal position of the resultants of the track reactions R_h and R_l. When $M_{nf} = 0$, there is no displacement; the resultants of track pressure are at the center of bearing of each track and each track experiences uniform pressure along its length. For nonzero values of front moment, the reactions are displaced:

$$e = M_{nf}/V \tag{5.36}$$

Because of eccentricity e, the track pressure diagram will take either a trapezoidal or a triangular shape (Fig. 5.19).

The length l of the triangular pressure diagram is found by solving equilibrium equations for vertical load and front moment. For either track

$$R = (p_{\max})wl/2$$

$$e \cdot R = \frac{(p_{\max})wl}{2}(d_l/2 - l/3)$$

which yields

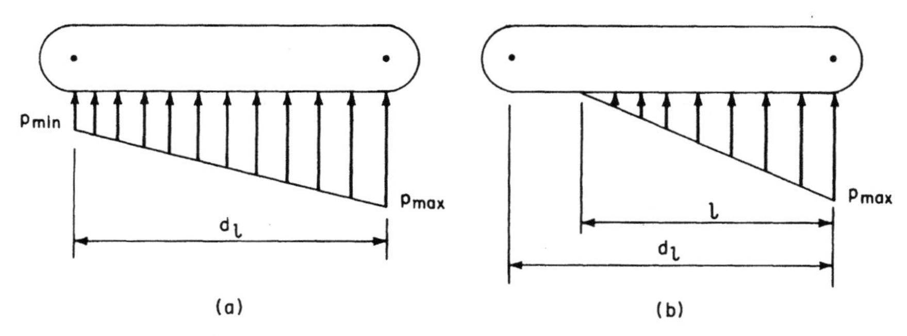

Figure 5.19 Crawler crane track pressure-diagram forms: (*a*) trapezoidal and (*b*) triangular.

$$l = 3(d_l/2 - e) \qquad l \leqslant d_l \tag{5.37}$$

For the triangular pressure diagram it follows that the maximum pressure is

$$p_{\max} = \frac{2R}{wl} \tag{5.38}$$

where R represents either R_h or R_l, depending upon the track being studied.

When $l > d_l$, this indicates that the pressure diagram is trapezoidal (Fig. 5.19*a*). The difference between the pressures at the ends of the track comes about because of the front moment. Using the equilibrium condition $\Sigma M_{\text{end}} = 0$, the pressures at the ends of the track are

$$p = \frac{R}{wd_l}(1 \pm 6e/d_l) \tag{5.39}$$

This analysis assumes that the crawler frames and carbody are absolutely rigid, a not unreasonable assumption for most machines. But, because of the rigidity assumption both tracks will have a common value for e and therefore will always have the same shape of pressure diagram, that is, both triangular or both trapezoidal. Actually, because of elastic effects, e will be the average displacement of the two tracks and small variations from the calculated pressures will be imposed on the ground.

Figure 5.20 shows a large crawler crane placed above subway ventilation gratings. The support structure was designed using the method given above.

Figure 5.20 Traffic restrictions dictated that this crawler crane be positioned above subway ventilation gratings. The steel beams maintain open areas for the vents and span from the building foundation wall to a spread footing on the street surface well beyond the vents. This footing is comprised of pontoons to spread the beam loads and a thin concrete course as a leveling medium. (*XLO Concrete Corp.*)

Example 5.8 (*a*) Find the pressures under the tracks of a crawler crane operating over the side while picking up a load, including hook block and slings, of 227.3 kips (103.1 t). The load is on the main hook at 24 ft (7.3 m) radius, but the 130-ft (39.6-m) boom has a 50-ft (15.2-m) jib mounted at 0° offset. Crane data are as follows:

W_u = 206.88 kips (93.84 t) d_u = 8.46 ft (2.58 m)

W_c = 129.31 kips (58.65 t) $d_c = 0$

W_b = 26.46 kips (12.00 t) L_b = 60.04 ft (18.30 m)

$\theta_b = 2.15°$

W_j = 1.50 kips (680 kg) J_j = 19.10 ft (5.82 m)

$\mu_j = 6.01°$

W_r = 1.60 kips (726 kg)

$w = 4.0$ ft (1.22 m) $\quad d_l = 24.0$ ft (7.31 m)

$d_t = 17.08$ ft (5.21 m) $\quad x_0 = 0 \quad t = 4.0$ ft (1.22 m)

Equation (5.1) is used to find the boom angle

$$24 = 4 + 130 \cos \theta \qquad \theta = 81.15°$$

and the boom moment is given by Eq. (5.21)

$$M_b = 26.46[4 + 60.04 \cos (81.15° + 2.15°)]$$
$$= 291.2 \text{ kip} \cdot \text{ft } (394.8 \text{ kN} \cdot \text{m})$$

The moment with jib weight included is then, from Eq. (5.22),

$$M_{bj} = 291.2 + 1.5[4 + 130 \cos 81.15° + 19.10 \cos (81.15° + 6.01°)]$$
$$= 328.6 \text{ kip} \cdot \text{ft } (445.5 \text{ kN} \cdot \text{m})$$

Equation (5.23) is used to calculate superstructure moment

$$M_u = 328.6 + (227.3 + 1.6)24 - 206.88(8.46)$$
$$= 4072.0 \text{ kip} \cdot \text{ft } (5520.9 \text{ kN} \cdot \text{m})$$

while the vertical load, from Eq. (5.24), is

$$V_u = 26.46 + 1.5 + 227.3 + 1.6 + 206.88$$
$$= 463.7 \text{ kips } (210.3 \text{ t})$$

The moments over the front and side are given by Eqs. (5.32) and (5.33) with $\alpha = 90°$:

$$M_{nf} = 0$$

$$M_{ns} = 4072.0 \text{ kip} \cdot \text{ft } (5520.9 \text{ kN} \cdot \text{m})$$

and, from Eq. (5.34), the total machine vertical load is

$$V = 463.7 + 129.3 = 593.0 \text{ kips } (269.0 \text{ t})$$

Equation (5.36) tells us that $e = 0$, and from Eq. (5.37)

$$l = 3(24.0/2 - 0) = 36.0 \text{ ft } (10.97 \text{ m})$$

$l > d_l$; therefore the pressure diagram is trapezoidal. Using Eq. (5.35)

$$R_l = (593.0/2 - 4072.0/17.08) = 58.1 \text{ kips } (26.3 \text{ t})$$

$$R_h = (593.0/2 + 4072.0/17.08) = 534.9 \text{ kips } (242.6 \text{ t})$$

Equation (5.39) is used to find track pressures. With $e = 0$, pressure will be uniform along each track ($p_{\max} = p_{\min} = p$). For the counterweight side track

$$p = \frac{58.1}{4.0 \times 24.0}\left(1 \pm \frac{6 \times 0}{24.0}\right) = 0.61 \text{ kips/ft}^2 \ (29.2 \text{ kN/m}^2) \qquad \textit{Ans.}$$

and for the boom side track

$$p = \frac{534.9}{4.0 \times 24.0} = 5.57 \text{ kips/ft}^2 \ (266.7 \text{ kN/m}^2) \qquad \textit{Ans.}$$

The pressures calculated are for the loaded case only, as asked.

(*b*) Find the track pressures when the horizontal operating angle $\alpha = 30°$.

Rotation of the superstructure causes a redistribution of upper moments and of track loadings. From Eqs. (5.32) and (5.33)

$$M_{nf} = 4072.0 \cos 30° = 3526.5 \text{ kip} \cdot \text{ft } (4781.3 \text{ kN} \cdot \text{m})$$

$$M_{ns} = 4072.0 \sin 30° = 2036.0 \text{ kip} \cdot \text{ft } (2760.5 \text{ kN} \cdot \text{m})$$

Using Eq. (5.35)

$$R_l = \frac{593.0}{2} - \frac{2036.0}{17.08}$$

$$= 296.5 - 119.2 = 177.3 \text{ kips } (80.4 \text{ t})$$

$$R_h = 296.5 + 119.2 = 415.7 \text{ kips } (188.6 \text{ t})$$

From Eqs. (5.36) and (5.37)

$$e = 3526.5/593.0 = 5.95 \text{ ft } (1.81 \text{ m})$$

$$l = 3(24.0/2 - 5.95) = 18.16 \text{ ft } (5.53 \text{ m})$$

$l < d_l$; therefore the pressure diagram is triangular. For the more lightly loaded counterweight side track, using Eq. (5.38)

$$p_{\max} = (2 \times 177.3)/(18.16 \times 4.0)$$

$$= 4.88 \text{ kips/ft}^2 \ (233.9 \text{ kN/m}^2) \qquad \textit{Ans.}$$

and for the boom side track

$$p_{max} = (2 \times 415.7)/(18.16 \times 4.0)$$
$$= 11.45 \text{ kips/ft}^2 \ (548.3 \text{ kN/m}^2) \qquad Ans.$$

The pressure diagrams for parts (*a*) and (*b*) are illustrated in Fig. 5.21.

Rough terrain cranes

Outrigger loadings for rough terrain cranes can be calculated using exactly the same procedures outlined for truck cranes, assuming that the data are at hand, of course. But, the data for telescopic booms is more complicated and difficult to use. For most operations such elaborate procedures are not called for, however.

Compared with truck and crawler cranes, rough terrain cranes are light in weight, although larger and heavier models are continually being introduced. For these machines, a rule of thumb can often be used for approximating maximum outrigger loads. Between 70 and 90% of gross machine plus load weight can be assumed to affect one outrigger (the lower figure for short-radius lifts) when the boom passes over that outrigger. For fully over-the-side or over-the-rear work, the percentage will be shared by two outriggers.

The upper limit for outrigger loads may be available as part of the technical documentation for telescopic cranes of both truck and rough

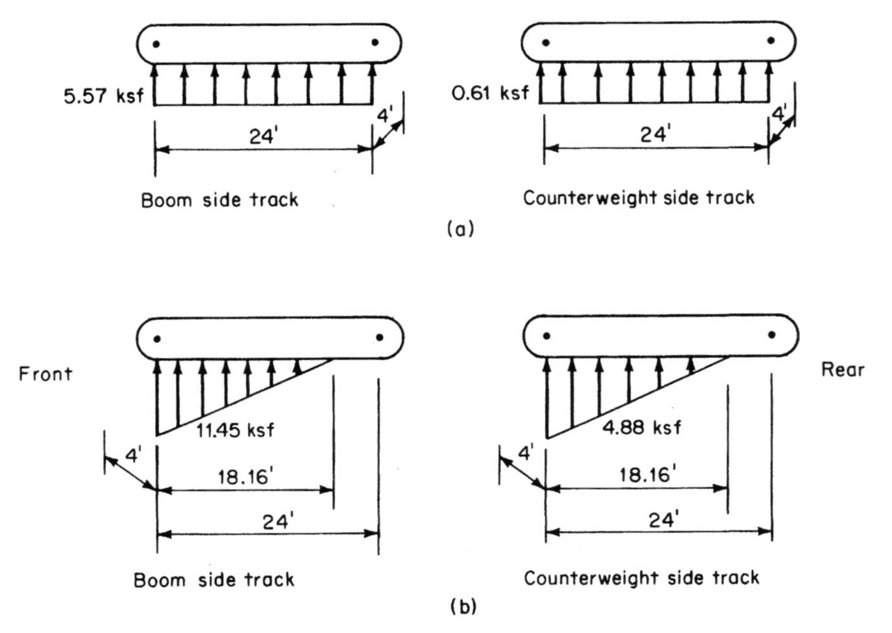

Figure 5.21 Solutions to parts (*a*) and (*b*) of Example 5.6.

terrain types. A technical committee† of the Society of Automotive Engineers (SAE) has adopted a recommendation that crane literature include a statement in the form "no single outrigger pad load exceeds the sum of the lifted load plus machine gross vehicle weight (GVW) times *xx*%, not to exceed a maximum of *xxxx* lb."

All terrain cranes

All terrain cranes (with telescopic booms) have been introduced with sizes and capacities approaching the largest truck cranes. Outrigger loadings, of course, can be calculated using the procedures given above, but the weight and CG data for various boom-length combinations will be much more difficult to acquire. Although the rule of thumb given for rough terrain cranes is also applicable, for the larger machines the pad loads can be very high so that more accurate values are preferable for safety and efficiency.

Some all terrain crane manufacturers include tables in their operator's manuals that give outrigger float loads for rated loads at all working configurations and radii. Means to adjust the data for smaller loads are also given.

5.6 Supporting the Crane

A note common to all mobile crane rating charts states, one way or another, that the "ratings have been derived assuming that the machine is standing on a firm, uniform, level supporting surface". Translated into practical language, this means that the ratings given are appropriate to use only if the outriggers, tracks, or tires are properly supported so that throughout operations the crane will remain level to within a specified tolerance. Most manufacturers specify $\pm$ 1%, but some manufacturers limit deviations to $\pm 1/2$%. An unstated but important further requirement is that the support arrangement must not permit the crane to rock or sway during operation. A truck, rough terrain, or all terrain crane can rock if three of the outrigger floats, or two diagonally opposite floats, are supported on hard points while the others are on a yielding material. An example would be a truck crane with three outriggers on concrete pavement and the fourth on soil. Similar effects occur with crawler cranes when a portion of one track is on a hard point.

A hard point is a local zone or area that will not settle or displace under load as much as the general support area will settle. Some ex-

†Subcommittee 31 of the Off-Road Machinery Technical Committee in its Recommended Practice SAE J1257, Rating Chart for Cantilevered Boom Cranes.

amples of hard points include bedrock, a massive concrete footing or block, a bed of brickbats, or a stiff steel member. A crawler crane with a long boom was discovered to have rocked and subsequently overturned because, during travel with a load, it passed over a buried timber with one track while on a general ground surface that was soft.

Many crane operators develop a sense of the ability of various soils in their "home" areas to support load and for the nominal cribbing needed to sufficiently spread outrigger float loads to a larger bearing area. But, then again, tipping accidents and near accidents occur too frequently for comfort or complacency. Consideration for crane support should be a part of the planning for just about every crane operation.

As a rule of thumb, for ordinary open-field operations on reasonably good ground, where the load can be grounded at any point during the work, crane supports can be left to the operator. Ground can only be considered "reasonably good" if there is knowledge or experience with the ground characteristics, or if some sort of testing has been done. Where those conditions do not exist, you will do well to arrange for crane support in your operation plans.

Ground support capacity

Presumptive soil-bearing capacities used for building foundation design are often overconservative when applied to crane supports. Buildings present long-term loads to the ground, so that allowable soil capacity must reflect a degree of settlement control not necessary for crane use.

Unless the crane is to remain installed for weeks or months, the initial settlement on first bringing the crane into place will be the only settlement of note; its effect will be overcome by initial leveling of the crane. For longer-term installations, periodic releveling may be required. Thus, long term settlement is of no practical consequence in mobile-crane support design if attention is paid to keeping the crane level at the site, and if ground pressure is appropriate for the soil.

When evaluated on the basis of strength alone, most soils can support higher crane loads than the building design values would seem to indicate. Although cranes are often operated on the ground on uncribbed floats with apparent success, there is actually great risk involved in doing this on many soils. Standard outrigger floats are limited in size by weight and stowage considerations and therefore can induce very high [15 tons or more per square foot (1.44 MN/m^2)] bearing pressures. Soil shear failure is sudden and results in almost certain tipping. Cribbing of some sort is needed in most instances.

The soil at any location is a unique blend of constituents that may include very fine particles such as silts or clays, fine to medium-sized sand particles, coarse sands, gravels, and boulders, or organic materials, as well as water or even ice. Soil makeup and content are important factors in soil bearing strength, as is initial density or degree of compaction.

Organic material in soil makes the mass resilient or spongy and is therefore an unwanted component of support soil. All surface soils contain some organic plant materials which ordinarily do not have an important effect on crane support properties, but as the proportion of such material increases, bearing capacity drops and the soil becomes unsuitable.

Fine-grained soils tend to shear at relatively low values. As soil particle size and the variation in size within the sample increase, soil strength will increase as well, until the highest capacities are reached in compact, well-graded sand-gravel mixtures made up of rough uneven particles.

A measure of soil capacity is its angle of repose: the maximum angle of slope that the soil can naturally maintain. As the angle of repose increases, shear strength (and therefore usable bearing capacity) rises as well. For any one soil mix, the angle of repose will improve as compaction, measured by density, is increased. For a number of soil types, Table 5.1 lists typical building-code presumptive bearing capacities as well as our own suggested crane-support bearing capacities. These are offered to help in installation planning; evaluation of the actual soil by an experienced person, preferably a licensed professional engineer, is certainly advisable if any doubt exists about soil capacity.

Be particularly wary of setting outriggers on recently backfilled soil, especially alongside foundations, since it is seldom as well-compacted as one might like it to be. Furthermore, the surface may appear to be reasonably compacted while the soil immediately below may be quite loose. It is a good idea, on backfilled soil, to level the crane and slowly swing the superstructure with the boom at about 60° and no load as a first check on the supports. Relevel if necessary, then repeat the procedure until the crane remains level. As a final check, with the crane facing the pick-up direction and then the load placement direction, run the boom from minimum to maximum radius permissible. If the crane remains level, the soil is ready for work, but if the crane settles, then the level should be rechecked shortly after operations start.

Caution should also be used when placing outriggers on some paved surfaces, particularly sidewalks. On older pavements, cracks or an uneven surface are an indication that the soil below may have settled or been washed out leaving voids below the surface. New sidewalks or existing sidewalks near excavations may well be on inadequate fill or undermined soil.

What would otherwise seem to be a firm surface may well be arching over a void that an outrigger will easily crush. If this happens while

TABLE 5.1 Soil Bearing Capacity Data†

Soil type	Density or state	Approx. unit weight, lb/ft³	Presumptive bearing capacity, tons/ft²: Typical building code	Presumptive bearing capacity, tons/ft²: For cranes	Angle of internal friction, degrees
Rock (not shale unless hard)	Bedrock		60	60	
	Layered		15	15	
	Soft		8	8	
Hardpan, cemented sand or gravel			10	10	
Gravel, sand and gravel	Compact	140	6	8	45
	Firm	120		6	40
	Loose	90	4	4	34
Sand, coarse to medium	Compact	130	4	6	42
	Firm	110		4.5	38
	Loose	90	3	3	34
Sand, fine, silty, or with trace of clay	Compact	130	3	4	34
	Firm	100		3	30
	Loose	85	2	2	28
Silt	Compact	135		3	30
	Firm	110		2.5	28
	Loose	85		2	26
Clay	Compact	130	3	4	30
	Medium	120	2	2.5	20
	Soft	90	1	1	10

†1 lb/ft³ = 16.018 kg/m³; 1 ton/ft² = 95.76 kN/m².

setting up a short-boom crane, the crane will drop until the outrigger has firmer support; most likely, no damage will be done. With a long-boom crane or with a substantial load on the hook, even a minor loss of outrigger support can cause the load to swing out or to the side and cause the crane to fail. The outcome in the former case is hopefully no more than a "shook up" crew; in the latter case it is disaster.

Examine all paved surfaces before the crane gets to the site. There are surface manifestations that can give a fairly good idea of subsurface conditions. If in doubt, drive a heavy truck over the area or drill a hole to probe beneath. When voids are found, it may be best to cave in the pavement and level off with fill.

Rough terrain cranes

The floats of some rough terrain cranes are flat steel plates without turned-up edges. Those cranes should not be operated on soil without some form of cribbing under the outriggers. In case of soil failure, the plate will dig in and the float may tilt on its pivot. The sharp plate edges will then cut into the soil, increasing the possibility of tipping. Turned-up edges, on the other hand, tend to help relevel the float and may allow the crane to recover and avoid tipping.

The floats on rough terrain cranes, in any case, are usually rather small, as little as 16 in (400 mm) square. Thus, even though relatively light in weight, these cranes develop among the highest unit pressures under their floats, particularly the larger models.

One difficulty in ensuring that a rough terrain crane will be properly cribbed by the crew in the field is that the crane is constantly moved to new work areas around the site. For models of 18-ton (16.3-t) capacity or less, productivity and safety can be enhanced by adding a 2-ft-square (600-mm) steel plate (with turned-up edges) to each float. In this way, the need for cribbing is precluded for all but the worst soils. With the elimination of the need for cribbing, crane setups are faster, and production will increase when operators are free of worry over the security of their floats. On models where the addition of plates will cause excess width for road transit, the plates can easily be made removable for stowage on the machine.

For machines of greater capacity, a steel spreader, such as shown in Fig. 5.22, can be designed for most cranes. It temporarily attaches to the outrigger float so that it can be carried by the crane from one working position to another. It assures proper cribbing and is easy to handle.

Truck cranes

Installations not requiring cribbing are the exception rather than the rule. Even on paved areas, failure to use cribbing often results in cracked concrete or in a clear impression of the float left in blacktop. An 82-ton (74-t) truck crane can easily have outrigger reactions of 85,000 lb (378 kN) on 30-in by 30-in (762-mm) floats, making the pressure on the supporting surface about 7 tons/ft^2 (670 kN/m^2).

Once the outrigger loads have been calculated and soil bearing capacity has been determined, the design of cribbing can proceed. The bearing area needed is obtained by dividing outrigger load P by soil capacity s, keeping units consistent. For timber cribbing, assume a trial timber size, such as 6 × 6 (150 × 150), 6 × 8 (150 × 200), or 12 × 12 (300 × 300). Only as many timbers can be used as can fit under, and be loaded by, the float (Fig. 5.23). The dimension $b = ne$, where n is the number of timbers to be used. When timbers overhang the float in width, a cross layer of planks should be added to make up for minor inaccuracy in centering the timbers. Timber lengths are cho-

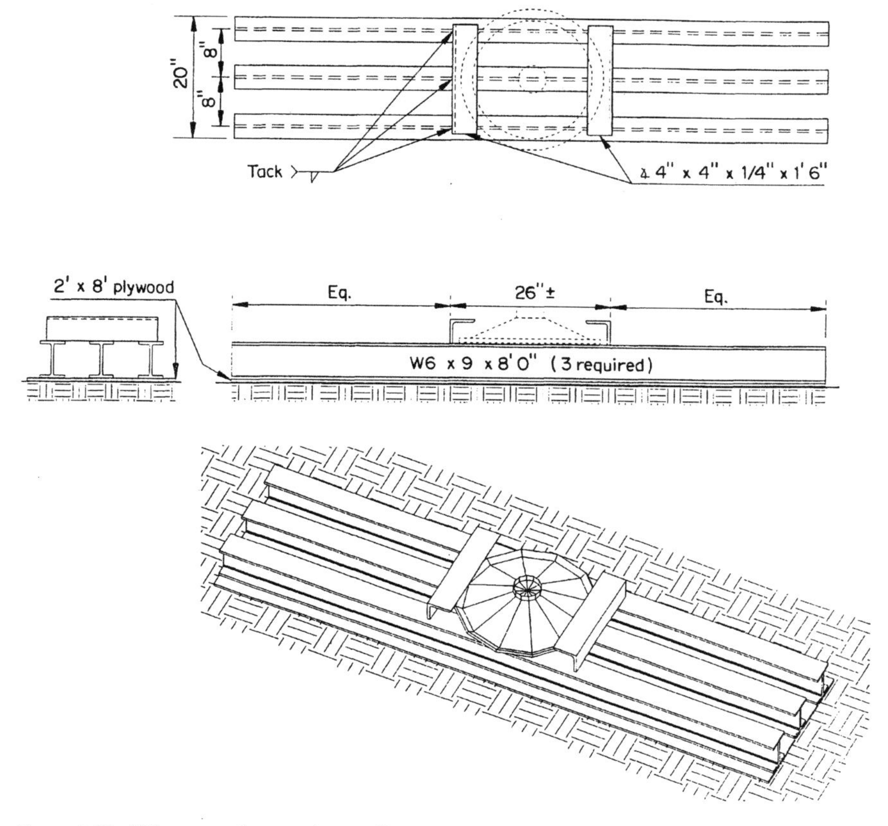

Figure 5.22 When rough terrain or all terrain cranes are to be moved about a site for a number of work assignments, prefabricated outrigger load spreaders can be used. A spreader of the type indicated here, which has been designed for a 35-ton (32-t) crane, can be lifted by the outrigger jacks and moved with the crane.

sen so that $bc \geqslant P/s$, the needed bearing area, and with c usually taken as an even, practical length.

The effective cribbing width b can be increased, where it is essential to do so, by placing a steel plate between float and cribbing. Cross timbers are not practical for this purpose if the combined height of the cribbing exceeds the outrigger ram lift clearance, which it usually does. (If unavoidable, this arrangement can be used by planking out under the crane wheel path to raise the machine.) With the use of plates the steel-to-steel contact can, under some operating conditions, permit sliding and displacement of the crane. To avoid this, a thin sheet of plywood can be inserted between float and plate to increase friction and stabilize machine position.

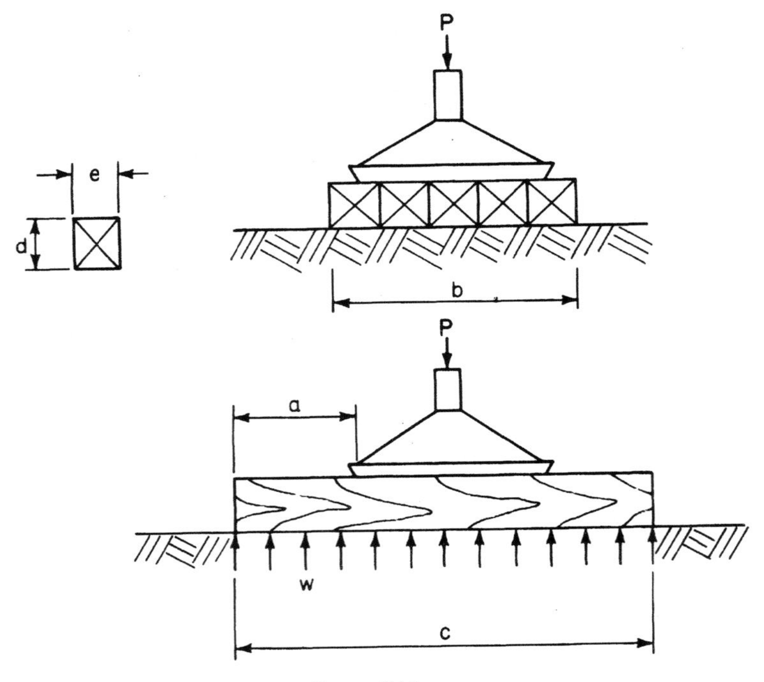

Figure 5.23 Timber outrigger float cribbing.

The applied soil pressure q under the cribbing is then $q = P/bc$ and, of course, $q \leqslant s$. The timbers must be checked for bending strength. The bending moment is given by

$$M = \frac{qba^2}{2} \tag{5.40}$$

and for n timbers of actual (as opposed to nominal) dimensions d and e, the bending stress will be $f = My/I$ or

$$f = \frac{3qa^2}{d^2} \tag{5.41}$$

with attention paid to keeping units consistent. For timber typically available at construction sites, the authors use 1200 lb/in^2 (8274 kN/m^2) as an allowable bending stress unless it has been confirmed that material with a higher strength grade will be used.

Another check must be made on the strength of the timbers, a check of horizontal shear stress which often controls cribbing design. If horizontal shear stresses in timbers are permitted to become excessive, the timber will crack horizontally at midheight, sharply reducing resistance to bending and inviting failure. The horizontal shear stress v is conservatively given by

$$v = \frac{1.5qa}{d} \tag{5.42}$$

with units kept consistent.

Allowable shear stress for typically available timber can be taken as 100 lb/in^2 (690 kN/m^2). Current timber design codes allow substantial increases in allowable stress for short-duration loads. The authors' practice has been to forgo those increases because outrigger loads are calculated without consideration for wind or dynamic effects.

Steel plates can be used for cribbing outrigger floats. As there is no longer a limit to dimension b, bending stress must be checked in two directions using Eq. (5.40). Equation (5.41) shows that moments in both directions will be equal and that maximum design efficiency will be obtained when $b = c$. Shear will not be important for steel plates, and the maximum permissible bending stress can be taken as 27,000 lb/in^2 (186 MN/m^2) for common mild steel.

Example 5.9 (*a*) Design timber support cribbing for a 32 by 32 in (813 by 813 mm) float that must carry 150 kips (667.2 kN) on soil with a capacity of 5 tons/ft^2 (478.8 kN/m^2). Assume new lumber, grade 1500f, and allowable shear stress of 125 lb/in^2 (860 kN/m^2).

Try 10 × 10 (250 × 250) timbers; the actual dimensions are $e = d = 9.5$ in (241 mm). Four timbers can be placed under a 32-in float so that $n = 4$ and $b = 4(9.5) = 38$ in $= 3.17$ ft (965 mm). The minimum permitted bearing area is

$$A_{\text{min}} = \frac{150 \text{ kips}}{(5 \text{ tons/ft}^2)(2 \text{ kips/ton})} = 15 \text{ ft}^2 \ (1.39 \text{ m}^2)$$

$$c \approx \frac{15}{3.17} = 4.73 \text{ ft } (1.44 \text{ m})$$

Use $c = 5.0$ ft (1.52 m), which gives $a = (5.0 - 32/12)/2 = 1.17$ ft (357 mm) and $q = 150$ kips/[(3.17 ft)(5.0 ft)] $= 9.46$ kips/ft^2 (453 kN/m^2). The bending stress is then given by Eq. (5.41)

$$f = \frac{3(9.46 \text{ kips/ft}^2)(1.17 \text{ ft})^2}{(9.5 \text{ in})^2} 1000 \text{ lb/kip}$$

$$= 430 \text{ lb/in}^2 \ (2.96 \text{ MN/m}^2) < 1500 \text{ lb/in}^2 \ (10.3 \text{ MN/m}^2) \qquad \text{OK}$$

Equation (5.42) gives the horizontal shear stress

$$v = \frac{1.5(9.46 \text{ kips/ft}^2)(1.17 \text{ ft})(1000 \text{ lb/kip})}{(9.5 \text{ in})(12 \text{ in/ft})}$$

$$= 146 \text{ lb/in}^2 \ (1.01 \text{ MN/m}^2) > 125 \text{ lb/in}^2 \ (860 \text{ kN/m}^2) \qquad \text{not acceptable}$$

Try 10 × 12 (250 × 300) timbers; e = 9.5 in (241 mm) and d = 11.5 in (292 mm). Rechecking shear, we get

$$v = \frac{(146 \text{ lb/in}^2)(9.5 \text{ in})}{11.5 \text{ in}} = 121 \text{ lb/in}^2 \ (832 \text{ kN/m}^2) \qquad \text{OK}$$

Use 4 timbers 10 × 12 × 5 ft. *Ans.*

(*b*) Using the same conditions as in part (*a*), design a steel-plate float support.

The minimum bearing area needed is 15 ft² (1.39 m²), so that a plate 4 ft (1.22 m) square will be adequate. The actual soil pressure then becomes

$$q = \frac{150 \text{ kips}}{(4 \text{ ft})^2} = 9.38 \text{ kips/ft}^2 \ (449 \text{ kN/m}^2)$$

and the dimension

$$a = \frac{4 - 32/12}{2} = 0.67 \text{ ft (203 mm)}$$

When the full allowable bending stress for steel is used, Eq. (5.41) can be rearranged to solve for the unknown plate thickness directly

$$d = \left(\frac{3qa^2}{f}\right)^{1/2}$$

$$= \left[\frac{3(9.38 \text{ kips/ft}^2)(0.67 \text{ ft})^2}{(27 \text{ kips/in}^2)}\right]^{1/2}$$

$$= 0.68 \text{ in (17.4 mm)}$$

Use one plate 4 ft (1.22 m) square by ¾ in (18 mm) thick. *Ans.*

No harm can follow if larger or thicker plates or longer or larger timbers are used than the design calls for. An intimate interaction between the soil and the cribbing takes place. Overlength cribbing will experience excessive elastic bending deflection, which will tend to make the soil resist the load over an area somewhat smaller than the cribbing. The soil will react by compressing and undergoing small settlements that will redistribute loading toward the periphery of the cribbing. A balance will be reached, soil settlements and cribbing deflections combining to produce an equilibrium state.

Overthick support members, because of their stiffness, will deflect less and cause a more even distribution of pressure beneath the cribbing. However, some care needs to be taken to assure that the cribbing members bear uniformly on the ground. Filling, leveling, or blocking may be necessary. It becomes obvious, then, that the assumption of uniform pressure beneath the cribbing, as used in the design procedure, is a myth. It is a useful myth, however, and one that has long been used successfully in foundation design.

Both steel plates and overlength timbers were used under the outriggers of the crane at the confined site shown in Fig. 5.24. The planks were used to drive onto and raise the floats high enough to clear the timber supports.

Crawler cranes

Cranes were originally mounted on crawler tracks so that they could move about construction sites without hindrance. They still can, and often do, when carrying short booms, particularly for clamshell and

Figure 5.24 Every aspect of crane and load positioning had to be planned in advance at this confined power plant renovation site. Planking had to be laid on the travel path to raise the retracted floats above the timbers and steel plates used to spread the outrigger loads during removal of the 80-ton (72.6-t) girder landed at the right of the crane. Oversized pontoons and plates were available and were used; there was no reason to cut them to size. (*Nab Construction Corp.*)

dragline work. Track pressures are relatively low, and even the largest machines with long booms exert maximum pressures of roughly 6 tons/ft^2 (575 kN/m^2). The wide tracks, low pressures, closely spaced track rollers, and immense power allow these cranes to slog through mud, climb over irregularities, and negotiate grades of 30% more or less. It is only when long booms are mounted that special care and very low speeds are needed to restrain excessive boom movements.

Except when placed on very poor soils, crawler cranes are not likely to develop enough pressure beneath the tracks to induce soil failure. Excessive settlement, which implies the crane may go too far out-of-level, must be addressed and guarded against, however. Often, the solution requires nothing more than placing small sized scrap lumber beneath the tracks and using crane weight to press the lumber into the soil. It may take a few repetitions before the spreading effect of the scrap will furnish a firm, level base, and releveling in the same manner may be required again from time to time. Good support under the ends of the track is essential to the stability of the crane and should be checked frequently.

When soils are generally good, it is possible that reliable support, and a level crane, can be attained with the tracks resting directly on the ground. For marginal or poor soils, a one to 1½ ft (300 to 450 mm) thick layer of course sand, gravel, road mix, or crushed stone will often furnish a satisfactory crane base.

Crawler crane tracks can arch over minor void spaces or low points if they are situated near the center of a track. But, all four track ends need to be well supported. Crawler bases are very rigid; therefore any three points of firm support (three of the four track ends) will define the support plane on which the crane will rest. The fourth corner may not be contacting the ground at all—there could be a gap below it. However, that unsupported track end will not show a gap beneath the track shoes. Instead, the shoes will rest on the ground, but the first track roller (and perhaps more) will not be in contact with the upper face of the shoe. Then, should the crane lift or land a load, boom in or out, or swing, the balance will change and that fourth corner will come down and make contact. The crane will rock. With a long boom, or a heavy hook load, this could impose excessive side loading on the boom or even conceivably overturn the crane.

A similar situation can exist if the fourth corner is on a soft spot. Likewise, positioning the crane where there is a hard point in the mid area of one track will produce similar unwanted results because the crane will tend to rock fore and aft about the hardpoint. The hardpoint can be rock or concrete, a manhole or precast pit, highly compacted soil in an area of generally loose soil, or even a piece of timber that the crane has run over and pressed into the ground. Anything that prevents the rigid crane base from setting up the four track ends in uniform contact with the support surface, all in the same plane, will cause rocking.

These support pitfalls seldom occur because the ground is usually sufficiently resilient to provide full support. However, on a number of occasions the authors have found such support subtleties explaining mysterious overturnings and close calls.

On poor soils and where a more definite support condition is required, timber pontoon mats can be used under the tracks. A pontoon is a group of timbers, usually four 12 × 12s (300 × 300), through-bolted together to form a unit. Design of pontoon supports is similar to design of outrigger cribbing if a simplified model is used. By assuming the pontoon to be discontinuous in length (Fig. 5.25), the support of each track on each segment of pontoon can be treated as a separate problem similar to the outrigger float. At each track, the load on each individual pontoon is equal to the average track pressure on that pontoon multiplied by the track area in contact with the pontoon. When

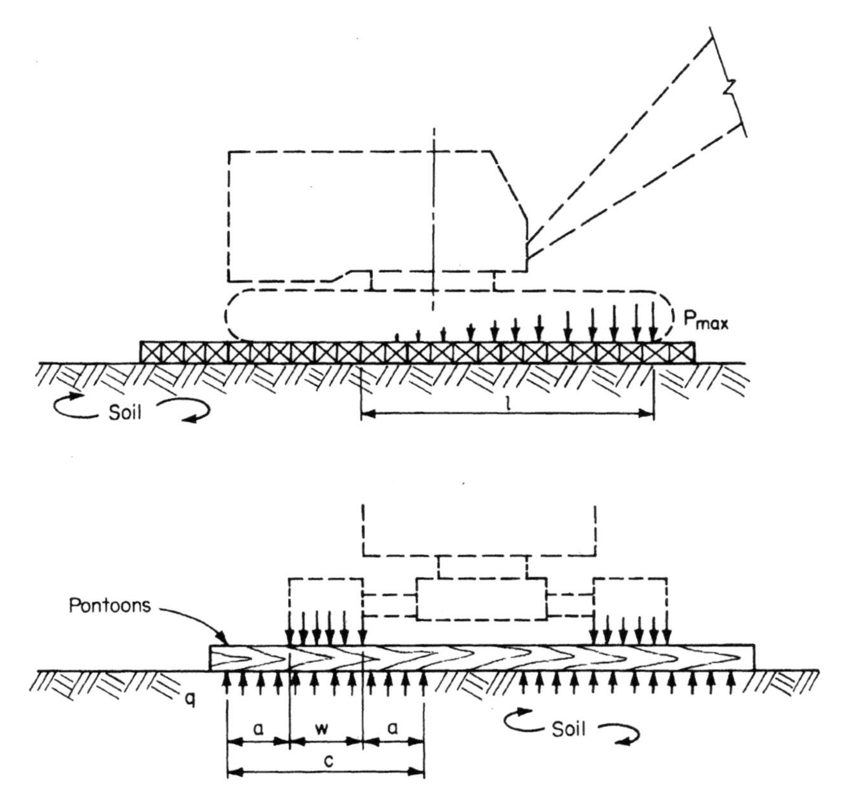

Figure 5.25 Timber pontoon mat for crawler crane support.

this assumption is used, the soil pressure areas (dimension c) are not permitted to overlap. If they do, the simplified method cannot be used.

It is necessary to explore several combinations of loading position and condition to be sure that the pontoons will not be overstressed. Under extreme conditions, a double layer of pontoons may be needed to carry the load, or single pontoons laid parallel to the tracks on top of the mat may satisfy stress limits. Steel plates can also be used to spread track pressures on the pontoons. Figure 5.26 shows a crawler crane on a pontoon support spanning to the foundation wall.

Example 5.10 Using the crane of Example 5.8 and the track pressures of part (*b*), design a timber pontoon mat for soil of 2 tons/ft² (192 kN/m²) capacity. Assume that old pontoons will be used.

The worst possible crane position on the mat will occur if the high end of the pressure diagram is placed at the edge of a pontoon. The boom side track has maximum pressure of 11.45 kips/ft² (548.3 kN/m²) diminishing to zero in 18.16 ft (5.53 m) (Fig. 5.21*b*) or at the rate of 0.63 kips/ft² per foot (99.2 kN/m²/m). Each pontoon is 4 × 11.5 in = 46 in = 3.83 ft (1.17 m) wide so that the pressure at the lower pressure edge is 11.45 − 3.83 × 0.63 = 9.04 kips/ft² (432.7 kN/m²). The load on this part of the pontoon is then

$$P = (4\ \text{ft})(3.83\ \text{ft})\,\frac{11.45 + 9.04}{2}$$

$$= 157.0\ \text{kips}\ (698.1\ \text{kN})$$

This requires a minimum soil bearing area of 157.0 kips/(2 tons/ft² × 2 kips/ton) = 39.25 ft² (3.65 m²), and a minimum value for c of 39.25/3.83 = 10.25 ft (3.12 m). The minimum pontoon length needed, with the crane centered, is then 10.25 ft + 17.08 ft = 27.33 ft (8.33 m), so that a 28-ft (8.53-m) pontoon can be used. The final value for c will then be 10.92 ft (3.33 m) and a = (10.92 − 4)/2 = 3.46 ft (1.05 m). Ground pressure will then be 157 kips/(4 ft × 10.92 ft) = 3.59 kips/ft² (172.1 kN/m²) which is lower than the given limit.

Checking bending stress in the pontoon with Eq. (5.41) gives

$$f = \frac{3(3.59\ \text{kips/ft}^2)(3.46\ \text{ft})^2(1000\ \text{lb/kip})}{(11.5\ \text{in})^2}$$

$$= 975\ \text{lb/in}^2\ (6722\ \text{kN/m}^2) < 1200\ \text{lb/in}^2\ (8274\ \text{kN/m}^2) \qquad \text{OK}$$

The horizontal shear stress is then given by Eq. (5.42):

$$v = \frac{1.5(3.59\ \text{kips/ft}^2)(3.46\ \text{ft})(1000\ \text{lb/kip})}{(11.5\ \text{in})(12\ \text{in/ft})}$$

$$= 135\ \text{lb/in}^2\ (931\ \text{kN/m}^2) > 100\ \text{lb/in}^2\ (690\ \text{kN/m}^2) \qquad \text{NG}$$

If two layers of pontoons are used, the shear stress will be halved. There-

Figure 5.26 A timber pontoon mat was used here to level out a crane working platform on the sloping street while also spreading track loads over a wider bearing area and permitting a small amount of travel. (*North Berry Structures, Inc.*)

fore, use two layers of 28-ft (8.53-m) minimum length pontoons for the crane support mat. *Ans.*

Operations near cellar walls

When a crane is installed in close proximity to a cellar wall, the installation design is not usually controlled by the soil bearing capacity but by the pressure the wall can safely sustain. Such walls are built to support the lateral pressure of the earth plus a nominal allowance for adjacent traffic. Crane weight can crack or even collapse a cellar wall.

The best solution is to keep the crane far enough away (Fig. 5.27) to preclude an increase in pressure on the wall. When this cannot be done and the crane must be positioned closer than the suggested distance, a complex design situation arises. This is clearly beyond the scope of this book, however. In such cases we urge that the services of a licensed professional engineer be retained. Not only will the engineer be able to provide the needed design but his services are required by the laws of most states and provinces.

During the course of construction, it is often found advantageous to place cranes near foundation walls either before or after the first floor has been placed. In some instances the walls are free-standing without backfilling, and in others they are backfilled and temporarily braced.

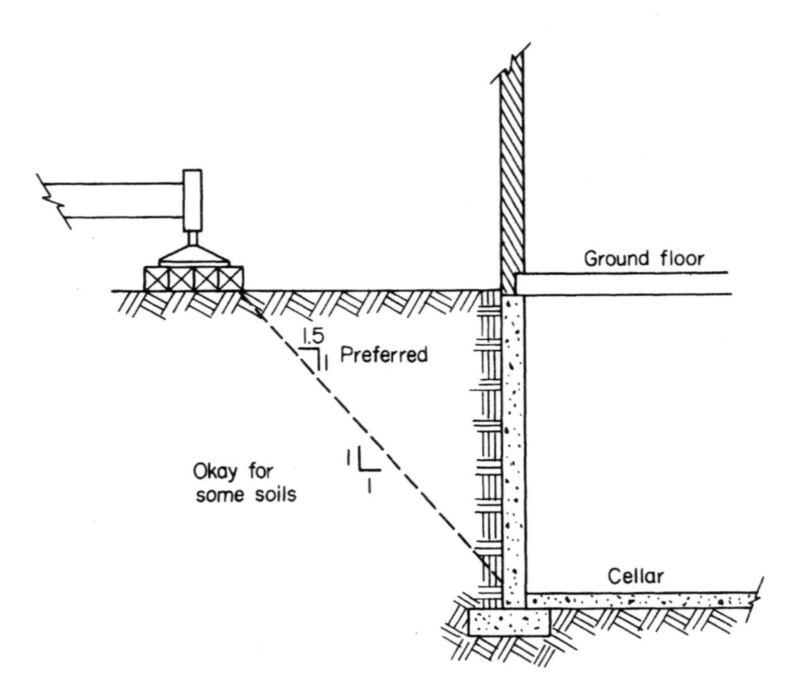

Figure 5.27 Positioning a crane near a cellar wall.

In either case, it is often possible to take advantage of wall capacity for supporting vertical load.

The wall can sometimes be used by installing timber pontoons, or similar structural material, to form a temporary platform for the crane to use to drive into place as well as to support the crane while operating (Figs. 5.28 to 5.30). The strength and stability of the wall should be checked by an engineer. The platform must be designed to have sufficient strength to accommodate wheel or track loads while driving into place; it also must be capable of carrying operating loads if the wall cannot be used for direct support. The pontoons are arranged to act as beams with one end supported on the wall and the other on the ground. The ground supports must be far enough from the wall so as to be stable (Fig. 5.29) or so as to avoid excessive pressure (Fig. 5.27). Two or more layers of pontoons are often required. If the crane must be positioned close to the wall, braces may be needed (Fig. 5.30).

For constructing buildings of between 100 and 350 ft (30 and 105 m) in height, the tower attachment on crawler or truck cranes has become the preferred equipment at urban sites. Since the boom luffs from the top of the tower, it can place loads anywhere on the topmost floor (up to the elevation of the top of the tower) without interference or clearance problems.

Figure 5.28 By using a pontoon deck and placing the outriggers on the wall, it was possible to bring this 250-ton (227-t) truck crane close enough to place the 32-ton (29-t) precast girders shown. The line of steel plates was needed to spread wheel loads while moving the crane from one working position to the next at this project over the right-of-way of a commuter rail line. The installation was designed by the authors' firm.

In urban areas, it is frequently necessary to place cranes as close as possible to the building face to minimize obstructions to traffic. A crane with a tower attachment is suited for this condition since the boom-to-building clearance is not restricted by proximity until building height exceeds tower height. Of course, minimum radius conditions must be considered since this can blank out access to a portion of the building. Unless the platform is long enough for the crane to travel to an access position, the minimum radius will define an arc within which loads cannot be placed.

If a portion of the deck cantilevers past the support (as in Figs. 5.28 and 5.29), the cantilevered area must be treated with caution. The tipping fulcrum of the crane must not be placed on the cantilever. To do so can cause the crane to tip at less than rated load because the machine tipping fulcrum will no longer be in control; tipping will occur about the base of the cantilever unless the other end of the pontoon is anchored down.

When timbers are used to form the deck and outrigger float base, large bearing pressures may be produced between the timbers and the concrete wall top—pressures great enough to crush the timbers. Again, a steel plate can be used to spread the float load, but when this is done, a sheet of plywood should be inserted between the plate and the float to increase friction and prevent sliding.

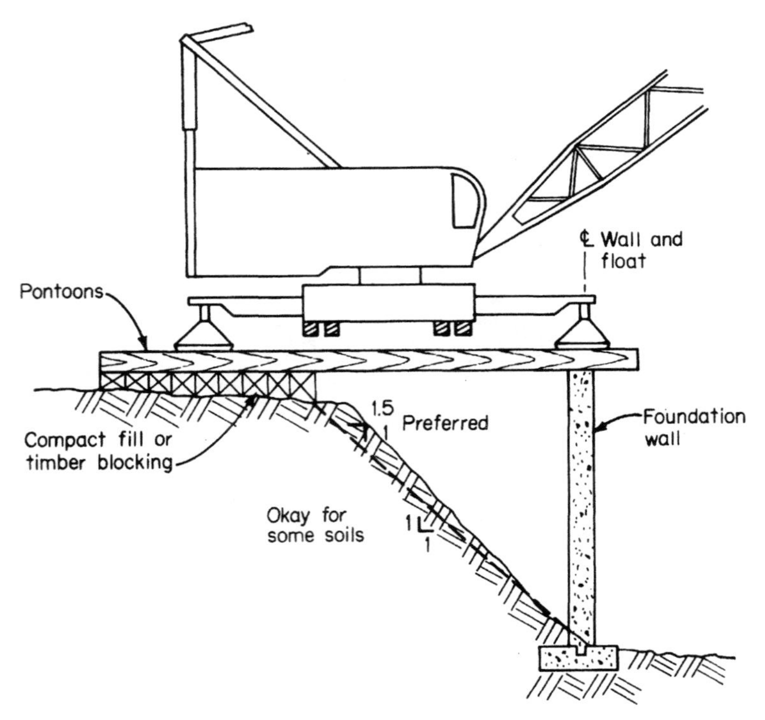

Figure 5.29 Installation of a truck crane on a foundation wall during construction.

Crawler crane tracks are not precision instruments, thus runway and platform arrangements must allow for some deviation in position and for maneuvering. Moving and positioning the crane often involves locking or reversing one track relative to the other. When this occurs, the lateral forces the tracks impose can shift pontoons about, destroying the integrity of the deck. Steel plates can be used to ameliorate the condition, but on occasion it may be necessary to grease the plates to reduce friction. Another alternative would be to nail down a "sacrificial" plywood wearing surface.

At urban jobsites, when the crane is positioned adjacent to, or partly on, a narrow street, it may be necessary to raise the crane high enough for traffic to pass below the counterweight, or for tower attachments, below the lowered gantry. This entails building a platform whose surface may be 4 ft (1.2 m) or more above street grade.

A high platform requires a temporary access ramp to enable the crane to climb into position. More often than not, the ramp cannot be placed to provide the straight run shown in Fig. 5.31; crane access may be from the roadway, and the ramp angled to the deck. The junc-

Figure 5.30 The timber pontoons supporting this crawler crane rest on steel beams spanning between the wall and ground supports. The free-standing wall has been backfilled and must also resist ground pressure from the crane, hence the timber braces. Note the steel clip angles bolted to the wall to resist uplift forces at the wall face. (*Tarrant USA, Inc.*)

ture of ramp and platform needs special attention. As the crane approaches the top of the ramp, the leading edge of the tracks will be in the air above the deck. At one point, the full weight of the crane will be concentrated on the head of the ramp.

When the ramp approaches the platform at an angle, the meeting of the two presents some difficulty in order to avoid having to sawcut expensive pontoons to make the angle. The crane's rigid side frames will permit one track to span a gap of 2 to 3 ft (0.6 to 1 m), beyond which timber blocking can be set up to receive the airborne track.

For deck heights of only one or two pontoon thicknesses, the ramp can be left to field crews to arrange as long as sufficient material is available. For higher decks, ramps should be designed. Doing so will save field time and money while reducing risk.

Lastly, give the crane operator a break. If the crane has to be maneuvered into operating position after climbing onto the deck, extra deck length is needed to do so. Be generous in sizing the deck to allow those movements to take place without hard cuts and too much backing and forthing. Not only will the operator appreciate a fully sized platform, but there will be less tendency towards pontoon displacement and surface damage. The extra work may be offset by avoiding the need for steel maneuvering plates.

Figure 5.31 A timber ramp had to be constructed for this crawler crane to get up onto its operating runway 3 ft (1 m) above the roadway at the low end. (*American Steel Erectors, Inc.*)

Operations near slopes and retaining walls

When operating adjacent to sloping ground, the crane must be placed at a proper distance away from the slope. If the crane is set closer than suggested (Fig. 5.32), it will be in an area of unstable soil due to

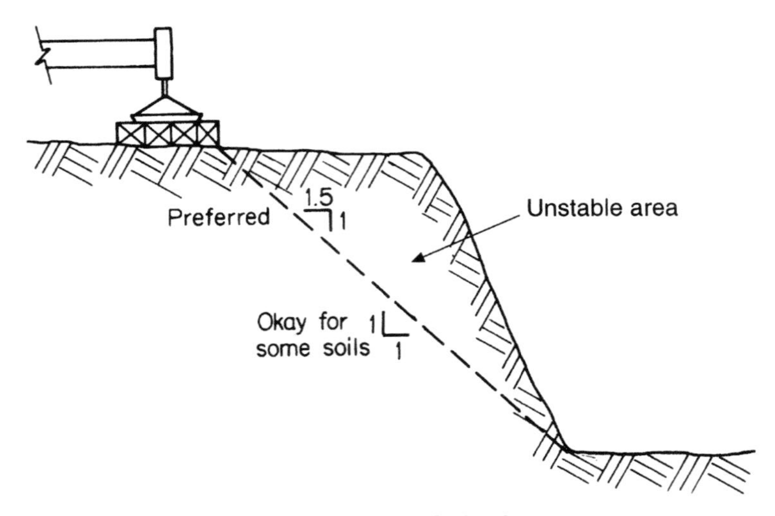

Figure 5.32 Placing a crane near the head of a slope.

significantly reduced shear strength, where soil failure and crane overturning are a possible outcome.

Figure 5.33 shows suggested safe positioning for work near retaining walls. Should it become necessary to set the crane closer to the wall, there are practical ways to do this. Although many retaining walls are designed for surcharge loads (loads in addition to that of soil to the level of the top of the wall), few can be expected to sustain crane loads safely unless special measures are taken.

Again, we urge that the services of a licensed professional engineer be retained for the final design of an installation close to a wall. There are several possible practical solutions to the problem; two of them are shown in Fig. 5.34. When the solution in Fig. 5.34*a* is used, wheel or track effects during the approach to the wall may prove to be the most severe loading on the pontoon.

Figure 5.35 is an installation design drawing for a large crawler crane reflecting a situation similar to that shown in Fig. 5.34*b*.

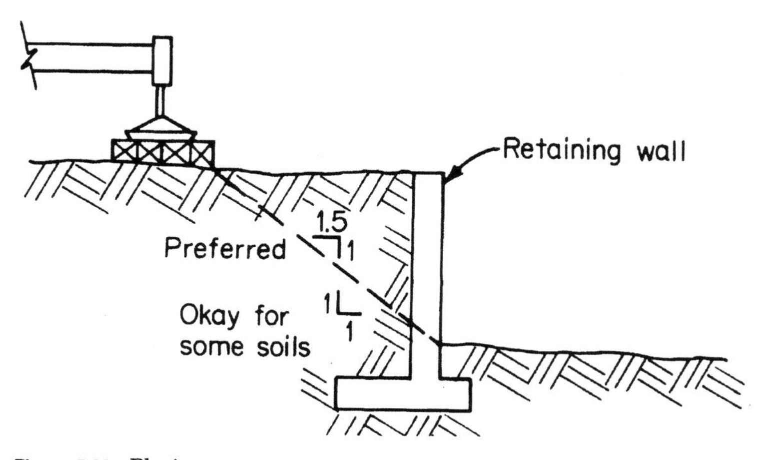

Figure 5.33 Placing a crane near a retaining wall.

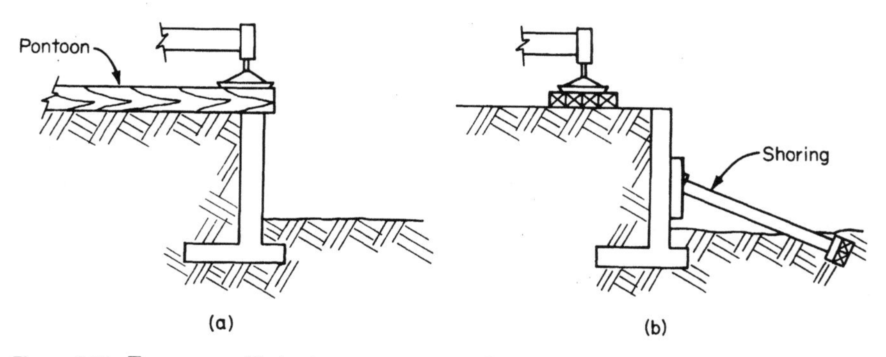

Figure 5.34 Two ways of bringing a crane very close to a retaining wall.

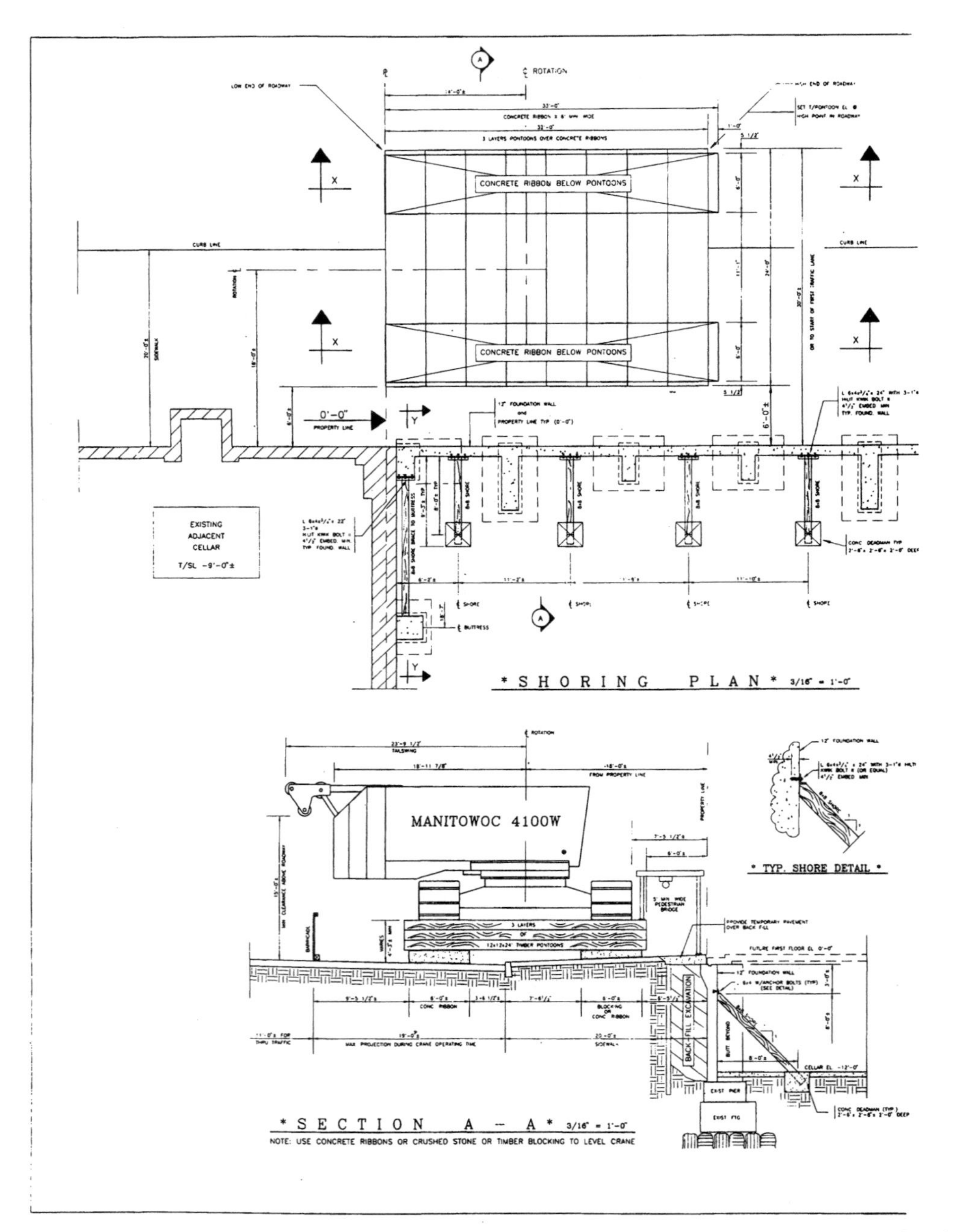

Figure 5.35 An installation drawing for a crawler crane with tower attachment that was placed on a sidewalk adjacent to a back-filled free-standing wall. The wall had to be braced to resist the crane-induced pressure. Note also that several layers of pontoons were used to raise the gantry and provide headroom clearance for traffic at this tight urban site.

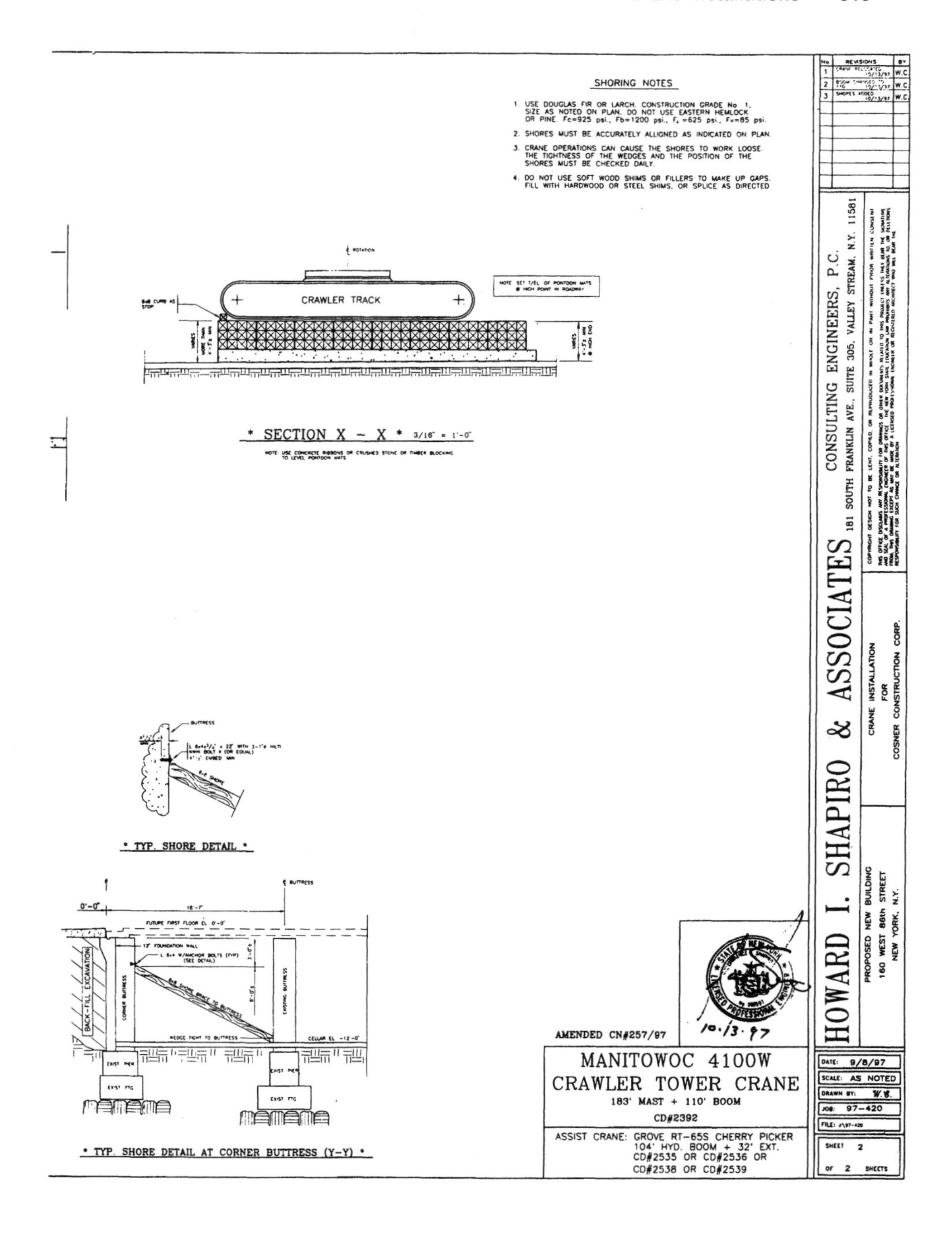
SHORING NOTES
1. USE DOUGLAS FIR OR LARCH. CONSTRUCTION GRADE No. 1. SIZE AS NOTED ON PLAN. DO NOT USE EASTERN HEMLOCK OR PINE. Fc=925 psi., Fb=1200 psi., F⊥=625 psi., Fv=85 psi.
2. SHORES MUST BE ACCURATELY ALLIGNED AS INDICATED ON PLAN.
3. CRANE OPERATIONS CAN CAUSE THE SHORES TO WORK LOOSE. THE TIGHTNESS OF THE WEDGES AND THE POSITION OF THE SHORES MUST BE CHECKED DAILY.
4. DO NOT USE SOFT WOOD SHIMS OR FILLERS TO MAKE UP GAPS. FILL WITH HARDWOOD OR STEEL SHIMS, OR SPLICE AS DIRECTED
CRAWLER TRACK
* SECTION X – X * 3/16" = 1'-0"
* TYP. SHORE DETAIL *
BUTTRESS
8x8 SHORE
FUTURE FIRST FLOOR EL. 0'-0"
12" FOUNDATION WALL
BACK-FILL EXCAVATION
CORNER BUTTRESS
EXISTING BUTTRESS
8x8 SHORE BRACE TO BUTTRESS
WEDGE TIGHT TO BUTTRESS
CELLAR EL. -12'-0"
EXIST PIER
EXIST FTG
* TYP. SHORE DETAIL AT CORNER BUTTRESS (Y–Y) *
REVISIONS
BY
W.C.
W.C.
SHORES ADDED
W.C.
HOWARD I. SHAPIRO & ASSOCIATES
CONSULTING ENGINEERS, P.C.
181 SOUTH FRANKLIN AVE., SUITE 305, VALLEY STREAM, N.Y. 11581
CRANE INSTALLATION
FOR
COSNER CONSTRUCTION CORP.
PROPOSED NEW BUILDING
160 WEST 86th STREET
NEW YORK, N.Y.
AMENDED CN#257/97
10.13.97
MANITOWOC 4100W
CRAWLER TOWER CRANE
183' MAST + 110' BOOM
CD#2392
ASSIST CRANE: GROVE RT-65S CHERRY PICKER
104' HYD. BOOM + 32' EXT.
CD#2535 OR CD#2536 OR
CD#2538 OR CD#2539
DATE: 9/8/97
SCALE: AS NOTED
DRAWN BY: W.S.
JOB: 97-420
SHEET 2
OF 2 SHEETS

Cranes operated close to walls induce surcharge loading on the walls. The surcharge effect decreases, of course, as the loading moves farther from the wall. For a surcharge of uniform pressure q starting at distance x from the wall and applied over width b (Fig. 5.36), the pressure against the wall at any distance y below the top of the wall for $\alpha = \psi + \eta/2$ is given by

$$\sigma_s(y) = \frac{2q}{\pi}(\eta + \sin\eta \sin^2\alpha - \sin\eta\cos^2\alpha) \tag{5.43}$$

where $\psi = \tan^{-1}(x/y)$ and $\eta = \tan^{-1}[(x + b)/y] - \psi$, both in radians. For total pressure on the wall, the effect of the weight of the soil against the wall must be included. This is given conservatively by

$$\sigma_w(y) = \gamma y K_a \tag{5.44}$$

where γ = soil unit weight (see Table 5.1)
$K_a = (1 - \sin\phi)/(1 + \sin\phi)$
ϕ = angle of internal friction (see Table 5.1)

The total pressure at any depth y is then

$$\sigma_t(y) = \sigma_s(y) + \sigma_w(y)$$

Taking pressures at a series of depths y gives the wall pressure diagram (Fig. 5.36). The wall must be checked for its ability to sustain the applied pressures; if it is inadequate, shoring or other means must be provided to assure stability.

This method is generally somewhat conservative for crane support because it is based on an infinitely long strip load whereas the crane cribbing is clearly finite in length. Using the following rational approach, the surcharge can be reduced. For example, if the cribbing length is taken as L, the surcharge pressure on the wall can be uniformly reduced by a coefficient equal to $L/(L + x)$, where x is as defined in Fig. 5.36.

The method above is useful for estimating and initial planning, but not always for final design. It is a simplistic approach to pressure calculation for active stress in a uniform cohesionless soil; it does not take into account other soil characteristics that may need to be considered. Further information is available by referring to a soils textbook, but for any situation where life and limb are at risk, or where collapse may destroy the crane, the services of a licensed professional engineer are needed.

Operations on structural decks and bridges

Even rough terrain cranes can produce outrigger loads that may be too great for elevated roadways or bridges to sustain safely without special provisions. Larger cranes certainly pose a potential for damage or worse. But there are many ways to permit travel in the operating configuration and operations themselves. Mobile cranes have been installed on such structural surfaces as bridges, airport terminal access

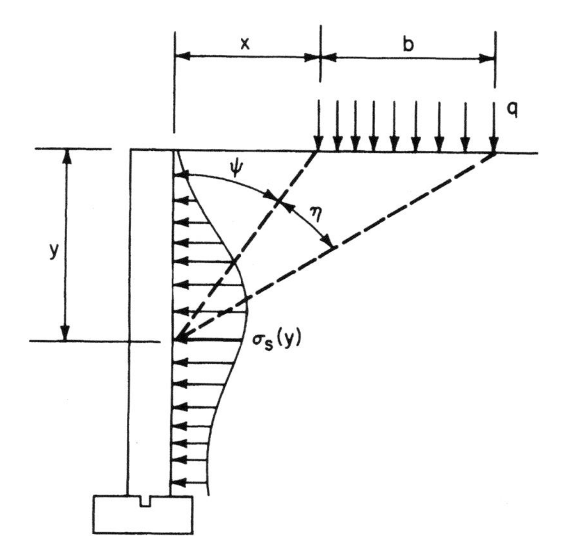

Figure 5.36 Pressure induced on a wall by an adjacent crane (surcharge) load.

ramps, parking decks, building floors, and plazas with cellars below (Figs. 5.37 and 5.38). The cranes accommodated varied from rough terrain models to the largest truck and crawler cranes.

For both economic and safety reasons, the first choice of operating area should always be on solid ground. Deck installations must be controlled operations, in almost every instance, where travel paths must be predetermined and closely adhered to. Operating positions must likewise be closely controlled; loads, operating radii, and horizontal angles should be known and allowed for in advance. There is little flexibility once the installation design has been set.

Deck operations usually require shoring from below, where the shoring is out of the view of the crane operator and operating personnel. Accidental damage to the shores or even their removal could go unnoticed, with catastrophic results.

It goes without saying that the structural aspects of all deck operations should be designed by, or under the direction of, a licensed professional engineer. It is also important that all installed supports and reinforcements be inspected before the crane is brought onto the deck and that operating personnel be thoroughly briefed on required and permitted procedures before the operation begins.

Deck operations can be and are regularly performed safely, but extra supervisory effort and attention are needed. Structural supports, both permanent and temporary, should be checked periodically for distress.

Elastic deflections need particular attention when deck installations are designed. Deflections whose magnitudes in themselves may appear acceptable can be excessive if they permit the crane to rock dur-

Figure 5.37 The timber mats under this crane rest on building steelwork reinforced for crane working loads. The crawler crane will be erecting a tower crane (the boom is in the background) which, together with a second tower crane, will erect the framework for this 800 ft (244 m) tall office tower at New York City's Times Square. (*Midwest Steel, Inc.*)

ing swing motions. If one is unsure, it is always best to choose the stiffer support arrangement.

5.7 Crane Loads

The planning process begins by assessing the loads to be handled. Rated loads include the weight of the hook block, slings, and other lifting accessories; an estimate of those weights may be needed at the outset. Sometimes the payload itself is provided with lifting lugs or specified support points that will dictate the sling or lifting-beam arrangement to be used.

Job planning sometimes includes the development of an entire scheme for handling the load, requiring the use of special equipment and unique procedures such as tripping a wall panel into a vertical position. The methods needed to handle the loads may impose added loads and specific conditions that will affect the choice of crane and its installation.

Loads with large surface areas act as wind sails. The effect, covered in Chap. 3, can be the equivalent of added load on the hook and may even require a strength derate as well (see Chap. 8). The ratings of available equipment may then dictate that wind limits be placed on the operation, a potential cost for lost time that must be reflected in bids or in job budgets.

Some situations impose impact loading on the crane, either by nature of the load or of the operations themselves. Concrete placement by bucket is one such situation. Efficient placement requires that cycle time be minimized, which can be accomplished in part by proper crane selection and siting. Nonetheless, rapid crane movements are needed; the loaded bucket is quickly lifted and swung and then quickly returned and lowered for another load. Most crane manufacturers recommend (and some specify) a 20 to 25% reduction in rated load because of the severity of this service.

Many difficult loading situations develop when a crane is used for demolition work. This is particularly true where load weights must be estimated on the spot and the crane becomes committed to holding the load suspended immediately once the piece is cut from the structure. The danger here is that the weights can be seriously underestimated or that after a while field personnel will replace pencil and paper with the eye. Load-indicating devices can serve well in demolition work for loads that can be test-lifted. Where this is not the case, good load estimates are the only means available. This implies that competent supervisory personnel must be present throughout the work.

A fairly common demolition operation that offers few problems when loads are kept small becomes a significant source of concern with large

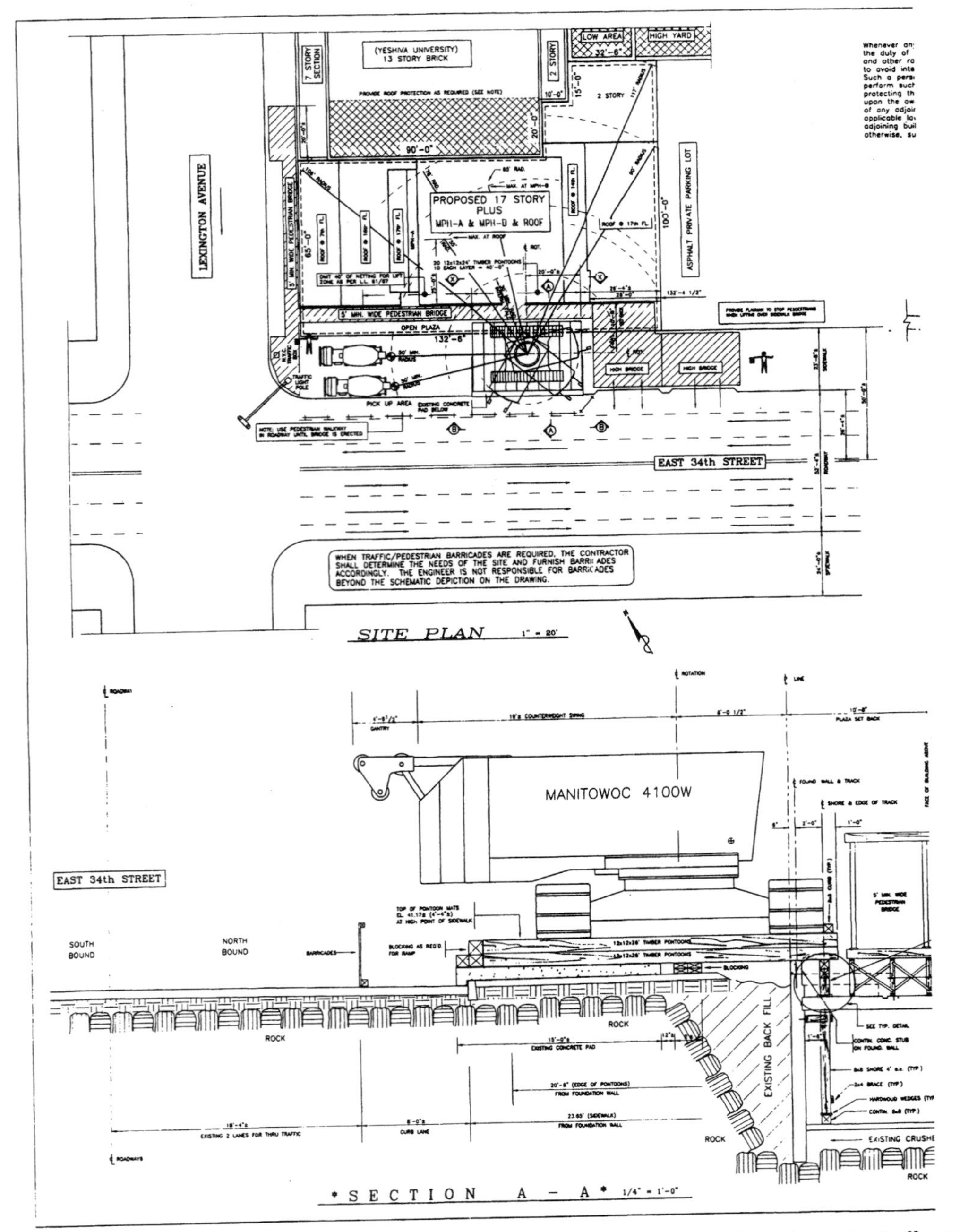

Figure 5.38 Installation drawing for a crawler crane with tower attachment at a typically constricted New York City site. One track of the crane is partly supported on the ramp down to the parking garage of this reinforced concrete residential building. The ramp is shored, and in turn the shores run up to support the ends of the pontoons.

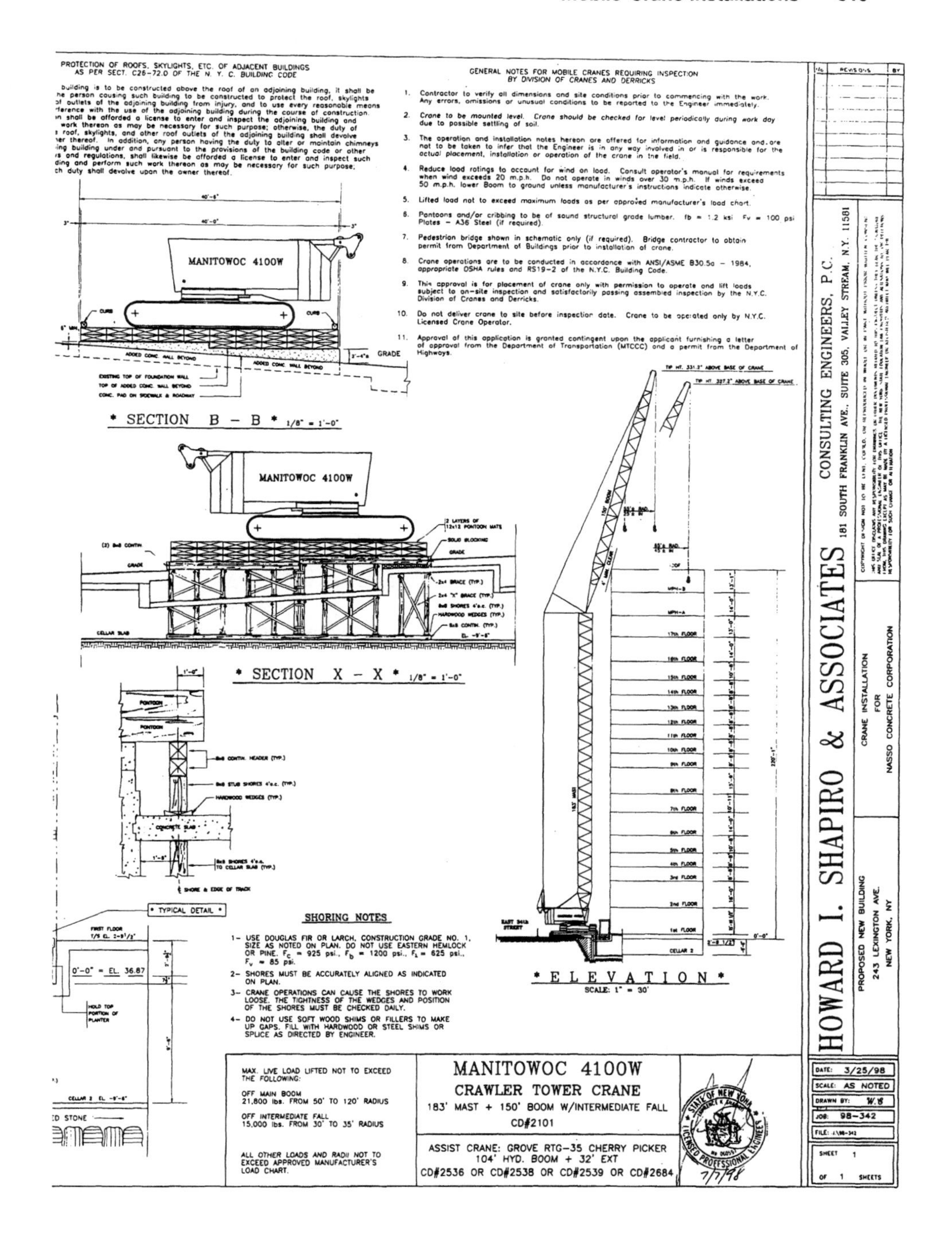

PROTECTION OF ROOFS, SKYLIGHTS, ETC. OF ADJACENT BUILDINGS AS PER SECT. C26-72.0 OF THE N. Y. C. BUILDING CODE
GENERAL NOTES FOR MOBILE CRANES REQUIRING INSPECTION BY DIVISION OF CRANES AND DERRICKS
1. Contractor to verify all dimensions and site conditions prior to commencing with the work. Any errors, omissions or unusual conditions to be reported to the Engineer immediately.
2. Crane to be mounted level. Crane should be checked for level periodically during work day due to possible settling of soil.
3. The operation and installation notes hereon are offered for information and guidance and are not to be taken to infer that the Engineer is in any way involved in or is responsible for the actual placement, installation or operation of the crane in the field.
4. Reduce load ratings to account for wind on load. Consult operator's manual for requirements when wind exceeds 20 m.p.h. Do not operate in winds over 30 m.p.h. If winds exceed 50 m.p.h. lower Boom to ground unless manufacturer's instructions indicate otherwise.
5. Lifted load not to exceed maximum loads as per approved manufacturer's load chart.
6. Pontoons and/or cribbing to be of sound structural grade lumber. fb = 1.2 ksi Fv = 100 psi Plates – A36 Steel (if required).
7. Pedestrian bridge shown in schematic only (if required). Bridge contractor to obtain permit from Department of Buildings prior to installation of crane.
8. Crane operations are to be conducted in accordance with ANSI/ASME B30.5a – 1984, appropriate OSHA rules and RS19-2 of the N.Y.C. Building Code.
9. This approval is for placement of crane only with permission to operate and lift loads subject to on-site inspection and satisfactorily passing assembled inspection by the N.Y.C. Division of Cranes and Derricks.
10. Do not deliver crane to site before inspection date. Crane to be operated only by N.Y.C. Licensed Crane Operator.
11. Approval of this application is granted contingent upon the applicant furnishing a letter of approval from the Department of Transportation (MTCCC) and a permit from the Department of Highways.
MANITOWOC 4100W
GRADE
* SECTION B – B * 1/8" = 1'-0"
* SECTION X – X * 1/8" = 1'-0"
* TYPICAL DETAIL *
SHORING NOTES
1– USE DOUGLAS FIR OR LARCH, CONSTRUCTION GRADE NO. 1, SIZE AS NOTED ON PLAN. DO NOT USE EASTERN HEMLOCK OR PINE. Fc = 925 psi., Fb = 1200 psi., Fv = 625 psi., Fv = 85 psi.
2– SHORES MUST BE ACCURATELY ALIGNED AS INDICATED ON PLAN.
3– CRANE OPERATIONS CAN CAUSE THE SHORES TO WORK LOOSE. THE TIGHTNESS OF THE WEDGES AND POSITION OF THE SHORES MUST BE CHECKED DAILY.
4– DO NOT USE SOFT WOOD SHIMS OR FILLERS TO MAKE UP GAPS. FILL WITH HARDWOOD OR STEEL SHIMS OR SPLICE AS DIRECTED BY ENGINEER.
* E L E V A T I O N *
SCALE: 1" = 30'
MAX. LIVE LOAD LIFTED NOT TO EXCEED THE FOLLOWING:
OFF MAIN BOOM 21,800 lbs. FROM 50' TO 120' RADIUS
OFF INTERMEDIATE FALL 15,000 lbs. FROM 30' TO 35' RADIUS
ALL OTHER LOADS AND RADII NOT TO EXCEED APPROVED MANUFACTURER'S LOAD CHART.
MANITOWOC 4100W
CRAWLER TOWER CRANE
183' MAST + 150' BOOM W/INTERMEDIATE FALL
CD#2101
ASSIST CRANE: GROVE RTG-35 CHERRY PICKER
104' HYD. BOOM + 32' EXT
CD#2536 OR CD#2538 OR CD#2539 OR CD#2684
HOWARD I. SHAPIRO & ASSOCIATES CONSULTING ENGINEERS, P.C.
181 SOUTH FRANKLIN AVE., SUITE 305, VALLEY STREAM, N.Y. 11581
CRANE INSTALLATION FOR NASSO CONCRETE CORPORATION
PROPOSED NEW BUILDING 243 LEXINGTON AVE. NEW YORK, NY
DATE: 3/25/98
SCALE: AS NOTED
JOB: 98-342
SHEET 1 OF 1 SHEETS

loads. The operation is that the crane must hold the load while it is being cut free from the rest of the structure. To do this safely, the operator must take a strain on the hoist line; i.e., the hoist line must be preloaded by the winch until it approximately matches the load to be lifted. When this is done, on being cut loose, the load will be freely suspended. On the other hand, if too little strain is taken, the load will suddenly drop upon being cut free; too great a strain will cause the load to jump up.

The difference between initial load-line strain and load weight can be called *impact*. Its effect on the crane has been discussed in Sec. 4.6 under the heading Inertial Forces Affecting Stability, and in Sec. 3.3 under Linear Motion. Use of a load-indicating device will permit the operator to preload the hoist line and avoid ill effects but only if the load has been reasonably well estimated in advance. As load weight increases relative to the tipping load, the accuracy of the estimate must likewise increase.

5.8 Positioning the Crane

In deciding between two prospective crane positions or more, discretionary tradeoffs need to be weighed, with costs, risks, and collateral effects at stake. Cost considerations might include the relative efficiency of operating from the alternate positions. Risk factors (see Chap. 8) could include the operator's ability to see the picks or the exposure of the public. Collateral effects could include the impact on various trades and on access in the vicinity of the crane position.

Figure 5.39 shows two operating positions on the same site for a large truck crane. One position requires a pontoon deck for access and to support outriggers on one side, and timber cribbing on the other. The second position needs timber and steel plate blocking for all four outriggers. Surely, the first position is more costly to build than the second. But support costs are not the only or overriding costs; faster load placement for long term or repetitive operations could be the governing consideration.

The cost of placing one load is a function of the time required for the operation, which in turn depends upon the crane motions needed to move the load from its initial to its final position. Crane motions, in the order of the fastest to the slowest are:

1. Hoisting
2. Swinging
3. Booming in or out
4. Traveling

Needless to say, for multiple-lift operations this list will help choose crane positions most likely to be operationally cost-effective. As a further refinement, duration of motions should be minimized, and finally, potential operating costs must be weighed against installation costs for each likely crane position. When only a few lifts are to be made, installation costs will be the overriding selection factor.

Boom extension, for telescopic cranes, has not been included in the motion list because of the uncertainty whether it can actually be performed under load. For that reason, it should not be part of a planned cycle operation. Boom retraction can always be done under load. Both it and extension without load are slow operations, about equal to item 3 on the list.

Swing motion tends to cause pendulum action in the load in the direction of swing. Although it is lateral motion having little effect on stability, time may be lost in landing the load. (Experienced operators have the ability to control acceleration and deceleration, minimizing pendulum effects.) Small swing arcs can be traversed at low speeds with little load disturbance and minimal elapsed time. Small swing arcs offer the further advantage of placing both ends of the operation within the operator's field of vision. When this is true, the operator can start a cycle with the assurance that the destination point is clear and ready to receive the load, a condition sure to reduce cycle time.

When radius change is needed during an operation, the work will be considerably slowed. Booming in or out are intrinsically slow procedures; attempts to speed up the motion by using high-range gearing and high engine speed will only increase radial pendulum action and decrease control of the load which could make the crane "tippy". For production work, radius change should be avoided.

Traveling with load (pick and carry), when permitted, is a very slow procedure, partly limited by the smoothness of the travel path. But use of a pick- and-carry procedure may enable a very small crane to do work that would otherwise require a large machine. It is within this context that the usefulness of pick and carry should be evaluated. Section 5.10 covers pick and carry work.

Risks are lower for those crane operations employing few controls and control engagements, low functional motion speeds and durations, and shorter boom lengths and operating radii. This is simply a matter of limitations inherent in machines and the humans who operate and work with them. As those crane and operating parameters change, risks tend to increase in subtle and unquantifiable ways. This is not to say that fast work with long booms or at long radii is unsafe. On the contrary, it merely reflects that low speeds and short booms and radii make a crane less sensitive to inadvertent occurrences and environmental changes. When planning operations, we should prefer to

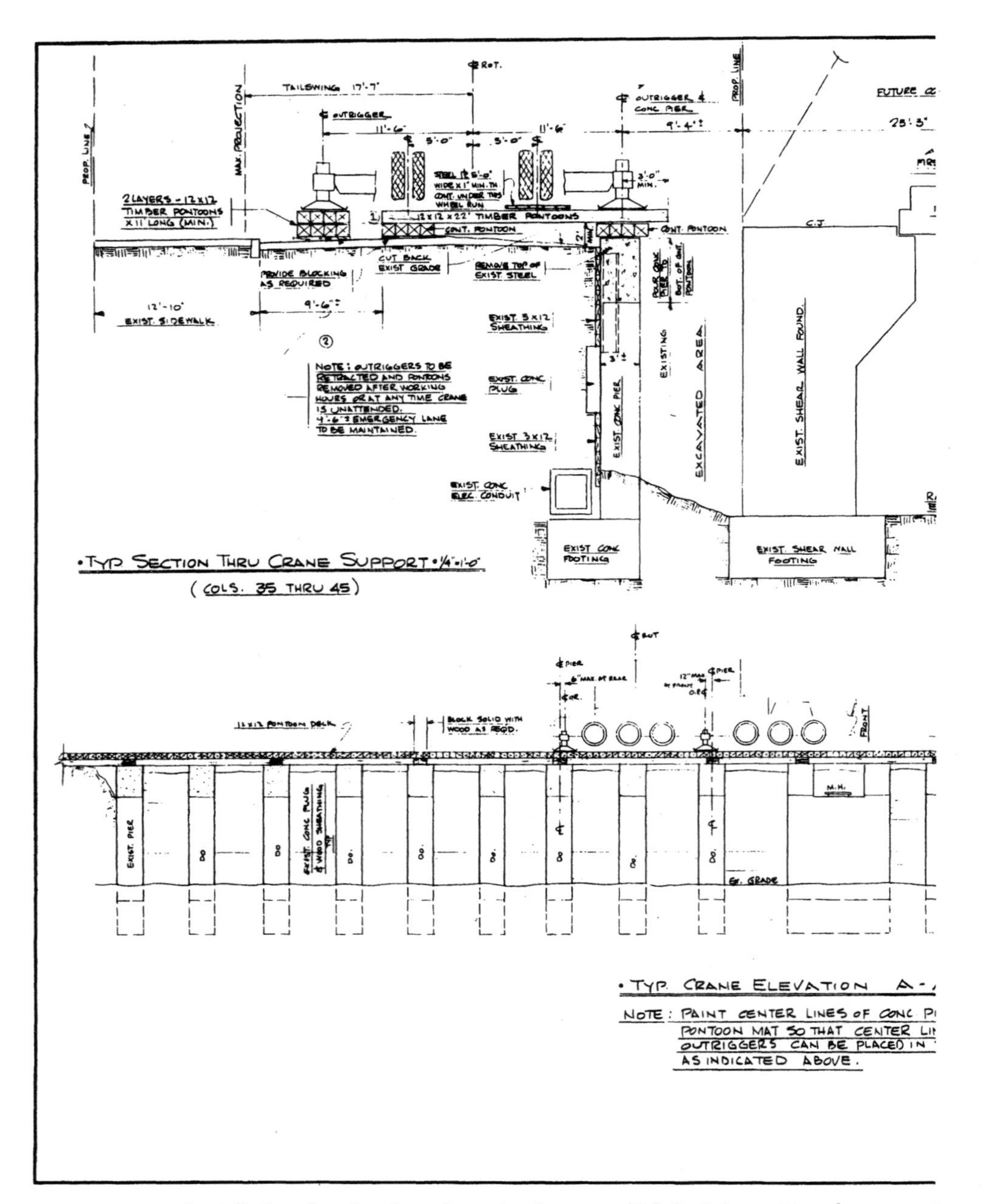

Figure 5.39 Installation drawing for a large truck crane. At left, it is positioned on a pontoon mat with the outriggers supported on a temporary retaining structure. At right, the crane is placed close to a heavy retaining wall capable of sustaining the surcharge.

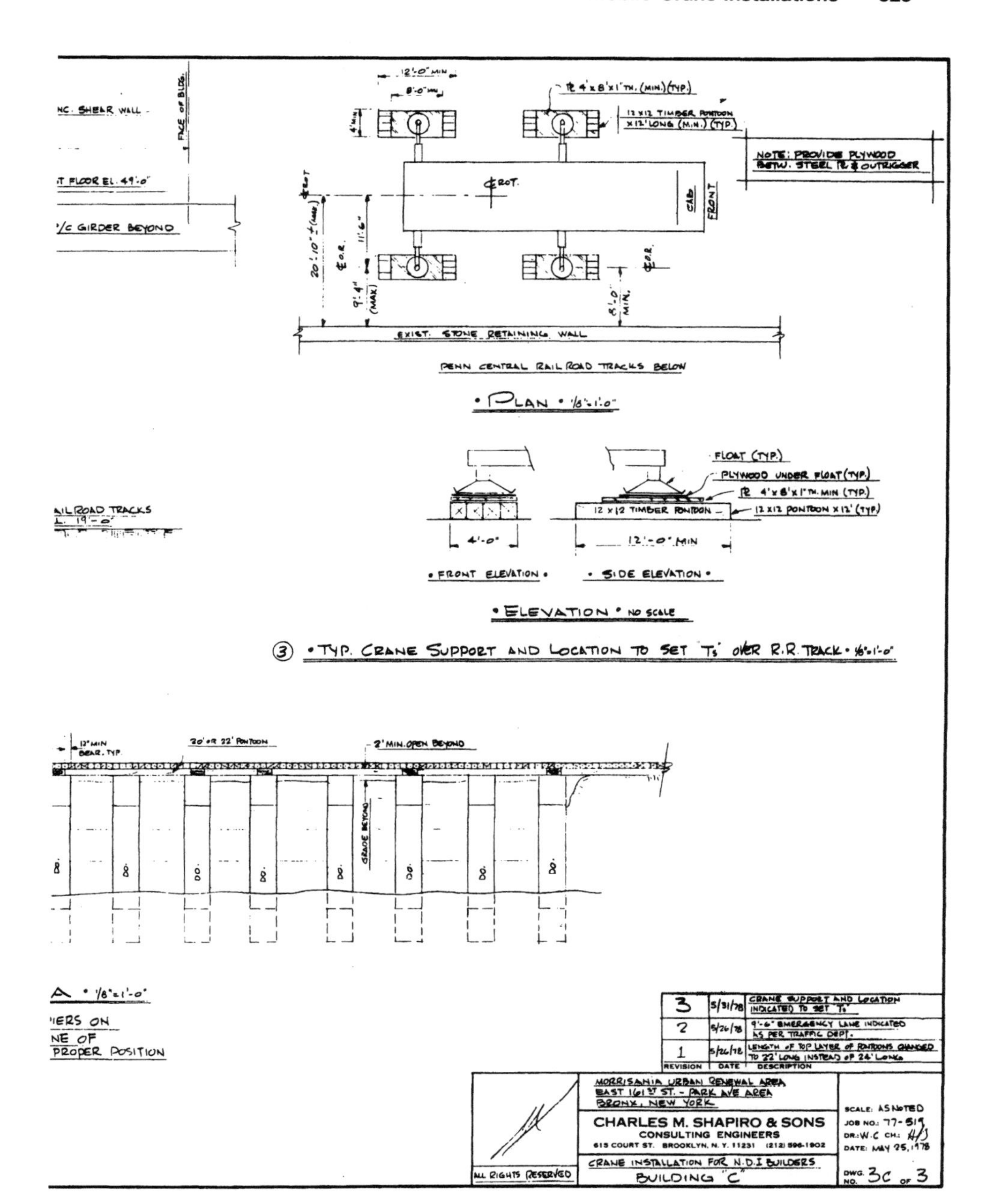

NOTE: PROVIDE PLYWOOD BETW. STEEL PL & OUTRIGGER
EXIST. STONE RETAINING WALL
PENN CENTRAL RAIL ROAD TRACKS BELOW
• PLAN • 1/8"=1'-0"
• FRONT ELEVATION •
• SIDE ELEVATION •
• ELEVATION • NO SCALE
③ • TYP. CRANE SUPPORT AND LOCATION TO SET T's OVER R.R. TRACK • 1/8"=1'-0"
MORRISANIA URBAN RENEWAL AREA
EAST 161ST ST. - PARK AVE AREA
BRONX, NEW YORK
CHARLES M. SHAPIRO & SONS
CONSULTING ENGINEERS
CRANE INSTALLATION FOR N.D.I BUILDERS
BUILDING "C"
ALL RIGHTS RESERVED

work the cranes in their least sensitive configurations but should not fear using any permitted arrangement when necessary.

5.9 Crane Selection

Many will say that it is an art to select the crane most suitable to perform a particular task. We don't agree. Once the technical requirements of load capacity, reach, clearances, and siting have been studied, crane selection is no more an art than deciding which automobile to buy. There are, first of all, economic considerations that dictate the type of crane to be used; second are purely subjective elements that may lead one person to choose brand X whereas another will select brand Y. Nowadays, even a computer can select the crane once the parameters have been established.

It is true that there are some optional features that are unique to individual brands or models, such as self-contained mechanisms for extending and retracting crawler side frames or for undecking superstructures, and that these features may have important cost advantages for some users. These factors must certainly be considered in the selection process, but few of them are important to the ordinary installation. Practical, desirable features are quickly adopted by all manufacturers.

Basic economic factors

When only a few loads must be placed in an operation that will last from several hours to perhaps 2 days or so, a telescopic crane is called for. With these cranes little time is lost in setup, and minimal crews are required, as virtually no assembly work is needed except for the largest machines. For short transit distances, the lower-cost rough terrain models are most economical. For longer transit distances, truck or all terrain cranes become economical.

Rough terrain and all terrain cranes are often used as service cranes on large worksites because of their quick setup, maneuverability, broad range of capabilities, quick movement from assignment to assignment at the site, and one-man crew. They serve well for any task from pick-and-carry work to precise placement of fabricated elements or machinery.

For longer-term operations of from several days to several months, latticed boom truck cranes will prove to be economically viable. Their purchase prices and rental rates are lower than for telescopic truck cranes of similar capacity, but the added labor for assembly and disassembly tends to rule them out for some short-duration work. This may not be so in areas where booms of 100 ft (30.5 m) or so are allowed to be carried folded during transit moves. Folded booms can be erected by a few workers in about one hour.

Crawler cranes are low in cost and rental rate compared with latticed boom truck cranes, but moving them into and out of a jobsite can be expensive; moving and assembly costs for very large machines run into five figures. These machines are therefore usually used for long-term service. In many instances the assembly costs are about the same as for an equivalent truck crane, so that the difference between rental rates clearly signals the minimum time that calls for use of a crawler machine.

If the nature of the work and the soil conditions make it necessary to operate the machine from a pontoon mat and from a number of different work positions, siting costs must be inserted into the decision process. The cost of placing a truck crane will usually be less than that of placing and leveling a pontoon mat.

For repetitive operations, such as concrete bucket work, we have always felt that an advantage is to be gained by using a boom that is a little longer than the minimum needed. Coupled with a 25% derating, this tends to relieve the operator of tipping and clearance anxiety and reduces cycle time. The small added cost in cranage should be more than offset by added productivity.

Some rigging supervisors have expressed similar feelings about crane selection for placing few but difficult loads. In addition, they tend toward cranes with extra capacity. The heavier crane seems to be less sensitive to such things as wind variations and gives the rigging crew both peace of mind and a steadier load. Here the added crane cost is an insurance premium.

5.10 Pick and Carry

The ability to travel with a load on the hook is a great productivity enhancement; crawler, rough terrain and all terrain cranes can do pick and carry work. Because wheel mounted cranes are supported on their tires during pick and carry, and dynamics due to motion are involved, ratings are reduced as compared to those for static work. Crawler cranes may also have lower ratings for travel with load, but their mountings are less sensitive to dynamic effects. To gain maximum advantage from pick and carry capability, it is necessary to understand how the cranes and loads behave and the inherent limitations on this activity. The operator's manual must be consulted for specific limitations for each particular model.

Wheel mounted cranes

Dual cab all terrain and truck cranes, when permitted by the operating instructions to pick and carry, transport lifted loads over the rear. Rough terrain cranes carry their loads over the front. All wheel mounted cranes must travel at very low speeds, generally 1½ mi/h (2.4 km/h) or less. Irregularities in the travel path cause the load to bounce, and this effect is exacerbated by the reaction from the tires. Excessive bouncing, a short duration dynamic action, would not ordinarily overturn the crane, but could, and therefore should be avoided. This motivates two rules for pick and carry;

1. Modulate travel speed to the condition of the travel path, and
2. Keep loads close to the ground, just high enough that they do not hang up from bounce and swing.

The condition of the travel path refers to the uniformity of elevation, or "levelness", of all points along the line of travel (high and low spots induce bouncing), and uniformity of compaction (soft spots induce bouncing too). But, if the side-to-side elevations vary, the load will tend to sway sideways pendulum fashion. This could make side stability tenuous, and should be controlled by use of tag lines. Travel path grade must be considered too. Entering onto a grade brings the crane and load combined CG closer to the front (or rear) tipping fulcrum, reducing the stabilizing moment. Downgrades reduce stability for loads carried in front, upgrades for loads over rear. The load, suspended from the boom tip, moves further from the tipping fulcrum, increasing the effective radius and hence the overturning moment. Coupled with bounce, this can prove troublesome.

Fore and aft load movement can be prevented by tying back or snubbing the load to the crane carrier. A chainfall or Cum-a-long is useful for this purpose because it permits the line to be taken up, pulling the load toward the carrier. Enough take up should be applied to keep the load from swinging and thus avoid shock loading of the snubbing line. Bringing the load closer to the carrier improves stability by reducing the overturning moment, but this should not be done for the purpose of increasing capacity—only for eliminating load swing.

Crawler cranes

Pick and carry limitations are less stringent for crawler cranes than for wheel mounted cranes. The crawler base offers a more rigid and stable platform that is inherently suited to carrying loads during travel. However, limitations vary from manufacturer to manufacturer and model by model; it is necessary to refer to the operator's manual for explicit data. Some crawler cranes are permitted to travel with full rated load, loads over front, rear or side, and up or down stipulated grades, but typically there are restrictions on travel speed and travel path levelness, and perhaps boom (or jib) length as well.

As with wheel mounted cranes, loads should be snubbed to the car-body to prevent fore and aft load swing when the boom is over front or rear. But when loads are over the side, there will be a tendency for load swing in the direction of travel that cannot be restrained by snubbing to the crane. Load swing will then side load the boom which could pose an unsatisfactory structural risk.

When a load approaching rated load must be carried over the side, and the travel path surface is not ideal, another means for snubbing should be used. For example, front and back tag lines, parallel to the travel direction, can be used for loads capable of being restrained by one or two persons at each end. For heavier loads, it may be necessary to snub the taglines over the bumpers of parked trucks or lugs on bulldozers or other such machines, giving the tagmen better purchase for control. Those anchorages may have to be moved as travel advances. The object, of course, is to eliminate or minimize load swing in all directions so that both stability and structural integrity of the crane can be maintained.

Generally, crawler cranes should travel with the load over the rear, that is, backwards with reference to the travel direction, and the drive sprockets should also be positioned at the rear. The exception would be travel with very light loads or on very firm support surfaces. When traveling with the load at rear, the operator does not see the travel path and must be controlled by a signalperson. But the rear of the tracks will carry heavier loading than the front—the front may in fact be unloaded. This will make the travel movement smoother on yielding travel surfaces as the tracks will readily advance over the ground surface.

With the load in front, and therefore heavier track loading there, the front end will tend to dig into the earth, making for greater resistance to movement and promoting less uniform motion. Moreover, when turns are attempted, the heavily loaded portions of the tracks will gouge their way through the earth, sometimes fouling the treads and track rollers. Carrying loads over the rear makes for smoother, cleaner travel. Operation planners have to decide whether the advantage of easier travel outweighs the disadvantage of the operator not seeing forward to anticipate what is coming.

Changes in grade along the travel path can be troublesome for crawler cranes doing pick and carry. It is not a question of power needs (crawlers have plenty of power) but one of load control; the longer the boom, the greater the potential problems. With their long rigid side frames, when crawler cranes crest a "hill", the CG can break over the crest and tilt the crane forward quite rapidly. For example, travelling across a 1% change in grade for a crane with 20 ft (6.1 m) bearing length and a 300 ft (91.4 m) boom, will cause the boom tip to move about 3 ft (910 mm) without considering elastic displacements. With the load either front or rear, the snubbing line could experience a strong shock and break. With the load over the side, excessive side loading could result, collapsing the boom in extreme cases. Also, the side load-swing could cause the superstructure to swing, putting the load where it may not be wanted. This tells us that the crane must break over crests very slowly, even for slight grades and small grade changes. Better yet, crests should be cut down to furnish a gradual transition whenever this is feasible.

5.11 Multiple Crane Lifts

When two cranes are used to lift a single object, the risk is more than double that of a single crane lift. The use of more than two cranes in a single lift causes a disproportionate increase in risk. The reason for the disproportionality is the interaction of the cranes on each other.

Consideration of risk makes clear why multiple crane lifts should be avoided if possible. Multi-crane lifts are critical operations requiring formal planning. The scope of planning and the risk control measures that need to be employed depend on the circumstances of each operation and are matters of judgment and experience. Although most decisions must be made case by case, there are a few rules that should be considered as absolute. The purpose of the discussion that follows is to provide guidance in planning multiple crane lifts and in making risk control decisions.

Risk control measures, as used herein, refer to all of the actions taken to keep the parameters of the lifting operation within limits established by the crane manufacturer, the operation planners, and industry or governmental authority standards. Those actions could include such things as crew training, briefing and "dry runs", establishment of communication methods and protocols, preparation of lift plans and monitoring of their implementation, site, soil and level checks, means used to coordinate and control crane movements, load weight verification and verification of weight sharing between cranes, establishment of wind limits and pre-operation weather checks, risk analysis and contingency planning, equipment inspection or certification, and barricading and personnel controls.

The Absolute rules

1. A single crane lift at full capacity is usually better than a multiple crane lift at less than full ratings.

2. On multiple crane lifts, the planners, field supervisors, and crane operators must be experienced and seasoned individuals.
3. If a multiple crane lift cannot be avoided, use as few cranes as possible. Avoid four-crane lifts on rigid loads unless two of them can be coupled or equalized to create the equivalent of three cranes. Four cranes or more should not be attempted unless full engineering support and rigorous risk control measures are employed.
4. A formal, written lift plan is advisable for multiple crane lifts. The plan, at a minimum, should set out the position of each crane and the load at the start of the operation, as well as the crane lift radii and applicable lift loads, the movements to be made by each crane including sequence and radius changes, and the operational and risk control measures to be used.
5. The weight of the lifted load, and its center of gravity (CG), must be known with reasonable accuracy early in the planning process. The distribution of load to each crane, and the resulting effect on their supports, should be determined for each phase of the operation. The weights of hook blocks and lifting accessories must be considered.
6. Crane load lines must be kept plumb within set tolerances at all times throughout the course of every multiple crane lift. The selection of crane movements to be made during the operation and the risk control measures chosen to protect the operation must be considered together with that requirement.
7. Normally, complex motions should not be performed. Avoid hoisting and swinging at the same time and hoisting and luffing. No motion should be combined with travel. The cranes can, of course, hoist together as needed, and any time that one crane swings, travels, or luffs it will be necessary for the other crane(s) to move synchronously in order to keep the load lines plumb.
8. Cranes can be used to near full capacity on multiple crane lifts that are fully engineered and where rigorous operational control and risk control measures have been established. In more typical circumstances, loads should be 75% of rated load more or less.

Two-crane lifts

The ideal situation to expect when using two cranes would be a lift of uniform weight, such as a bridge girder, by identical cranes which are symmetrically attached, that is, hooked on the same distance from each end of the girder. Each crane would be equally loaded. With their load lines kept plumb, they would remain equally loaded while the load is in the air; they will be safe provided the cranes are both kept within their rated radii, all other normal crane operation measures having been taken (Fig. 5.40). The load should be kept generally level, although only nominal attention to this is necessary since the load will not shift from one crane to the other if the load goes a little out of level.

When the load is landed, the weight distribution between cranes will not change if both cranes land their loads at the same time. But, if the crane hook-up points and the landing contact points are not the same, load can shift to the other crane if one end is landed first. Load shifts of that nature are both predictable and quantifiable during the planning process. The point is that for even the most ideal two-crane lifts, planning is necessary to manage the added risks, those that are not present when a single crane is used alone.

For the typical two-crane lift neither the load nor the attachment points are symmetrical. The cranes often also differ in either model or configuration. Distribution of the load between the cranes will follow a ratio that follows from the relative position of the load CG between the two crane hooks. The crane closest to the CG will carry more of the load than the farther crane. In Fig. 5.41, the horizontal distance from the CG of the load to the hook of the larger crane A is a and the distance from the CG to the hook of crane B is b. The horizontal distance between the two hooks is then $a + b$. The load on crane A can be found by taking moments about the hook of crane B.

Figure 5.40 A double truss 290 ft (88 m) long and weighing 148 tons (134 t) being lifted by two Series 2 Manitowoc 4100W crawler cranes with 160-ft (49-m) booms prior to traveling several hundred feet with the load. Including lifting attachments, each crane is carrying 84 tons (76 t), 87.5% of rated load, during this fully engineered and controlled lift. (*Douglas Steel Fabricating Corporation.*)

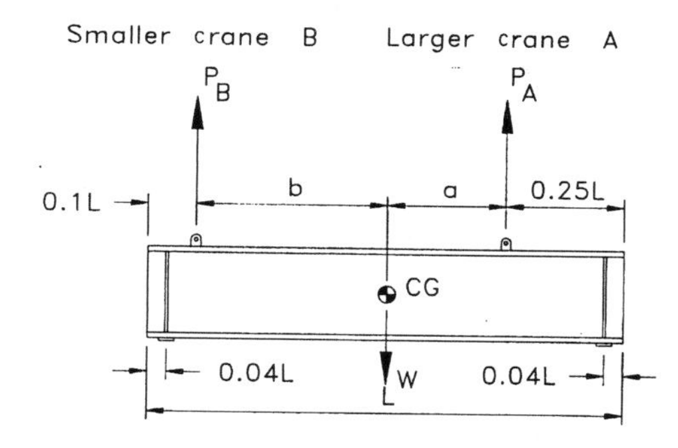

Figure 5.41 A girder dimensioned for determining the load distribution during a two-crane lift.

The moment introduced by the load W must be balanced by the moment of the load P_A on crane A, or

$$Wb = P_A(a + b)$$

$$P_A = Wb/(a + b)$$

Inserting numbers into the relationship given above, suppose a is 10 ft (or meters) and b is 30 ft (or meters). The hooks are then 40 ft (or meters) apart. The load on crane A will be 30/40 or ¾ of the lifted load; the rest of the load, or ¼ of it, will be on crane B.

If the load is permitted to go out of level, the horizontal distances will become shorter, but their proportions will remain the same. For example, if the load is tilted to the point where a measures 9 ft (or meters), it will be at an angle of about 26° to the horizontal and b will become 27 ft (or meters); the distance between the hooks will then be 36 ft (or meters). The load on crane A will be 27/36, or ¾ of the lifted load as before. This is true, or true for practical purposes, for nearly all loads. The only situation that will cause a significant shift in hook load, when the lifted load goes out of level, occurs when the vertical position of the load CG is very much above or below the hook-up point levels with respect to the distance between the hooks or if the hook-up points are considerably offset from one another vertically. For example, if the girder CG is 1.5 ft (or meters) below the hook-up points, crane A will actually carry just over 73% of the lifted load and not ¾, a difference of less than 2% for the extreme case of 26° off level.

Suppose crane A landed its end first. Using Fig. 5.41 but changing the dimensions causes new relationships to develop. Taking a as equal to $0.25L$ (which places the CG at the center of the girder), b becomes $0.4L$. When the right end of the girder is landed at its bearing $0.04L$ from the girder end, crane A becomes unloaded. Taking moments about the right end bearing, the moment of the load W must be balanced by the moment of the load on crane B:

$$W(0.5L - 0.04L) = P_B(L - 0.04L - 0.1L)$$

$$0.46WL = 0.86P_BL$$

$$P_B = 0.53W$$

Whereas crane B carried only 38% of the load while the girder was in the air [0.25/(0.4 + 0.25) = 0.38], landing the larger crane's load first increases the load on crane B by almost 40%. Had crane B been near capacity during the lift, this landing sequence could have caused overload and failure. The same danger would occur should crane B try to lift the load first.

The opposite procedure, landing the crane B end first, would only increase the load on crane A from 62% of girder weight to about 65%. Clearly, the load lifting and landing procedures must be studied as part of planning and the lift plan must specify the sequence whenever a meaningful load shift can occur.

When the load line goes out of plumb, the effect on the crane could be far more severe than the eye would indicate. Strength, stability, or both could be affected depending on the direction of the misalignment. If the hook has been pulled out past the boom tip, which is called *off-lead,* the effect is equivalent to adding load on the hook or booming-out to a greater radius—stability is reduced. Multiple crane lifts are typically heavy lifts at relatively short radii where a small increase in radius corresponds to a large loss of stability and hence of lift capacity. What may appear as a small off-lead at the start of a lift can become quite significant as the load is raised. With a rising load, stability loss increases along with the possibility of overturning, loss of the cranes, and destruction of the load.

As part of planning, out-of-plumb tolerances can be examined. If, for example, the crane shown in Fig. 5.42 has an off-lead of magnitude e, the out-of-plumb load line will exert a vertical force component on the boom tip equivalent to load W and a horizontal component of pW where $p = e/h$ and h is the distance from the boom tip to the hook. Taking moments about the boom foot pin F,

$$M_F = Wr + WpL \sin \theta$$

where θ is the boom angle and L is the length of the boom. The vertical load effect producing an equivalent moment can be found by dividing through by r, the radius measured from the boom foot:

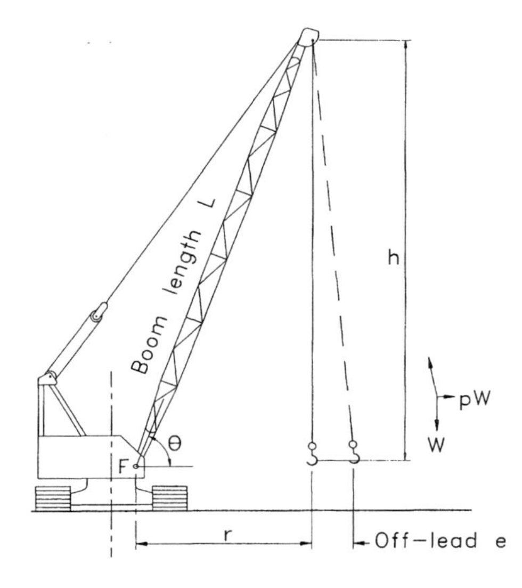

Figure 5.42 A schematic view of a crawler crane annotated for calculating off-lead effects.

$$W_{eff} = W + W(pL \sin \theta)/r$$

Substituting $r = L \cos \theta$,

$$W_{eff} = W(1 + p \tan \theta)$$

The term in parentheses is the load multiplier that reflects the effects of off-lead. For a 75° boom angle, and several values for p, for example, the term in parentheses becomes

1.019	when $p = 0.005$
1.037	when $p = 0.010$
1.075	when $p = 0.020$
1.149	when $p = 0.040$

If a two-crane lift is being conducted using two identical cranes facing each other with long booms at 75° boom angles, and if the load is initially suspended 200 ft (61 m) below the boom tips with an off-lead of 1 ft (305 mm), $p = 1/200 = 0.005$. The equivalent vertical load will be

$$W_{eff} = W(1 + 0.005 \times \tan 75°) = 1.019W$$

which reflects about a 2% increase in load effect.

Should the load be hoisted to 25 ft (7.6 m) below the boom tips, the off-lead will remain 1 ft, but p becomes $1/25 = 0.040$. The equivalent load becomes $1.149W$, an increase of about 15%; a load that started out as 75% of rated load becomes effectively 86% after hoisting. Had the boom angle been 80° instead, the increase in effective load would be about 23%. That is because the tangent function increases with boom angle; the higher the boom angle, the more pronounced the load increase effect. Also, the plumbness parameter p directly controls the load increase effect. For any off-lead, that parameter will increase as the load is hoisted.

Operating procedures and plumbness checks can be set up to assist the crane operators in adjusting their booms to bring the load lines to within the predetermined plumbness tolerance, but more importantly, the cranes must possess adequate lift capacity to safely support the effective load.

These concepts show the relationship between crane capacity and out-of- plumb tolerance that must be maintained at the jobsite. Unlike a single crane lift where the load CG *will* align itself below the hook, in a multiple crane lift the load line *will* be out of plumb. With the load grounded, as the load line is drawn in on the winch drum, the crane starts to become loaded; the boom hoist and pendant ropes start to stretch, the boom shortens under compressive load, the tolerance in the swing bearing is taken up, and the superstructure and the undercarriage frames elastically distort, all of which result in an increase in radius. Normally, operators try to compensate for this by starting out with the boom at a shorter radius, but they do this by judgment. The result will be an out-of-plumb load line unless judgement is perfect.

The other effect that could result from an out-of-plumb load line is side loading. This occurs when the deviation from plumb is near right angles to the boom, such as when one crane swings and is permitted to pull the second crane with it. Side loading is a structural situation that could cause boom collapse without warning. In the United States, mobile-crane booms are designed to sustain a side load of 2% of rated load (3% for telescopic cranes), which is intended to account for the minor dynamic effects of a freely suspended load. Experience has shown that this allowance is adequate for single crane hook work. However, side loading causes appreciable, nonlinear increases in boom stress. Lateral, or sideways, out of plumb in excess of $p = 0.020$ is therefore very hazardous and can result in boom collapse. It would be imprudent to set the out-of-plumbness tolerance too close to $p = 0.020$, since this would leave no allowance for dynamic side loading, wind, or for the crane to be out of level.

The reader may have started to think that perhaps undue emphasis has been placed on out-of-plumb load lines. However, the reason for highlighting this hazard is that almost everything that is done during a multiple crane lift could lead to the load lines going out of plumb. Therefore, most risk control measures are concerned, in the end, with maintaining the lines plumb.

In a two-crane lift, for example, assume that the crane hooks are at a set distance from one another when the load is in the air and that the load lines are plumb; each crane is at a set radius. If one crane lowers (or raises) the load while the other holds fast, the load goes out of level and the horizontal distance between hooks becomes shorter. This means that the hooks are pulled toward one another—the load lines go out of plumb. But, an out-of-level load is really not critical in this regard. Monitoring by eye or by simple means can certainly keep the load within 5° of level, which will usually have an insignificant effect on the load lines. The greater the distance between hooks, the more significant levelness becomes. For a distance of 100 ft (30 m), 5° out of level brings the hooks almost 5 in (122 mm) closer, whereas 10° would bring them 1.5 ft (460 mm) closer.

When one of the cranes swings, it is usual practice to leave the superstructure of the second crane free to follow the other crane. But even with the low-friction swing bearings used on mobile cranes, the second crane will resist swinging and must be pulled around by the first crane. That frictional resistance will cause both cranes to be side loaded as will be evidenced by their load lines going out of plumb after the swing commences. When loading levels are high enough for this to be important, both cranes must power swing together at very slow speed, or each crane must swing separately in small increments that will not cause out-of-plumb tolerances in the load lines to be exceeded.

If two cranes with initially plumb load lines are facing one another and one of the cranes booms-up, the distance between boom tips will increase, but the load will keep the hooks at their original horizontal separation. The load lines, however, will now be out of plumb with an off-lead that will act as though the load was at a greater radius than the boom tip for each crane. The radius change effect will only be equal if both cranes were originally sharing the load equally and if the boom tips were at the same height. However, the stationary crane, the one that did not boom-in, will now be at an increased effective radius. For any other condition, each crane will experience a radius change of different magnitude, but the stationary crane is usually the one that will suffer an effect greater than its original radius. The crane operators cannot perceive the radius change effect from their boom angle indicators; however, these changes can be calculated using principles of equilibrium, and tolerances can then be established as part of the planning and risk control measures.

If two cranes are going to travel with a common load, there are three ways in which the movement can cause the load lines to go out of plumb. If the travel path of either crane is not level, or if there are "soft spots" that will permit a crawler track or tire to "dig in" and place the crane out of level, the boom tips will change position relative to one another and the load lines will go out of plumb, as they will if one crane travels faster than the other and runs too far ahead. Lastly, the same effect occurs if the cranes are permitted to travel on paths that are not truly parallel (Fig. 5.43). If reduced loading levels are specified when travel is required, the more critical effect of these travel irregularities is often side loading of the booms and the danger of structural collapse. Each of the travel conditions may be analyzed, however, and control tolerances established accordingly.

Now, after having discussed the ways in which hoist lines go out of plumb, it has become more evident why loads should be held to 75% (more or less) of rated load for most multiple crane lifts. Although the 75% figure is an arbitrary number, it is nonetheless practical for the general case, but certainly not an absolute limitation. If we assume that at full rated load all of the allowable boom strength is being utilized (including an allowance for some wind and a 2% side load), no meaningful spare strength will remain for the inevitable side loading or off-lead that occurs with the restrained loads which are characteristic of multi-crane lifts. At 75% of rated load, however, spare strength will be made available for those additional effects, because the reduced hook load will subject the boom to a lower level of direct stress. This is clearly true for latticed booms, but less so for most telescopic booms.

The configuration of a telescopic boom makes it behave in an entirely different way than does a latticed boom, in part because there

Figure 5.43 The travel path for the operation shown in Fig. 5.40 has been made of leveled and compacted crushed stone ballast. At right are distance flags used to keep the cranes abreast of one another. A track centerline has been sprayed on the ballast for maintaining alignment, as seen at left center. (*Douglas Steel Fabricating Corporation.*)

are no pendants supporting the boom tip. That difference makes telescopic booms more susceptible to side loading than latticed booms; the pendants of a latticed boom counteract part of the side load. That is why telescopic booms are designed for a 3% side load rather than the 2% used for latticed booms. With freely suspended loads, telescopic boom crane working motions do not produce greater side loading than other cranes.

Finally, we reiterate that the 75% limit can be exceeded; it is not by any means absolute. However, a note common to mobile crane rating charts stipulates that the ratings given apply when loads are freely suspended. During multi-crane lifts, loads are not freely suspended; because they are attached to a common load, the cranes interact with each other. Therefore, lifted loads must be less than rated loads, but the amount of reduction is a matter of judgment.

Reductions can be less than 25% of rated load when the nature of the operation and site conditions are such that there will be little tendency for the hoist lines to go out of plumb, or when the operation is suitably controlled. A suitably controlled operation is one in which meaningful, realistic risk control measures are in place to keep conditions and crane motions within predetermined limits. Predetermined limits apply to the portion of the lifted load going to each crane as the operation proceeds, and to the off-lead or side loading to which each crane will be exposed, a function of how much the hoist lines will be permitted to go out of plumb. Risk control measures include establishing engineering requirements, or field monitoring procedures that will be followed, or a combination of them, to keep crane loads to known values and to maintain hoist lines plumb to within stipulated tolerances.

In short, the capacity reduction used for a multi-crane lift should be a realistic measure of the inability to achieve the "freely suspended" load condition, which is a requirement for lifting full rated loads specified on all mobile crane rating charts.

Tailing operations

Vessels, or for that matter many loads that are longer than wide, are shipped lying down and must be raised to vertical at final destination. This uprighting procedure is often called *tripping*. When loads are tripped, much attention is paid to the main crane attached to the top of the vessel, but the smaller crane at the bottom, the *tailing* crane, is too often given short shrift and undersized. The tailing crane is also critical to the success of the lift, and there are subtleties to tailing work that should be understood.

Earlier in this section, there was a discussion of how load can shift between a pair of cranes lifting a symmetrical load, and how relatively insensitive the cranes will be to a moderately out-of-level load. Now, however, we must consider a load going out of level in the extreme; the load must be rotated 90°. What may be minor effects when lifting an essentially horizontal load can become significant during tripping.

For tripping operations, typically the main crane hoists the vessel while the tailing crane lifts the bottom a little and moves horizontally to keep the horizontal distance between hooks close to ideal and thereby minimizing off-lead or side loading. As mentioned earlier, the distribution of load between the cranes will remain constant as the load is tripped if the CG and crane attachment points are on a common line. This could be called the ideal case, but it suffers from one shortcoming: for tripping, it doesn't work in practice. As the load approaches vertical, the boom tip of the tailing crane must go inside the load if the hoist line is to remain plumb. There are very few loads indeed for which that is possible. The tailing crane attachment point should therefore normally be at or above the upper edge of the load, while the load is still horizontal, so that the boom tip will remain clear when the load is vertical.

For vessels, the main crane is commonly attached to lift points, usually trunnions, which are mounted on each side of the load along the longitudinal centerline. The tailing crane would be attached to a lift point at the upper edge of the load, but close to the bottom of the load when vertical. With this arrangement, at the start of the lift the load distribution will be defined by the horizontal distance of the load CG from each of the lift points. As the angle of the load centerline to the ground increases, the portion of the load carried by the main crane will gradually increase. However, as the longitudinal centerline of the load approaches vertical, the main crane burden starts to increase at a faster and faster rate.

This load sharing between the cranes can be demonstrated by referring to Fig. 5.44. With the vessel horizontal, as in Fig. 5.44*a*, the load M on the main crane is found by taking moments about the tailing crane attachment point

$$M = Wb/(a + b) \qquad T = W - M$$

where W = weight of the vessel
M = the part of the load on the main crane
T = the part of the load on the tailing crane
a = the horizontal distance from load CG to main hook-up point with the load horizontal
b = similar distance to tailing hook-up point

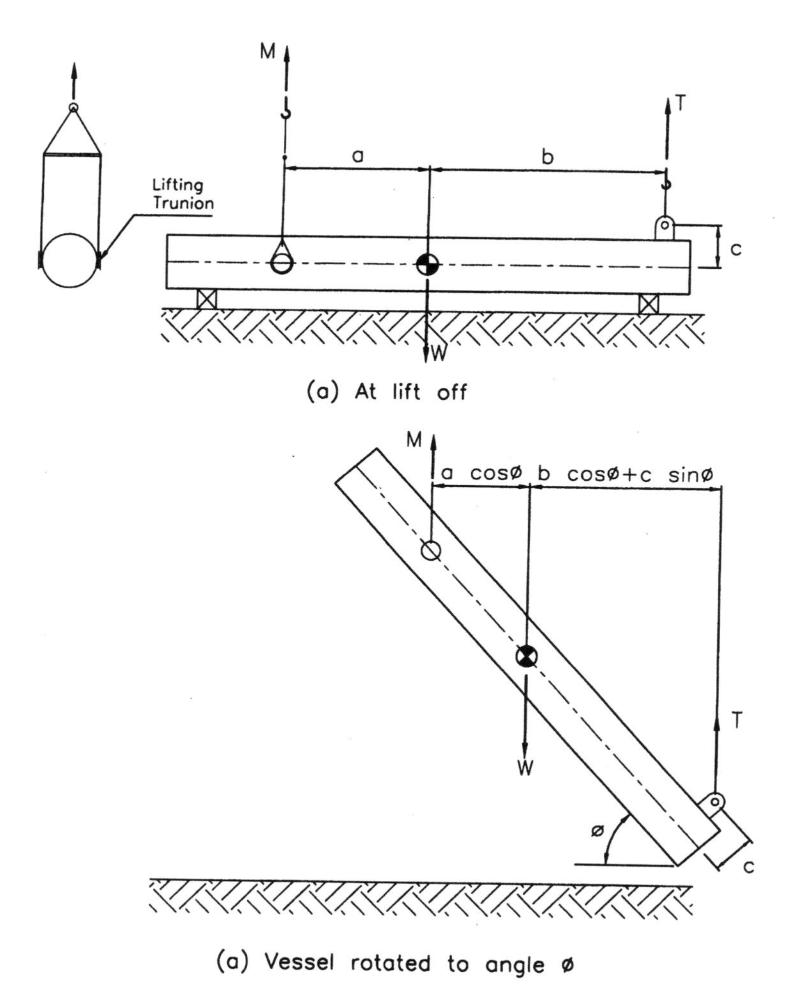

Figure 5.44 A vessel annotated to demonstrate load distribution changes that take place between a main and tailing crane as the load is tripped. The main, or upper, crane lifts straight up while the tailing crane must move toward it as the load is lifted.

The main crane hoists the vessel as in Fig. 5.44*b* until the centerline makes an angle ϕ with the horizontal. Taking moments again creates an expression for M

$$M = W(b \cos \phi + c \sin \phi)/(b \cos \phi + c \sin \phi + a \cos \phi)$$
$$= W(b + c \tan \phi)/(a + b + c \tan \phi) \quad (5.45)$$

where c is the vertical offset of the tailing hook-up point from the load CG when the load is horizontal.

When $\phi = 0°$, $\tan \phi = 0$ and Eq. (5.45) becomes identical to the expression derived for a horizontal vessel. When $\phi = 90°$, $\tan \phi = \infty$, making the terms in parentheses in the numerator and denominator both infinite and therefore equal, and indicating that $M = W$, or that the main crane takes the entire load. The tangent function increases rapidly approaching 90°. This means that as the vessel approaches vertical, the load on the main crane will quickly approach vessel weight, while concurrently tailing crane load rapidly drops. Therefore, for smooth load transfer without causing the load lines to go too much out of plumb, the last 5° or so of rotation should be accomplished by slowly booming out with the tailing crane, and slowly lowering the tailing hoist line when and if necessary.

Another possible lift point arrangement would have the lift attachments for both cranes at the upper edge of the load when horizontal as shown in Fig. 5.45*a*. The crane loads P_a and P_b can be found by taking moments as done previously.

$$P_a = \frac{W(b \cos \phi + c \sin \phi)}{(a \cos \phi - c \sin \phi + b \cos \phi + c \sin \phi)}$$

$$= \frac{W(b + c \tan \phi)}{a + b} \tag{5.46}$$

From Eq. (5.46) we find that $P_a = W$ when $\tan \phi = a/c$ which tells us that $\phi < 90°$. Although perhaps obvious from examination, that value for ϕ is actually the angle at which the crane attachment at the high end will be directly above the load CG, or in other words, the angle of the load when the tailing crane is released.

This situation, however, poses problems and potential dangers. The tripping operation cannot be easily completed unless the corner is lowered to the ground, which may cause damage. In order to complete the tripping in the air, it will be necessary to use horizontal (or near horizontal) lines at the top and bottom of the load as pulling and holdback lines. There are three cases that need to be examined. Case 1 is when the CG lies on the diagonal line from the high lift attachment to the lower corner of the load. When that line is vertical, the CG will be directly above the corner and the load will be in balance. The load can now be tripped only if pushed or pulled further, but the second crane must then furnish a balancing force to rotate the load to vertical.

Case 2 is when the load CG is above the diagonal line. When the CG is directly below the high attachment point, rotation can continue no further unless the corner of the load is placed on the ground, or helper lines installed to complete the tripping in the air. But, if placed on the ground, the load will tend to fall into a vertical position unless

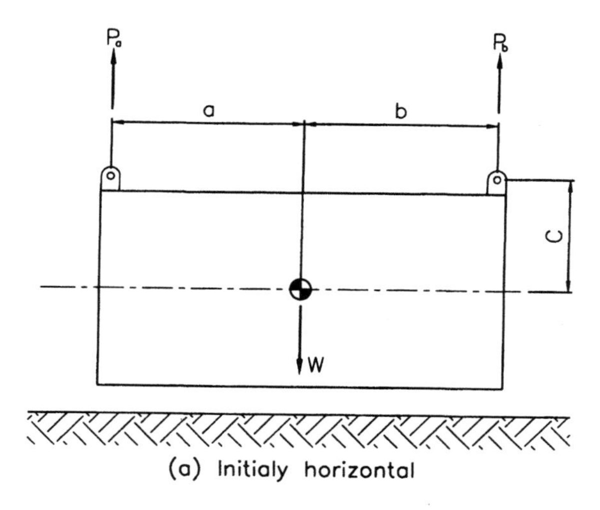

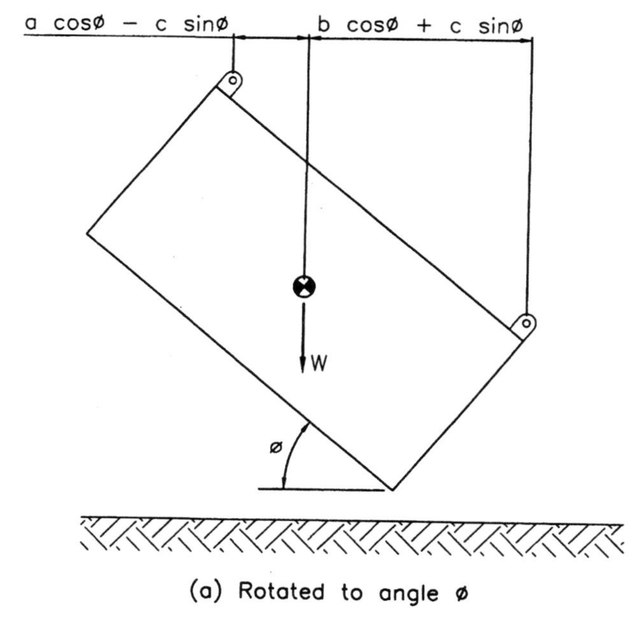

Figure 5.45 An arbitrary load annotated for calculating load distribution changes as the load is tripped by two cranes with lift points at or near the top surface.

restrained and controlled by the second crane. For Case 3, the CG of the load lies below the diagonal line. When the load CG has been rotated until directly below the high attachment point, and the corner placed on the ground, the load will tend to fall back to the horizontal position. The second crane cannot prevent this. Therefore, additional equipment is needed: a horizontal pulling line to force rotation to continue toward vertical, and a restraining line to limit side load or off lead. But, when the load CG passes the corner, the load will tend to fall forward to vertical. It must be restrained and controlled by the second crane or the helper lines.

Another possible lift point arrangement must be avoided unless special precautions are taken. When the lift points are positioned such that the load CG is higher than both lift points, the load will tend to be unstable and could roll over.

The last potential lift point arrangement is the best; it is shown in Fig. 5.46. When load configuration and other conditions make this arrangement feasible, it has three main advantages to offer

1. Neither crane will be required to support the entire load weight during any part of the operation,
2. The load can be tripped in the air, and
3. Both the main and tailing crane will always be well loaded, and therefore the load will be stable and controllable.

Using the same procedures as earlier employed

$$P_a = \frac{W(b \cos \phi + c \sin \phi)}{b \cos \phi + c \sin \phi + a \cos \phi + d \sin \phi}$$

but the only points of interest are for $\phi = 0°$ for a horizontal load and $\phi = 90°$ for a vertical load. Inserting those values

$$P_a = Wb/(a + b) \text{ when } \phi = 0°$$

$$P_a = Wc/(c + d) \text{ when } \phi = 90° \qquad P_b = W - P_a \tag{5.47}$$

During tripping, P_a will not exceed the larger of the values from the equations above, therefore no other values need to be calculated.

Example 5.11 A vessel weighing 150 tons (136 t) must be tripped to vertical. With the vessel horizontal, the main crane will be attached to the load 90 ft (27.4 m) from the CG, while the tailing crane will be 125 ft (38.1 m) from the CG. The vessel CG is on the longitudinal centerline. It is possible to place attachment points either 6 ft (1.8 m) above the centerline or 4 ft (1.2 m) below. How much load will the main and tailing cranes have to lift for various attachment point arrangements?

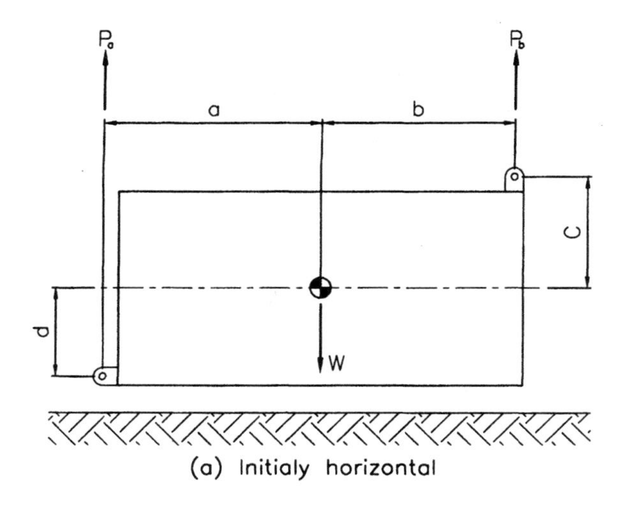

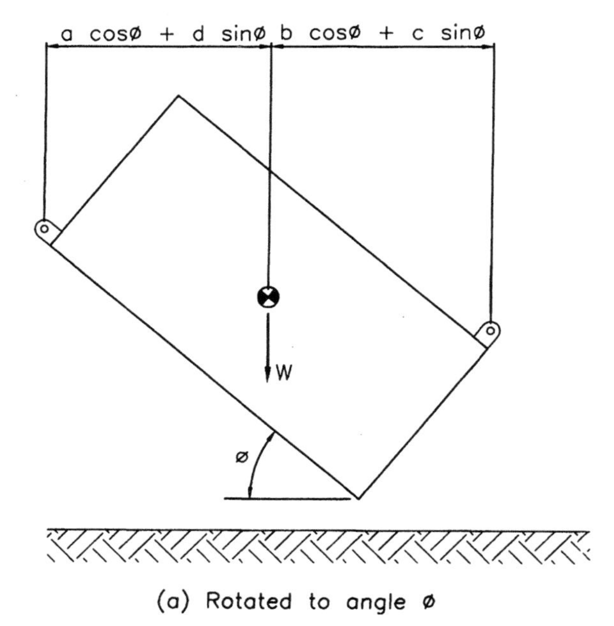

Figure 5.46 An arbitrary load annotated for calculating load distribution changes as the load is tripped by two cranes one lifting from at or near the top surface and the other from a low lifting point.

(a) Try an arrangement as shown in Fig. 5.44, with the main crane attached on the centerline and the tailing crane above it. Using Eq. 5.45, when the vessel is horizontal the tangent terms zero out and

$$M = 150 \times 125/(125 + 90) = 87.2 \text{ tons } (79.1 \text{ t})$$

$$T = 150 - 87.2 = 62.8 \text{ tons } (57 \text{ t})$$

When the vessel is vertical, the main crane will support the entire load and the tailing crane will be unloaded.

(b) Check the set up of Fig. 5.45, with both lifting attachments at the top. From Eq. 5.46, with the vessel horizontal, the tangent term will drop out and the results will be as above. The vessel CG will be directly below the main crane attachment point when

$$\phi = \tan^{-1}(90/6) = 86.2°$$

At that angle, the base will not be below the CG (125/tan 86.2 = 8.3 > 6), or in other words, the load CG is below the diagonal line from the upper lifting point to the lower corner. The vessel cannot be tripped to vertical without additional help. A holdback will have to be attached at the bottom and a pulling line in the opposite direction near the top. The maximum load on the main crane will be vessel weight; on the tailing crane the maximum will occur while the vessel is horizontal.

(c) Try the arrangement of Fig. 5.46, where the main crane is attached below the centerline and the tailing crane above. From Eq. 5.47, when the vessel is horizontal the cranes will share the vessel weight as in the examples above. But when the centerline is vertical

$$M = 150 \times 6/(6 + 4) = 90 \text{ tons } (81.6 \text{ t})$$

$$T = 150 - 90 = 60 \text{ tons } (54.4 \text{ t})$$

Summary of results: maximum crane loads

	Main crane	Tailing crane
Part (a) Fig. 5.44	150 tons (136 t)	62.8 tons (57 t)
Part (b) Fig. 5.45	150 tons (136 t)	62.8 tons (57 t)
Part (c) Fig. 5.46	90 tons (81.6 t)	62.8 tons (57 t)

Three crane lifts

There are three ways in which three cranes can be used to lift together. The manner in which the cranes are attached to the load and the positions of the cranes with respect to one another are the chief issues which must be addressed, although the characteristics of the load itself may leave little choice.

The first arrangement is one we have already discussed, tripping a long slender vessel or other load, but in this case two cranes are used to lift the heavy end, usually the top. However, normally the pair of cranes will be positioned so that their booms will be perpendicular to the long axis of the load. Therefore, when the hoist lines go out of plumb, these cranes will be side loaded. In all other respects, their loading will be as previously mentioned. And, of course, side loading of the main cranes implies off-lead on the tailing crane.

Figure 5.47 illustrates the lifting arrangement. Using the dimensions in the figure, if we assume a vessel weighing 196 tons (177.8 t), the load on each crane can be calculated. Taking moments about the tailing crane attachment, the upper cranes will each lift

$$1/2 \times 196 \text{ tons} \times 50/70 = 70 \text{ tons (63.5 t)}$$

so that the tailing crane must carry

$$196 - 2 \times 70 = 56 \text{ tons (50.8 t)}$$

During the course of the lift, when the vessel has been raised until it

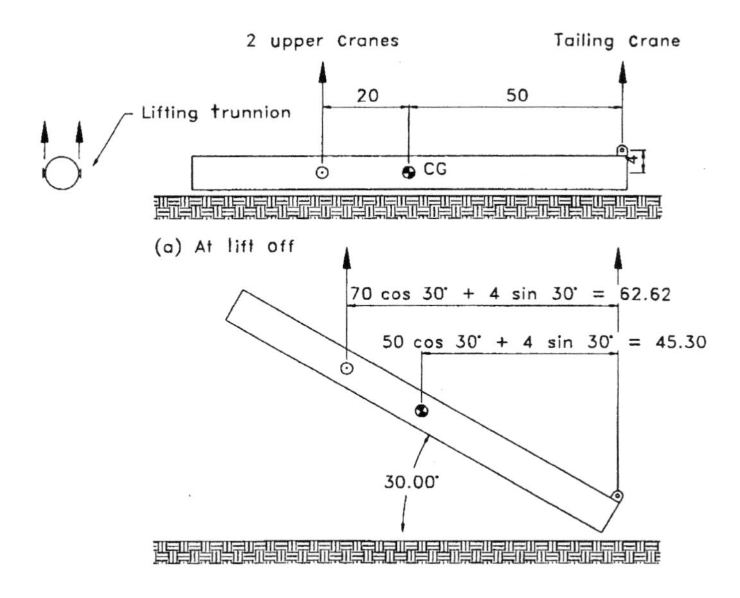

Figure 5.47 A vessel annotated for a typical three crane tripping operation. It is similar to Fig. 5.44, except that two cranes are at the top end.

is 30° to the ground (Fig. 5.47*b*), the distribution to the upper pair of cranes becomes

$$1/2 \times 196 \times 45.30/62.62 = 70.9 \text{ tons } (64.3 \text{ t}) \text{ each}$$

and the tailing crane now carries

$$196 - 2 \times 70.9 = 54.2 \text{ tons } (49.2 \text{ t})$$

The shift of load from the tailing to the upper cranes has nothing to do with the load going out of level; it is simply a consequence of the tailing crane attachment point being located above the trunnions. It is easy to verify that with lift points at the same elevation there will be no shift in load. To do that, eliminate the 4 sin 30° terms in the dimensions of Fig. 5.47*b*.

A similar arrangement may be used to raise and position a large girder, but the similarities are superficial. For two cranes hoisting from trunnions, only nominal synchronization is needed; if one hook runs a little ahead of the other, the load will turn in the trunnions, but the weight distribution will not change materially. Girder lifts, however, are not made with trunnions. The lift attachments will be at the girder flange, well above the girder CG. If one hook gets ahead of the other, there will be a shifting of load. Therefore, an equalizer beam should be used. The distribution of the load between the single crane at one lift point and the pair of cranes at the other will then be similar to the two-crane lift.

The second three-crane arrangement is used for loads that are wide with respect to their length. Here, three discrete attachment points are used—in plan view the three hook-up points make a triangle regardless of the shape of the load. The distribution of load among the three cranes is determined from the position of the load CG with respect to each of the attachment points. During the lift, the load should be kept nominally level, but no meaningful load shift will occur unless the load is seriously out of level, the CG of the load is considerably above or below the attachment points, or the attachment points are at different levels.

The distribution of loading using a triangular three-crane lift can be found for the machine shown in Fig. 5.48 by using the conditions of equilibrium; the sum of the vertical forces and of the moments must equal zero. Assuming that the machine weighs 141,000 lb (64 t), the vertical force condition gives

$$P_A + P_B + P_C - 141 \text{ kips} = 0$$

Moments to the right of lift point C yield

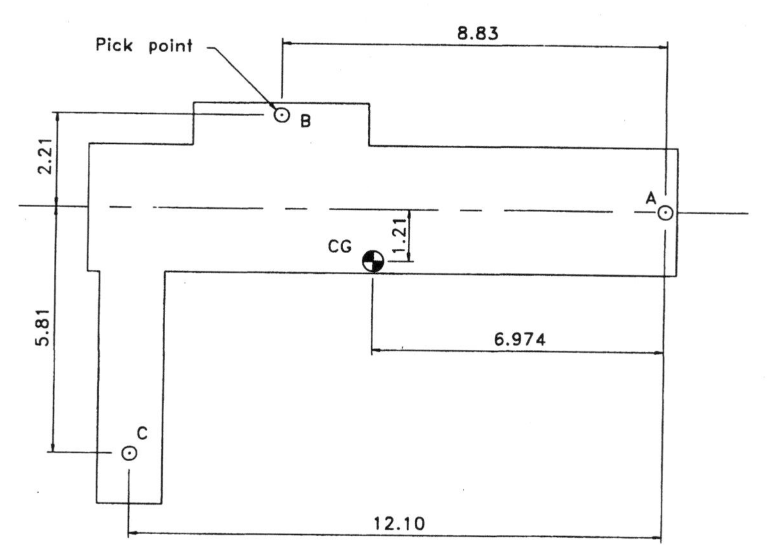

Figure 5.48 Plan view of a machine and its pick points A, B, and C, dimensioned for calculating load distribution.

$$(12.10 - 8.83)P_B + 12.10P_A - 141(12.10 - 6.974) = 0$$

$$3.27P_B + 12.10P_A - 722.77 = 0$$

and moments above point C produce

$$(5.81 + 2.21)P_B + 5.81P_A - 141(5.81 - 1.21) = 0$$

$$8.02P_B + 5.81P_A - 648.60 = 0$$

Those three equations containing three unknowns can be solved simultaneously and will yield equal hook loads of 47,000 lb (21.3 t) at A, B, and C.

The third three-crane arrangement is linear with the cranes positioned along the length of a long load such as a girder or section of pipe. Unlike the other two arrangements, the distribution of the load between the cranes is now indeterminate unless the load is very limber. For the most part, the load on each crane will depend on the operator (or on a load indicator if the crane is so equipped). The load on the middle crane can vary from nothing to almost the entire load weight. When this arrangement must be used, the cranes should be:

1. Rated well above the expected load
2. Equipped with load or load-moment indicators in working order

3. Reeved with multiple parts to permit slow raising and good control
4. Operated very slowly with the indicated load kept within a predetermined range

For three-crane lifts, all of the precautions given for two-crane lifts are applicable. And of course, if load lines go out of plumb, one or more of the crane booms will be side loaded.

Four-crane lifts

The problem with a four-crane lift, where each crane is attached at its own corner of the load, can best be visualized by comparison with a common kitchen table. If a four-legged table is rigidly built, it will almost always rock back and forth because most floors are not dead level and true. That cannot occur with a three-legged table—it always sits firmly.

When four cranes lift a rigid symmetrical load, they cannot be perfectly synchronized and the load will tend to rock like a four-legged table. But the implication for the cranes is that two of the cranes, diagonally opposite to one another, will get substantially all of the load. If two cranes must be capable of sharing the whole load, why use four cranes? If neither pair of diagonal cranes can hold the entire load, sharing can only take place if one or more cranes partially tips or severely deflects.

In order to get four cranes to predictably share a rigid load, when each crane is at a corner, two of them must lift by means of an equalizer. In that way the cranes and load will interact as a three-legged table; the load on each crane will become determinable even if the lifted load is unsymmetrical. Without an equalizer, the crane loads are indeterminate. If this situation cannot be avoided, each crane should be chosen so as to possess excess capacity, equipped with a load indicator, and reeved with extra parts. The lift must then be carried out slowly so that the operators have time to manipulate their hoist controls and keep each crane loaded to within a range of preselected values.

For some loads, pairs of cranes can lift by means of equalizers and the operation then resembles a two-crane lift in many respects. But, when four cranes are arranged in a line for lifting, the same problems prevail as for three cranes similarly arrayed. For this setup, and any time equalizers are not practical, the operating rules given under three-crane lifts should be followed. The same is true for all lifts requiring more than four cranes.

All unequalized arrangements of three or more cranes are indeterminate, that is, the distribution of load between the cranes cannot be calculated in advance. Those lifts therefore require strict adherence to planned and rehearsed procedures tailored to the equipment selected for use, the equipment arrangement, and the specifics of the work that needs to be done.

Most of the dangers discussed above either do not exist or are of limited importance for single crane operations. Any of them, and certainly their combinations, comprise the envelope of risk that must be addressed for multiple crane lifts. During planning the movements and operational steps required to accomplish the particular lift are determined and the risks associated with each step identified. Once found, control measures are established to bring each risk, and the set of risks, under management. With defined procedures, rationally set tolerances controlled in the field, and a clear supervisory chain of command and reporting, multiple crane lifts are viable, productive, and efficient lifting procedures.

Chapter

6

Tower Crane Installations

Tower cranes have become symbols of urban development. Though European in origin; they are now manufactured on most continents and have penetrated virtually all construction markets.

In the cities of Europe, most tower cranes are installed freestanding on simple aboveground, ballasted, knee-braced support bases. These basic installations require minimal site engineering and suffice to erect buildings of a dozen or so stories.

North American contractors seldom use tower cranes in the manner of their European counterparts. Here mobile cranes are used to erect buildings unless there is a distinct disadvantage in doing so or unless there are constraints that prevent its use such as building height or constricted sites. These tower crane uses demand expertise for planning and installation design.

This chapter provides a fairly detailed view of common tower crane installation methods and an overview of some solutions to more difficult installation problems. The intention is not to present cookbook solutions but rather to give sufficient background for resourceful engineers and planners to implement their own solutions.

There is an emphasis on storm winds in this chapter because they so often control what can be done. Also, we feel that the geographic variation of wind is an issue that manufacturers and suppliers of tower cranes have not adequately dealt with. Tower crane users in hurricane-prone places like Florida and the Gulf Coast should take special note of this. Chapters 3 and 6 combined provide the analytical tools needed to adapt tower cranes to high storm-wind areas.

6.1 Introduction

On a job not too long ago, the construction superintendent, a graduate civil engineer, thought he would save his company some money by designing the foundation for a freestanding tower crane by himself. In order to provide hook coverage over the entire job, the crane had to be placed in such a position that building footings restricted the space available for the crane base. He therefore designed an eccentric footing block to fit in the remaining space.

Fortunately, before the crane was erected, the superintendent got cold feet and decided to have the design confirmed. When one of the authors checked out-of-service stability, it was evident that a wind of relatively low velocity would overturn the crane because of the short fulcrum distance on one side.

The most cost-efficient way we could discover to add ballast to this in-place concrete block required some steelwork and mass concrete. When our design arrived at the jobsite, however, the superintendent called frantically with the news that an erection crew and mobile crane were due to start work in 2 days' time but the steelwork specified would take about a week to procure and fabricate. The crane installation could not be delayed, as it was the key to future job scheduling and progress.

Now cost effectiveness had to be thrown to the winds and an alternative scheme, previously explored but rejected as too costly, was put to use. This plan required the inefficient use of large masses of concrete but had the advantage of needing no special materials or processes. After explaining the concept to the superintendent and being informed that there was a backhoe at the job-site, we instructed him to "set up the backhoe and start digging a trench on the long end of the footing. We'll call you back in about two hours and tell you how wide and deep to make it. By then the design will be finished."

And that is exactly how it happened. After two hours he was given the details of the ballasting, and the work was then completed in sufficient time for the crane erection to go ahead on schedule.

But that is hardly the recommended procedure for design and construction of tower crane bases. Although the superintendent was at fault for waiting so long to have his work checked, we have always felt strongly that the design engineer must understand and make every effort to accommodate the time pressures of the construction site. This should never be done, however, at the expense of thorough and proper design.

The design of a tower crane installation includes several elements besides support of the loads. First, the crane must be positioned where there will be adequate space to lay out the crane components before erection and to position a mobile crane to do the erection work. When erected, cranes that must be free to weathervane need 360° of clear space without obstruction to jib slewing. Furthermore, when in position the crane must provide hook coverage and adequate load capacity at all required points. Finally, at the completion of the work there must be access for a mobile crane or other means to disassemble the crane.

Where two or more tower cranes are to operate at the same site, the cranes must be arranged so that there is no possibility that the jibs will collide (Fig. 6.1). Obviously, when luffing jib cranes are used, potential collision problems are easier to solve (Fig. 6.2).

Loads acting on tower cranes

Tower cranes are subject to two loading regimes: the in-service condition includes the effects of lifted load, impact, moderate wind, and slewing torsion while the out-of-service condition incorporates only dead-weight effects and exposure to storm winds. The design for the installation of a crane with a high tower located on the Gulf Coast

Figure 6.1 A pair of Richier climbing tower cranes constructing the Tour Nobel, Paris. Keeping the cranes at different heights avoids collision of the jibs, but the operator of the lower crane must take care to keep away from the hoist lines of the other crane. (*H. Baranger et Cie, Paris.*)

Figure 6.2 Three luffing boom tower cranes atop the roof framing of the American Express Operations Center in New York City. The luffing motion helps avoid boom and hook interferences. Coordination is much easier than in Fig. 6.1, but communication among operators is still important. (*A. J. McNulty & Co., Inc.*)

obviously will be governed by out-of-service loading, whereas a crane with a stubby tower surrounded by tall buildings in Times Square will not. For almost everything in between, both loading regimes must be investigated.

An out-of-service crane in calm air exerts only dead-load effects on its supports. This includes the vertical load itself and a moment that results because the CC of the rotating portion is located toward the counterweights. When tower cranes are left unattended, the jibs normally are left free to slew. With the introduction of wind, the jib will then weathervane. The counterweight end will point into the wind; therefore the dead-weight moment must be subtracted from wind moment to yield the net moment felt by the supports. A weathervaning crane swings freely and therefore does not impose torsion on its tower.

An in-service crane imposes lifted-load effects on the supports as well as dead weight. This will include an increase in vertical load and a moment in the direction of the jib. Wind may come from any direction, and although wind from the rear will directly add to load moment, this condition may not be the most critical. Instead, the large wind exposure areas of the side of the crane may make wind from that direction govern.

The in-service wind moment and the load moment are then normal to each other for that particular loading case, and their resultant is the moment the supports must resist. The resultant can take any orientation with respect to the tower and footing.

Given site conditions that require that the swing be locked when the crane is not in service, such as an adjacent taller structure interfering with weathervaning, both side and front wind effects must be checked. Storm-wind forces acting on the large side area create a moment normal to the dead-weight moment, but frontal wind acting on a much smaller area induces moment that adds directly to the dead-weight moment. Both cases must be investigated to verify that the tower can withstand the combined axial and bending effects and to determine reactions on the support structure.

For all tower cranes, both wind from the side and slewing inertia forces introduce torsional effects in the horizontal plane, and, of course, there are horizontal shearing forces from the wind as well.

Crane manufacturers provide support reactions to be used for installation design, although practice varies among them, some giving only maxima and others giving all data for each permissible configuration. Installation designers should insist on being given the data for the configuration they will be working with. A complete set of data is needed for in-service and out-of-service conditions so that the designer can evaluate the particular requirements that will govern in the context of actual site conditions. For example, site restrictions may dictate the use of guying to satisfy out-of-service loads. During operation the guys may be removed; therefore, in-service data are required for the unguyed condition.

When out-of-service conditions control any phase of the design, the question of permissible stress levels must be faced. Most American design codes permit ordinary allowable stresses to be increased by as much as one-third when wind loading is included. We have always preferred to ignore this provision when dealing with in-service wind, taking the position that an increase is not usually justified for this case; there are situations, however, where rational considerations indicate otherwise, such as when in-service design wind speed is high for the operating location and rated loads are infrequently lifted (an increase of about one-sixth, similar to European practice, then seems reasonable). But, for out-of-service loading, a full one-third increase in the allowable stress makes sense. Design storm-wind effects, including gusts and dynamic amplification, impose a transient stress state that may occur on average only once in 25 years, that is, only once in the life of the crane. The low probability and the transient nature of the loading justify high design stress levels provided that they are not high enough to induce damage. For further guidance, see Sec. 3.4.

The out-of-service problem is complicated by the fact that tower crane manufacturers design cranes and calculate support-reaction data based on a universal storm-wind level, that is, without recognition that there is geographic variation. This level may be chosen from the design code of the country of crane origin, or it may be an arbitrary choice, but it may not be at all appropriate for the place where the crane is to be installed. If the design wind is higher than the installation wind, no harm can result other than an overconservative and a bit more costly crane setup. But if local conditions are more severe than accounted for in the design, it may be necessary to limit crane height or to deal with increased support reactions. The mast and anchorage must be investigated if the overturning moment at the crane base will exceed the standard design value. Such investigations should be done by the manufacturer unless mast design data are available.

Users of tower cranes that are to be located in high-exposure areas, such as coastal belts (particularly those in hurricane zones) and some mountain regions, should raise these issues with their crane suppliers. We believe that it is only the statistical rarity of full storm winds and the relatively infrequent use of tower cranes in high-wind areas that have combined to limit the actual number of storm-wind failures. Installation designers, tower crane users, and manufactures will have to face this problem squarely if they wish to be assured of successful and safe installations.

In some instances where design-level winds derive from hurricane conditions, tower cranes may have to be secured, climbed down, or partially dismantled on the basis of storm warnings. As a measure utilized to avoid the cost penalty of hurricane loading, this must be done only with great forethought and preparedness. When it comes time to implement the preplanned hurricane measures, resources must be available to do so without reluctance and without undue haste.

6.2 Selection and Positioning

Picking a suitable model and configuration for a tower crane goes hand-in-hand with its placement on a site. By nature, a tower crane is "nailed down" to a spot or, if the crane is rail mounted, confined to a path. A crane that has characteristics that are remarkably suited to the work may have those characteristics foiled by a poor choice of crane position.

Some considerations in selecting a suitable crane model are:

User preference. Comfort with a vendor and previous experience with a particular crane are subjective criteria, but hardly to be ignored. A tower crane is a key piece of equipment on a project; downtime is costly. A crane user should count on a history of reliable performance from the machine and strong support from the vendor.

Availability. Fleets of tower cranes are usually small and geographically dispersed, while individual machines are mostly committed to long term assignments. Thus availability of a preferred crane is hit-or-miss. A machine coming from another site is a hostage to that site's progress: with a prospect of delay, a contingency plan should be made to bring in another tower crane, or to make a short-term mobile crane substitution. Time is needed between installations for the crane to be prepared and for scheduled maintenance. That lag time is variable depending upon the condition of the tower crane; a decision to defer needed maintenance for the expedience of meeting a schedule will frequently come back to haunt the user.

The question of availability applies as well to accessories such as mast sections, ties, base frames and climbing gear. During busy times, these items need to be reserved along with the crane.

Capability. Its load chart describes the primary characteristic of a tower crane, the lifting capacity. The next important trait is the speed of its various motions and the degree of control that the operator has over these motions. Means of adapting to the needs of the site—for example, using accessories such as climbing gear or a travelling base—round out a tower crane's capabilities.

A sales brochure is a superficial tool for evaluating an unfamiliar crane's abilities and limitations; "kicking the tires" is no better. A savvy user can sift important nuances from a close physical study of the equipment and a review of its documentation. Watching a crane in operation can be informative, but questioning reliable people who have experience with the crane is the best way of all.

Expecting fast work cycle speed from a crane and simultaneously expecting precise control would seem to be mutually exclusive features. Whether electric or hydrostatic, however, modern drives do permit the same crane machinery to offer both, but not usually simultaneously. Efficient concrete bucketing, for example, requires high work cycle speed but little precision; requirements for steel and precast concrete erection are the opposite.

Work cycle speed is a combination of the speeds of all crane motions: hoisting, swinging and either trolleying (on a hammerhead) or derricking (raising and lowering a luffing boom). Most tower cranes can execute simultaneous motions within the limitations of overall demand on the prime power source. The speeds of most motions are stepped or variable; for instance, a full load would be raised in low speed mode and the empty hook would be lowered to the ground in high speed.

Power source. Most tower crane models are electric; thus there is not usually a choice about an alternate power source. If adequate power is not available at the site, or if the available power is unreliable, a generator is needed. Some tower cranes, particularly North American and Australian models, have diesel prime movers and hydrostatic drives; a few have electric motors powering the hydrostatic machinery. When there is a choice, objections to noise and fumes might motivate selection of an electric prime mover. Alternately, inadequacy of electric power is an incentive to use diesel power.

Configuration. Many urban and industrial sites favor luffing boom tower cranes to avoid the interference of adjacent structures. This boom type is also a suitable choice when multiple cranes operate in close proximity (Fig. 6.2), though arrangements of overlapping hammerhead crane booms are not unusual (Fig. 6.1). The greater complexity of luffing systems does make them more expensive to manufacture and maintain; they are also more difficult to erect and dismantle than hammerhead models. The latter are thus preferred where luffing booms offer no advantage.

Ease of erection and dismantling. Some tower cranes are manufactured with ease of erection paramount. These models have many pin-connections instead of bolts. Components come apart in neat packages and erection weights are manageable for medium size mobile cranes to handle. These cranes typically have thoughtfully placed rigging aids such as lifting lugs, access platforms and alignment holes for spud wrenches at clevis connections. At the opposite extreme, a few tower cranes are so complex and awkward to erect that they seem to have been designed as if for permanent installations.

Cost. Selection should consider overall costs and benefits, not just purchase or rental price. Overall costs include trucking, erection/dismantling, climbing, rope replacement and downtime. Replacement part and service technician availability can reflect on cost. The relative impact of the crane's motion speeds and cycle time on productivity should be factored into its cost evaluation too.

Choosing a tower crane may be easier than finding the optimal place for it on the site. From selection to positioning, one goes from having a few choices to facing many; more often than not, each potential location involves conflicts and compromise.

Initially, looking at a blank slate, positioning often seems like a "nobrainer" issue. Complexity creeps into the picture when the construction process is fully mapped out, and erection and dismantling issues are studied. A complete mapping includes the three spatial dimensions plus the dimension of time. By nature, a construction project is constantly changing; the tower crane should be viewed in the context of the evolving project landscape.

Seasoned crane planners repeat the same exercise in hundreds of jobsite shanties. With a set of plans flipped open on the table, an "X" is penciled where the project superintendent thinks the tower crane should go. A scale is placed on the drawing and used to sketch a series of defined arcs centered on that mark to see what the crane would reach. The load chart for the proposed crane is perused; the superintendent points out pick zones, laydown areas and whatever out-of-the-ordinary loads that the crane will be expected to handle. On a few fortuitous occasions, crane planning ends with a session back at the office confirming and fine tuning the preliminary results of this field conference and recording the information on a drawing. More often than not, the field session is the first of many rounds as various complicating factors start to muddy the picture.

The considerations in pinning down a suitable tower crane location can be numerous and varied, but they can also be corralled into a handful of categories:

Reach and capacity. There can be a hidden dimension to reach and capacity requirements. The crane expert and the general superintendent together do not necessarily possess all of the information needed for a satisfactory layout; subcontractors and vendors might need to be queried. For example, oversized gusset plates make some fabricated steel sections heavier than contract drawings would imply. The fabricator and steel erection subcontractor should be consulted. In reinforced concrete construction, crane requirements are often governed by the contractor's choice of forms; flying tables, for instance, must be carried free and clear from the perimeter of the building by the crane.

The reach of the crane needs to include pick zones and areas for laydown, shakeout or makeup. The "end game" needs to be looked at, too. The top of a building sometimes carries appurtenances such as cooling towers, chillers, generators, signs, water tanks, antennas and architectural "gingerbread". These building features can interfere with or dictate removal means and methods.

Reach and capacity requirements might also include one tower crane dismantling another.

While a mobile crane might be loaded to full rated capacity only with reluctance and under controlled circumstances, there should be no similar trepidation lifting with a tower crane. That is not to suggest that overloading, recklessness or cavalier use is tolerable. Unlike its mobile cousin, a tower crane is usually on a more substantially constructed mounting and at rated capacity it is not as close to tipping. Its load chart is more straightforward and has few, if any, deductions or adjustments. Under routine working conditions, tower cranes can work up to full rated capacity.

Efficiency. For bucketing concrete, cycle speed is a paramount consideration. Swing and boom motions are slower than load hoisting; an efficient crane location is one that keeps these slower motions to a minimum.

When erecting steel or precast concrete, efficiency is not much affected by cycle time. More important for efficiency is an ability to erect pieces in the largest practical sizes so that field assembly is minimized. For example, a project with large trusses is probably better served by a crane big enough to erect these whole than by a smaller one that can only assemble individual truss pieces on falsework.

Conflict. A tower crane can interfere with part of the new construction, an existing structure or with another temporary structure. When a tower crane is placed within the perimeter of a new building, it requires deck penetrations and perhaps temporary support elements (shores and braces) as well. Deck openings prevent the building from being closed in from the weather; both the penetrations and the support elements can interfere with the various systems and finishes on the penetrated floors. A crane that is mounted externally and tied at intervals to the building will have perimeter penetrations that may hold up completion of the wall system and some interior work.

Interference with an existing structure is the problem in reverse: as the building was there first, the effect is to restrict the tower crane. The crane might be forced into an inefficient work cycle, or it might be prevented from properly weathervaning.

Interference with other temporary structures can come about from a lack of coordination or communication. The conflict can be with scaffolding, hoists, concrete formwork or with other cranes. As the tower crane is normally higher than hoists and scaffolding, the problem does not usually occur until a climbing crane starts to jack down prior to removal. Conflict with other cranes typically is an issue of obstruction to weathervaning or avoidance of clashing loads. The former might occur if a mobile crane raises its boom inside the swing arc of a tower crane boom, in which case the mobile crane must lower its boom to establish a clear path for weathervaning. Establishing reliable communications and coordination protocols among the cranes solves the latter.

Conflict can also occur at the tower base. The tower and its base mounting might interfere with a roadway. Excavation for the footing might impede the placement of buried utilities or clash with existing utilities or foundations.

Support. Cost and complexity are not much of an issue for a static mounted tower crane on sound soil. For a crane mounted on less than desirable support materials, engineering and the crane support base become more involved and costs can be expected to escalate. Likewise, a tower crane that climbs through a building or that is laterally braced will require an engineered solution. When crane installations become complex, evaluation of alternative solutions is not always straightforward. Each solution has its own costs and difficulties. Sometimes the issues are easily identified and quantified at the outset of the planning process but this is not always so. Support issues—those that rise above stock solutions—need engineering input during planning as well as engineering follow-up with design and analysis.

Erection and dismantling. The planning of a tower crane installation is not complete unless due consideration has been given to the logistics and costs of erection and dismantling.

6.3 Static-mounted Cranes

The static mount is the most common installation configuration used in the United States. Furthermore, it is the initial condition used in most climbing, braced, and guyed installations. When tower height is short, the design will be governed by in-service conditions, but as height above base increases, a point will be reached at which out-of-service loads become more critical. Storm winds typically limit the freestanding tower height.

For most North American static-mounted installations, a mass of concrete is used to provide the ballast needed to resist overturning and to provide a margin of stability. The concrete footing block also transmits vertical loads to the ground and must resist shear forces from wind and torsional effects. The shear forces are generally small and it is only under exceptional conditions that the design of the footing block will be affected by them, but shear may have to be taken into account when the connection between the mast and the block is designed.

Older American design codes specified that a footing block must be able to accommodate from 1.33 to 1.5 times the applied overturning moment.† Thus, not only were stability requirements inconsistent within our own design codes, but neither figure matches the 1.20 minimum value of the codes in force where the machines are manufactured (see Sec. 4.3). However, an American tower crane code‡ calls for the 1.5 value.

If the out-of-service wind-selection procedure we have suggested (see Sec. 3.4) is followed, the question becomes moot. Using a design storm wind based on an overload-level wind and the stability margin means that each of the three code provisions results in the same required footing resisting moment and therefore the same design. But with present uncertain requirements in effect, we prefer to use the

---

†The Uniform Building Code, International Conference of Building Officials, 1973, sec. 2308(i) specified 1.5, while the Basic Building Code, Building Officials and Code Administrators International, Inc., 1975, sec. 717.1, specified 1.33. Current editions of both codes do not include similar provisions.

‡American National Standard for Construction Tower Cranes, ASME B30.3-1996, American Society of Mechanical Engineers, New York.

1.5 ratio for both in-service and out-of-service conditions, since it is more consistent with the strength margins used in the design. In this way, should winds beyond the design value occur, stability will still govern; at a small addition to footing cost a great increase in system reliability has been provided.

The ratio of footing resisting moment to applied moment is the main parameter controlling crane base design. Assuming that the crane applies net moment M_o to the base, that the crane vertical load is Q, the weight of the footing block itself is W, and that the footing is a square of sides b, then

$$1.5M_o = \frac{(Q + W)b}{2} \qquad b = \frac{3M_o}{Q + W} \tag{6.1}$$

where 1.5 is the stability factor referred to above. The design process is begun by assuming a starting value for W at 1 to 2 times Q. Equation (6.1) can then be used to select first trial values for the footing parameters or for rough planning or cost estimates. For final design more accurate figures must be used.

The unit weight of reinforced concrete w is about 150 lb/ft^3 (2400 kg/m^3). In an accurate check of stability, the overturning effect of the horizontal shear force V should not be neglected (Fig. 6.3). For a square footing, stability will be maintained if

$$1.5(M_o + Vd) \leqslant \frac{(Q + W)b}{2} \leqslant \frac{Qb}{2} + \frac{wb^3d}{2}$$

$$d \geqslant \frac{3M_o - Qb}{wb^3 - 3V} \tag{6.2}$$

Equation (6.2) can be solved for the footing depth d needed for any trial side dimension b. The depth should be nominally $b/6$ minimum, however. As d decreases, footing volume and therefore concrete cost decrease as well. This is because the increasing values of b accompanying decreasing d provide a greater fulcrum distance. But with decreasing depth, footing bending moments and reinforcing requirements may increase; then again soil pressure will decrease. On top of this, there are minimum depth requirements, related to anchorage of the crane to the footing block, and local building-code stipulations governing placement of footing bottoms below the frost line. The designer has to consider all these factors and then produce the most economical and practical design. Anchorage will be covered later in this section, but first soil pressure must be investigated.

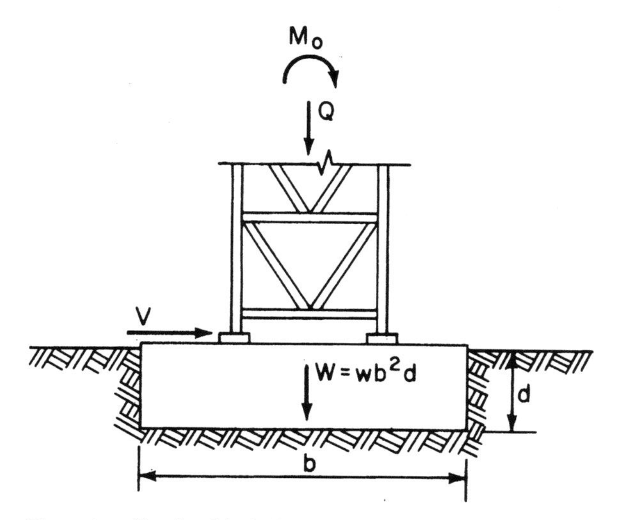

Figure 6.3 Footing block for a freestanding tower crane.

Induced soil pressures

Uniform settlement of the footing block is of little consequence; its small magnitude cannot have a significant effect on crane height. Differential settlement can be critical and will be dealt with later. Settlement is a minor consideration when selecting permissible soil-bearing values because of the crane's relative insensitivity to it. The values used can therefore be somewhat greater than those typically used for building design. Because of tower crane footing size and its usual placement with the footing bottom below ground surface, the crane values suggested in Table 5.1 may be conservative. A licensed professional engineer can offer better guidance for the specific soil at the site.

The maximum pressure exerted on the soil by the footing is the combined effect of the vertical loads and the moments. The component of pressure contributed by the vertical loads v is given by

$$v = \frac{W + Q}{b^2} = \frac{wb^2d + Q}{b^2} = wd + \frac{Q}{b^2} \tag{6.3}$$

Using a beam analogy ($\sigma = My/I$), the moment is found to contribute pressure f, given by

$$f = \pm \frac{(M_o + Vd)b/2}{b^4/12} = \pm \frac{6(M_o + Vd)}{b^3} \tag{6.4}$$

If $v > |f|$, the resultant pressure under the footing (Fig. 6.4) will follow a trapezoidal pattern and the maximum pressure p_{max} is then

$$p_{max} = v + |f| \tag{6.5}$$

But if $v \leqslant |f|$, the pattern will be triangular (Fig. 6.5). Taking the length of the triangle as t and writing the expression for applied and resisting vertical forces, we get

$$W + Q = \frac{p_{max}bt}{2} \tag{a}$$

Then the expressions for applied and resisting moments in equilibrium are

$$(M_o + Vd) = (W + Q)\left[\frac{b}{2} - \frac{t}{3}\right]$$

$$t = 1.5b - \frac{3(M_o + Vd)}{W + Q} = 1.5b - \frac{3(M_o + Vd)}{vb^2} \tag{6.6}$$

Rearranging Eq. (a), we get

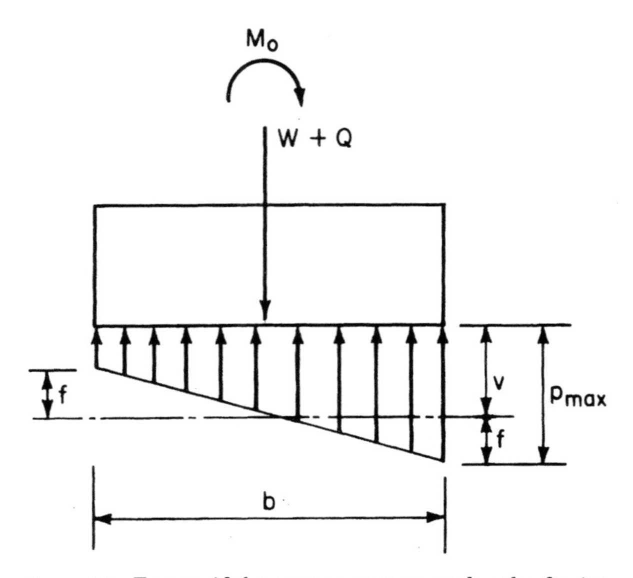

Figure 6.4 Trapezoidal pressure pattern under the footing block of a freestanding crane.

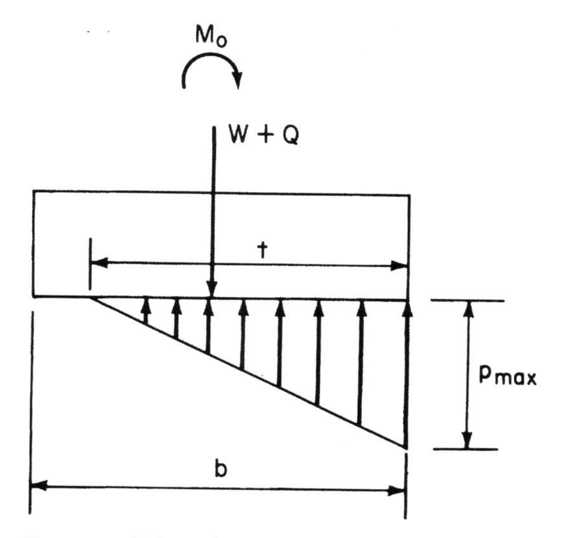

Figure 6.5 Triangular pressure pattern under the footing block of a freestanding crane.

$$p_{\max} = \frac{2(W + Q)}{bt} = \frac{2vb}{t} \tag{6.7}$$

Wind can, of course, blow from any direction, and the magnitude of the wind moment will vary as direction varies with respect to a mast side. There are two conditions of interest, the two extremes of wind blowing normal to a mast face and on the diagonal. The crane documentation should give separate values for the two cases in order to facilitate efficient footing design. Ordinarily, however, only one value is given with the crane data, and it is most likely to be the value for wind on the diagonal.

The soil pressure calculations for wind on the diagonal involve an equation derived from use of the calculus. This equation is an exact solution giving maximum pressure at the footing corner:

$$p_{\max,\text{diag}} = \frac{(v/2)[\text{b}/(A\sqrt{2}) - 1]}{b/(3A\sqrt{2}) + (2A/b)(1/\sqrt{2} - A/3b) - 1} \tag{6.8}$$

where $$A = \frac{M_o + Vd}{W + Q} = \frac{M_o + Vd}{vb^2}$$

The term maximum pressure is used here in a manner consistent with the assumption commonly made by civil engineers in simplifying footing design. Their assumption is that pressure is uniform across the width of the footing, when in fact it drops off near the edges and so must increase toward the central region. The calculated maximum pressures may in fact be exceeded, but the values for permitted soil pressures take this into account.

Exact solutions for bearing pressure under odd-shaped footings would be very difficult using traditional methods. Some structural analysis computer programs, however, are capable of modeling these footings and generating solutions for various loadings.

With Fig. 6.6 the calculus can be used to derive an expression for maximum soil pressure with wind on the diagonal. The differential area across the footing can be expressed as

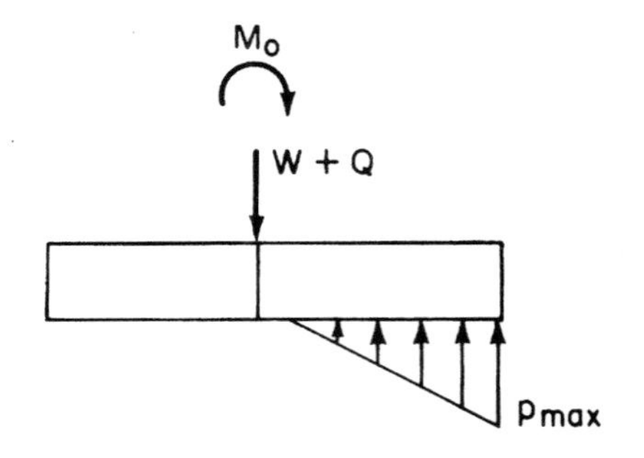

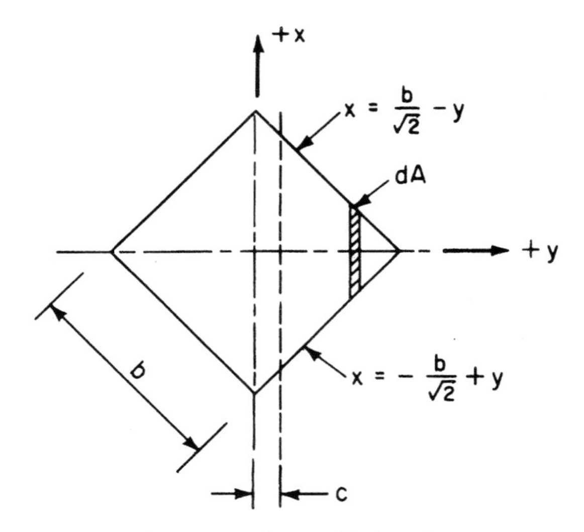

Figure 6.6 Tower crane footing block with moment applied on the diagonal.

$$dA = 2x\, dy \quad \text{where } x = \frac{b}{\sqrt{2}} - y$$

If we assume that pressure will diminish to zero at $y = c$, the pressure at any point y is then

$$p(y) = \frac{p_{\max}(y - c)}{b/\sqrt{2} - c}$$

The equilibrium equation for the veritcal forces is

$$\Sigma V = 0 = W + Q - \int p(y)\, dA = W + Q - 2\int p(y)x\, dy$$

$$= W + Q - \frac{2p_{\max}}{b/\sqrt{2} - c}\int_c^{b/\sqrt{2}} (y - c)\left(\frac{b}{\sqrt{2}} - y\right) dy$$

$$\frac{(W + Q)(b/\sqrt{2} - c)}{2p_{\max}} = -\frac{y^3}{3} + \frac{(b/\sqrt{2} + c)y^2}{2} - \frac{cby}{\sqrt{2}}\Bigg|_c^{b/\sqrt{2}}$$

$$\frac{b/\sqrt{2} - c}{2p_{\max}} = \frac{1}{W + Q}\left(\frac{b^3}{12\sqrt{2}} - \frac{b^2c}{4} + \frac{bc^2}{2\sqrt{2}} - \frac{c^3}{6}\right) \qquad (b)$$

which is a cubic equation in the unknown c. Likewise, the moment equilibrium equation can be written

$$\Sigma M = 0 = M_o + Vd - \int p(y)y\, dA = M_o + Vd - 2\int p(y)xy\, dy$$

$$= M_o + Vd - \frac{2p_{\max}}{b/\sqrt{2} - c}\int_c^{b/\sqrt{2}} y(y - c)\left(\frac{b}{\sqrt{2}} - y\right) dy$$

$$\frac{b/\sqrt{2} - c}{2p_{\max}} = \frac{1}{M_o + Vd}\left(\frac{b^4}{48} - \frac{b^3c}{12\sqrt{2}} + \frac{bc^3}{6\sqrt{2}} - \frac{c^4}{12}\right) \qquad (c)$$

Since the left sides of Eqs. (*b*) and (*c*) are identical, the right sides must be equivalent. Setting them equal to each other yields a fourth-order equation in c. Simplified, it reads

$$c^4 - 2\left(A + \frac{b}{\sqrt{2}}\right)c^3 + \frac{6Ab}{\sqrt{2}}c^2$$

$$- \left(3Ab^2 - \frac{b^3}{\sqrt{2}}\right)c + b^3\left(\frac{A}{\sqrt{2}} - \frac{b}{4}\right) = 0 \qquad (d)$$

where A is as previously given.

One or more roots will be imaginary and can be ignored. The remaining roots will be identical solutions to the location of the zero-pressure line. Substituting the solution value into either Eq. (*b*) or (*c*) will give the maximum footing corner pressure p_{max}. But through good fortune and the application of pure logic to the physical aspects of the problem, an exact expression for the solution root has been found. When $c = 2A - b/\sqrt{2}$ is substituted into Eq. (*b*) or (*c*), the result is Eq. (6.8).

Example 6.1 (*a*) Design a footing block for a tower crane weighing 230 kips (104 t) that will subjected to a storm-wind net moment of 3600 kip · ft (498 t · m) across the mast faces and 4450 kip · ft (615 t · m) on the diagonal. The wind shear forces are 18.8 and 22.6 kips (8.5 and 10.3 t), respectively. Soil bearing capacity is 4 tons/ft² (383.0 kN/m²).

Since the component of the diagonal wind that acts across the side is $4450 \cos 45° = 3147 < 3600$, the side moment will be used to satisfy stability criteria. From Eq. (6.1), try

$$W = 1.1 \text{ (230 kips)} = 253 \text{ kips (115 t).}$$

$$b_1 = \frac{3(3600)}{230 + 253} = 22.36 \text{ ft} \quad \text{try 22 ft (6.71 m)}$$

and from Eq. (6.2) the trial depth is

$$d_1 = \frac{3(3600) - 230(22)}{0.15(22^3) - 3(18.8)} = 3.73 \text{ ft}$$

Check: $$\frac{b}{6} = \frac{22}{6} = 3.67 \quad \text{try } d = 3.75 \text{ ft (1.14 m)}$$

The trial footing is then 22 by 22 by 3.75 ft (6.71 by 6.71 by 1.14 m). Soil pressure must now be checked. The vertical component is given by Eq. (6.3)

$$v = 0.15(3.75) + \frac{230}{22^2} = 1.04 \text{ kips/ft}^2 \text{ (49.8 kN/m}^2\text{)}$$

The footing side overturning moment, $M_o + Vd$, is $3600 + 18.8 \times 3.75 = 3670$ kip · ft (4976 kN · m). Equation (6.4) then gives the moment contribution to ground pressure

$$f = \pm \frac{6(3670)}{22^3} = \pm 2.07 \text{ kips/ft}^2 \text{ (99.1 kN/m}^2\text{)}$$

Since $|f| > v$, the soil pressure pattern will be a triangle whose length is given by Eq. (6.6)

$$t = 1.5(22) - 3 \frac{3670}{1.04(22^2)} = 11.13 \text{ ft (3.39 m)}$$

The maximum soil pressure from side moment, Eq. (6.7), is then

$$p_{\max,\text{side}} = \frac{2(1.04)(22)}{11.13} = 4.11 \text{ kips/ft}^2 \text{ (196.8 kN/m}^2\text{)}$$

Now the diagonal pressure must be checked. Using diagonal moment and shear, the parameter A takes the value

$$A = \frac{4450 + 22.6(3.75)}{1.04(22^2)} = 9.01 \text{ ft (2.75 m)}$$

Equation (6.8) gives the maximum diagonal pressure

$$p_{\max,\text{diag}} = \frac{\dfrac{1.04}{2}\left(\dfrac{22}{9.01\sqrt{2}} - 1\right)}{\dfrac{22}{3(9.01\sqrt{2})} + \dfrac{2(9.01)}{22}\left[\dfrac{1}{\sqrt{2}} - \dfrac{9.01}{3(22)}\right] - 1}$$

$$= 8.81 \text{ kips/ft}^2 \text{ (421.8 kN/m}^2\text{)}$$

The diagonal pressure is more than twice the overside pressure. Furthermore, it exceeds the permitted value of 4 tons/ft². If we try a larger footing with $b_2 = 22.5$ ft (6.86 m), $d_2 \geqslant 3.40$ (1.04 m) but $b_2/6 = 3.75$ ft (1.14 m). The new trial footing is then 22.5 by 22.5 by 3.75 ft (6.86 by 6.86 by 1.14 m). Only the diagonal pressure need by checked.

$$v = 0.15(3.75) + \frac{230}{22.5^2} = 1.02 \text{ kips/ft}^2 \text{ (48.8 kN/m}^2\text{)}$$

$$A = \frac{4450 + 22.6(3.75)}{1.02(22.5^2)} = 8.81 \text{ ft (2.69 m)}$$

$$p_{\max,\text{diag}} = \frac{\dfrac{1.02}{2}\left(\dfrac{22.5}{8.81\sqrt{2}} - 1\right)}{\dfrac{22.5}{3(8.81\sqrt{2})} + \dfrac{2(8.81)}{22.5}\left[\dfrac{1}{\sqrt{2}} - \dfrac{8.81}{3(22.5)}\right] - 1}$$

$$= 7.68 \text{ kips/ft}^2 \text{ (367.7 kN/m}^2\text{)} < 8 \text{ kips/ft}^2 \quad \text{OK}$$

(*b*) using the same data, design a footing with the crane positioned as shown in Fig. 6.7.

The diagonal moment now acts about the side of the footing. For b try $W = 1.25\,Q$, $b = 25.80$ ft rounded to 25.75 ft (7.85 m).

$$d \geqslant \frac{3(4450) - 230(25.75)}{0.15(25.75^3) - 3(22.6)} = 2.98 \text{ ft} \quad \text{try 3 ft (0.91 m)}$$

Although $b/6 = 4.29 > 3$, this orientation will produce smaller bending moments in the footing and thinner sections may prove to be practical. This

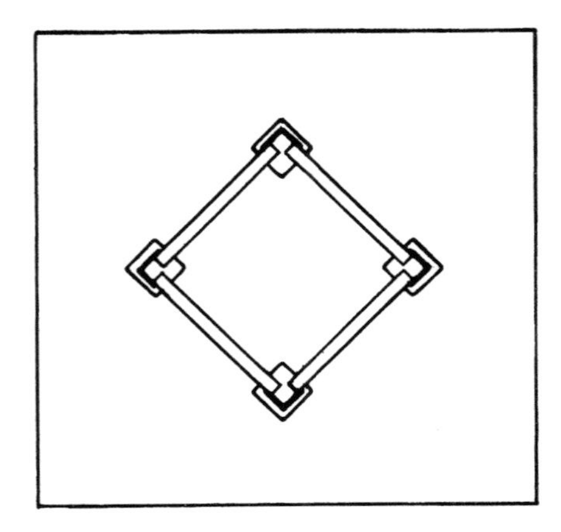

Figure 6.7 Tower crane footing block with the mast rotated 45°.

trial footing contains about 5% more volume and weight than the footing of part (*a*).

$$v = 0.15(3) + \frac{230}{25.75^2} = 0.80 \text{ kip/ft}^2 \text{ (38.2 kN/m}^2\text{)}$$

$$f = \pm \frac{6[4450 + 22.6(3)]}{25.75^3} = 1.59 \text{ kips/ft}^2 \text{ (76.0 kN/m}^2\text{)}$$

Since $f > v$, the pressure pattern is triangular with

$$t = 1.5(25.75) - 3\,\frac{4535}{0.8 \times 25.75^2} = 12.98 \text{ ft (3.96 m)}$$

$$p_{\text{max,side}} = \frac{2(0.8)(25.75)}{12.98} = 3.17 \text{ kip/ft}^2 \text{ (152 kN/m}^2\text{)}$$

Checking the diagonal, we see that the applicable moment was the side moment before.

$$A = \frac{3600 + 18.8(3)}{0.8(25.75^2)} = 6.89 \text{ ft (2.10 m)}$$

$$p_{\text{max,diag}} = \frac{\dfrac{0.8}{2}\left(\dfrac{25.75}{6.89\sqrt{2}} - 1\right)}{\dfrac{25.75}{3(6.89\sqrt{2})} + \dfrac{2(6.89)}{25.75}\left[\dfrac{1}{\sqrt{2}} - \dfrac{6.89}{3(25.75)}\right] - 1}$$

$$= 3.11 \text{ kips/ft}^2 \text{ (148.9 kN/m}^2\text{)}$$

The side and diagonal ground pressures are almost identical, and both are considerably below those of part (a). Rotating the crane 45° with respect to the footing therefore is a viable arrangement for sites where permissible soil bearing values are low.

Mast anchorage

The crane mast must be anchored to the footing block so that the vertical load, moment, and shear can be transferred from mast to block. If we assume a square mast measuring s between the centroids of the mast legs on each face, the diagonal distance will be $s\sqrt{2}$. With the moment on the diagonal, the force on the legs affected by the moment will be

$$F_{\text{diag}} = -\frac{Q}{4} \pm \frac{M_o}{s\sqrt{2}} \tag{6.9}$$

the negative sign denoting compressive force. For a moment applied parallel to a mast side, the legs would carry

$$F_{\text{par}} = -\frac{Q}{4} \pm \frac{M_o}{2s}$$

The diagonal case can be seen to produce greater leg loads both in tension and compression. For most cranes the portion of leg load contributed by moment is sure to leave net tensile forces in one leg.

The absolute magnitude of the compressive load will be $Q/2$ greater than the tensile load with moment on the diagonal. Therefore, compression will certainly govern the design of the mast legs, but anchor bolts must be designed for a greater load than Eq. (6.9) would suggest.

Good design requires that the ultimate strength of the bolts be no less than that of the mast legs in compression or alternatively that they be capable of transmitting 1.5 M_o to the footing block without failing. Then in the event of an overload wind it will be overall crane stability that will control failure. Inasmuch as peak wind effects will be from gusts, the crane will stand a good change of recovering before overturning completely. If the bolts were the weakest link, their failure might lead to collapse.

The anchor bolts themselves should be heavily greased or sleeved with some material that will prevent bonding to the concrete. Bonding should be prevented because of the fluctuating nature of the leg loading, both in and out of service. The fluctuations would destroy the bond in any case, but in the process the concrete around the bolt could spall, damaging the compressive support area and loosening the bolts. Unbonded bolts, on the other hand, will stress uniformly over their length and apply loading to the concrete only at the end bearing zones. Consequently, it is necessary to have adequately sized bearing plates anchoring the bottom ends of the bolts.

Common practice is for use of bolts of higher quality than common carbon steel (ASTM A307 or ISO 4.8) and initial pretensioning should be specified by the engineer. Anchor bolts that are preloaded to at least their maximum applied loading will not permit the tower leg baseplate to loosen or to drift.

In steel-erection practice, it is not unusual to install leveling nuts on some of the bolts at each anchorage. Then, when the baseplate is set on the nuts, it is at proper elevation and ready for placement of grout. However, this procedure must not be used when the bolts are to be preloaded because the bolts with leveling nuts cannot be preloaded; the entire preload will be confined to that small part of the bolt which lies between the two nuts and will therefore be ineffective.

To assure proper preloading, the baseplates should be leveled on steel shim stock and grouted. When the grout has reached sufficient strength, the bolts can be tightened to the specified preload. Bolt torque should be checked a week or so after the crane has been put into operation, since the dynamic loading could cause further bond breaking or tighter seating of the baseplate on the footing. If the bolts do require additional tightening at that time, they should be checked again after another week or so.

As already explained, the anchor bolts should terminate at suitably sized plates bearing in the concrete. These should be stiff plates. Though their thickness can be determined by calculation, by rule-of-thumb the thickness should be approximately equal to the diameter of the anchor bolts. Length and width of these embedded plates should usually be at least that of the leg baseplate.

The transfer of tensile load is then through the bolts to the plate mechanically anchored in the concrete mass. The shear, or diagonal tension, strength of the concrete must lock the plate in place. For a square plate of sides p buried to a depth d_0 in the concrete (Fig. 6.8), the imaginary surface that must experience diagonal tension comprises planes rising from each edge of the buried plate at 45° to the direction of the applied force. In plan view, the projected area of this surface is

$$A = (p + 2d_0)^2 - p^2 = 4d_0(p + d_0) \tag{6.10}$$

and the tensile stress on the area is then (using 1.5 times the applied moment, discussed earlier)

$$\sigma_t = \frac{F^+_{\text{diag}}}{A} = \frac{\frac{1}{4}[-Q/4) + 1.5M_o/(s\sqrt{2})]}{pd_0 + d_0^2} \tag{6.11}$$

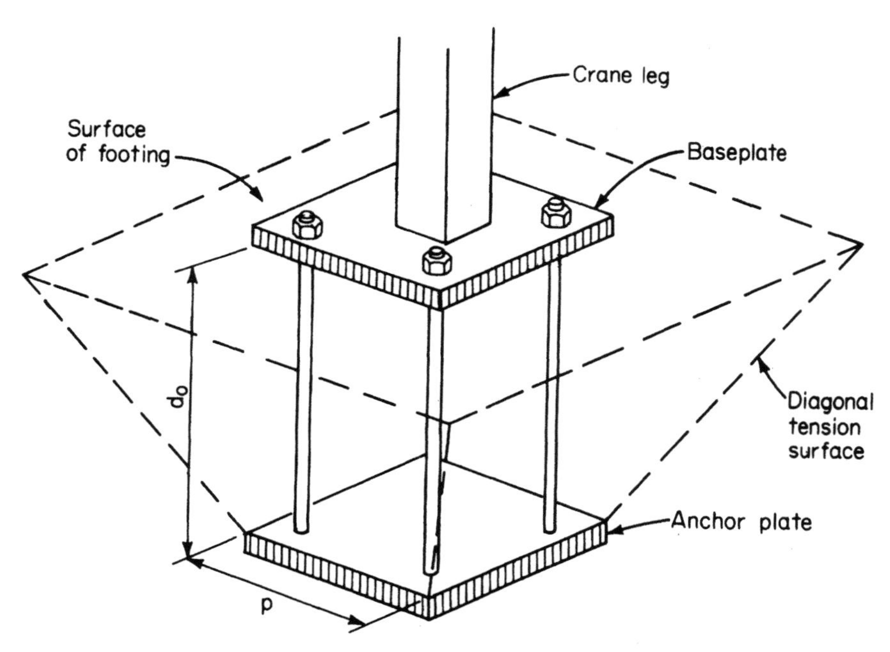

Figure 6.8 The imaginary concrete surface assumed to act to secure the anchor bolt plate against pullout forces.

The limiting value† (in working stress design) for σ_t is $\sigma_a = 2\sqrt{f'_c}$, where f'_c is the 28-day compressive strength specified for the footing concrete. Usual practical values for f'_c can be specified from 2500 to 5000 lb/in² (17 to 35 MN/m²).

Equation (6.11) will give the diagonal tension stress for a plate with sides p at depth d_0 for given loads. If p is taken to match leg baseplate size, then d_0 is the unknown value when full permitted stress is used. Rearranging Eq. (6.11) and solving for d_0 with the quadratic formula gives

$$d_0 = \frac{1}{2}\left[\left(\frac{p^2 + F^+_{\text{diag}}}{\sigma_a}\right)^{1/2} - p\right] \tag{6.12}$$

as the minimum plate depth needed to satisfy both the loads and concrete strength.

A similar effect takes place under the leg in compression, the stressed surfaces spreading downward at 45° from the bearing-plate

† Building Code Requirements for Reinforced Concrete (ACI 318), American Concrete Institute, Detroit, Mich., 1983.

edges to the bottom of the footing depth d, but d must be greater than d_0, of course. Assuming the bearing plate with sides p, we have

$$\sigma_a \geqslant \frac{-F^-_{\text{diag}}}{A} \geqslant \frac{\frac{1}{4}[(Q/4) + M_o/(s\sqrt{2})]}{pd + d^2} \tag{6.13}$$

and to find the minimum acceptable value for d we use

$$d = \frac{1}{2}\left[\left(p^2 - \frac{F^-_{\text{diag}}}{\sigma_a}\right)^{1/2} - p\right] \tag{6.14}$$

When the edge of the bearing (or anchorage) plate is closer to the footing edge than footing depth d (or anchorage plate depth d_0), the full diagonal-tension-resisting surface area will not develop and Eq. (6.13) or (6.11) will no longer be applicable. In this case, Eq. (6.10) must be replaced and the true projected area A determined for the actual footing-edge conditions. The corrected value for A can be used in Eq. (6.13) or (6.11), but neither Eq. (6.12) nor (6.14) can be used.

Another means for anchoring a crane mast to a footing block is by casting a section of mast into the concrete. When this method is used, a special mast bottom section, called an *expendable base,* is employed as the lowermost section of the mast.

At the upper end of the expendable base, standard mast-connection arrangements are provided for erection of ordinary mast sections, but at the bottom end of each leg a stiffened plate is provided. The plate connection to the leg is arranged so that it will be capable of transmitting the maximum tensile and compressive forces developed.

Since the leg baseplates are buried in the concrete, tensile loads are transmitted to the concrete through diagonal-tension-resistance surfaces that rise from the plate edges at 45° to the footing surface in exactly the same way that the anchor-bolt-plate loads were transmitted. Equations (6.10) to (6.12) are therefore applicable for determining the anchorage depth needed, subject to the limitations mentioned for distance to the edge of the footing, of course.

In like manner, the compression loads are resisted on diagonal-tension planes projecting downward from the leg baseplate. Equations (6.13) and (6.14) are therefore applicable, with the limitations mentioned above. Now, however, total footing thickness can be no less than $d_0 + d$, and the distance d is the depth of footing remaining below the leg baseplate (Figs. 6.9 and 6.10). Thus, when an expendable base is used, overall footing thickness may be controlled by pullout and punching-strength requirements rather than by bending-resistance needs. However, expendable inserts, such as in Fig. 6.11, can be designed to minimize footing thickness by providing top plates for punching forces and bottom plates for pullout forces.

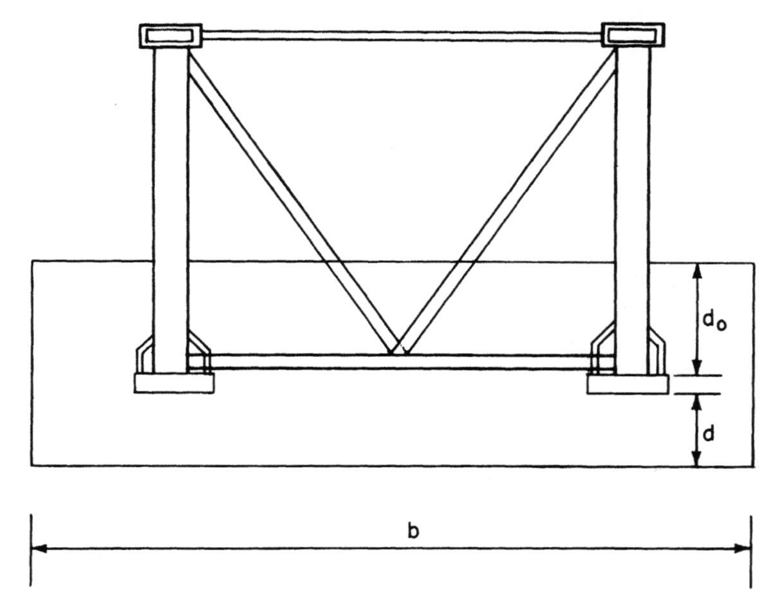

Figure 6.9 Section through a footing block for a crane with an expendable base section.

Some tower crane manufacturers furnish individual expendable legs which must be installed with a template to ensure correct alignment (Fig. 6.11). These are designed to utilize overlapping tension and compression pullout areas, thereby minimizing footing thickness.

A tower section is frequently the best template for setting expendable legs. Though much heavier than a purpose-built template, it has several important advantages:

1. There is no question of tolerances or fit as the tower section is already mated to the expendable legs.
2. A tower section is convenient and reliable for checking plumbness before and after the concrete is poured.
3. Climbing cranes require the mast to be correctly oriented; this is easily checked before the pour when a tower section is in place.
4. A template such as that shown in Fig. 6.11 can leave a "rocking table" effect if it is not perfectly leveled. This cannot occur when a tower section is used as a template.

Needless to say, the determination of footing depth for anchorage should be the first step in tower crane footing design, a step performed before stability is considered. Then with minimum footing depth known, it can be tested with Eq. (6.2) for stability requirements.

Figure 6.10 Richier model 1425 crane on a footing block founded on solid rock. The high bearing capacity of the support surface permitted use of a short-sided but deep footing mass to clear an obstruction at the left (not in view).

With footing dimensions set, the footing reinforcement needed to resist the moments and shears can be designed using the ordinary principles of concrete design.

Example 6.2 (*a*) What anchorage depth is needed for a crane with leg centroids 9 ft (2.74 m) apart and with 26-in-square (660-mm) anchor plates? Use the loads of Example 6.1 and concrete with f'_c = 4000 lb/in^2 (27.6 MN/m^2).

Equation (6.9) gives the uplift or tensile leg load that must be accommodated

$$F^{+}_{\text{diag}} = -\frac{230}{4} + \frac{1.5(4450)}{9\sqrt{2}} = 467.0 \text{ kips (211.8 t)}$$

The allowable concrete stress σ_a is

$$\sigma_a = 2\sqrt{4000} = 126.5 \text{ lb/in}^2 \text{ (872.1 kN/m}^2\text{)}$$

and the minimum anchorage depth is given by Eq. (6.12)

Figure 6.11 Expendable legs for a Pecco tower crane. These legs are in position for the footing to be poured. A template has been bolted to the tops of the legs so that they will mate to the mast section that will later be mounted. Each leg has been carefully blocked up to a uniform elevation; otherwise, shims will be needed to plumb the mast and avoid a rocking table effect. (*Photo by Alex Zajac.*)

$$d_0 = \tfrac{1}{2}\left[\left(26^2 + 467.0\,\frac{1000\text{ lb/kip}}{126.5}\right)^{1/2} - 26\right]$$

$$= 20.0\text{ in} = 1.67\text{ ft }(0.51\text{ m}) \qquad \textit{Ans.}$$

(*b*) What minimum footing depth will be needed if an expendable base is used?

The anchorage depth is as calculated in part (*a*). For resistance to punching through the footing, the compressive load is given by Eq. (6.9)

$$F^{-}_{\text{diag}} = -\frac{230}{4} - \frac{4450}{9\sqrt{2}} = -\ 407.2\text{ kips }(184.7\text{ t})$$

and the depth below the bearing plate is given by Eq. (6.14)

$$d = \tfrac{1}{2}\left[\left(26^2 + \frac{407{,}200}{126.5}\right)^{1/2} - 26\right]$$

$$= 18.2\text{ in} = 1.52\text{ ft }(0.46\text{ m})$$

which makes a total depth required of

$$1.67 + 1.52 = 3.19\text{ ft }(0.97\text{ m}) \qquad \textit{Ans.}$$

The footing size chosen in Example 6.1 would satisfy for an anchor bolt base or for an expendable base.

Ballasted or knee-braced bases

Knee-braced bases with external ballast have the advantage of requiring less on-site preparation work (Fig. 6.12). Typically, four poured-in-place footing blocks of relatively small size are used, be-

Figure 6.12 Both tall hammerhead tower cranes are on ballasted knee-braced bases. The shorter crane, called a self-erecting crane, is not often seen in North America. (*Morrow Equipment Co., L.L.C.*)

cause the ballast provides most of the overturning resistance. The principles used for the design of this arrangement are similar to those used with a massive footing block. As given earlier, the resisting moment M_r should be 1.5 times the applied moment M_o. Then

$$M_r = 1.5M_o = (Q + B + W)b/2 \tag{6.15}$$

where B is the weight of the prefabricated ballast blocks, W is the combined weight of the four identical corner footing blocks, and b is the distance between footing block centers each way. For footing blocks measuring $c \times c$ in plan, and an allowable soil pressure of p_{allow}

$$c = \left(\frac{Q + B + W}{2p_{\text{allow}}}\right)^{1/2} \tag{6.16}$$

The footing blocks of depth d and unit concrete weight w will then weigh

$$W = 4c^2dw \tag{6.17}$$

For practical problem solving, Eq (6.15) is rearranged to give

$$B + W = 3M_o/b - Q$$

and then Eq. (6.16) is solved for a trial value of c; d is chosen by judgement in consideration of c, permitting W and then B to be calculated. But B represents the weight of available prefabricated blocks, and may not match the calculated value. Therefore, blocks are chosen to make two symmetrical piles of ballast blocks weighing not less than the trial calculation value. It is necessary for the CG of the ballast blocks to be on the mast centerline, and this is accomplished with two equal weight stacks on opposite sides of the mast. Now, the footing blocks may need recalculation.

The anchor bolts used to fasten the knee brace legs to the footing blocks need to be of sufficient strength to lift the weight of the block at the allowable tension force for the bolt size. Those bolts can never be exposed to a greater force, but they must also resist wind shear forces from the entire structure.

Example 6.3 Using the same data as for Example 6.1, and with $b = 24$ ft (7.3 m), design the footing blocks and determine the required ballast weight. The combined weight of ballast and footing blocks can be calculated from the rearranged form of Eq. (6.15)

$$B + W = 3 \times 3600/24 - 230 = 220 \text{ kips (978.6 kN)}$$

Equation (6.16) will then give the footing block plan dimensions

$$c = \left(\frac{230 + 220}{2x8}\right)^{1/2} = 5.3 \text{ ft (1615 mm)}$$

but use 5.33 ft or 5′ − 4″ (1625 mm)

For a footing of this size, a depth of 3 ft (915 mm) would be appropriate. Equation (6.17) now gives footing weight

$$W = 4 \times 5.33^2 \times 3 \times 0.15 = 51 \text{ kips (227 kN)}$$

which requires ballast weight to be 220 − 51 = 169 kips (752 kN). The ballast selected will therefore weigh somewhat more when available blocks are chosen.

Variations on the static mount

The basic ballast block type of footing is simple to conceptualize and almost as simple to implement, but it is not suited to all conditions. The soil may be inadequate, or rock that is too expensive to excavate en masse may be present. Sometimes the space available is not suitable for a spread footing, or to compensate for a small available plan area, an uneconomically deep excavation would be required. In such cases, the best solution is often to mimic the method used for the building footings or to utilize other foundation methods that are common in local practice. Caissons, piles, mats, rock anchors, and steel base frames have all been used successfully to support tower cranes (Fig. 6.13).

Caissons and piles are economical only when the placing equipment is already at the site. In crane practice, batter piles are not ordinarily used to resist torsional and lateral loads. Instead, the surrounding soil mass is utilized for lateral restraint, but the backfill must then be placed in a manner suitable for this purpose. Caissons and some types of piles can be installed so as to possess uplift capacity which can be exploited to reduce footing mass.

Oftentimes a crane base can be improvised by augmenting structural elements of the building. For example, a mat foundation can be extended (Fig. 6.13*d*) or several pile caps can be interconnected. A tower crane need not necessarily be mounted on grade; some circumstances favor a setup on a deck or on roof framing (Fig. 6.13*b*). A roof-mounted crane might be used, for instance, to erect cladding on a completed building frame as in Fig. 6.2.

Caution must be exercised when a crane footing is to be interconnected with spread footings or friction piles that are part of the building structure. One must assume that the dynamic action of the crane will cause some differential settlement. Control joints might be con-

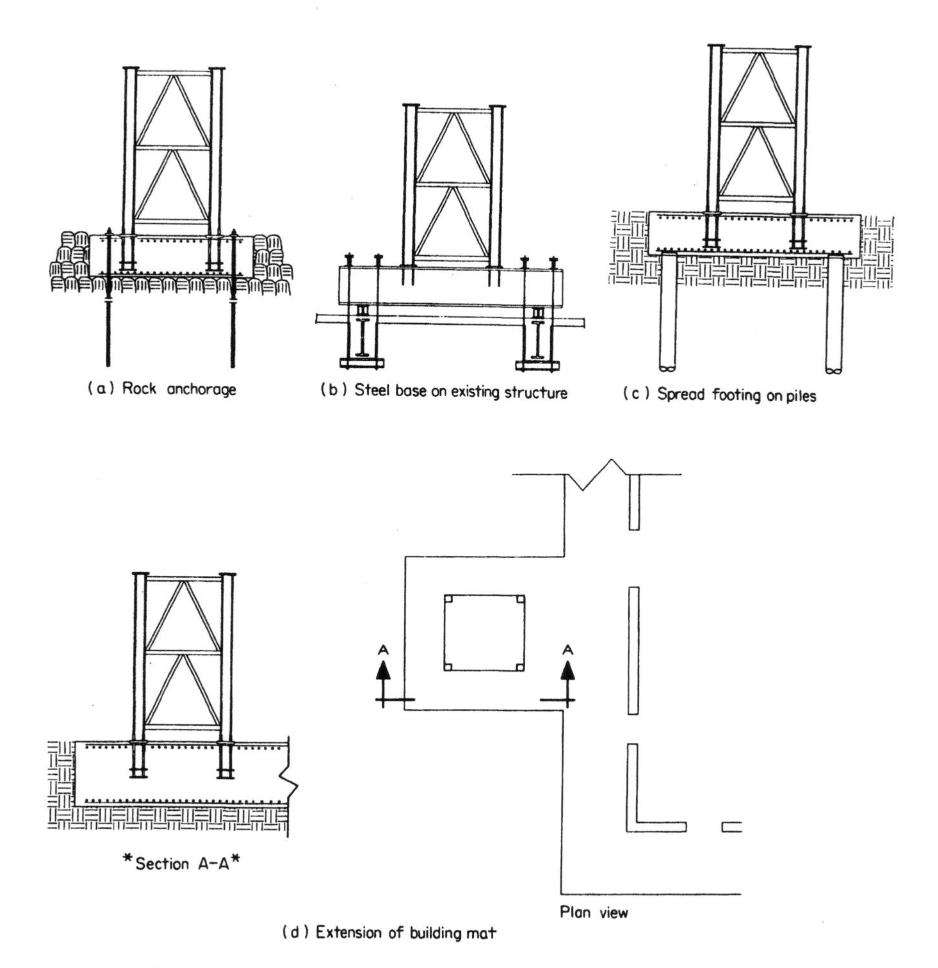

Figure 6.13 Several methods of supporting a freestanding tower crane (*a*) Rock activation posttensioned anchors may be economical where rock excavation is difficult. (*b*) A base frame distributes the crane loads to adequate points of support on a building frame. (*c*) Piles may be suitable where the pile-driving equipment is already on the job. (*d*) A mat foundation may be well-suited to double as a tower crane base.

sidered as a remedy to prevent cracks from developing, or the combined footing mass could be made so robust that cracking would be precluded.

If the tower crane base cannot be built into the building foundation, it can sometimes be built over it by mounting the crane on a steel frame that spans footings. The frame can be secured by ballast, anchors, or some combination of these.

A steel base frame is also usually the most viable means for mounting a tower crane on a deck above grade such as a roof. The larger the plan area of the base, the smaller are the corner loads imposed on the building. If the deck is not adequate to support the crane by itself, the loads can be shared by several tiers of framing or transferred to stronger parts of the building. Combinations of shores, high-strength tie rods, and load-transfer framing are often economical means for distributing vertical crane loads. Horizontal loads from wind and torsion might be distributed through a soundly connected decking system, or alternately carried by plan braces installed for this purpose (Fig. 6.14).

Tower crane leg loads reverse repeatedly from tension to compression. Lateral loads also go through reversals from shifting wind and alternating swing motion. Details of base support structures should be designed with cyclical dynamic loading in mind. Connections should be designed to transmit shear forces through friction.

When drilling equipment is available, rock anchors may be used to great advantage for tower crane mounting on rock. The anchors can tie down a steel base frame, the individual tower legs, or a concrete footing block (Fig. 6.13*a*). The base frame has the advantage of attenuating the reactions by spreading them out but then later must be removed from an area that may have difficult access. The tie-down of individual legs is simplest but requires the most rock anchors and

Figure 6.14 This crane is supported on a steel base built into the building framing. Note the plan bracing installed to accommodate horizontal forces from torsion and wind and to permit the crane to operate before the concrete deck is poured. (*Canron Construction Corp.*)

presents the greatest difficulty should an anchor be unsuccessful. The tied-down footing block requires the most excavation but the least number of anchors. It also offers the greatest safety through redundancy—with this method, tie-downs simply replace ballast mass and the footing is reduced to the smallest size required for anchorage (Fig. 6.15).

Design of rock anchorage requires familiarity with the materials, specialized hardware, and local conditions. Each anchor should be proof tested to at least 33% more than the maximum working load and locked off at a tension at least 15% above working load. The anchors must be allowed to stretch under the influence of the pretensioning load; a free stressing length of at least 15 ft (4.5 m) is a recommended practice.

Tower bases and anchors should be part of the daily inspection rounds of the operator or oiler. Connections should be checked for distress. Rock anchors cannot be permitted to loosen. Each can be checked by tapping gently with a hammer. A well tensioned anchor will emit a high pitched ping, while a loose anchor will answer back with a dull thud.

Differential settlements

Using nothing more than a surveyor's level, one can establish a regular procedure for monitoring differential settlement. The intervals between monitoring readings are a function of soil type and the observed settlement activity, but guidance on this point is straightforward.

As soon as possible after completion of crane erection, the initial, or control, elevation readings are taken. On relatively soft soils follow-up readings to monitor movement should be taken weekly for 4 to 6 weeks. After that reading intervals can be expanded to 2 weeks if little movement is noted. Later, monthly intervals can be used if warranted by the records.

The initial and monitoring readings are taken at the four corners of the footing block (Fig. 6.16). The points chosen should be marked with a masonry nail or other permanent means and numbered to assure that all readings will be taken at exactly the same points and properly identified.

The relationship between individual initial readings is of no importance. Let us call the initial elevation readings $R_1(I)$, $R_2(I)$, $R_3(I)$, and $R_4(I)$; subsequent readings will be denoted $R_1(n)$, $R_2(n)$, $R_3(n)$, and $R_4(n)$, where n represents the sequential sets of monitoring readings, $n = 1, 2, 3, \ldots$. The initial readings and each set of monitoring readings should be recorded in a log book and dated.

Figure 6.15 Corner of a rock anchor footing similar to the scheme is Fig. 6.13*a*. The anchor, in the center, carries a preload of 140 kips (623 kN). A thick plate washer is needed to maintain the bearing stress on the concrete to within the permissible level. (*Photo by Alex Zajac.*)

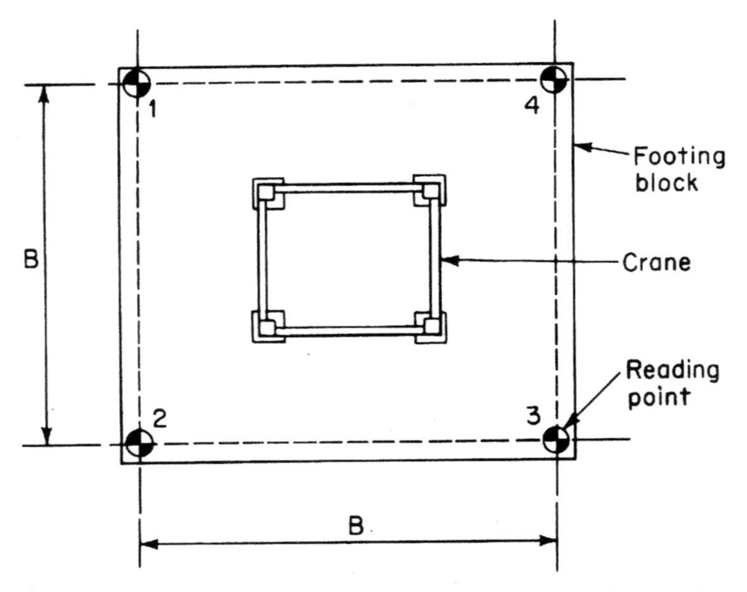

Figure 6.16 Locating control points for monitoring footing-block differential settlements.

Although 1:400 is accepted as an out-of-plumb limit, a properly installed crane mast should not lean more than 1:1000 in calm air with the boom pointed in the axis of the observer. The same plumbness limit must therefore be the guiding criterion during mast erection. Initial mast lean together with lean direction should be measured and recorded in the log book immediately after erection.

Changes in mast lean are determined from the monitoring elevation readings. The change in elevation from initial conditions of any corner is given by

$$\Delta_3(n) = R_3(I) - R_3(n) \tag{6.18}$$

for corner 3 and similarly for each of the other corners. The change in lean between two adjacent corners is then

$$C_{34} = |\Delta_4(n) - \Delta_3(n)| \frac{1000}{B} \tag{6.19a}$$

for corners 3 and 4, with similar expressions for the other adjacent corners. The vertical lines indicate *absolute difference* (i.e., the greater value minus the smaller value). Lean will be in the direction of the greater value. Dimension B and the elevations must be in consistent units.

Equation (6.19a) is modified somewhat for monitoring changes in lean across the diagonals

$$C_{24} = |\Delta_4(n) - \Delta_2(n)| \frac{1000}{B\sqrt{2}} \tag{6.19b}$$

The C values determined from Eqs. (6.19a) and (6.19b) are the number of units of lean per 1000 units of vertical mast height and should not be permitted to exceed 1. C values calculated after each set of monitoring readings are made should be recorded in the log book.

We have not found any crane manufacturer who issues mast-lean criteria for either erection or subsequent settlement. The criteria that follow are therefore our own; we have based them on ordinary good practice in steel erection and on an analysis of the effect of lean on crane operation.

If any C value is found to be 1 or greater, it must be compared with initial mast-lean readings. If the net lean in any direction exceeds 1:400 (2½ units of lean), corrective action should be taken. If leans exceeds 1:500 (2 units of lean), corrective action is suggested.

Differential settlements can be corrected by parking overnight in such a way that the jib will point in the direction of the lean. The slewing brakes can be locked only if there is assurance that low-level winds will prevail during the unattended period. When this is done, the counterweight moment will act opposite to the lean and tend to correct it. It is unlikely that any loaded moment will exceed the counterweight moment, so that loading the jib is not a viable alternative. If no clear corrective progress is evident, more positive measures must be taken, e.g., ballasting the footing block. Monitoring readings should be taken frequently during corrective actions.

Given the same soil subgrade, tower crane footings can usually tolerate higher bearing pressure than buildings. The reasons for this are several:

1. Allowable bearing pressures for buildings are usually governed by settlement, and tower cranes are more tolerant of uniform settlement.
2. The crane is a temporary structure.
3. Peak pressures are ephemeral, whether induced by wind or operations, whereas settlement is time dependent.
4. Peak pressures tend to be randomly distributed with respect to direction due to the weathervaning of the crane.
5. The excavated mass of soil may weigh nearly as much as the crane footing. The net pressure of the crane installation on deep strata of soil is thus small.

With due caution, a tower crane can sometimes be supported on soil that is considered unsuitable for building footings.

Example 6.4 Check the net lean of a tower crane having the base-elevation monitoring marks 25 ft (7.6 m) apart. Initial and monitoring readings were made as follows:

n	R_1	R_2	R_3	R_4
0	132.42	132.38	132.41	132.41
3	132.41	132.37	132.38	132.39

Initial mast was measured, and the following values, in units per thousand, were calculated:

$$\text{Mast lean} = \begin{cases} 0.25 & \text{from side 1-2 toward side 3-4} \\ 0.33 & \text{from side 2-3 toward side 1-4} \end{cases}$$

The diagonal mast-lean values can be immediately determined from the measured data. The lean from 2 toward 4 is $(0.25^2 + 0.33^2)^{1/2} = 0.41$, while from 3 toward 1 it is $(0.33^2 - 0.25^2)^{1/2} = 0.22$.

From Eq. (6.18), the corner elevation changes are

$$\Delta_1(3) = 132.42 - 132.41 = 0.01$$

$$\Delta_2(3) = 132.38 - 132.37 = 0.01$$

$$\Delta_3(3) = 132.41 - 132.38 = 0.03$$

$$\Delta_4(3) = 132.41 - 132.39 = 0.02$$

The settlement lean of each side is given by Eq. (6.19*a*)

$$C_{21} = 0 \qquad C_{23} = \frac{(0.03 - 0.01)1000}{25} = 0.80$$

$$C_{43} = \frac{(0.03 - 0.02)1000}{25} = 0.40$$

$$C_{14} = \frac{(0.02 - 0.01)1000}{25} = 0.40$$

while the diagonal leans are given by Eq. (6.19*b*)

$$C_{24} = \frac{(0.02 - 0.01)1000}{25\sqrt{2}} = 0.27$$

$$C_{13} = \frac{(0.03 - 0.01)1000}{25\sqrt{2}} = 0.57$$

The net leans are determined by adding the initial and settlement leans, giving appropriate consideration to direction:

$$2\text{-}1 = 0.33 + 0 = 0.33 \quad \text{OK}$$

$$2\text{-}3 = 0.25 + 0.80 = 1.05 \quad \text{OK}$$

$$4\text{-}3 = -0.33 + 0.40 = 0.07 \quad \text{OK}$$

$$1\text{-}4 = 0.25 + 0.40 = 0.65 \quad \text{OK}$$

$$2\text{-}4 = 0.41 + 0.27 = 0.68 \quad \text{OK}$$

$$1\text{-}3 = -0.22 + 0.57 = 0.35 \quad \text{OK}$$

The leans are all less than 2½ units (1:400), for which corrective action is required, and less than 2 units (1:500) for which action is suggested. It should be noted that the calculated net leans are not geometrically consistent. This is not a cause for alarm, as the difference results from reading the elevations to the nearest 0.01 ft (3 mm).

6.4 Braced and Guyed Towers

Storm winds limit the permissible tower height of the static-mounted crane. When this height is insufficient for completion of the work, the tower must be braced or guyed to permit greater height. The use of one versus the other is dictated by the jobsite environment; guying requires open space on all sides and bracing demands close proximity to a major supporting structure. Almost always, this structure is the one being erected.

Guying and bracing systems given rise to large forces that must be resisted by both the framework of the tower and the supporting structure. Guys must be anchored either to the earth or to suitably massive elements. While brace loads can easily be sustained by the wind-resisting systems of most high-rise buildings, the local area of brace attachment may have to be reinforced to accommodate these loads. On steel-framed structures that rely on fully made connections and concrete floors to complete the structural system, temporary bracing schemes may be more elaborate.

Guying or bracing operations are commonly coordinated with the addition of tower sections to increase crane height. The combined work of raising and securing the crane is outlined in a *climbing schedule* which also links the construction schedule to the operation (Fig. 6.17). If the building is permitted to move ahead of the crane, the crane will be obstructed by the rising structure. If the crane moves too far ahead of the building, there will not be structure high enough to receive the bracing. Devising a climbing schedule requires knowledge of crane limitations and familiarity with the pace and practices of construction.

Top climbing

To increase the height of guyed or braced cranes, tower sections are added in a process called *top climbing*. The process varies, but all methods require the crane to be in a balanced condition. Generally, hydraulic rams raise the upper portion of the crane in conjunction with placement of a new section of tower above the existing tower. The climbing operation is repeated as many times as necessary to achieve the specified height.

Each manufacturer has a proprietary top climbing system, but they are all of two types. In the most common setup, a *climbing frame* forms a sleeve around the tower (Fig. 6.18). There is an opening on one face of the climbing frame to permit the insertion of a tower section (and conversely its removal). A trolley beam, platform, or other means is furnished on the open side to temporarily hold a tower section that has been hoisted on the crane hook. Hydraulic rams then raise the climbing section until it engages the turntable and picks up the weight of the superstructure of the crane. Extension is continued until the height is sufficient to slide the new section of tower into place. Last, the climbing frame is lowered to bring the weight of the superstructure onto the new tower section, and it is secured.

The other type of top climbing system has the superstructure mounted on an *inner tower* that telescopes from an *outer tower* with the help of hydraulic rams. New tower sections are hoisted in U-shaped halves that are joined together in place. As with the other type, the climbing apparatus can then engage the added section to continue the climb.

Top climbing requires skill and attentiveness since improper procedures can lead to disaster. An experienced and knowledgeable crew is recommended; there must always be at least one key person on the crane with those qualifications.

The crane must be balanced before the climb, which usually requires that a specified hook load be held at a specified radius, although some cranes are balanced by setting counterweight or boom positions. Field adjustment of the radius is almost always needed to achieve a fine balance.

Top climbing should not be done in high winds. If the manufacturer does not recommend an upper limit for wind, 20 mi/h (9 m/s) should be used.

Often bracing or guying cannot be installed until after the crane has been top climbed. Thus the tower may temporarily exceed the unbraced height limitation. This does not present a problem as long as there will be no high winds while the crane is in this vulnerable condition. Weather forecasts must be checked before climbing to excess height.

Braced towers

The bracing of a tower crane requires attachment to a structure much more massive than the crane itself (Fig. 6.19). The mathematical assumption is that the building is rigid and that the tower flexes. Although this simplification is suitable for calculation of bracing loads, it is contradicted by the motion one feels standing on a slender building to which a working tower crane is attached. All buildings will flex and twist in varying degrees under the influence of braced towers. The magnitude of side sway will depend on the flexural rigidity of the structure; the extent of twisting will be affected by both torsional ri-

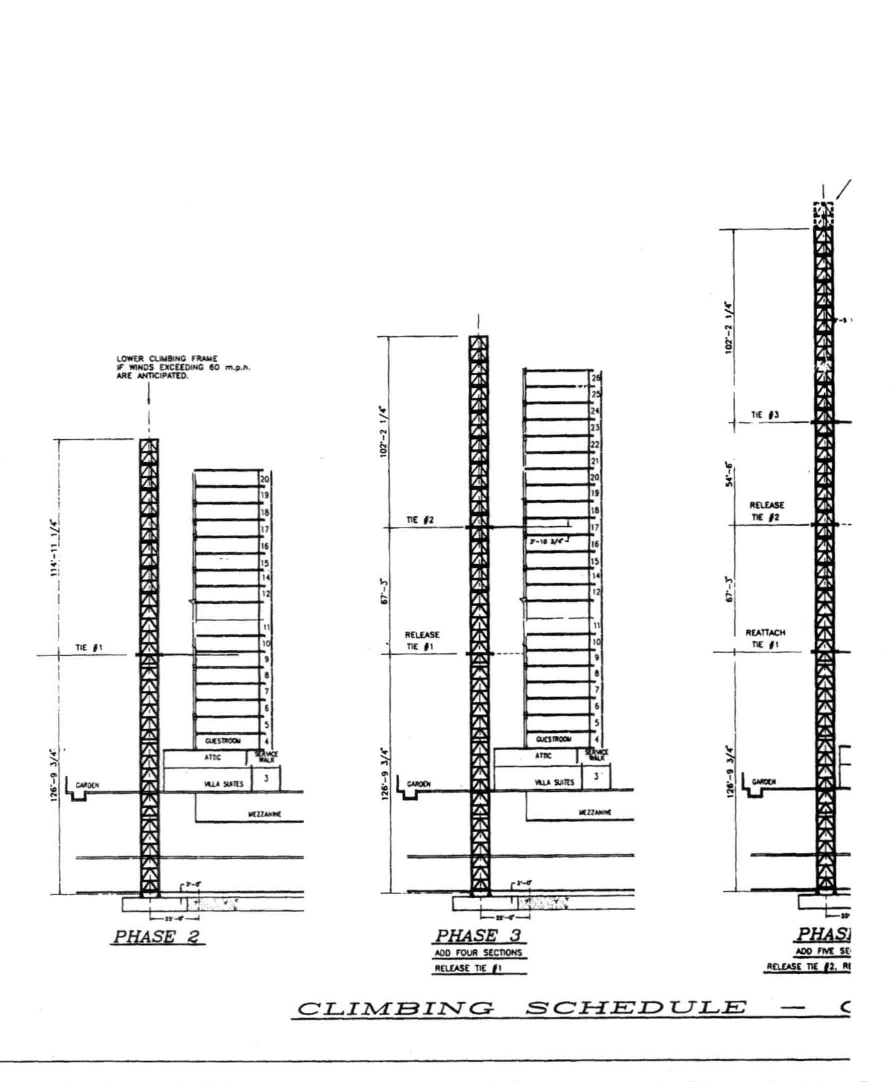

Figure 6.17 Climbing schedule for an externally mounted tower crane building a highrise Las Vegas Hotel. Wind controlled how high the crane could rise above the top tie.

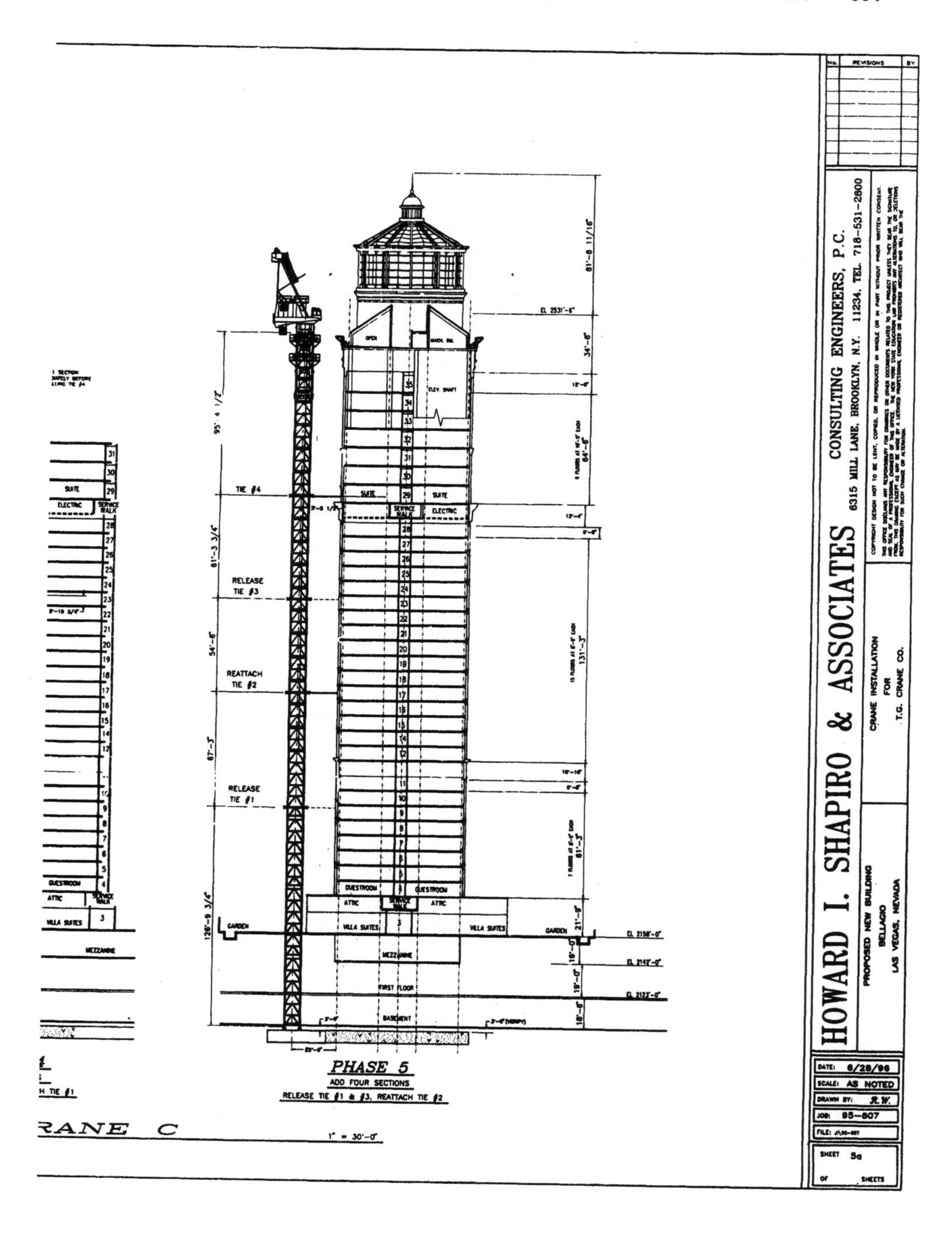
PHASE 5
ADD FOUR SECTIONS
RELEASE TIE #1 & #3, REATTACH TIE #2
1" = 30'-0"
HOWARD I. SHAPIRO & ASSOCIATES
CONSULTING ENGINEERS, P.C.
6315 MILL LANE, BROOKLYN, N.Y. 11234, TEL. 718-531-2800
CRANE INSTALLATION FOR T.G. CRANE CO.
PROPOSED NEW BUILDING
BELLAGIO
LAS VEGAS, NEVADA
DATE: 6/28/96
SCALE: AS NOTED
JOB: 95-607
SHEET 5a
TIE #4
RELEASE TIE #3
REATTACH TIE #2
RELEASE TIE #1
EL 2531'-6"
EL 2158'-0"
EL 2142'-0"
EL 2123'-0"

Figure 6.18 Climbing an externally mounted Pecco SN355 crane. After the climbing frame raises the upper works of the crane, the loose mast section (on the platform) is rolled into position and bolted to the mast below and to the turntable above. (*Delro Industries, Inc.*)

Figure 6.19 A collar and brace support an externally mounted tower crane. The collar is very rigid to prevent the tower from distorting and to take advantage of building stiffness. (*Delro Industries, Inc.*)

gidity and by the location of the crane with respect to the shear center of the building.

When large or sharp motions are observed, investigation would be prudent. The sway of the building can be measured with surveyor's instruments during crane operation and the results checked by the structural engineer.

Almost all buildings require some preparatory work before brace loads can be imposed. For cast-in-place concrete structures, this may entail test cylinder breaks or in-situ testing to assure that the material anchoring the brace has reached sufficient strength. Added local reinforcing or posttensioning may be required to distribute brace loads into the structure. Steel framed buildings will necessitate completion of connections in the vicinity of the crane brace and perhaps stiffening or reinforcement of the area and temporary braces to levels below. All of these measures must be clearly defined in advance, subjected to the scrutiny of the structural engineer, and coordinated with the climbing schedule.

On a steel framed building, the extent of temporary bracing needed to support crane lateral loads is frequently determined by the progress of construction. Braces might be required to provide strength where connections have not been completed or metal decking not installed. The subcontractor using the crane typically controls this follow-up work and therefore has a measure of control over the extent of temporary bracing. On the other hand, the extent of bracing needed might be governed by the progress of shear walls or concrete decking—work performed by another trade.

A single brace or tie may suffice to raise the height of a crane for completion of a building of 15 stories or so, whereas 60 stories may require a half dozen braces or more (Fig. 6.20). The spacing usually works out to be from 40 to 90 feet (12.2 to 27.44 m), but it should always be made as large as practical for economy and to minimize counterflexure forces.

Two factors control brace spacing: the tower alignment to the building and the permissible tower height above the top brace. A point on the tower suitable for attachment of a brace must be found that aligns with a suitable elevation for attaching a brace to the building. On most towers, braces can be attached only where horizontal cross-members come into the tower legs. Coincidence of suitable tower and building elevations rarely happen at every floor; therefore braces are hardly ever located at optimum spacing (Fig. 6.21). Maximum spacing is limited by the height of the building already erected, which in turn is limited by the tower crane height.

Braces should be spaced apart as generously as conditions allow. Motivation for this is based both on economics and engineering. First,

Figure 6.20 A 450-ft-high crane braced to the outside of the Broadway Crowne Plaza Hotel in New York City. During the jacking-down operation, the crane must clear the curtain wall at the lower part of the building. (*North Berry Structures, Inc.*)

braces are relatively expensive to install. Second, close spacing of braces tends to magnify the forces they carry, as one brace "fights" against the next.

When trying to optimize tie spacing, consideration should be given to the preparatory work that must be completed before the brace is installed and loaded. For instance, on a concrete building one must allow time for the concrete to cure to a stipulated strength, while on a steel framed building connections, reinforcements, etc., have to be made up. Often, the last floor erected does not have adequate strength to support bracing.

The crane vendor sometimes provides bracing hardware, but oftentimes it is the responsibility of the user. Vendor-supplied hardware is generally modular and adjustable. A user is more apt to furnish "rude and crude" braces that will be discarded after one use.

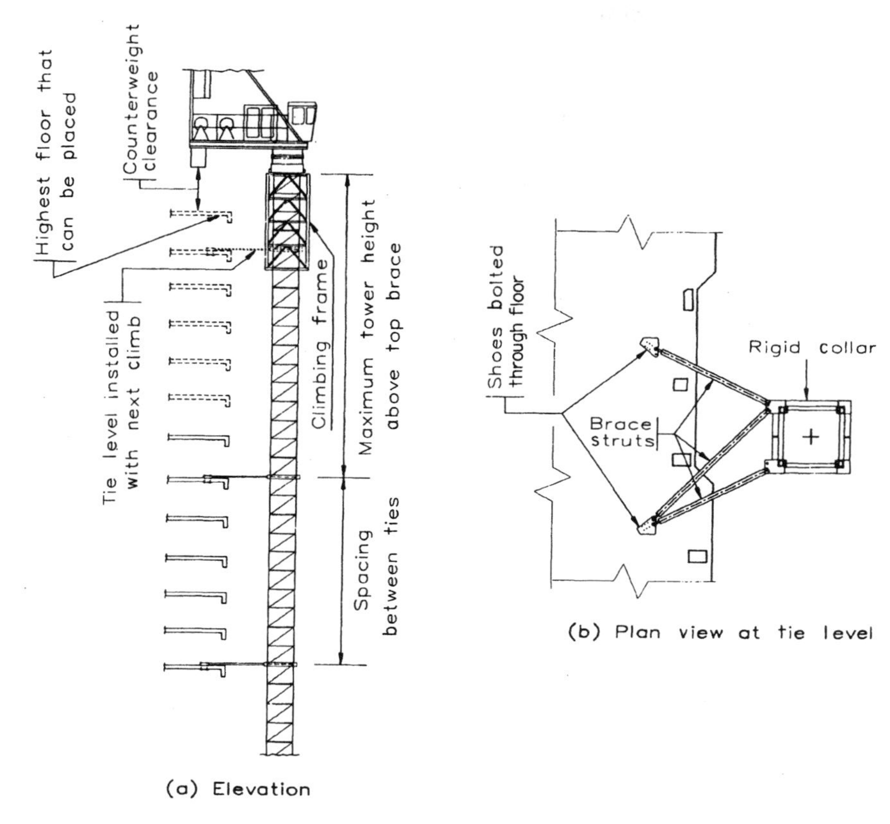

Figure 6.21 Top climbing crane mounted outside a building. (*a*) Tie spacing is controlled by the crane clearance above the building and by the requirement that tie elements must align to suitable elevations on both the mast and the building. (*b*) Schematic view of a brace such as in Fig. 6.19. The reinforced concrete floor acts as a diaphragm to distribute crane forces into the building.

Braces are not typically connected directly to the crane mast. Usually they frame into a collar that has been framed tightly around the legs of the mast (Fig. 6.19). The collar is a stiff "picture frame" that distributes brace forces concentrically to the legs and inhibits racking distortion of the tower. Most tie strut arrangements have three struts because this is statically determinate and economical. The struts can be welded, bolted or pinned at the end connections.

Vendors often supply a set of "shoes" that allow the tie struts to be bolted down to a floor. On a cast-in-place concrete building, the floor is normally the place of choice for securing tie struts because of its diaphragm strength and ease of access. The loading on the floor diaphragm is eccentric—an aspect that the designer tries to minimize—and the reinforcing of the slab must be made adequate for this. Walls and columns have been used for tie attachment points as well.

On a steel building, tie struts are typically framed into connections to the framing system that have been detailed and fabricated for that purpose.

The permissible tower height above the top tie is controlled by out-of-service wind loading. Consequently, this height is affected by the intensity of wind and by the effective exposure area. It follows then that, all other things being equal, a hammerhead crane can be raised higher than a luffing boom crane, and an inner-city crane can be raised higher than a crane at a coastal installation.

It also follows that under carefully controlled conditions, the installation engineer might permit the tower height to be increased over a short period of time. Installation of a tie is time consuming—typically an all-day operation—and causes an interruption of the work cycle. A temporary mast height increase might "buy time" for the tie installation to be scheduled in synchrony with that cycle. However, the crane user must be prepared to install the tie on short notice if high winds are forecast.

Guyed towers

For situations where local storm-wind levels exceed design values for the needed freestanding mast height, a guying system may be used to restrain the excess wind load. In some instances, the guys cannot be left in place while the crane is in service. In that event, the guys must be installed on the mast and left slack or disconnected at the anchorage ends. The anchorages, or deadmen, must be in place ready to receive the guys, and definite procedures are needed to make certain that work crews will be available to complete the guying should high winds threaten at any time of day or night. A *deadman* is a temporary anchorage point, either existing or built for the purpose.

Guys, when installed, will hang in a catenary shape, the amount of sag depending on the tension in the rope. If loaded by wind, the tower will lean with the wind and load one or more guys, causing a reduction in catenary sag. This has the same effect on the mast as lengthening the rope would. But loading the rope will also induce constructional and elastic stretch, resulting in further elongation. The combined increases in rope length cannot be permitted to let the mast deflect excessively, since excessive mast deflection can lead to collapse.

Rope elongation can be controlled. Constructional stretch can be minimized by using prestretched rope, while elastic stretch can be limited by using oversized rope and catenary stretch reduced by preloading the guys with turnbuckles or other devices. Preloading must be done carefully so that the crane remains plumb afterward.

After setting up a guying system, preloading it, and replumbing the mast, it is necessary to monitor both the preload forces and mast plumbness. This should be done every few days initially and after any significant high wind, but the intervals can be extended as preloads stabilize (i.e., when the constructional stretch has been fully removed from the rope). At the same time fittings, such as rope clips, should be checked and retightened.

For guys intended to carry excess wind loads, the excess wind moment M_e can be approximated from the design wind moment M_d and the wind velocities

$$M_e = \left[\left(\frac{V_s}{V_d}\right)^2 - 1\right] M_d \tag{6.20}$$

where V_s is the site wind velocity and V_d the design velocity.

A simple guying arrangement (Fig. 6.22) would have four guys radiating from the mast corners. The guys are affixed to the mast at a guying frame supplied by the crane manufacturer or in accordance with details supplied by the manufacturer (Fig. 6.23). If neither frame nor details are available, the guys should not be attached before a rigorous stress analysis of the mast under storm-wind conditions has been made. Unless they are held within acceptable limits and transmitted to the mast properly, guying loads can collapse a mast leg.

The guying connection should be made high on the mast between half and two-thirds of the distance up to the slewing circle. The guy

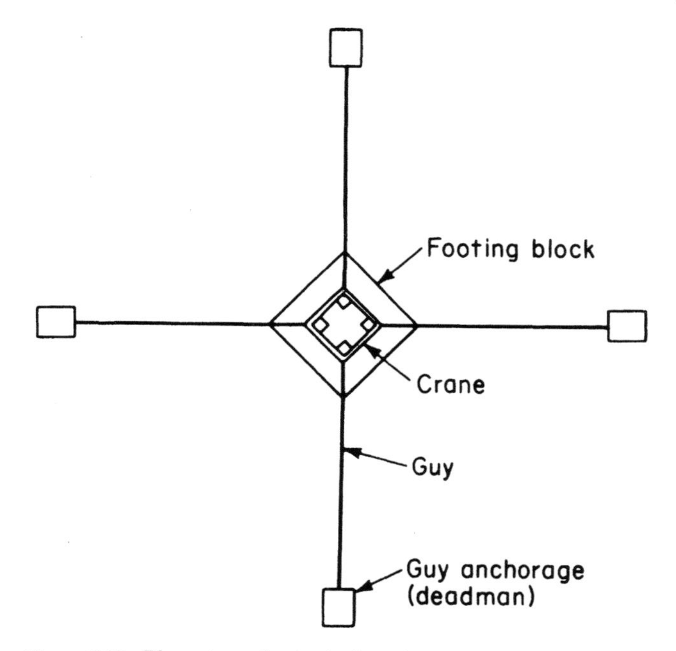

Figure 6.22 Plan view of a typical guying arrangement.

Figure 6.23 Guying frame for a Richier model 1372 crane. The frame is designed to distribute guy forces to the mast members without distorting the mast cross section. *(Photo by J. D. Telenko.)*

angle θ should be kept somewhat flat, however, preferably about 45°, and angles in excess of 60° should be avoided if at all feasible.

To restrain the excess wind given in Eq. (6.20) each guy must be capable of supporting a horizontal force (Fig. 6.24)

$$F_h = \frac{M_e}{h}$$

or a load in the rope of

$$F = \frac{M_e}{h \cos \theta}$$

Although this appears to be mathematically correct, nevertheless in practice the difference between the elastic properties of the rope and the mast will cause the rope to take more load. Approximately, then, take

$$F = \frac{(V_s/V_d)^2 M_e}{h \cos \theta} \tag{6.21}$$

When selecting the rope to support this load, a strength margin of not

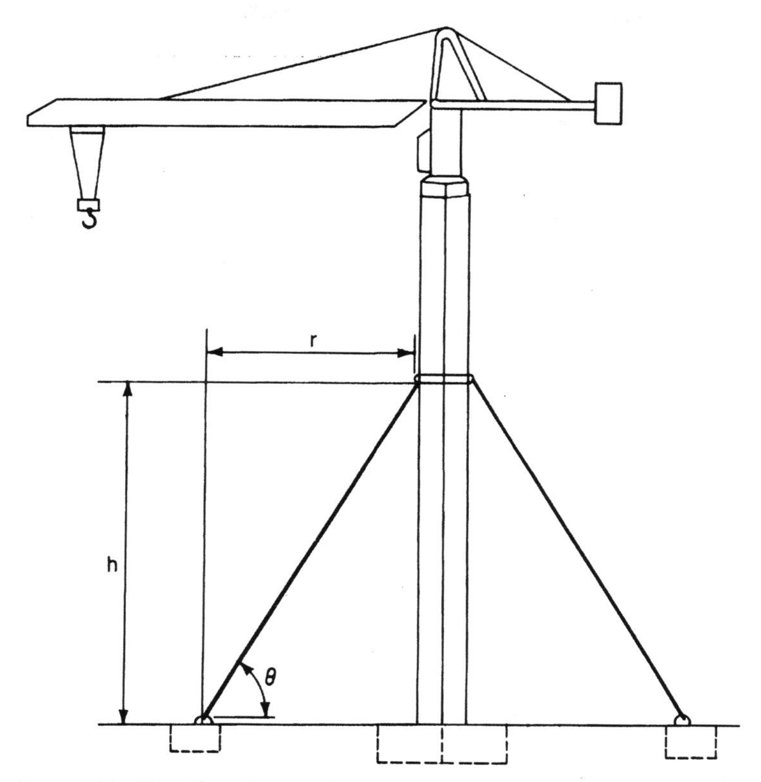

Figure 6.24 Elevation of a guyed tower crane.

less than 5 should be used to minimize elastic stretch and to provide additional safety because of the crude approximation used, an approximation that still tends to underestimate guying loads. To avoid ropes of very large diameter, two or more parts per guy can be used. When this is done, sheaves or blocks must be placed to assure equal loading of the parts. Guy anchorages and fittings must be sized with regard for rope breaking strength in order to provide an adequate strength margin against the true load.

This simple guy-selection method should be used only for planning or estimating purposes. For the working design of the system the following exact method should be employed.

Exact analysis of a guyed mast

Don't let the title of this subsection fool you. The results achieved by the method that follows will not be exact but will rather be an acceptably accurate prediction of conditions to be expected in practice. What *is* exact are the mathematical concepts, but inaccuracies are introduced by the wind calculations, the variations in the properties of ropes, and the approximations made to adapt the real crane to a practical mathematical model.

As mentioned earlier, the guy ropes hang in a catenary, and the changes in this curve as rope load changes are important to the design. A catenary rope loaded by a horizontal end force F_h and by its own weight (Fig. 6.25) will transmit that horizontal force to the anchorage at its other end undiminished. If the rope is assumed to be perfectly flexible, there can be no bending moment at any point in it. The only applied load within the rope length, i.e., the rope weight, is vertical, so that the applied end load F_h constitutes the only horizontal force and therefore must be present throughout the rope. Taking moments about b (Fig. 6.25), we get

$$\Sigma M_b = F_h h - R_{ay} r - \frac{w' r^2}{2} = 0$$

$$R_{ay} = F_h \tan\theta - \frac{w' r}{2} \tag{a}$$

where $w' = w/\cos\theta$ is the unit rope weight projected onto a horizontal plane. Taking moments about c of the loads to the left of c gives

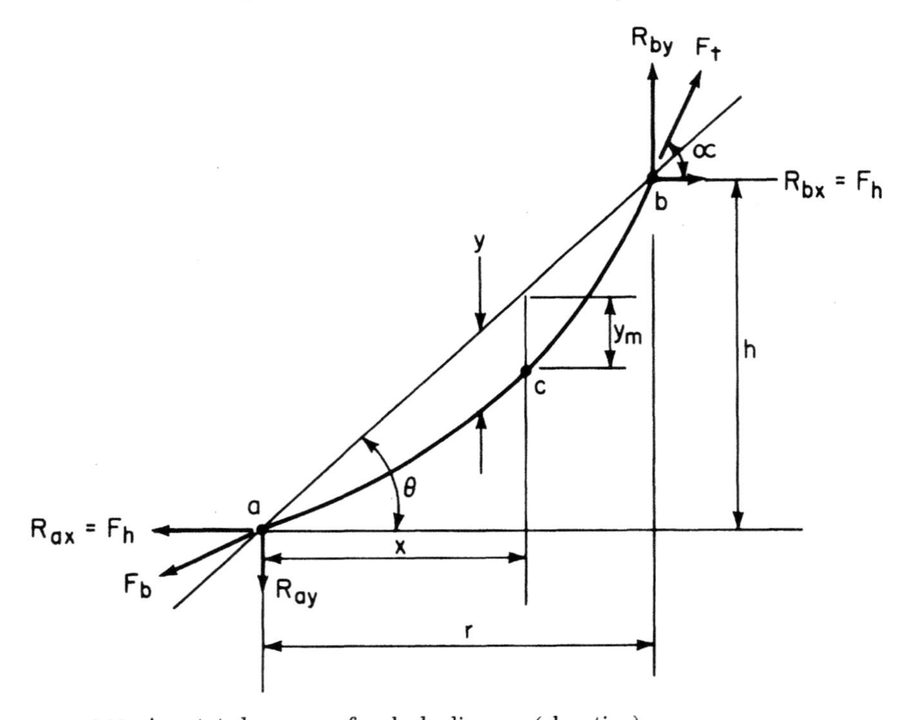

Figure 6.25 Annotated guy-rope free-body diagram (elevation).

$$F_h(x \tan \theta - y_m) - R_{ay}x - \frac{w'x^2}{2} = 0$$

where y_m is the difference in height between the chord and the rope. Substituting R_{ay} from Eq. (*a*), we get

$$F_h y_m = \frac{w'rx}{2} - \frac{w'x^2}{2} \tag{b}$$

The right side of Eq. (*b*) is seen to be the bending moment in a simple beam with uniform load w' about a point a distance x from one end. If the same sort of procedure were to be used with a series of concentrated vertical loads along the rope, the same form of result would ensue, giving rise to the general theorem for a nonvertical wire-rope load-supporting member: at any point on a wire rope that is loaded only vertically, the horizontal force in the rope multiplied by the vertical distance from the rope to its chord is equal to the moment that would be produced by those same vertical loads about the same point on a simple beam of the same span as the horizontal span of the rope.

Going to the center of a span carrying only rope dead load and calling the difference in height between chord and rope at this point y_0, we get

$$F_h y_0 = \frac{w'r^2}{8} \qquad F_h = \frac{w'r^2}{8y_0} \tag{6.22}$$

Proceeding, we can derive expressions for tension at each end of the rope.† For an inclined rope, tension will be greatest at the top, and

$$F_t = F_h(1 + 16S^2 + \tan^2 \theta + 8S \tan \theta)^{1/2}$$

$$F_b = F_h(1 + 16S^2 + \tan^2 \theta - 8S \tan \theta)^{1/2} \tag{c}$$

where $S = y_0/r$ = sag ratio
F_t = tension in rope at upper end
F_b = tension at lower end

The maximum rope force can be restated as

$$F_t = \frac{F_h}{\cos \alpha} \tag{6.23}$$

where $\alpha = \cos^{-1}[(1 + 16S^2 + \tan^2 \theta + 8S \tan \theta)^{-1/2}]$

The true length off the rope can be represented, for practical purposes, by the approximate expression

†See J. B. Wilbur and C. H. Norris, "Elementary Structural Analysis," McGraw-Hill, New York, 1948, p. 250.

$$L_0 = r\left(\frac{1}{\cos\theta} + \frac{8}{3}S^2\cos^3\theta\right) \tag{d}$$

which takes only the curvature of the rope into account.

In addition, the rope will stretch under load, but as shown by Eq. (*c*), the force in the rope varies through its length. The average force can be taken as

$$F_{av} = \frac{F_h r}{L_0}\left(1 + \frac{16}{3}S^2 + \tan^2\theta\right) \tag{e}$$

From Hooke's law, the elastic increase in rope length can be expressed in terms of the modulus of elasticity E_r as

$$\Delta L_0 = \frac{FL_0}{A_r E_r} \tag{f}$$

where A_r is the sum of the cross-sectional areas of the wires in the rope. Combining (*e*) and (*f*), we have

$$\Delta L_0 = \frac{F_h r}{A_r E_r}\left(1 + \frac{16}{3}S^2 + \tan^2\theta\right)$$

which when added to the change in length indicated by (*d*) gives the true length change of the rope, including curvature and elastic effects.

$$\Delta L = r\left\{8/3(S_1^2 - S_2^2)\cos^3\theta + \frac{1}{A_r E_r}[(F_{h2} - F_{h1})(1 + \tan^2\theta) + 16/3(F_{h2}S_2^2 - F_{h1}S_1^2)]\right\} \tag{6.24}$$

where S_1 is the sag ratio at the initial horizontal load F_{h1} and S_2 is the sag ratio at the final load F_{h2}. It has been assumed that the rope had been preloaded to F_{h1} to reduce initial sag and subsequently loaded by wind, for example, to F_{h2}. The rope-length change given by Eq. (6.24) takes place between those two loading levels. Of course, the equation provides change in length between any two loading levels, a negative change indicating a shortening of the rope.

Change in rope length, as loading increases, must be reflected in elastic lean of the crane mast. In turn, the elastic lean is resisted by the springlike action of the guy. The horizontal displacement, in terms of rope parameters, is expressed by (Fig. 6.26)

$$\Delta_r = [(L_c + \Delta L)^2 - h^2]^{1/2} - r \tag{6.25}$$

where L_c, the initial chord length, is $(r^2 + h^2)^{1/2}$.

Using the method of elastic stability (see Sec. 4.3) and the model in Fig. 6.27, we can develop expressions relating mast elastic deflection and guying restraint.

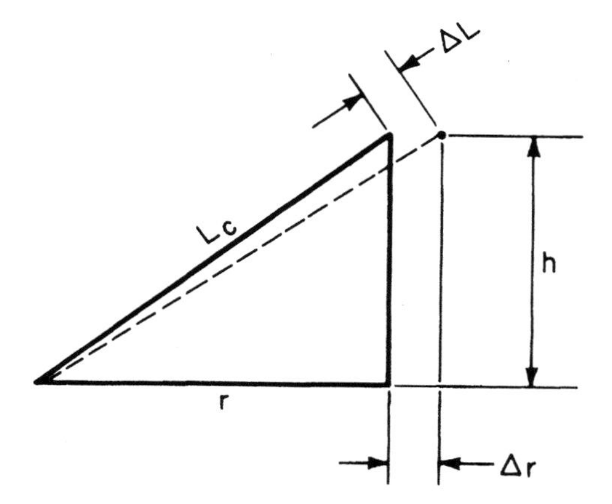

Figure 6.26

In the upper section of the mast, $0 \leqslant x \leqslant h_1$, the moment in the mast is given by

$$M_x = W_w x + \frac{wx^2}{2} - M_c + Qz$$

where w is now used to denote wind force on the mast per unit of mast length. The differential equation for this section then becomes

$$z'' + k^2 z = \frac{M_c - W_w x - wx^2/2}{EI}$$

where $k^2 = \dfrac{Q}{EI}$

$$Q = Q_0 + \frac{uH}{3}$$

u = dead load per unit of mast length

The solution to this equation is

$$z = A \cos kx + B \sin kx + \frac{M_c - W_w x - wx^2/2}{Q} + \frac{w}{Qk^2}$$

where A and B are unknown.

From the boundary condition $z(0) = 0$

$$0 = A + \frac{M_c}{Q} + \frac{w}{Qk^2} \qquad A = -\frac{M_c}{Q} - \frac{w}{Qk^2}$$

and at the lower end of the section $z(h_1) = d$, which is an arbitrary name for displacement at this level.

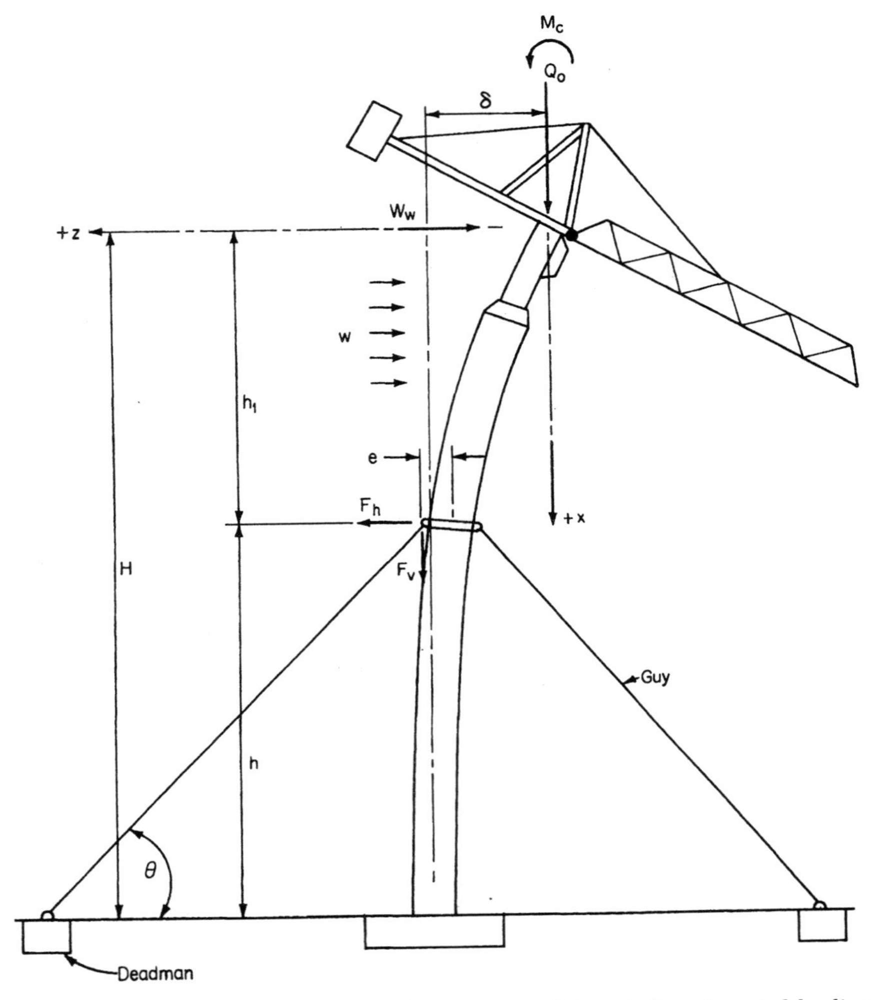

Figure 6.27 Analytical model of a guyed crane exposed to out-of-service wind loading (bending deflection exaggerated).

$$d = -\left(M_c + \frac{w}{k^2}\right)\frac{\cos kh_1}{Q} + B \sin kh_1$$

$$+ \frac{M_c - W_w h_1 - wh_1^2/2}{Q} + \frac{w}{Qk^2}$$

$$B \sin kh_1 - d = \frac{1}{Q}\left(M_c + \frac{w}{k^2}\right)(\cos kh_1 - 1) + \frac{W_w h_1 + wh_1^2/2}{Q} \tag{g}$$

The slope at the base of the section will be given an arbitrary value s; $z'(h_1) = s$

$$z' = \frac{k}{Q}\left(M_c + \frac{w}{k^2}\right)\sin kx + Bk\cos kx - \frac{W_w + wx}{Q}$$

$$s = \frac{k}{Q}\left(M_c + \frac{w}{k^2}\right)\sin kh_1 + Bk\cos kh_1 - \frac{W_w + wh_1}{Q} \qquad (h)$$

For the second section of the mast, $h_1 \leqslant x \leqslant H$, the moment is given by

$$Mx = W_w x + \frac{wx^2}{2} - M_c - F_h(x - h_1) - F_v e + Qz$$

motivating the differential expression

$$z'' + k^2 z = \frac{M_c + F_h(x - h_1) + F_v e - W_w x - wx^2/2}{EI}$$

for which the solution is

$$z = C\cos kx + D\sin kx + \frac{M_c + F_h(x - h_1) + F_v e - W_w x - wx^2/2}{Q} + \frac{w}{Qk^2}$$

For continuity with the previous section, $z'(h_1) = s$ must be satisfied

$$z' = -Ck\sin kx + Dk\cos kx + \frac{Fh}{Q} - \frac{W_w + wx}{Q}$$

$$s = -Ck\sin kh_1 + Dk\cos kh_1 + \frac{F_h}{Q} - \frac{W_w + wh_1}{Q}$$

The arbitrary s can now be eliminated by equating this expression with (h), giving, after simplification,

$$B + C\tan kh_1 - D = \frac{F_h}{Qk\cos kh_1} - \left(M + \frac{w}{k^2}\right)\frac{\tan kh_1}{Q} \qquad (i)$$

As a further condition of continuity, $z(h_1) = d$

$$d = C\cos kh_1 + D\sin kh_1 + \frac{M_c + F_v e - W_w h_1 - wh_1^2/2}{Q} + \frac{w}{Qk^2}$$

$$C\cos kh_1 + D\sin kh_1 - d = -\frac{F_v e}{Q} - \frac{M_c + w/k^2}{Q} + \frac{W_w h_1 + wh_1^2/2}{Q} \qquad (j)$$

The boundary condition at the base of the crane, $x(H) = \delta$, gives

$$\delta = C \cos kH + D \sin kH + \frac{M_c + F_h h + F_v e - W_w H - wH^2/2}{Q} + \frac{w}{Qk^2}$$

$$C \cos kH + D \sin kh - \delta = -\frac{F_h h + F_v e}{Q} - \frac{M_c + w/k^2}{Q} + \frac{W_w H + wH^2/2}{Q} \qquad (k)$$

and finally, the slope at the mast base must be zero, $z'(H) = 0$

$$0 = -Ck \sin kH + Dk \cos kH + \frac{F_h}{Q} - \frac{W_w + wH}{Q}$$

$$-C \sin kH + D \cos kH = \frac{W_w + wH}{Qk} - \frac{F_h}{Qk} \qquad (l)$$

Since we now have five equations *g*, *i*, *j*, *k*, and *l*, for the five unknowns B, C, D, d, and δ, the system of equations can be uniquely solved. In matrix form

$$\begin{bmatrix} \sin kh_1 & 0 & 0 & -1 & 0 \\ 1 & \tan kh_1 & -1 & 0 & 0 \\ 0 & \cos kh_1 & \sin kh_1 & -1 & 0 \\ 0 & \cos kH & \sin kH & 0 & -1 \\ 0 & -\sin kH & \cos kH & 0 & 0 \end{bmatrix} \begin{bmatrix} B \\ C \\ D \\ d \\ \delta \end{bmatrix} = \begin{bmatrix} C_1 \\ C_2 \\ C_3 \\ C_4 \\ C_4 \end{bmatrix}$$

where $C_1 = \frac{1}{Q}\left(M_c + \frac{w}{k^2}\right)(\cos kh_1 - 1) + \frac{W_w h_1 + wh_1^2/2}{Q}$

$$C_2 = -\left(M_c + \frac{w}{k^2}\right)\frac{\tan kh_1}{Q} + \frac{F_h}{Qk \cos kh_1}$$

$$C_3 = \frac{1}{Q}\left(W_w h_1 + \frac{wh_1^2}{2}\right) - \frac{M_c + w/k^2}{Q} - \frac{F_h}{Q} e \tan \propto$$

$$C_4 = \frac{1}{Q}\left(W_w H + \frac{wH^2}{2}\right) - \frac{M_c + w/k^2}{Q} - \frac{F_h}{Q}(h + e \tan \propto)$$

$$C_5 = \frac{W_w + wH}{Qk} - \frac{F_h}{Qk}$$

where $F_h \tan \alpha$ is substituted for F_v.

The matrix expression $U = A^{-1}C$, where U is the column vector of unknowns, C is the column vector of constants, and A^{-1} is the inverse of the matrix of coefficients, can readily be solved by computer. The solution yields values for each of the unknowns and hence the expressions for deflections and moment at any point in the mast for any value of the rope reaction F_h, but a unique value for F_h has not yet been found.

Using a manual solution method, Kramer's Rule, the values for displacement d at guy height and δ at mast top are found for any F_h; the difference $\delta - d$ is then the mast lean at guy height. This must correspond to rope stretch under load F_h.

$$\delta - d = N_1 - \frac{F_h}{Q}(N_2 + N_3 \tan \alpha)$$

where
$$N_1 = \frac{1}{Q}\left[\frac{W_w + wH}{k}\left(\frac{\sin kH - \sin kh_1}{\cos kH}\right) + \frac{wh_1^2}{2} - \frac{wH^2}{2} - W_w h - \frac{(M_c + w/k^2)(1 - \cos kh)}{\cos kH}\right]$$

$$N_2 = \frac{\tan kh - 2 \tan kh_1(1/\cos kh - 1)}{k(1 - \tan kh \tan kh_1)} - h$$

$$N_3 = e\left(\frac{1/\cos kh - 1}{1 - \tan kh \tan kh_1}\right)$$

By rearranging, we can obtain an expression for final guy force as a function of guy level displacement

$$F_h = \frac{Q[N_1 - (\delta - d)]}{N_2 + N_3 \tan \alpha} \tag{6.26}$$

When Eq. (6.26) is used, dimensional units must be kept consistent and trigonometric arguments are in radians. Final mast lean is dependent upon the net horizontal load F_h. As wind load is applied, the mast will lean, causing an increase in rope loading on the windward side and a decrease from the preload condition on the leeward side. Approximately, if preload $F_{h1} \leqslant F_h/2$, the leeward guy will go to nearly slack when F_h is applied and $F_{h2} = F_h$ on the windward side. But if $F_{h1} > F_h/2$, the leeward guy will remain loaded and the final guy loads on both sides will have to be such as to be consistent with mast lean. In that case, the difference between final guy loads must be F_h.

Guy loading will affect crane mast stiffness and natural frequency of vibration; therefore the gust factor can be affected. The horizontal spring rate of the loaded guy will be

$$K_r = \frac{F_{h2} - F_{h1}}{\Delta r} \qquad \Delta r = \delta - d \tag{6.27}$$

Equation (3.10) will reflect the effect of the guys, for any K_r, on the effective mast spring rate, which is used to calculate natural frequency. A reasonable initial approximation for K_r should yield results close enough that a new gust factor should not be needed after the true value of K_r is found.

All of the information needed to solve the guying problem has now been developed. The solution procedure is:

1. Calculate a trial value for F_h using a guestimate for $\delta - d$ (½ to 1% of h is a good start) and θ as an approximation for α. Use Eq. (6.26).
2. Choose a preload value F_{h1}; a suggestion is $F_{h1} = F_h/4$ then $F_{h2} = F_h$.
3. Select a trial guy rope using F_{h2}, Eq. (6.23), and strength margin $(\mathrm{SM}) = 3$.
4. Determine Δr from Eqs. (6.24) and (6.25), noting that S changes with rope load ($S = y_0/r$).
5. Compare Δ_r with $\delta - d$; if they are satisfactorily close, F_{h2} has been found; if not, repeat the procedure using the mean of Δr and the initial value for $\delta - d$, or Δr alone, in Eq. (6.26) along with a calculated value for α.
6. Verify rope SM using F_{h2} and Eq. (6.23).

Mast top displacement should also be determined, as it may be necessary to control this parameter in some instances. Its value is given by

$$\delta = N_4 - F_h N_5 \tag{6.28}$$

where
$$N_4 = \frac{1}{Q}\left[(W_w + wH)\frac{\tan kH}{k} - W_w H - \frac{wH^2}{2} - \left(M_c + \frac{w}{k^2}\right)\left(\frac{1}{\cos kH} - 1\right)\right]$$

$$N_5 = \frac{1}{Q}\left[\frac{\sin kH - \sin kh_1}{k \cos kH} + e \tan \alpha \left(\frac{\cos kh_1}{\cos kH} - 1\right) - h\right]$$

If this deflection is excessive, it may be reduced by using more parts of line, rope of larger diameter, or rope with higher modulus of elasticity—or by increasing preload. Each of these measures, or combinations of them, will reduce Δr and therefore $\delta - d$ with an attendant increase in F_h and decrease in δ.

The crane in Fig. 6.28 is guyed at three levels; at each level, each corner is guyed in two directions with two-part guys.

Preloading guys

The change in length under guying loads results partly from taking up the slack from the unloaded state. To reduce initial sag, and thereby reduce movement under load, the guys must be preloaded, but for preloading to be effective the preloading force must be measured and controlled.

Figure 6.28 A guyed tower crane mast at the center of a partly completed cooling tower shell at Satsop, Washington. This installation, engineered by the authors' firm, has three guy levels, as dictated by the 400-ft (122-m) mast height. Two-part guys in two directions at each corner of each level were necessary to control mast-top movements because of coastal winds. The same installation inland might have required one less guy level.

A calibrated spring device, together with a turnbuckle, can be used to give measured preloading (the turnbuckle to take up on the rope and the calibrated device for measuring preload). The device is easily made up using a spring of known spring rate built into a housing marked at several intervals of compression with the force required to produce that spring movement (Fig. 6.29). Depending on rope size and length, preload forces may be as small as 1000 lb (4.5 kN) or as high as 10,000 lb (45 kN). It would not be practical to try to accommodate such a large range with one preloading device.

Preload can also be monitored with load cell, strain gauge, or other stress- or force-measuring devices. In any case, the crane top moment should be balanced and the crane checked for plumbness in calm air after preloading. It may be necessary to make minor adjustments to the guys to bring the mast into plumb regardless of the accuracy of the preloading.

It is important that adequate preloading levels be employed. If insufficient force is used, the slack guys will whip under wind-gust action, throwing shock loadings into the rope and the mast and perhaps permitting excessive mast lean. The large stress variations in the rope

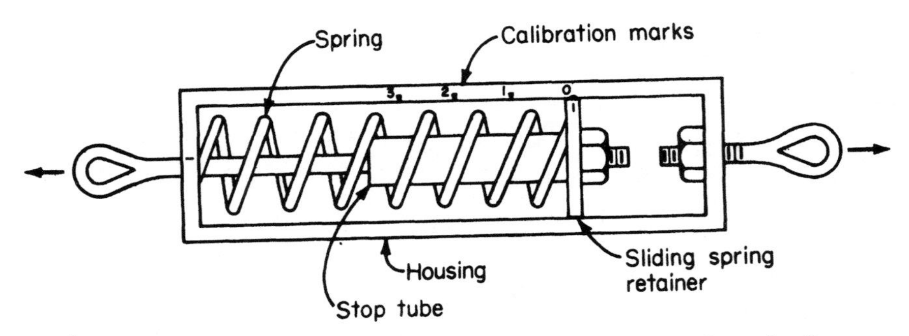

Figure 6.29 Schematic drawing of a calibrated spring takeup device for preloading guys. The housing and rope-attachment parts must be designed to full guy working loads, but the stop tube is needed to prevent the spring from being crushed.

may also contribute to fatigue failure. High preload forces will stabilize the system, reduce mast movement, and minimize or eliminate shock loading.

Rope parameters

The change in length that accompanies rope loading is a function of constructional stretch and the elastic properties of the rope. Both are reflected to a degree in the modulus of elasticity of the rope; the constructional stretch is caused by looseness in the placement of individual wires and diminishes with loading. As the constructional stretch works out of the rope, the modulus increases.

The constructional stretch can be minimized, and the modulus of elasticity thereby increased, by using prestretched rope. Prestretching must be specifically ordered when purchasing rope, and a delay in delivery must therefore be anticipated. Ropes are ordinarily prestretched at the factory by loading them to some 55% of the break strength, but on order any level of loading can be provided.

Approximate values of modulus of elasticity for several rope types can be taken as the following:

Type	*kips / in²*	*GN/m²*
6 × 19, fiber core	12,000	82.7
IWRC	15,000	103.4
Prestretched	17,500	120.7
6 × 37, fiber core	11,000	75.8
IWRC	14,000	96.5
Prestretched	16,500	113.8
Bridge rope, prestretched	20,000	137.9
Bridge strand, prestretched	24,000	165.5

Except for the bridge rope and strand, the manufacturer's literature ordinarily does not include data on modulus of elasticity or listings of metallic cross-sectional areas for the various styles and rope sizes. This is largely because the data are seldom requested. Approximate values for metallic area are given in Table 6.1.

Example 6.5 (*a*) A tower crane is designed for freestanding installation to a height of 215 ft (65.5 m) to the hook. Out-of-service wind at 50 mi/h (22.4 m/s) at a base height of 30 ft (10 m) induces a wind moment about the base of 6415 kip · ft (8698 kN · m). Using the approximate method, estimate the guy break strength needed for 60 mi/h (26.8 m/s) winds if the guys are fastened to the mast 121 ft (36.9 m) above the base. Assume that the guying distance r = 125 ft (38.1 m) (Fig. 6.24).

The increased wind moment will be given by Eq. (6.20)

$$M_e = \left[\left(\frac{60}{50}\right)^2 - 1\right](6415) = 2823 \text{ kip} \cdot \text{ft } (3827) \text{ kN} \cdot \text{m})$$

Geometry gives us the guying angle θ as

$$\theta = \tan^{-1}\frac{121}{125} = 44.07°$$

TABLE 6.1 Approximate Metallic Area of Ropes, in² †

Rope diameter, in	*6 × 7 fiber core*	*6 × 19 and 6 × 37*		*8 × 19 fiber core*
		Fiber core	*IWRC*	
¼	0.024	0.025	0.029	0.022
5⁄16	0.037	0.039	0.045	0.034
⅜	0.054	0.056	0.065	0.049
½	0.095	0.10	0.12	0.088
⅝	0.15	0.16	0.18	0.14
¾	0.21	0.23	0.26	0.20
⅞	0.29	0.31	0.35	0.27
1	0.38	0.40	0.46	0.35
1⅛	0.48	0.51	0.58	0.44
1¼	0.60	0.63	0.72	0.55
1⅜	0.72	0.76	0.87	0.66
1½	0.86	0.90	1.0	0.79
1¾		1.2	1.4	
2		1.6	1.8	
2½		2.5	2.9	
3		3.6	4.1	

†1 in² = 6.4516 mm²

Equation (6.21) then gives the force in the guy rope

$$F = \left(\frac{60}{50}\right)^2 \frac{2823}{121 \cos 44.07°} = 46.75 \text{ kips (208 kN)}$$

For a strength margin of 5, the required break strength is

$$\text{BS} = 46.75(5/2) = 117 \text{ tons (1041 kN)} \quad Ans.$$

(*b*) Design the guying system using the exact method and the following additional data (Fig. 6.27):

$$M_c = 1000 \text{ kip} \cdot \text{ft (1356 kN} \cdot \text{m)} \qquad Q = 207.5 \text{ kips (94.1 t)}$$

$$W_w = 16.2 \text{ kips (72.1 kN)} \qquad w = 0.23 \text{ kip/ft (3.4 kN/m)}$$

$$e = 4.25 \text{ ft (1.30 m)} \qquad H = 221.5 \text{ ft (67.5 m)}$$

The elastic parameter $k = 0.000215$ rad/in (0.00848 rad/m).

Equation (6.26) can be used to find a trial value for the horizontal component of the fully loaded guy force, but first the N parameters need to be evaluated. Using radians for all trigonometric arguments, we have

$$h_1 = 221.5 - 121 = 100.5 \text{ ft (30.63 m)} \qquad \theta = 44.07° = 0.769 \text{ rad}$$

$$\sin kH = \sin[0.000215 \text{ rad/in}(221.5 \text{ ft})(12 \text{ in/ft})] = 0.5409$$

$$\sin kh_1 = \sin[0.000215(100.5)(12)] = 0.2564$$

$$\cos kH = \cos[\sin^{-1}(0.5409)] = 0.8411$$

$$\cos kh = \cos[0.000215(121)(12)] = 0.9517$$

$$\tan kh = \tan[\cos^{-1}(0.9517)] = 0.3227$$

$$\tan kh_1 = \tan[\sin^{-1}(0.2564)] = 0.2653$$

$$N_1 = \frac{1}{207.5}\left\{\frac{16.2 + 0.23(221.5)}{0.000215}\left(\frac{0.5409 - 0.2564}{0.8411}\right) + 12 \text{ in/ft } (0.23)\right.$$

$$\times \left(\frac{100.5^2}{2}\right) - 12(0.23)\left(\frac{221.5^2}{2}\right) - 16.2(121)(12)$$

$$\left. - \left[1000(12) + \frac{0.23/12}{0.000215^2}\right]\left(\frac{1 - 0.9517}{0.8411}\right)\right\}$$

$$= 18.53 \text{ in } (471 \text{ mm})$$

$$N_2 = \frac{0.3227 - 2(0.2653)(1/0.9517 - 1)}{0.000215[1 - 0.3227(0.2653)]} - 121(12 \text{ in/ft})$$

$$= 52.48 \text{ in } (1.333 \text{ m})$$

$$N_3 = 4.25(12 \text{ in/ft})\left[\frac{1/0.9517 - 1}{1 - 0.3227(0.2653)}\right]$$

$$= 2.83 \text{ in } (72 \text{ mm})$$

For an initial value for $\delta - d$, try ½% of h, which is 7.26 in (184 mm); then

$$F_h = \frac{207.5(18.53 - 7.26)}{52.48 + 2.83 \tan 0.769}$$

$$= 42.35 \text{ kips } (188.4 \text{ kN})$$

when θ is taken to approximate α. Using a preload of $F_{h1} = 42.35/4 = 10.59$ kips (47.11 kN), $F_{h2} = F_h = 42.35$ kips (188.4 kN).

To select a trial guy rope, we must convert the horizontal component of load to maximum rope load using Eq. (6.23) and θ for α

$$F_t = \frac{42.35}{\cos 0.769} = 58.93 \text{ kips } (262.1 \text{ kN})$$

$$\text{BS required} = \frac{3}{2}(58.93) = 88.4 \text{ tons } (786.4 \text{ kN})$$

Try two parts of 1-in-diameter (25 mm) 6 × 19 IWRC IPS PRF prestretched rope which has a BS of 44.9 tons (399.5 kN) per part, a metallic cross sectional area $A_r = 0.46$ in^2/part (2.97 cm^2/part), and a dead weight of 1.76 lb/ft (2.62 kg/m); the modulus of elasticity is $E_r = 17{,}500$ kips/in^2 (120.7 GN/m^2).

The chord length of the guys L_c is $(121^2 + 125^2)^{1/2} = 173.97$ ft (53.02 m), and the rope dead weight projected onto the horizontal is then

$$u' = 1.76\left(\frac{173.97}{125}\right) = 2.45 \text{ lb/ft (3.65 kg/m)}$$

Note that the guy loads for each part are now $f_{h2} = 42.35/2 = 21.18$ kips (94.21 kN) and $f_{h1} = 10.59/2 = 5.30$ kips (23.58 kN). The sag ratios are then

$$S_2 = \frac{y_o}{r} = \frac{u'r}{8f_{h2}} = \frac{2.45(125)}{8000(21.18)} = \frac{0.0383}{21.18} = 0.00181$$

$$S_1 = \frac{0.0383}{5.30} = 0.00723$$

Equation (6.24) gives the change in guy rope length induced by the loading

$$\Delta L = 125\left(\frac{8}{3}(0.00723^2 - 0.00181^2)\cos^3 0.769 + \frac{1}{0.46(17{,}500)}\right.$$

$$\times\left\{(21.18 - 5.30)(1 + \tan^2 0.769) + \frac{16}{3}[21.18(0.00181^2)\right.$$

$$\left.\left. - 5.30(0.00723^2)]\right\}\right) = 0.48 \text{ ft (146.3 mm)}$$

and from Eq. (6.25)

$$\Delta r = [(173.97 + 0.48)^2 - 121^2]^{1/2} - 125 = 0.67 \text{ ft}$$

$$= 7.99 \text{ in (203 mm)} \neq \delta - d = 7.26 \text{ in (184 mm)}$$

Going back to Eq. (6.26) with $\delta - d = (7.99 + 7.26)/2 = 7.63$ in (194 mm) and a calculated value for α,

$$\alpha = \cos^{-1}\{[1 + 16(0.00181^2) + \tan^2 0.769 + 8(0.00181)(\tan 0.769)]^{-1/2}\}$$

$$= 0.773 \text{ rad}$$

$$F_h = \frac{207.5(18.53 - 7.63)}{52.48 + 2.83 \tan 0.773}$$

$$= 40.94 \text{ kips (182.1 kN)}$$

$$f_{h2} = \frac{40.94}{2} = 20.47 \text{ kips (91.1 kN)}$$

$$f_{h1} = \frac{20.47}{4} = 5.12 \text{ kips (22.8 kN)}$$

from which $S_1 = 0.00748$ and $S_2 = 0.00187$. Continuing, $\Delta L = 0.47$ ft (143 mm) and $\Delta r = 0.65$ ft (198 mm), making the difference from $\delta - d$ only about 3/16 in (4.3 mm). This is satisfactory convergence, and the guy loading has been found with sufficient accuracy. Total rope load will be [Eq. (6.23)]

$$F_t = \frac{40.94}{\cos 0.773} = 57.2 \text{ kips (254.4 kN)}$$

and the actual rope strength margin

$$\text{SM} = 2(44.9)\left(\frac{2}{57.2}\right) = 3.14 > 3.0 \quad \text{OK}$$

Assuming that the preloading mechanism will be located near the base of the guy, the preloader force needed is

$$F_b = 5.12[1 + 16(0.00748^2) + \tan^2 0.769 - 8(0.00748)(\tan 0.769)]^{1/2}$$

$$= 7.0 \text{ kips/part (31.1 kN/part)}$$

(*c*) Calculate mast top deflection, guy horizontal spring rate, and the wind moment remaining at the mast base after guying.

Before Eq. (6.28) can be used, additional N values are needed.

$$\tan kH = \tan[\sin^{-1}(0.5409)] = 0.6431$$

$$\cos kh_1 = \cos[\sin^{-1}(0.2564)] = 0.9666$$

$$N_4 = \frac{1}{207.5}\left\{[16.2 + 0.23(221.5)]\frac{0.6431}{0.000215} - 16.2(221.5)(12 \text{ in/ft}) - 0.23(12)\left(\frac{221.5^2}{2}\right) - \left[1000(12) + \frac{0.23/12}{0.000215^2}\right] \times \left(\frac{1}{0.8411} - 1\right)\right\} = 45.67 \text{ in (1.160 m)}$$

$$N_5 = \frac{1}{207.5}\left[\frac{0.5409 - 0.2564}{0.000215(0.8411)} + 4.25(12 \text{ in/ft})\tan 0.773\left(\frac{0.9666}{0.8411} - 1\right) - 121(12)\right]$$

$$= 0.621 \text{ in (15.8 mm)}$$

It is interesting to note that without guys $\delta = N_4 = 45.67$ in (1.16 m) or about 1.7% or H. With the guys.

$$\delta = 45.67 - 40.94(0.621) = 20.25 \text{ in (0.51 m)} \quad \textit{Ans.}$$

or about 3/4% of H.

From Eq. (6.27),

$$K_r = \frac{40.94 - 2(5.12)}{0.65} = 47.2 \text{ kips/ft (689 kN/m)} \quad Ans.$$

The wind moment on the crane is

$$M_w = 16.2(221.5) + 0.23\left(\frac{221.5^2}{2}\right) = 9230 \text{ kip} \cdot \text{ft (12,510 kN} \cdot \text{m)}$$

so that the moment at the base of the crane will be

$$M_{\text{net}} = 9230 + 207.5\left(\frac{20.25}{12}\right) - 1000 - 40.94(121)$$

$$-40.94 \tan 0.773(4.25 - 0.65) = 3483 \text{ kip} \cdot \text{ft (4722 kN} \cdot \text{m)} \quad Ans.$$

(*d*) Calculate the effect of reducing the preload horizontal reaction to 2000 lb (8.9 kN).

$$f_{h1} = 1.0 \text{ kip (4.45 kN)} \qquad S_1 = 0.0383 \qquad \Delta L = 0.79 \text{ ft (241 mm)}$$

$$\Delta_r = 1.10 \text{ ft (335 mm)} = 13.2 \text{ in}$$

Try $\delta - d = (7.63 + 13.2)/2 = 10.42$ in (265 mm) $\alpha = 0.773$ rad, from which $F_h = 30.46$ kips (135.5 kN) and $f_{h2} = 15.23$ kips (67.75 kN). This gives $S_2 = 0.00251$, $\Delta L = 0.61$ ft (186 mm), and $\Delta r = 0.85$ ft $= 10.20$ in (258 mm) for reasonable convergence (¼ in or 5.5 mm). Mast lean at guy level has increased by a small amount, but the increase is about one-third nonetheless. At mast top

$$\delta = 45.67 - 30.46(0.621) = 26.75 \text{ in (0.68 m)}$$

an increase of some 32%. Spring rate decreases to 33.5 kips/ft (489 kN/m), while the net base moment increases by some 40% to 4906 kip · ft (6651 kN · m).

Some 10 kips (44.5 kN) of force that would have been taken by the guy ropes are instead resisted by the mast itself because of the added lean permitted by reducing the preload. *Ans.*

6.5 Internal Climbing Cranes

For attainment of height above the capability of a freestanding tower, the *internal climbing crane* is an alternative to a top climber. Whereas the top climber stands apart from the building it is erecting, the internal climbing crane mast is enveloped and supported by it (Fig.

Figure 6.30 This internal climbing crane has just topped off a high-rise hotel. It must now be disassembled and lowered to the ground from a difficult perch.

6.30). The building under construction forms the base for the crane mast. As new floors are created, the crane raises itself and is supported on them.

Commonly, the internal climbing crane is identical to the static version except for the addition of climbing gear. The crane may initially be set up freestanding and later converted to internal climbing after the building height approaches the top of the mast.

To accommodate the mast and to permit climbing, a series of floor openings must be provided through the height of the building. In some instances, an elevator shaft will suit this purpose, but as a rule, on high-rise construction this may cause an unacceptable delay of elevator installation. If existing openings are not to be used, temporary openings must be made in each floor. These are filled in after the crane passes, except for the last few that allow the crane to be jacked down a few floors at the end of the job to permit dismantling.

The crane imparts two distinct sets of loads to the building: vertical forces that derive from crane and load weight and lateral forces that derive from overturning moment, wind shear, and swing torsion.

A tower crane might weigh perhaps 200 kips (91 t) which cannot be supported on a typical floor. This problem is usually solved by distributing the weight to a number of floors through shoring. If, for example, each floor slab can carry 20 kips (9.1 t), and all floors can be considered equally stiff, nine sets of shores are needed to deliver the weight of this hypothetical crane to ten floor slabs. Fortunately, lateral forces are dispersed with comparative ease, as they are of smaller magnitude and are applied in the plane of the floor structure which acts as a diaphragm; the floors are much stronger in that direction.

The crane is wedged at two floor levels; this creates a couple to resist overturning moment. The wedge floors must be far enough apart to reduce the resultant forces against the crane legs to a manageable level. In addition to the couple force, the upper wedge level must sustain wind shear and slewing torsion.

Positioning a crane in a building is an art that recognizes the need to weigh several concerns, some that can be compromised and some that leave little room to maneuver. The position must be suitable for servicing the construction work and be practicable for erection and dismantling. Temporary openings, if used, must not be too close to the building edge or to other large floor openings. These temporary openings must not materially intrude upon major building structural members, and they must repeat, one directly above the other, throughout the height of the structure regardless of changes in framing arrangements.

Finishing trades may be disrupted by the shoring and temporary openings, especially when the crane is located in areas of plumbing, HVAC, or electrical chases. Owners and project managers are less tolerant of these disruptions than in the past and often demand that the crane be placed in less active space or outside of the building altogether.

Figure 6.31 is an installation drawing for a typical internal climbing crane on a high-rise structure.

Vertical loads

Vertical loads do not vary greatly during the operating cycle, as by far the greatest part is dead weight. Less than 25% of the maximum attainable load can be attributed to live load, and the effect of impact is almost nil.

Although details of support arrangements will vary from manufacturer to manufacturer, and some may even comprise user-supplied beams, nearly all have the common characteristic of transmitting the vertical loads to the structure through pairs of bearing points, normally four. They also share the requirement that the support beams be mounted level. Should they be out of level, the crane can be thrown out of plumb. If the crane is then plumbed by adjustments at the wedges, two undesirable consequences ensue.

First, the vertical-load bearing points will become unevenly loaded; instead of experiencing one-quarter of the vertical load, any one support could theoretically receive double or more of its share. As a practical matter, the redistribution of loading will cause added elastic deflection at the more heavily loaded parts, with the likelihood that only a portion of the excess load will actually occur. It is not unrealistic, however, for any one support to be burdened by as much as one-third of the vertical load, or 33% more than its share.

Second, the upper horizontal reactions could be increased by a force on the order of 5% or more of the vertical load. Under certain circumstances this can prove critical, as will be discussed in the next section.

It is unrealistic to assume that the concrete floor surface surrounding the crane opening will be level enough to receive the support frame properly. A difference in elevation of ½ in (13 mm) across the opening diagonal is a little extreme but should not be precluded from consideration; ¼ in (6 mm) difference between any two supports is almost a certainty. The support frame should be level to about 1:1000, or roughly to within ⅛ in (3 mm) between any two adjacent supports and $^3/_{16}$ in (5 mm) on the diagonals.

Steel shim stock is not usually a satisfactory leveling means unless grout is inserted after the leveling has been completed. The high vertical loads coming to the concrete surface through the shims will probably cause local spalling of the concrete and a loss of levelness. The grout will increase the bearing area and reduce surface stresses to below the critical level. Figure 6.32 shows a climbing frame. Note proximity of the bearing points to the edge of the floor opening; this is typical. Crane weight is transferred to the climbing frame through dogs (Fig. 6.33).

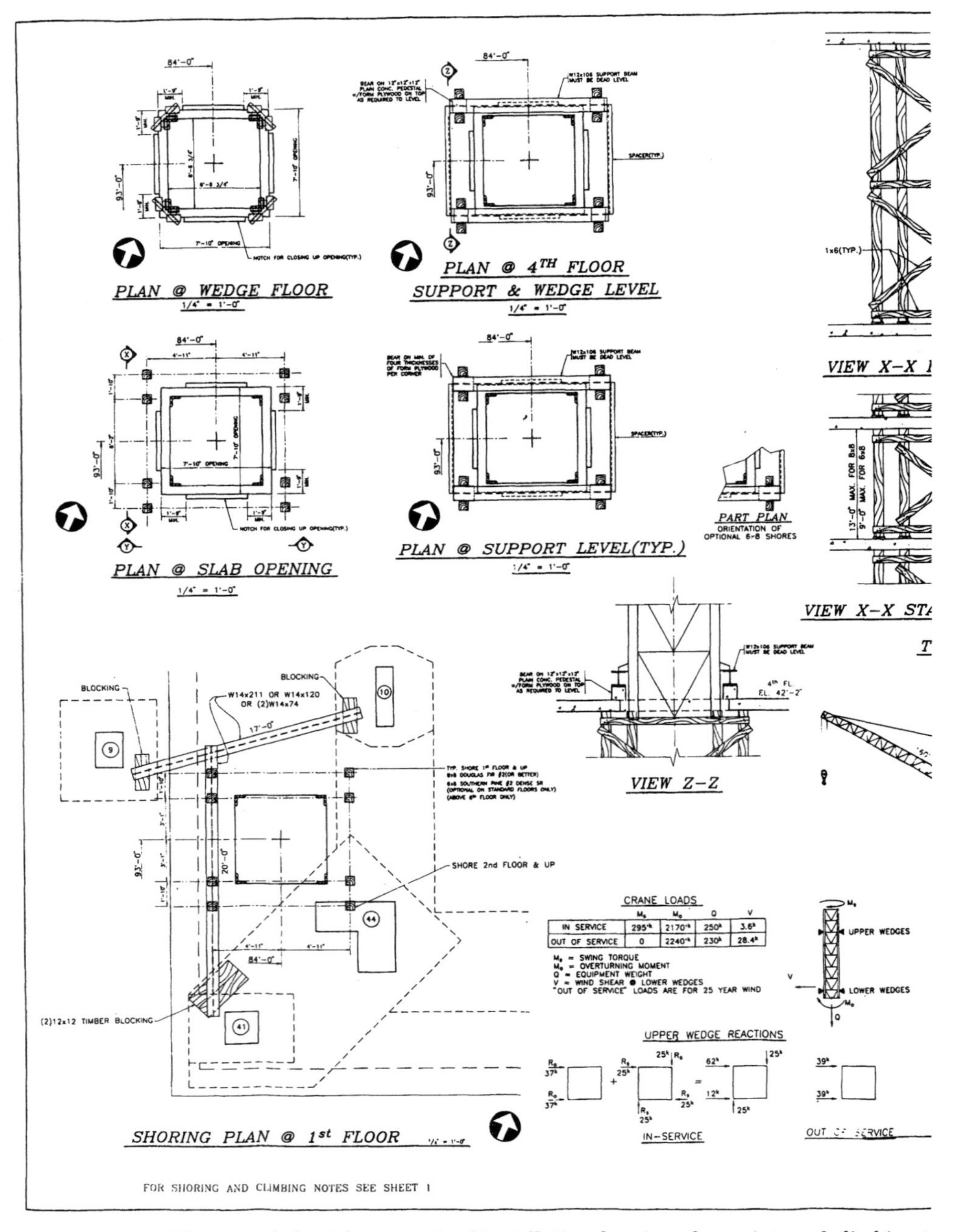

Figure 6.31 The second sheet from a pair of installation drawings for an internal climbing tower crane. Shoring is an important aspect of the installation, particularly when there are high floor to ceiling heights. Note the climbing schedule.

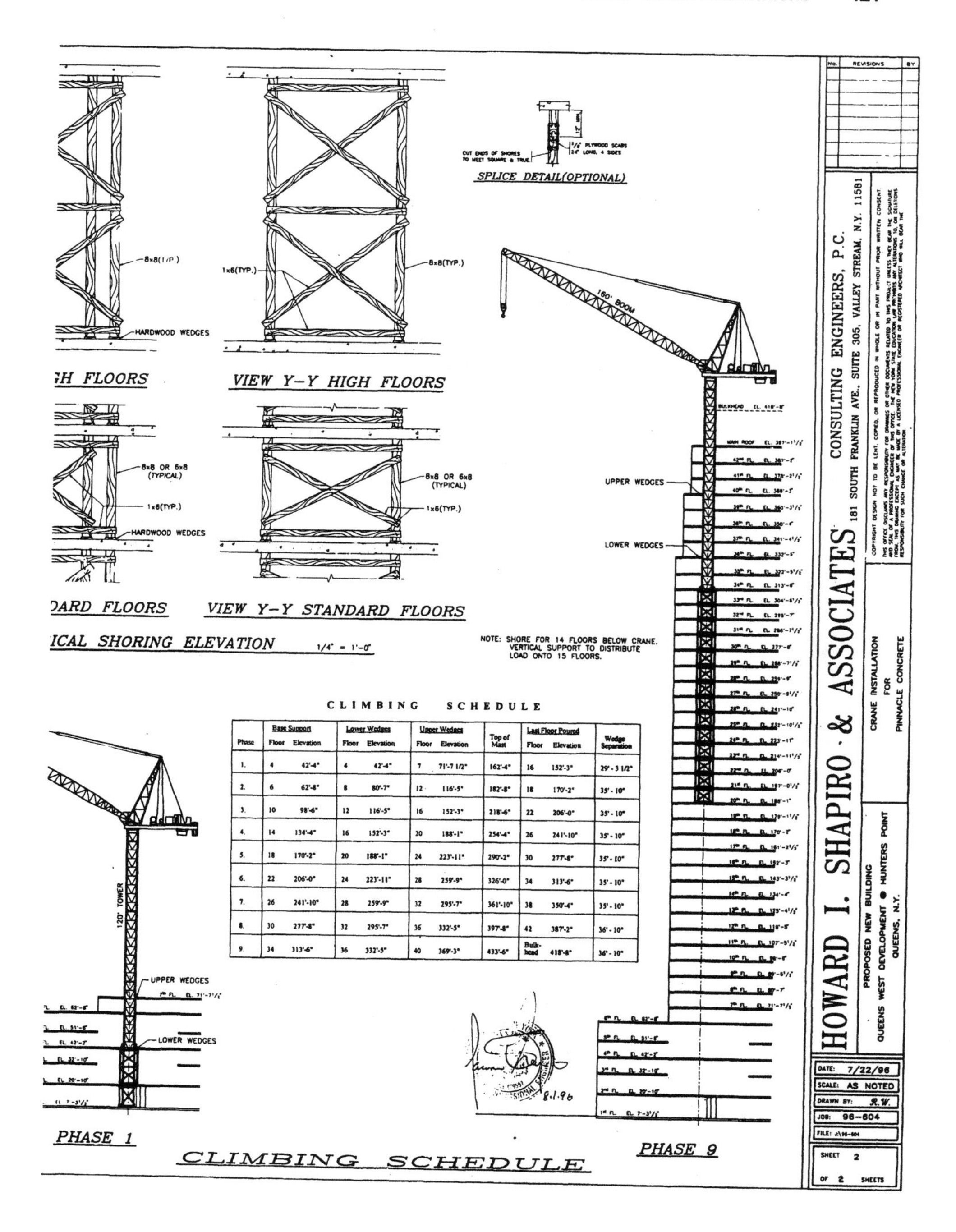

CLIMBING SCHEDULE

Phase	Base Support Floor	Base Support Elevation	Lower Wedges Floor	Lower Wedges Elevation	Upper Wedges Floor	Upper Wedges Elevation	Top of Mast	Last Floor Poured Floor	Last Floor Poured Elevation	Wedge Separation
1.	4	42'-4"	4	42'-4"	7	71'-7 1/2"	162'-4"	16	152'-3"	29' - 3 1/2"
2.	6	62'-8"	8	80'-7"	12	116'-5"	182'-8"	18	170'-2"	35' - 10"
3.	10	98'-6"	12	116'-5"	16	152'-3"	218'-6"	22	206'-0"	35' - 10"
4.	14	134'-4"	16	152'-3"	20	188'-1"	254'-4"	26	241'-10"	35' - 10"
5.	18	170'-2"	20	188'-1"	24	223'-11"	290'-2"	30	277'-8"	35' - 10"
6.	22	206'-0"	24	223'-11"	28	259'-9"	326'-0"	34	313'-6"	35' - 10"
7.	26	241'-10"	28	259'-9"	32	295'-7"	361'-10"	38	350'-4"	35' - 10"
8.	30	277'-8"	32	295'-7"	36	332'-5"	397'-8"	42	387'-2"	36' - 10"
9.	34	313'-6"	36	332'-5"	40	369'-3"	433'-6"	Bulk-head	418'-8"	36' - 10"

Figure 6.32 This support frame for a Richier tower crane rests on the floor slab and is supported below by eight levels of shoring. Each corner of the support frame is bearing on several thicknesses of plywood. In the middle foreground is a climbing ladder support.

An alternate leveling material. A surprisingly satisfactory material for leveling under support-frame bearing plates is exterior-grade structural form plywood. The shim pads should be made of several thicknesses of the plywood.

The obvious advantages of form plywood are its availability at construction sites, its low cost (it can usually be taken from the scrap pile), and its ease of fabrication. In addition, however, the plywood will not mar the concrete surface or disintegrate even under extremely high stresses, and it will adapt to small irregularities in the concrete surface.

Tests conducted a few years ago by one of the authors suggest that plywood has outstanding characteristics for shimming crane support beams. The test results indicated that pressure against the face of the plywood causes elastic behavior up to about 500 lb/in^2 (3.4 MN/m^2). Between that stress level and about 1500 lb/in^2 (10.3 MN/m^2), the stress-strain curve is almost flat, an indication that the cell walls

Figure 6.33 Working dog on a Kodiak tower crane. The dogs are designed to pivot out of the way under their own weight when the crane is jacked up. (*Photo by Lawrence K. Shapiro.*)

within the wood are crushing. Above 1500 lb/in^2 the slope of the stress-strain curve again sharply rises as the material stiffens following the full collapse of the walls. The tests were conducted up to a stress level of 13,350 lb/in^2 (92.0 MN/m^2), which caused the plywood to reduce in thickness from an initial ⅝ to ¼ in (16 to 6.4 mm).

Plywood functions best in this role when it is stressed in the crushing range of 500 to 1500 lb/in^2. The greater the thickness of the plywood shim packs, the greater is its capability to equalize the crane load among the support points. For instance, we were able to infer from the test data that 5-in (127-mm) shim packs loaded to 1250 lb/in^2 (8.6 MN/m^2) permit a self-adjustment of about ¼ in (6.4 mm) between support points. This should be sufficient to correct for a rather poorly finished floor. However, the plywood must be replaced each time the crane is relocated to another floor because the cells have been crushed.

The same characteristics that make plywood ideal for shimming the support beams make it entirely unsuitable for shimming shores.

Shoring. Only a very robust floor slab can safely manage the imposed support-frame reaction without shoring or other supplementary accommodations. The actual concentrated load a single floor can carry must be calculated by a licensed professional engineer, preferably by the building design engineer for the project in question. With this information, the number of floors needed to support the crane load is readily determined. For planning purposes, assume that 5000-lb (2-t) concentrated loads are permitted at each corner on each floor.

For distributing loading to a series of floors, timber shoring posts are readily installed (Fig. 6.34). They have the advantage of being easily fitted and wedged into place and are readily removed when the crane climbs to another level. Bracing, to assure that the shores will remain in position under fluctuating vertical loading, and barricades for protection of the floor opening, require only common lumber and very little labor to fit and place.

Extra shoring material is kept at hand to be installed before climbing to another level. When a new support frame is set at a higher floor in preparation for the climb, it must be shored in advance, but the previously installed shores are still loaded and must be left in place until after the climb. When the climb is completed, the excess shores can be removed and kept in reserve for the next climb.

The maximum loading on a single shore will occur immediately after the climb and before the crane is plumbed. It is prudent to assume that 33% of the dead weight of the entire crane will be on any one support point for the short time that it takes to plumb the crane. Inasmuch as this is a short time loading that occurs but once, it is

Figure 6.34 Shoring for a tower crane mounted inside a concrete frame building. The cross braces are used to prevent accidental dislodgement of the shores. The floor openings here have already been filled in. (*Photo by Lawrence K. Shapiro.*)

reasonable to permit allowable stresses to be increased by one-third under this load.

Unless plywood bearing pads or some other load equalization arrangement is used under the crane support beams, we suggest that normal loading be taken as 30% of crane dead and live load on a single support. The load going into each topmost shore is then the support reaction less the support value of one floor slab. Each successive level of shoring will carry the load from the shore above minus the support value of the intervening floor. In theory, smaller shores can be used at successive floors below the support frame. But this is impractical in most cases, as it complicates the work in the field.

Timber shores are designed using ordinary timber-column design formulas. Only sound structural-grade lumber should be used. Wedges must be inserted to assure contact and transmission of loads between the floor slabs and shores. They must be of hardwood, must be firmly driven into place, and must be checked periodically for tightness.

Timber has great end-bearing strength—about double what it can support against the side grain. Consequently, headers and sills cause a reduction in the load-bearing capacity of the shoring system. Any material used to fill out or extend the length of a shore must be suitably matched to it in strength, and consideration must be given to directional properties if the filler is of wood.

Improper fillers or loose shores can cause local distress of the support slabs. The distribution of the crane weight among load points is thrown out of balance, as some points receive loads that are much greater than the designer had intended. Cracks and large deflections in the slab are the telltale signs of such distress.

A common assumption is that the floors are equally loaded by the shores. This notion is usually reasonable but requires some caution. Its corollary assumption is that all floors deflect equally. Each shore in the vertical progression of shores, however, shortens under its compressive load. With some load dropping off at each floor level, each shore will carry less load than the one above and hence foreshorten less. The shortening in the shores must be taken up by deflection in the floor slabs so that the upper floors will deflect the most and carry a disproportional share of the load.

On a concrete building the effect of the shortening of the shores is counteracted somewhat by the aging of the concrete. With age, concrete gains in strength and stiffness; this added stiffness on the lower floors causes them to take on more load.

Changes in floor framing bring about changes in floor stiffness as well. These occurrences can affect the overall distribution of load among floors or local distributions on a floor. Simplified assumptions can usually be used in place of complex analyses, but not without careful consideration of the structural consequences.

Moments and horizontal loads

The wedges used to hold climbing cranes upright react to the horizontal loads that are induced by overturning moment, wind, and swing torsion as indicated in Fig. 6.35. When the crane is exposed to working—or in-service—conditions, the wind effect is small but the torsion reactions can be large. On the other hand, out-of-service loading is heavily influenced by wind exposure, but there is no torsion unless the jib is secured against weathervaning.

Any torsional loading will be resisted at the upper wedge level by two couples, as shown in Fig. 6.36*a*. The torsion reaction at any one wedge point is then

$$R_s = M_s/2d \tag{6.29}$$

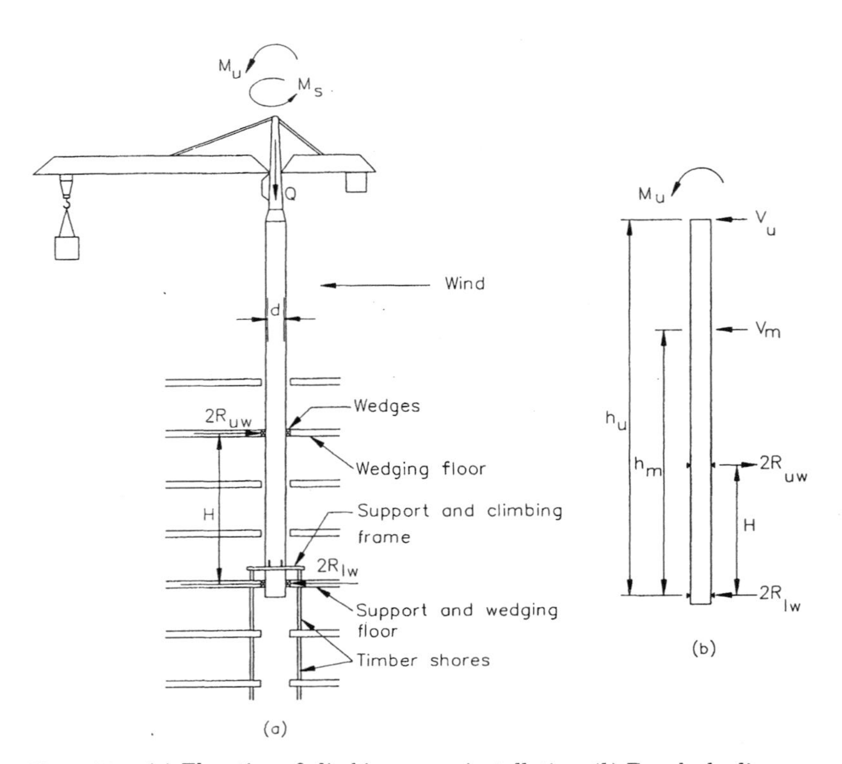

Figure 6.35 (*a*) Elevation of climbing crane installation. (*b*) Free body diagram for moments and wind forces.

where d is the horizontal distance between mast leg centroids. No part of the torsion can bypass the upper wedging level unless there is a large elastic deformation permitting the crane to rotate through the upper wedge level. Such deformations are unlikely.

Overturning moment and wind reactions are best evaluated by using a free body diagram to represent the crane mast (Fig. 6.35*b*). The terms M_u and V_u are the overturning moment and wind shear at the top of the mast and V_m is the wind force acting on the exposed portion of the mast. Reactions can be determined by taking moments about the lower wedges. The upper wedge reaction per corner (Fig. 6.36*b*) is then

$$R_{uw} = \frac{M_u + V_u h_u + V_m h_m}{2H} \tag{6.30}$$

The maximum wedge reaction at any one corner of the upper wedge floor (Fig. 6.36*c*) includes torsion and is given by

$$R_{\max} = R_s + R_{uw} \tag{6.31}$$

whereas the reaction at each corner at the lower wedges is

$$R_{lw} = \frac{R_{uw} - V_u - V_m}{2} \tag{6.32}$$

and is opposite to the direction of R_{uw}.

The crane manufacturer always specifies the minimum wedging distance H. This must be respected unless measures are taken to reinforce the crane mast under the directions of the manufacturer or a licensed professional engineer. Some masts are furnished with wedge point reinforcing (see Fig. 6.37), while others must be positioned so that the wedges line up closely to mast cross-framing. If the alignment does not occur, there is a strong possibility that the leg will collapse from the combined bending and compression loads.

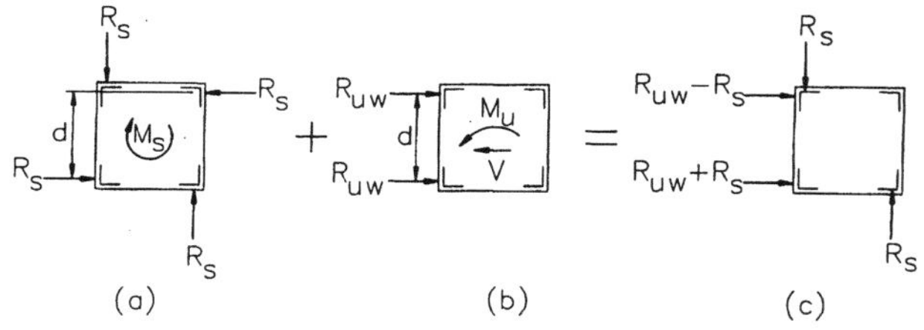

Figure 6.36 Distribution of upper wedging-level reactions.

Figure 6.37 This Richier mast is stiffened to resist wedge loads at any elevation. Reinforcing bars are bent up around the opening, a practice that has largely been abandoned because protruding bars can cause the crane to hang up during jacking.

Example 6.6 Determine the wedge reactions for a crane with an in-service overturning moment of 2100 kip · ft (2847 kN · m) at the top of the mast, a torsional moment of 220 kip · ft (298 kN · m), wind shear at the top of the mast of 3.1 kips (13.8 kN) acting 125 ft (38.1 m) above the lower wedges, and a resultant wind force on the exposed part of the mast of 1.5 kips (6.7 kN) acting 75 ft (22.9 m) above the lower wedges. The mast leg centers are at 7.56 ft (2.30 m) and the distance between wedging levels will be three full 8-ft 9-in (2.67-m) stories.

From Eq. (6.29)

$$R_s = \frac{220}{2(7.56)} = 14.55 \text{ kips } (64.72 \text{ kN})$$

Since H is 3(8.75) = 26.25 ft (8.00 m), h_u = 125 ft (38.1 m), and h_m = 75 ft (22.9 m), using Eq. (6.30) we get

$$R_{uw} = \frac{2100 + 3.1(125) + 1.5(75)}{2(26.25)} = 49.52 \text{ kips } (220.3 \text{ kN})$$

The maximum wedge reaction is at the upper wedge level and is given by Eq. (6.31):

$$R_{\max} = 14.55 + 49.52 = 64.07 \text{ kips } (285.0 \text{ kN}) \qquad \textit{Ans.}$$

while Eq. (6.32) gives the lower wedge reactions:

$$R_{lw} = [2(49.52) - 3.1 - 1.5]/2$$

$$= 47.22 \text{ kips } (210.0 \text{ kN}) \qquad \textit{Ans.}$$

Climbing procedures

A variety of systems have been devised by manufacturers to raise their tower cranes from one operating level to the next. All of the modern systems use one or more hydraulic rams; the differences among them are in the means employed to overcome the shortness of typical ram strokes. The systems known to the authors are described below.

Long-stroke rams. The hydraulic extension of the ram is sufficient to raise the crane one entire typical floor height. When overly high floors are encountered, an intermediate step must be arranged.

A variation of this system is the "three beam" arrangement. The rams push off from a telescoping beam centered under the mast and seated on a support level. At the end of the stroke, the ends of two support beams are telescoped out to a new support level. The cycle can be repeated as often as necessary to reach the required support elevation. A three beam arrangement is particularly suitable for a crane in an elevator shaft. Pockets can be made in the walls to receive the telescoping ends of the three beams. No shoring is needed in a shaft with masonry walls.

Climbing ladders. A pair of steel ladders is hung on opposite sides of the crane mast (Fig. 6.38), supported from a frame set above the new operating level. Pawls (dogs) on the ends of a floating cross-beam en-

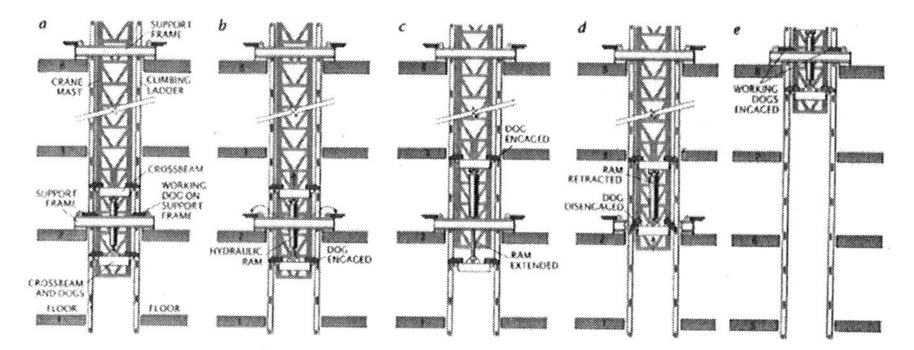

Figure 6.38 Sequence of an internal climbing operation. (*a*) Working dogs impose the crane weight on the lower support frame. (*b*) Climbing dogs linked to the hydraulic ram engage the climbing ladder and lift the crane. (*c*) A second set of climbing dogs is engaged after the crane has been raised by one rung. (*d*) The hydraulic ram is retracted to reengage its dogs on the next rung. (*e*) After a number of repetitions, the crane reaches a sufficient elevation to come to rest on the next support frame. (*Reprinted by permission of Scientific American.*)

gage the ladders. The ram raises the crane by pushing down on this cross-beam. At the end of the stroke, the crane is brought to rest by engaging another set of pawls on the ladders. Retraction and reengagement of the floating cross-beam pawls complete a cycle that is repeated until the crane has been raised to the new operating level.

Rod climbers. There are at least two variations on this system: one uses smooth square rods of finite length, and the other uses threaded high-strength reinforcing bars that can be coupled to most any length.

Cranes that lift themselves on square rods operate similarly to ladder climbers. In this case, a pair of ram and gripper mechanisms is mounted on opposite sides of a lifting frame installed on a floor at or above the new support level. The rods are fixed at the base of the mast and run upward through the grippers.

The ram gripper mechanisms consist of a base that sits on the lifting frame, a pair of jaw-type rod grippers at the base level, a hydraulic ram, and a pair of grippers above the ram. The gripper jaws are on an eccentric pivot so that weight on the rods causes the jaws to grip. It is only by relieving the weight that the jaws will release.

At the start, the rams are retracted, the upper jaws are lightly gripping the rods, and the working dogs are engaged. As the rams are extended, the upper jaws tightly grip the rods and the crane rises. The working dogs are then retracted and the rams are extended to full stroke.

At the top of the stroke, slight retraction causes the lower jaws to grip and the top jaws to release. The rams are then fully retracted and the cycle repeated.

The advantages of this lift-rod system include its simplicity, low cost, and fast lifting cycle. The disadvantage lies in the limited length of rod that is practical for jobsite use. This necessitates climbing after each new floor is placed (or at most, after two floors), a requirement that can prove inconvenient and reduces scheduling flexibility.

On the other hand, the threaded rod system can be coupled to permit a climb of any height. Aside from permitting couplings, the threads are used to support the weight of the crane and transfer it to the rods.

Tube climbers. A square or round tube column rests atop a cross-beam that engages the crane support frame at the old support level. This column is furnished with holes that are spaced vertically to match the ram stroke, for the purpose of receiving pins. A long pin is inserted through the lowest hole and projects through both walls of the column. The rams, connected to a floating yoke, push down on this pin to raise the crane. At the end of a stroke, a second long pin is inserted, and this engages a second yoke that is fixed to the crane. Retracting the rams ends one step in a cycle that is completed when the crane reaches its next resting level.

The tube climbing system is very easy to operate and quick to complete a climb, but the climb is limited to the height of the column, usually 25 to 30 ft (7.6 to 9.1 m).

Before and after the climb. Except for cranes that climb one floor at a time, each climbing operation must be preplanned and a climbing schedule prepared (Fig. 6.31). The schedule will indicate each floor level that will become a vertical-load support level and each level for horizontal support. In addition, it shows the maximum floor level that can be built before the next climb becomes necessary.

In this way, work crews will know which levels will require such things as cast-in-place inserts, climbing and support frames, and shores. Lumber and other requirements can be foreseen and provided for.

After completion of a climb, the floor below the one last placed, and sometimes the one last placed, will become the upper horizontal support level. Therefore, before a climb can be started, it must be known with certainty that the concrete at the support levels has reached sufficient strength to sustain the applied loads safely. This will be discussed in detail in the next section.

Preparations for climbing include placing all new support and climbing frames and the shoring needed to support them, but in addition it is necessary to determine that wind at the time of the climb will not exceed limits set by the manufacturer. The crane must also be balanced, either by retracting the counterweights to a balance radius or by lifting a balance weight on the hook. Climbing ladders or lifting rods must be mounted plumb and true, and, finally, horizontal wedges must be released immediately before the climb. An inspection must be made to be sure that there are no obstructions interfering with the climb, or preventing the mast legs from sliding through the loosened wedges.

On completion of the climb, while the crane is resting on preleveled supports, it must be brought into plumb by adjusting the wedges. All wedges are to be firmly set in order to lock the crane in a plumb posture that will be maintained during operation. Crane movements and elastic action in the crane legs can cause wedges to work loose. Wedges should be checked for tightness daily, or more often if necessary.

Supporting horizontal loads

Horizontal loads are transmitted to the building structure, more often than not, through contact with the edge of the floor slab at the side of the crane opening rather than at a more substantial mass of material. For smaller cranes, which impose relatively low-level forces, it is sufficient to drive hardwood wedges between the slab and the mast legs. Larger cranes need more elaborate wedging systems.

As horizontal forces increase with crane size, it becomes necessary to spread the loads to larger bearing areas. Some manufacturers require that steel inserts be cast into the floor slab for this purpose while others use wedging mechanisms that provide the larger area (Fig. 6.39). In either case, the object is to avoid failure of the concrete and loss of support.

Modern construction methods make it possible to place floors in concrete high-rise structures on a 2-day cycle (i.e., at the rate of one floor every 2 days). Furthermore, enclosure and space-heating methods have been devised that permit work to proceed through the winter in all but the most severe weather. These techniques were first developed by concrete construction contractors in New York City during the post-World War II housing boom almost 50 years ago. They have since been refined and have become more widespread.

Fast-cycle high-rise construction methods were developed using mobile cranes for hoisting the concrete and other materials. As the method caught on, fueled by the cost savings enjoyed, a need arose for equipment able to service taller and taller buildings. The mobile-crane industry responded by making larger cranes and longer and longer booms. Tower cranes, until recently, were slow to penetrate the United

Figure 6.39 Richier wedging arrangement has small steel wedges that bear against the mast leg but transfer load to a casting which bears on the concrete edge with a larger area. The bolts visible on top of the wedges are used for adjusting the wedges and locking them into final position.

States market. Guy derricks could erect the highest of buildings—which were all steel framed—and mobile cranes were able to reach the rooftops of any concrete structures that designers of the generation considered feasible.

Improvements in concrete technology have greatly extended its height limitations. At this writing, the tallest buildings in the world† are reinforced concrete. Constructing very tall concrete structures without the benefit of tower cranes would be impractical.

For construction to be economical, tall concrete buildings must progress in fast cycles—from two to four days per floor. Efficient movement of concrete and formwork are a key to sustaining these rapid cycles. A concrete pour might require 250 yd^3 (190 m^3) of concrete in one pour. Completion of a pour of this magnitude in one shift is possible only if the pour rate is in the order of 50 yd^3/h (38 m^3/h) or more. If the concrete is conveyed by bucket, such a high rate of placement can be sustained only if that bucket carries at least 3 yd^3 (2.3 m^3). The weight of the loaded bucket will be at least 13,000 lbs. (5900 kg).

Formwork for a high rise concrete building is often assembled into large gangs or tables for the sake of efficiency. These form assemblies, which are lifted from one placement to the next by crane, can weigh the same or more than the loaded concrete buckets.

Concrete strength. If the floor below the last-placed floor is to become the upper wedging floor (for 2-day-cycle work the last-placed floor cannot be used) the concrete will be about 2¼ days old at the time of climbing, assuming that climbing will start at the end of a normal work day. The concrete will be about 4 hours short of 3 days old when the crane is next put into service.

After 2¼ days, concrete test cylinders taken at the jobsite during winter can be expected to develop about 30% of the standard 28-day strength,‡ on average, unless accelerants are added to the mix.† But cores drilled from the slab itself can be expected to show only about 90% of cylinder strength or about 27% of 28-day strength. This leads to the suggestion that permissible bearing stress at the upper wedges be taken as no more than 10% of 28-day strength for out-of-service conditions.

After 3 days, concrete test cylinders will normally show about 40% of 28-day strength, and cores will develop about 36%, which leads to

†Petronas Towers in Kuala Lumpur, Malaysia.

‡See C. Berwanger and V. Mohan Malhotra, "Strength Development of Concrete Exposed in Winter," proc. paper 10489, *J. Eng. mech. Div. ASCE,* vol. 100, no. EM2, April 1974, pp. 305–321.

the suggestion that 10% of 28-day strength is appropriate for in-service conditions as well.

The expected strengths used in the discussion above were drawn from the source cited. The paper presented data gathered from as series of tests and, as with all experimental data, some scatter is inevitable; the same will be found to be true with tests taken at any jobsite. Moreover, the data might be considered conservative because modern cements, having been more finely ground than earlier cements, tend to hydrate (cure) faster.

To make certain that the concrete has reached adequate strength for climbing, we urge that field samples be taken and tested before the climb, say at age 2 days. If the cylinders are kept at the jobsite and taken to the testing laboratory a few hours before test time, cylinder strength can be assumed to approximate slab strength. If the cylinders test at 25% or more of 28-day strength, the climb may be executed. If not, the climb should be delayed.

Using our suggestions, permissible bearing stresses will be on the order of 400 lb/in^2 (2.7 MN/m^2), which is not very much when compared to the forces to be accommodated. We have usually found, therefore, that it is necessary to call for steel plates or wedge seats to be inserted between the mast leg wedges and the concrete to spread the load. Provision should be made to prevent the plates from falling if the wedges come loose.

It is the duty of the installation designer to adapt the crane to the particular jobsite conditions; the crane was designed in the sterile condition of a free body in space. The structure must be made suitable to support the crane loads—the lives of the construction workers and passersby depend on it. The installation designer cannot abdicate this responsibility.

On the other hand, field personnel must be made aware that tower cranes can impose intense dynamic loads that do not tolerate sloppy field practices. Floor openings must be accurately placed with square edges and wedge areas well formed and solid. Shores must be correctly aligned and made of approved materials.

6.6 Traveling Cranes

Cranes on traveling bases are erected freestanding and are ballasted by the user to accommodate in-service loads. When out of service they may need to be parked and anchored down to prearranged storm ballasting blocks or guyed to resist storm winds.

The bases travel on railroad-type rails set to a very wide gage. At each corner of the base one or more wheels are provided; when more than one wheel is used, they are mounted in a bogie that will equalize the load on all wheels at any one corner.

Some crane manufacturers offer options on the number of wheels to be placed at each corner (Fig. 6.40) of any one crane model. As the number of wheels increases, the weight of track and the number of track supports needed decrease. This can have significant ramifications for installation cost, particularly if soil conditions are poor.

Crane rails can be supported in a number of ways, including wooden ties on stone ballast (in this case the term ballast is used to refer to the bed of material placed between the tie and the native soil or sand base), a continuous steel beam on wooden ties and stone ballast or on concrete footings, or a continuous concrete footing or concrete sleepers on stone ballast (Fig. 6.41). The best system is that which will support the crane properly at least cost; this will be a function of crane wheel loads, soil conditions, and availability and cost of the materials at the jobsite. The crane manufacturer provides the wheel-load data, but the installation designer must make the decisions from that point on.

The spacing of sleepers or ties can be determined from rail strength and the wheel loads. For multiple-wheel arrangements, some continuity can be taken into account, but we suggest that supports outside of the bogie should be taken as simple. Deflection should also be checked to avoid lifting the ties off their beds.

Track splices are designed to carry only shear loads, so that splices must be centered between close-spaced supports or placed directly over a support. The spacing must be set so that the two rail ends do not differ in elevation (as a result of deflection or any other cause) as the wheel passes over the splice. This will prevent horizontal impact forces from occurring, forces that can be quite significant given the inertia of the tall crane above.

Rails must be laid to comply with the tolerances given by the manufacturer or specified by code (see Sec. 4.5). There are strict limits to variations permitted in gage, in elevation along the tracks and between the tracks, in straightness, and in slope.

Crane rails can be laid to curves but only if the bogies are designed to permit this. It must be noted that centrifugal forces which develop as the crane travels a curve can have an important effect on stability. The manufacturer must supply data for minimum radius of curvature that will permit safe travel at the speed the crane is capable of attaining.

Curved track as well as slewing forces, wind, and rail misalignment induce lateral forces on the rails. Rail strength and anchorage must be sufficient to restrain these forces. Magnitudes, however, are not easily determined; the crane manufacturer's recommendations should therefore be sought and followed.

On poor soils, track differential settlements can be a problem, as they may cause track elevations to go beyond permitted tolerances and endanger operations. It would be wise to monitor elevations at marked

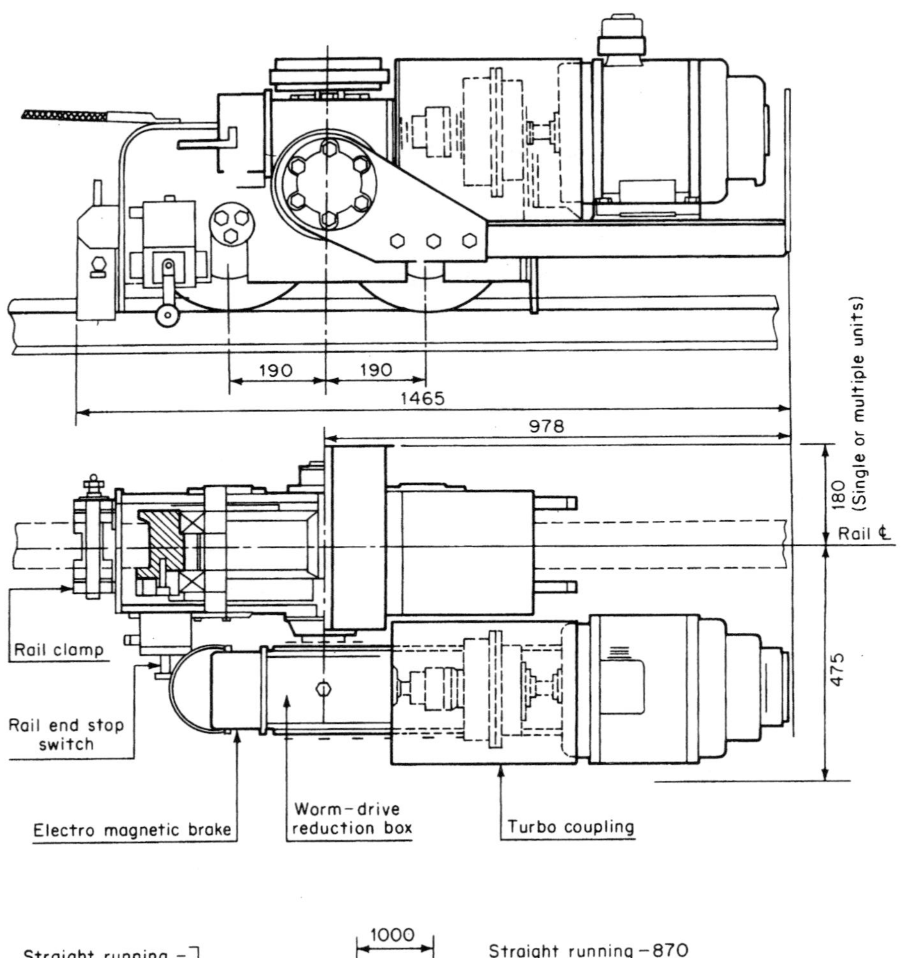

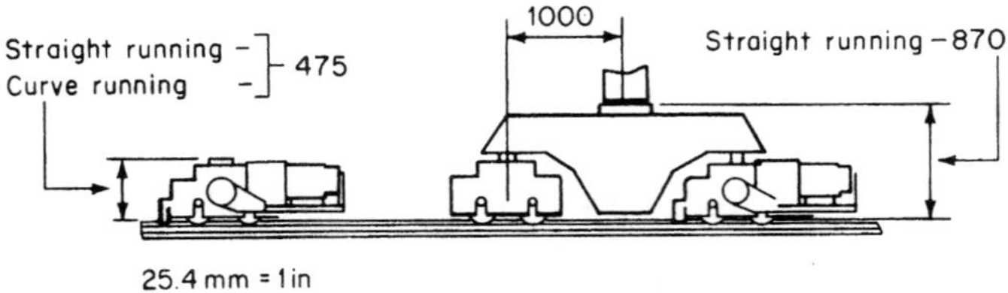

Figure 6.40 Optional arrangements of wheels showing single- and two-bogie sets; the dimensions are in millimeters. (*F. B. Kroll A / S.*)

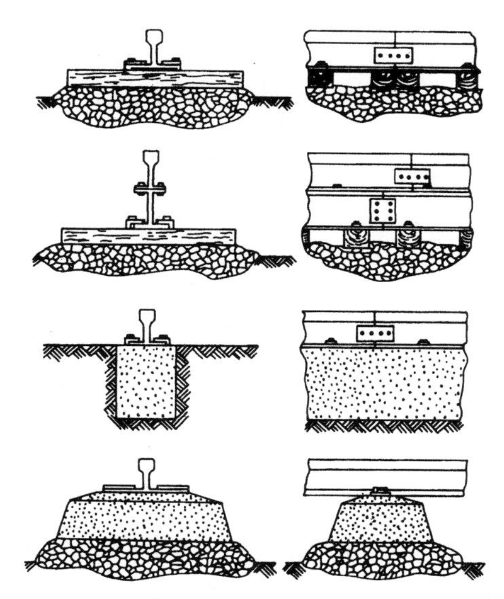

Figure 6.41 Several methods for supporting traveling crane rails. (*F. B. Kroll A / S.*)

points and record the data. This will quickly show whether settlement will be a continuing phenomenon or track supports have stabilized. With wooden ties, settlements can be corrected by jacking the rail and tie and resetting the stone ballast. For concrete supports it may be necessary to install steel shim plates with sufficient contact area to prevent the concrete from being crushed.

The parking area must be designed and constructed in advance of crane erection. It will consist, for most cranes, of an area with close support spacing for the rails that will be capable of resisting the storm-wind compressive wheel loads. In addition, there must be four buried ballast blocks to which the crane can be tied down by means of cast-in fittings. The buried ballast together with the traveling ballast must be capable of counterbalancing 1.5 times the maximum overturning moment (see Sec. 6.2).

At the ends of the tracks, trippers are set that will automatically cause the crane travel brakes to engage. At a distance somewhat beyond the crane stopping distance, end stops, or bumpers, are installed as a last means to prevent the crane from running off the rails.

6.7 Erection and Dismantling

A considerable proportion of tower crane accidents occur during erection and dismantling rather than during actual operations. For example, a British study found that some 24% of tower crane accidents and 42% of repair costs are associated with erection and dismantling.† Although site conditions can be difficult as Fig. 6.42 shows, the difficulties can be overcome by planning. Several other reasons can explain but not justify the high rate of such accidents.

Tower crane manuals are seldom written from a rigger's perspective. They typically present idealized erection conditions and schemes that can rarely be replicated under jobsite conditions. Often the manuals are literal translations from a foreign original; the intention can become cloudy or obscure, particularly where specialized terms or jargon are used.

Erection weights are not always as clear as they could be. The manual may give an extensive list of subassembly weights that in practice need to be compiled and added. Other times, weights of standard assemblies are listed, but in the field additions and subtractions must be made for subassemblies added or deleted to suit erection crane capacity or other jobsite restrictions. All too commonly, riggers work with an inaccurate weight list.

Centers of gravity are unsymmetrically located on some components. As a consequence, the placement radius of the erection crane may easily be miscalculated. During disassembly, picking eccentrically with respect to the CG is almost inevitable, and the ensuing lateral movement of the load must be anticipated and controlled.

The moments acting on the crane mast are naturally unbalanced. During erection and dismantling, the unbalance must be kept within permissible limits by following the correct sequence of work. Otherwise, bolts, pendants, or other structural components may become damaged or fail.

Wind and darkness are both enemies of the crane erector. Critical work, such as boom erection, must be scheduled with this in mind and ample time allowed to secure everything before dark. Wind cannot always be avoided, but it is prudent to consult with the local airport tower before proceeding with a critical operation.

Erection

The erection supervisor should not expect to see the site for the first time on the day the work is to begin. At the very least, the site should be visited a month or so beforehand for planning, a week before for

---

A. J. Butler, Crane Accidents: Their Causes and Repair Costs, *Cranes Today,* Edgeware, Middlesex, England, no. 62, March 1978, p. 24.

Figure 6.42 Stiffleg derrick removing a Kodiak KL300 tower crane from the roof of the Royal Concordia Hotel in New York City. At this difficult site, the roof was not large enough for the derrick, so the sill ends had to be supported on posts resting two stories below. A situation like this dramatically illustrates why clearances and reaches must be thought out step-by-step before the derrick is placed on the roof. This installation was designed by Malcolm Johnston of the authors' firm.

coordination, and a day before for final details. Close coordination with the tower crane technician should be maintained through this period to verify that the many important details are attended to.

Sufficient space must be allocated at the crane location to allow placement of the erection mobile crane and layout of the tower crane components within reach of the mobile crane. When inadequate space is provided, erections costs will increase disproportionately, as additional equipment will be required to bring the components to the mobile crane one by one or a larger mobile crane will be needed to reach to the components. During preplanning, space must be allocated for ground storage of each component, with consideration given to truck positioning and offloading activities as well as preassembly work.

A top climbing crane can be erected to minimum height and then jacked up to final height as the work progresses. This will minimize both the capacity of the mobile crane needed and the initial labor cost. Final labor cost will depend, however, on crane utilization. If climbing must be done at overtime rates, the final erection cost may be higher than for initial erection to full height.

There should always be a cost advantage to preassembling as many components as possible on the ground, but this must be weighed and balanced against the cost of the larger mobile crane that may be needed to place the heavier combined assemblies. Another consideration is that work on the ground is far safer and much faster than work in the air. Decisions must be based on cost, availability of equipment, and the level of experience of the erection crew.

In many instances both cost and safety have suffered from insufficient erection preplanning. Before the crews are sent to the jobsite, preassembly decisions must be made and each individual lift identified with respect to weight, necessary mobile-crane reach and height, and lifting accessories required. It can be dangerous to make do with whatever slings, shackles, and so forth happen to be in the gang box. With many sharp-cornered components, provisions to soften the bends of wire-rope slings need to be made or special gripping devices used. The tower crane will be exposed to substantial loads during operation so that nicks and dents in its members cannot be tolerated.

Accurate erection weights must be known in advance. If there is doubt, components should be weighed. Decisions concerning preassembly or partial disassembly of mechanical elements are part of the preplanning that must take place before the crews and equipment arrive at the site. In this way lift weights and crew instructions are set down on paper so that field work can proceed smoothly and quickly. Risks are then minimized.

Counterweights are not always supplied by the tower crane manufacturer although the distributor or rental agency may do so. If made on site, they must be made to strict specifications and be ready for use when erection begins. For many cranes, the load jib cannot be erected without part of the counterweighting in place.

The manufacturer's erection instructions must be carefully studied in planning jib placement procedures, as in all phases of the erection process. Usually, the counterweight jib must be in place with a stated portion of counterweights mounted before the load jib can be installed. This is to assure that the mast will not be exposed to excessive unbalanced moment at any time during erection.

Each crane model has a particular style of jib support pendants, and the erection procedure will depend upon the style in use. The simplest form employs fixed-length wire ropes with factory-installed end fittings that make up to pins on the crane top tower, sometimes called a *rooster* on hammerhead cranes, and on the jib. For this pendant and jib type, the jib is lifted from the ground at 20° to 30° to the horizontal and the jib foot connection pins inserted. With a come-along or similar puller, the pendants are pulled into place one by one and the pins set. With all pendants attached and the puller released, the jib can be lowered to its operating position.

Procedures for placing steel bar pendants are similar except that pullers of greater capacity are needed; chainfalls are usually used. In addition, the heavy bar sections may have to be temporarily tied in place on top of the jib until the puller is functioning. Since few tower cranes are equipped with service platforms at the tower top, a working scaffold of some sort may be needed to make the pendant connections.

Luffing jibs are the most difficult to erect. They must also be lifted at an angle to the horizontal. After the foot pins are installed, the jib may be hung on temporary pendants, or the boom hoist must be reeved while the jib remains suspended on the erection crane hook. The latter operation should not be attempted unless there is power to the luffing hoist drum, little wind, and sufficient remaining daylight.

When erecting jibs, top towers, and machinery decks or counterjibs, preplanners often overlook checking clearances. These checks are needed to verify that the erection crane boom is long enough to provide adequate drift, but also to position the erection crane and to provide clearance during the placement and alignment of components. When evaluating hook height requirements, allowance must be made for slings, for moving the piece into place, for trip settings of anti-two-block devices, and for the maximum height of the hook with respect to the boom tip (Fig. 6.43). The distance from boom tip to hook, while the hook is in position to place a load, is called *drift*. Adequate drift assures height clearance, the ability to raise the load to the needed height.

In addition to adequate drift, it is necessary to make certain that acceptable load clearance has been achieved as well. Load clearance is the distance from the closest part of the piece being lifted to the

Figure 6.43 Erecting the boom on an FMC TG-1900 tower crane. A small time allotment on a busy city street favored the use of a telescopic crane, but one can see the importance of checking load clearance and hook height requirements. (*Photo by Jay P. Shapiro.*)

closest part of the boom or any other obstruction, such as an adjacent building. Load clearance checks therefore often require several independent evaluations with respect to various obstacles near the load movement path.

The methods given in Chap. 5 for calculating lift and swing clearances can be applied during erection preplanning. The amount of clearance taken as acceptable for any specific situation is a matter for judgment but should be influenced by the experience of the erection crew, the height at which the work is taking place, the size and weight of the load, and the means to be used to control the load, such as tag lines. Clearance of 4 ft (1.2 m) is usually sufficient under any circumstances, and as little as 1 ft (300 mm) has proven to be satisfactory on erection jobs with highly experienced supervision and crews.

One of the most important aspects of erection on some tower cranes concerns bolting. This subject is of such critical concern and is so little understood that detailed treatment is called for.

Bolting

There is a basic philosophical difference between high-strength-bolting procedures used in the United States and those in use in Europe, although there is agreement on technological concepts and aims. In essence, the United States view is that bolt tension is the parameter controlling the success of the completed joint so that procedures must be geared to achieving required tension. The Europeans seem to agree on the need for controlling tension but allow that tension will result from applying torque to the bolt; their procedures are therefore centered on producing the proper torque.

In practice, American procedures require calibration of torque wrenches on the job to a tension-measuring device that will show the torque needed to secure the tension required under the ambient conditions. Frequent recalibration is required, but an alternate method, the turn-of-the-nut method, requires no calibration equipment whatsoever. When this highly reliable method is used, the nut is turned from one-half to three-quarters turn after a "snug tight" condition has been reached. The turning range depends on bolt size and the slope of the elements joined, but there is a tolerance of $\pm 30°$. "Snug tight" is defined as the point where an impact wrench starts impacting or as the tightest possible setting made by hand with a spud wrench.

In European practice, torque is prespecified for a particular set of conditions (such as plated bolts treated with wax before installation or galvanized bolts installed dry), and those conditions must be met at the jobsite. Quite simply, *we* calibrate to existing conditions at the jobsite, and *they* stipulate conditions that must be met in the field. When either procedure is followed properly, the results will be the same.

For European-made cranes with European-written documentation, American installation personnel are faced with bolts and procedures that are outside their experience and often not adequately explained.

Most European bolts and nuts are made, rated, and marked to International Organization for Standardization (ISO) specifications or to national standards such as the German DIN series, which are in substantial conformity with ISO. The bolts are graded for strength by numbers, such as 5, 6, 8, or 10, which represent one-tenth of the minimum tensile strength in kilograms force per square millimeter (1 $\text{kip/in}^2 = 0.70\ \text{kgf/mm}^2$);† a second number following indicates the ratio of the minimum yield point to the minimum tensile strength. A decimal point may be inserted between the numbers, producing designations that take the form 5.6, 8.8, 10.9, and so forth (Fig. 6.44). Nuts are similarly graded but by tensile-strength number only. The

†The kilogram force is not an SI unit, but this standard was written before the SI was widely adopted, and it has not yet been revised. The kilogram force is called a *kilopond* in European literature.

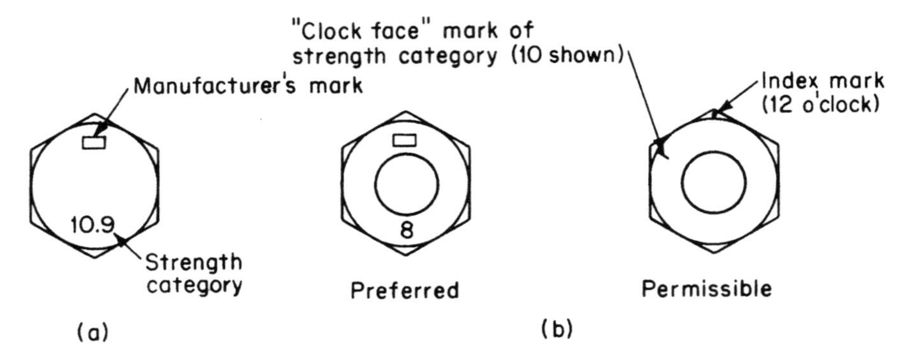

Figure 6.44 International Organization for Standardization (ISO) strength markings common to European-made nuts and bolts: (*a*) bolt-head marking; (*b*) nut marking.

tensile-strength number of properly matched nuts and bolts must agree.

For comparison, ASTM A325 (SAE grade 5) bolts are about equivalent to ISO 8.8 while ASTM A490 (SAE grade 8) are about the same as ISO 10.9.

High-strength bolts are installed by applying torque to secure a high level of pretensioning. American practice is to preload to 70% or more of bolt tensile strength with the general understanding that excess preload will do no harm. As Black and Adams† have put it,

> It can be reasoned that, if a well-designed bolt does not fail in hard tightening up, it will not fail in service. The reasoning is that, when the bolt is being tightened, the stress will be due to the tensile-load stress combined with the torsional stress due to the tightening torque. The latter stress disappears on removal of the tightening torque; hence the remaining stress is less than the maximum stress during tightening.

Several series of tests have proved that excessive preload does not diminish the ability of the bolt to perform and may even assure better performance. Therefore, when high-strength bolts of any manufacture—domestic or foreign—are installed, it is safer to err in the direction of overtensioning rather than to chance undertensioning.

Nuts are designed so that they and their threads are stronger than the bolts to which they are fitted. Overtorquing will therefore fail the bolt in tension rather than cause hidden nut damage and later failure during operation.

†P. H. Black and O. E. Adams, Jr., "Machine Design," 3d ed., McGraw-Hill, New York, 1968, p. 192; used by permission.

Mast bolts

Mast sections are usually joined together with butt joints fastened with high-tensile prestressed bolts. During operation and under wind loading the mast is exposed to bending moments that want to load and then unload the mast bolts. This is a classical situation of a detail exposed to potential fatigue damage, the severity of which will depend upon the range of the stress variation and the number of cycles of loading. The solution to the problem applied to tower crane mast legs is a classic example of engineering reasoning; it is a series of concepts and procedures that are quite effective but need particular attention at the jobsite.

The key to avoiding fatigue failure is in keeping tensile-stress variations within very close limits, particularly when overall tensile stress is high. This can reliably be accomplished by using concepts developed for prestressed joints. Although the best joint-design parameters should be set by testing, the theory underlying prestessed joints offers direct and serviceable insights.

The joining surfaces of the mast legs are constructed of large, thick steel blocks drilled to receive the bolts. The bolts are inserted and torqued to produce high bolt tensile stress that can extend beyond the bolt yield stress. This will compress the steel leg blocks.

If the bolt stress area is A and the block area is n times A, with bolt prestess load P the blocks will undergo compressive strain (Fig. 6.45)

$$\Delta t_c = \frac{Pt}{nAE}$$

i.e., the blocks will be under compressive stress while the bolt is in tension. E is the modulus of elasticity and t is the block thickness.

Should the mast leg become exposed to tensile load F, block compression will be relieved to that extent, and for $P > F$ the compressive strain will be reduced by

$$\Delta t_t = \frac{Ft}{nAE}$$

but Δt_t is a tensile strain that will be imposed upon the bolt and will add to bolt tensile stress. Calling the additional bolt tension F_b, we have

$$\Delta t_t = \frac{F_b t}{AE} = \frac{Ft}{nAE} \qquad F_b = \frac{F}{n}$$

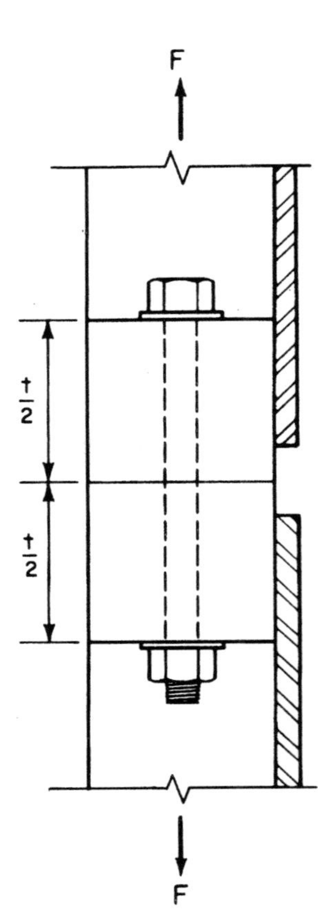

Figure 6.45 Schematic of mast-leg prestressed joint.

The additional stress on the bolt resulting from leg loading is therefore a function of the ratio of bolt to block area. This is true even if bolt tensile stress is above the proportional limit, but in that case the bolt stress will increase less than the block strain would seem to imply.

Large block areas can be effective only if thickness increases as well, so that local strain effects in the vicinity of the bolt head and nut, which act on a smaller effective area, become but a small part of total strain. Thick blocks are also needed so that they will not bend, causing prying action at the bolt head and the additional tensile stress in the bolt this action implies.

If $F \approx P/2$ and $n = 10$, the increase in bolt stress under tensile load will be only 5% if the stress remains below yield and less if above yield. But leg loads will vary from tension to compression. With the introduction of compressive force to the already compressed leg blocks, the blocks will undergo further compressive strain. This strain will be reflected in the bolt as a reduction in tensile load with the bolt now experiencing stress fluctuations that include components of both leg tension and leg compression.

Equation (6.9) was used to describe the design leg forces at the mast-footing interface. If M_o is redefined as the moment at any particular point along the mast height, Eq. (6.9) will give mast joint loads as well. This equation shows us that $Q/4$ comprises a compressive increment common to each mast leg.

If mast sections are preassembled on the ground with their bolts fully torqued, after final assembly each bolt prestress will be reduced by $Q/4$ because of the compressive strain that will be introduced into the leg block by crane dead weight. The same result will occur if the bolts are fully torqued as the mast sections are erected. A reduction in effective bolt prestress will introduce greater variation in tensile stress in the completed crane, and a heightened susceptibility to fatigue failure will ensue. The approximate bolt stress range will be $2M_o/(nAs\sqrt{2})$, including moment effects.

If the crane were fully assembled and the bolts torqued with the crane balanced, the stress range in the bolts would be reduced by $Q/4nA$ from the range induced by preassembly. A further reduction can be achieved by leaving the crane unbalanced. For an unloaded crane, the unbalanced moment in the direction of the counterweights is approximately the same as the unbalanced moment in the direction of the hook for a fully loaded crane. By aligning the counterweight jib over a mast leg and then torquing the bolts in that leg, the bolt stress range will be reduced to approximately $[M_o/(s\sqrt{2}) - Q/4]/nA$, the smallest range of the bolt prestressing conditions discussed.

For this reason, assembly instructions for most cranes specify that mast bolts should be finally torqued only after assembly has been completed and with the counterweight jib above the mast leg on which the bolts are being torqued. This procedure should be followed even if the instructions fail to mention it, as it materially reduces susceptibility to bolt fatigue failure during operation.

In the United States, tower cranes are usually erected by ironworkers, whose ordinary duties involve erection of steel building frameworks. The high-strength bolts they normally see in their work would be the ASTM A325 or A490 types used in building construction. It is most important, therefore, that erection crews be alerted to the ISO grading and marking systems so that European-made high-strength bolts will be properly placed and installed.

Dismantling

In the best of circumstances, dismantling a tower crane is essentially the reverse of erecting it. But in most actual cases, dismantling is more difficult. Conditions change during the time of the crane's use—pins and bolts corrode, the crane may have been raised to a higher elevation, and the new building is now around or alongside it.

A climbing crane is usually jacked down before removal. Clearances to the building must be investigated, and the final jack-down elevation determined in advance. Removal with a mobile crane or derrick suggests that elevation may be critical: If the tower crane is not jacked down sufficiently, the drift and clearance could become inadequate.

When external climbing cranes are jacked down, they will no longer be capable of free swinging because the building will be in the way. The dismantling procedure must allow for this. During erection, the boom and counterjib could be swung to an advantageous position; in dismantling, the dismantling hook must reach these components where they are.

The removal of a pendant suspended jib or counterjib is more difficult than its assembly because of several factors: The CG of the pick is hardly ever accurately known, and the interplay of pendant and footpin forces introduces the danger of dramatic load shifting as soon as these are released.

Before the slings to be used for removing the jib or counterjib are selected, the steps that need to be taken to release the pendants should be reviewed. Assume that a pair of slings is attached to the jib or counterjib from a ring at the hook so that the angle between the legs is 60°, the crane hook is positioned directly above the CG, and load weight is P (Fig. 6.46*a*). If the piece is then lifted to make an angle of 20° with the horizontal in order to unload the pendants, the forces in the sling legs will change from their values in the level position. For equilibrium of the horizontal components of the sling leg forces (Fig. 6.46*b*)

$$F_1 \cos 80° - F_2 \cos 40° = 0$$

and for the vertical forces including foot reaction R

$$F_1 \sin 80° + F_2 \sin 40° - P + R = 0$$

Taking moments about the foot pin, we get

$$F_1(d - e) \sin 80° + F_2(d + e) \sin 40° - Pd = 0$$

where d is the distance from the foot pin to the jib CG and e is the distance to the sling points on either side of the CG. Solving simultaneously gives $F_1 = P/D$ and $F_2 = P/4.41D$, where $D = 1.131 -$

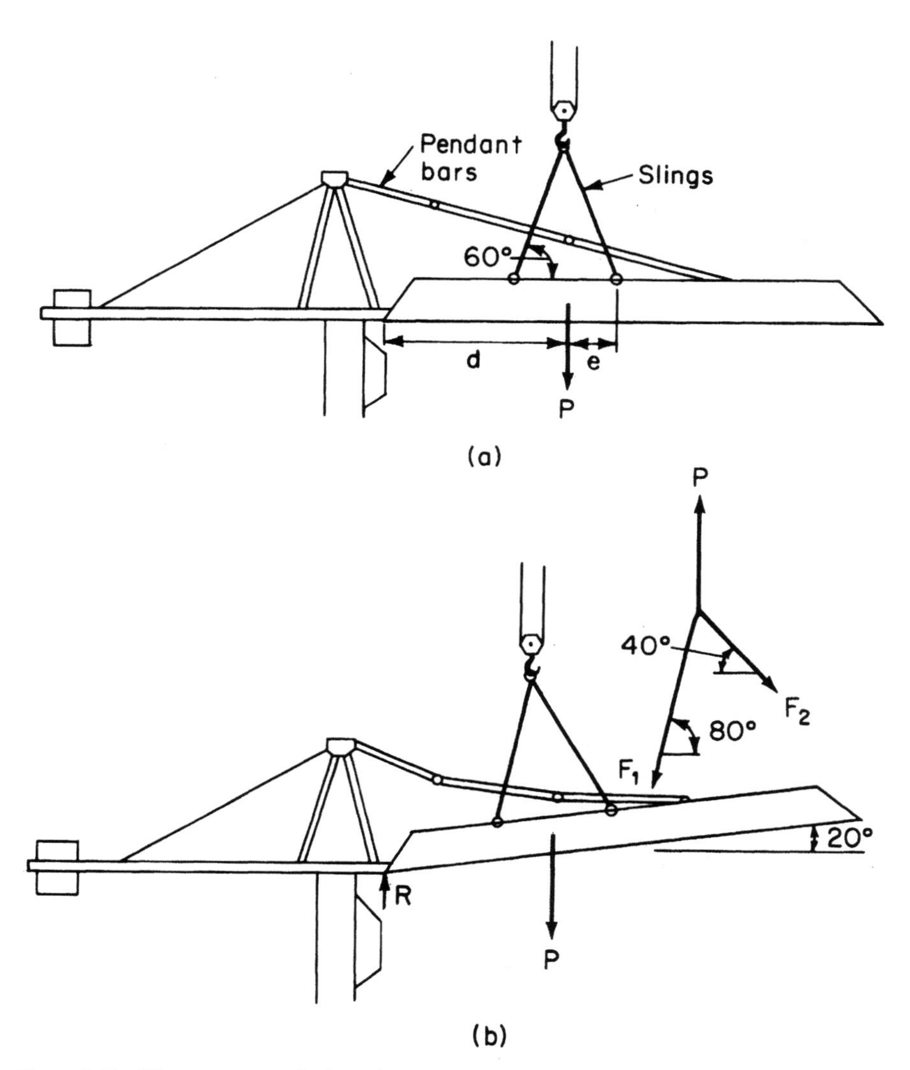

Figure 6.46 Sling geometry during tilting of the jib to slacken the load on the pendants.

$0.839e/d$. A reasonable value for e/d would be 0.1, which makes $F_1 = 0.96P$.

In the initial position, before tilting the jib or counterjib, each sling leg carried a load equal to $0.58P$, so that by tilting the piece 20° the maximum sling leg force has increased by some 66%. Note that tilting the piece 30° will place more than the entire load weight on one sling leg while the other leg will be unloaded.

A picture like Fig. 6.42 makes one appreciate the magnitude of the dismantling problem.

6.8 Tower Crane Operation

Operating a tower crane bears little resemblance to the experience of operating a mobile machine. Aside from the dramatic difference in vantage points from the cabs, there is a marked difference in the way that the machinery responds to the controls. A traditional mobile crane has internal combustion engine power that directly responds to the hand of the operator. Tower cranes are electric or hydraulic, and the operator's application of power is modulated and limited by the control system.

Fear tends to keep many from the cabs of tower cranes, but that fear is not well-founded. The mobile-crane operator who does not know the proper limits of the machine can exceed them; but the tower crane operator is protected by the control system, if it is functioning properly.

The control system is made up of various circuits and devices that do the following: prevent overloading and overacceleration, keep the various crane motions from exceeding prescribed travel limits, and cause the crane to return to stable or static condition when malfunctions occur. Electric and hydraulic machinery lend themselves to such intervention.

Line pull and load moment are monitored, and when a load limit is reached, a warning is signaled to the operator. An additional increment of load causes the hoist motion to stop.

Motors, brakes, and gears are sized to limit acceleration; hence the forces acting on the crane during operation are also limited. While the structure is protected from excessive stress, the operator is also restricted from having complete freedom of action. He cannot cause the load to freefall or quickly stop during swinging. He might be prevented, on occasion, from swinging against a moderate wind, especially when the load has a large surface area and is at a long radius.

The standard motion-limiting devices include those that prevent two-blocking and trolley overrun. More exotic devices are available on some models to prevent the jib from colliding with buildings or with other cranes or to restrict the crane to lifting within designated areas.

All tower cranes are furnished with deadman controls that return to neutral (and stop the motion) when the hand is taken off the lever or button. In addition, circuitry is designed to apply brakes to hoist, trolley, and boom motions when fluid pressure or power is lost.

The benevolent guidance of the control system should not be permitted to lull anyone into feeling that the operator can do no wrong. The intelligent operator (and supervisor) treats the various limits as though they are safety nets; they are there when needed, but they are not intended for jumping into. After all, devices do malfunction; the operator who is careless enough to depend on the vigilance of these devices is probably also too careless to check and monitor them.

Limiting devices can be defeated even without malfunction. Two-block and hook low-position limiters, for example, must be reset from time to time; incorrect settings can void the effect of the mechanism. Full throttle thrust backed by inertia or wind may overrun some motion-limiting devices. Shock load—a sudden application or release of tension on the load line—cannot always be stopped by the load limit device.

The operator would not be doing anyone a favor by tampering with the equipment to boost its performance. The small gain would hardly be worth the risk: the strong likelihood that the crane will be damaged and the potential danger of a mishap.

Shock loading is insidious because it has a direct link to fatigue failure even though any one episode may impart no visible sign of damage. But the damage accumulates and repetitive shock loading can cause fatigue failure during the course of a single installation. Construction tower cranes are not designed for duty cycle work.

Some examples of work that must not be performed with ordinary tower cranes are pulling formwork free from freshly poured concrete surfaces, dropping a demolition ball, and pulling piles or sheetpiling from the ground.

Chapter

7

Derrick Installations

An installation design engineer, faced with a derrick job, starts out with many questions but with few facts. As in crane installation design, the minimum data needed includes the loads that must be handled, where they are and where they must go.

But what comes next? Derricks come in several configuration types, as well as in varying dimensions and capacities. They can be placed on foundations on the ground, on raised supports, or anchored to the framework of a building, during construction or long after completion, on a static mounting or movable either horizontally or vertically.

In making derrick choice and placement decisions, mathematics comes into play, but experience carries the day; derrick selection and installation is an art. In this chapter, we offer such insights as we can to aid the installation designer, and we describe the physical and working characteristics of the major derrick types. But most of the following material will concern the practical details that need to be addressed after the designer has made the important installation decisions.

7.1 Introduction

A friend of ours, a rigging company executive who had worked his way up through the ranks, was once sounding off about how easy his hoisting riggers have it today. Their derrick installations are all worked out in advance, on paper, with the drawings indicating every bolt, turnbuckle, and shackle needed, as well as the details at anchorages and support points. Ropes are specified and lengths calculated; operating radii, swing arcs, and load pickup points are indicated. All the foreman has to do is to put the pieces together, make the picks, pack it all up, and then go home.

In the good old days (and that was not so long ago, as our friend is in his middle sixties now) the foreman had to do the whole thing by himself. The only support he received from the office was a pat on the back and a couple of cuss words from the boss for encouragement. Their equipment was a few steel-pipe and wood-pole masts, or perhaps a latticed angle-iron boom, and whatever was in the gang box. The job was done by guts, the strength of their backs, and their wits. But their hearts were in their mouths from the time they arrived at the job until the piece was finally in place.

The good old days were not so good after all. Not often mentioned in the tales told about that era of hoisting by guess and a prayer are the occasional injuries and deaths, the pieces dropped or damaged, and the projects delayed waiting for replacement equipment.

In the past, society may have seemed to accept the proposition that work is dangerous, and that injuries and deaths are the inevitable consequence of modern life. Not so anymore. The court of public opinion and courts of law impose harsh judgements on those responsible for failures. What's more, modern business people shun uncertainty, and accidents produce delay, demoralized staffs and wildly escalated costs. We can no longer afford not knowing in advance how the job is going to be done.

Preplanning does not make hoisting work a sure thing, but it certainly comes a lot closer than the old methods did.

Planning for derrick installations starts with contemplating the loads: where they start out and where they must end up. This, together with the physical aspects of the site, leads to choice of derrick type, size, and capacity. Often several types of derrick can do the job equally well. The selection must then be made on the basis of availability and installation and operational costs.

There are two general types of derrick installation, permanent and temporary. Permanently installed derricks are not nearly as common today as they used to be. They have largely been replaced by mobile cranes, heavy fork-lift trucks, and other mobile equipment that is far more flexible. Therefore, little will be said of permanent installations.

Temporary installations encompass the standard derrick forms and units built for a special purpose, which may be mounted either on the ground or on structures for performing specific lifts or operations. They can include unusual limited-use devices, such as a transformer derrick our office designed whose only function was to hoist replacements in the event of a transformer burnout. Adaptations of this derrick were used in all-electric office buildings that had the transformers distributed to several power rooms in various parts of the structure.

For this reason, the unit had to be made up of components easily assembled and disassembled for transport in an elevator and through corridors, but mounting provisions were permanently placed at each potential location for use. Because the device was expected to be needed only on rare occasions, it did not include a permanent power unit or winch but was arranged to make use of the winch on a mobile crane at ground level.

When a standard derrick form is used, its accessories and where and how it is to be mounted must be tailored to suit the installation site and the work to be performed. The possibilities are infinite, but careful study of the loads and their placement as well as labor requirements for both installation and operation will usually point quite clearly to the most practical type of solution. Several possible schemes should be devised and reviewed before a choice is made and final design started.

7.2 Chicago Boom Derricks

An important consideration in the installation and use of Chicago booms arises from studying where the loads must go once they have been hoisted. Several entirely different situations can occur. For example, when building facing panels are hoisted and placed, the CG of the panel will fall just outside the building when it is in its final position. The derrick must be capable of placing its hook directly over the CG position so that the stone setters can make up the connections without fighting against a tendency for the panel to fall away. This means that the derrick swing pivot must lie some distance beyond the building edge; a distance of as much as 4 ft (1.2 m) is used. Such great eccentricity can have a dramatic effect on the installation design and cost.

When hoisting relatively lightweight loads, the boom pivot can be placed as close as one-half the boom width from the edge. At full 90° swing, this will leave the CG of the load outside the building that same distance, but for light loads drifting or pulling with a chainfall or come-along will easily shift the CG inside the structure.

For heavy or bulky loads it is particularly important to set the swing pivot as close as possible to the structure to minimize the effects of eccentricity on the installation design. For 90° swing, placing the boom parallel to the building edge, one-half of the boom width will be the minimum. But the load CG is outside, and pulling or drifting such loads into the building can be hazardous. Therefore, a time-consuming load-shifting procedure must be used (Figs. 7.1 and 7.2). As an alternative, a load landing platform can be cantilevered from the building edge.

Figure 7.1 Loads cannot be landed until the CG is inside the structure. For heavy or bulky loads, this means that a load-shifting procedure, such as shown in Fig. 7.2, must be used. (*Gendelman Rigging and Trucking, Inc.*)

In Chap. 1 we learned how to distribute the effects of live load, lead-line forces, and the topping lift load and to determine the reactions to the boom fittings. From this point on, we must deal with the interactions between the derrick and the host building frame or other support structure. If the frame needs reinforcement, for buildings under construction costs will be minimized if this work is done as part of the fabrication process; in other words, it is most advantageous to have the derrick installation designed before the steel or concrete frame is built.

When selecting the floor levels at which the boom foot and topping lift anchorages are to be placed, it is preferable to avoid the floors at which steel column splices occur. The splices may be located from 1½ to 2½ ft (0.5 to 0.75 m) above finished floor level, thus forcing the bottom of the anchorage fittings to be placed well above that point in order to clear the splice plates. It is best to keep the fittings as close as possible to floor level, both to simplify installation and to minimize the bending moments (biaxial) introduced into the building column.

Look at the boom foot fitting in Fig. 7.3; the boom has been swung to 45° to the building face, and we are taking the boom vertical reaction as B_v and the horizontal reaction as B_h. With the load P representing the dead load plus some construction live load delivered to the

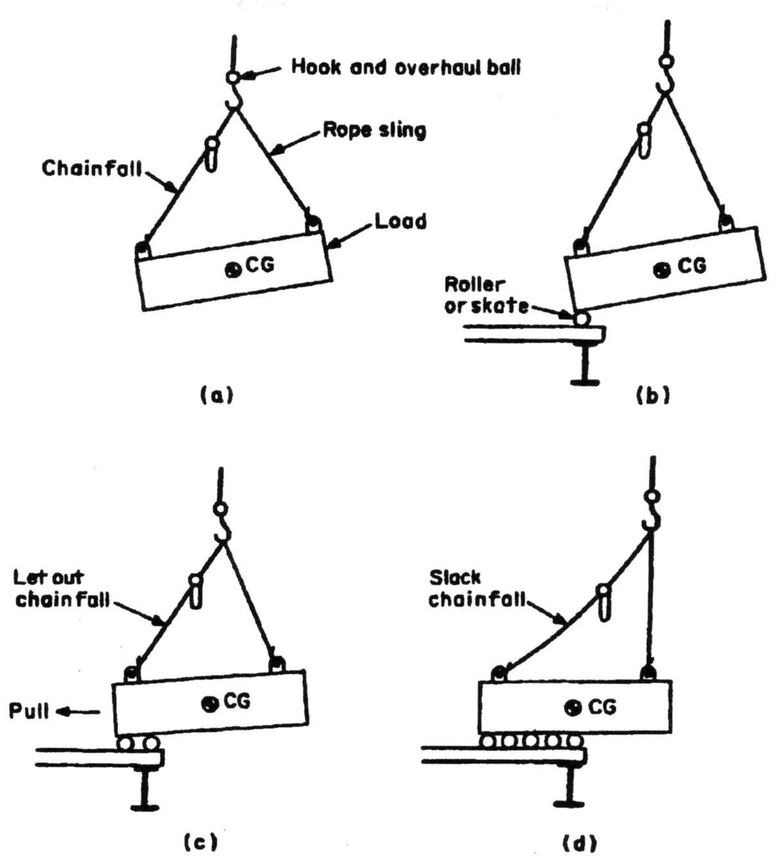

Figure 7.2 Procedure used for shifting loads suspended from crane or derrick hooks in order to move the CG into the building: (*a*) load is in hoisting position; (*b*) end of load is landed; (*c*) load starts in; (*d*) CG is inside building.

building column from all the floors above, column axial load below the fitting is then $P + B_v$. For strong-axis bending (Fig. 7.3*b*), the bending moment is $0.7B_h d(1 - d/h)$ by examination. Weak-axis bending comprises two terms, one from the boom vertical reaction and the other from the horizontal. By statics, the reaction at the column foot is

$$R = \frac{B_v e}{h} + \frac{0.7B_h(h - d)}{h}$$

which gives the maximum moment in that plane of

$$M = \frac{[B_v e + 0.7B_h(h - d)]d}{h}$$

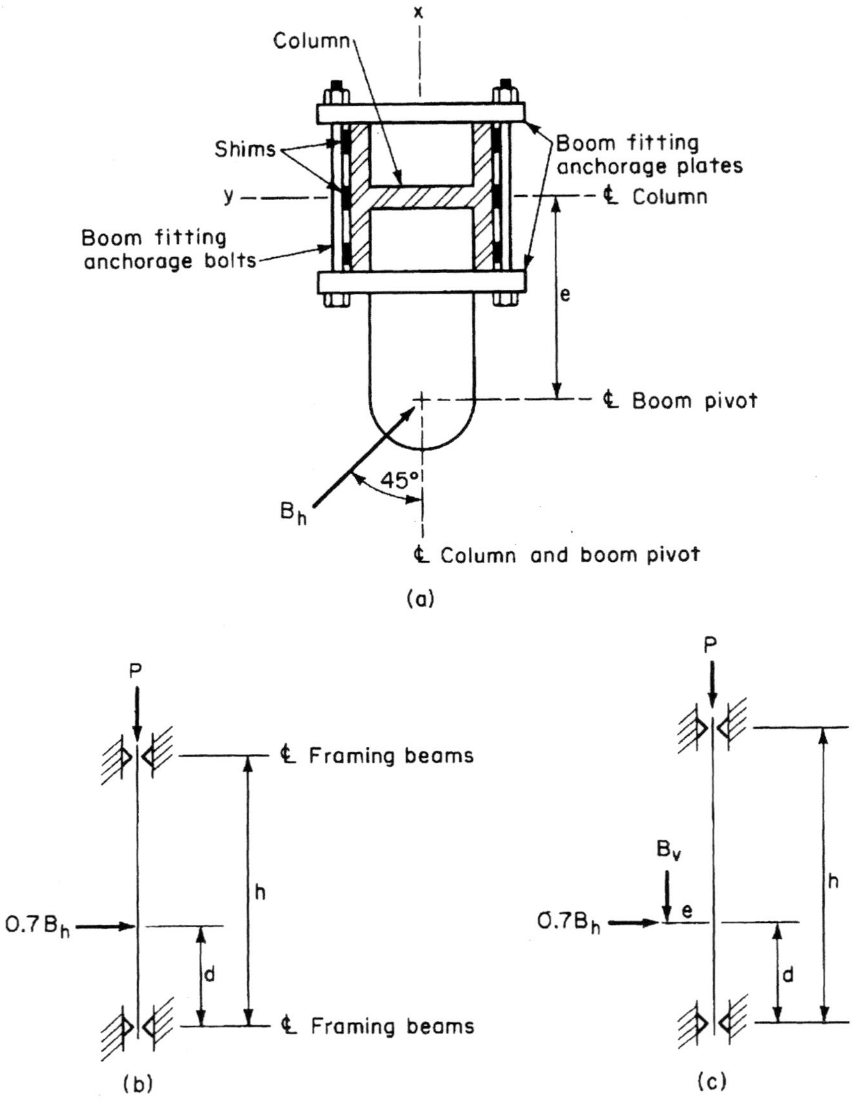

Figure 7.3 (*a*) Boom foot fitting and bending loads on (*b*) column strong axis and (*c*) column weak axis.

The axial load and the bending moments can then be converted into stresses by introducing the column section properties, but an additional moment must also be considered. The force $0.7B_h$ acting on eccentricity e imposes a torsional moment on the column. If the beams framing into the column at each floor are provided with rigid connection details able to absorb the torsion, this will introduce additional bending across the column flanges that must be evaluated. If not, the column will be free to twist and could distort considerably, perhaps permanently.

Of course, on any particular structure the columns may be positioned 90° from the orientation shown in Fig. 7.3. In addition, the loads imposed by the boom when it is parallel, perpendicular, or at any other angle to the building may be more severe than for the 45° position discussed. The installation designer must evaluate the structure for the actual loading conditions defined by the operations to be performed or those that may be inferred from reasonably expected field procedures. The job lift plan must then incorporate instructions that will limit induced loads to those provided for in the design. Where necessary, warnings against actions that will cause excessive loadings would be included.

The Chicago booms shown in Figs. 7.4 and 7.5 are clamped on the flange tips of the mounting columns, as illustrated diagrammatically in Fig. 7.3.

Torsion

Column torsional distortions can be troublesome: They can cause fireproofing to become loosened, damage beams and their connections, and raise the possibility of column buckling. When column ends are restrained from twisting, the added stresses induced by the restraint can cause yielding at the flange tips and a considerable loss of column strength (Fig. 7.6). For this reason, particularly on lighter weight columns, lateral ties are often installed to reduce or eliminate the torsional moment by transferring the force that produces it to adjacent members (Fig. 7.7). Torsion reducers are just barely visible in both Figs. 7.1 and 7.4.

It is obvious that wire rope is the easiest material to use for lateral torsion ties, but it is not so obvious that wire rope does not do the job well. Even when the rope ties are taken up very tightly, they will stretch considerably when the torsional load is introduced. As a consequence, the column will still be exposed to most of the torsional loading. For example, an IWRC rope will stretch some 3½ times more than a steel rod of equivalent diameter. If a 1¼-in-diameter (32-mm) mild-steel rod is needed to absorb the torsional moment, to get the equivalent distortion resistance we need seven parts of ⅞-in-diameter (22-mm) rope, and the rope will end up loaded to only about one-seventeenth of break strength. Rope, therefore, will often prove to be unsuitable for this purpose.

The effect of torsion in a column can be analyzed by using elastic theory. For the model in Fig. 7.8, where the column ends are restrained to permit neither

Figure 7.4 Chicago boom clamped to the weak axis of a column and arranged for landing loads to the left. Several parts of line are used in the right swing line to develop enough force to pull the boom away after unloading. (*Gendelman Rigging and Trucking, Inc.*)

warping nor twisting, the applied torsional moment T will be distributed to the column ends as reactions T_0 and T_h. In order to maintain consistency in angular twist, the greater part of T will be restrained by the shorter and torsionally stiffer segment of the column.

A column unrestrained at its ends will twist or warp evenly (Fig. 7.8*b*), causing shear stresses in the cross section. This is known as *St. Venant torsion* and is show in Fig. 7.6. If the column ends are restrained, warping becomes nonuniform as resistance to torsion induces bending moments in each of the flanges (also shown in Fig. 7.6). The applied moment will then be resisted by the combination of the two effects

$$T = T_v + T_w \tag{a}$$

where the St. Venant torsional resistance is

$$T_v = GJ\frac{d\phi}{dz} \tag{b}$$

G is the shearing modulus of elasticity and J is a torsional constant to be described later. The nonuniform warping resistance is

$$T_w = -Vs \qquad V = \frac{dM_f}{dz}$$

where the flange bending moment is

$$M_f = EI_f \frac{s}{2} \frac{d^2\phi}{dz^2}$$

giving

$$T_w = -EI_f \frac{s^2 d^3\phi}{2\, dz^3} \tag{c}$$

when I_f is taken as the moment of inertia of the cross section of one flange about the y axis and E is the modulus of elasticity. Substituting (*b*) and (*c*) into (*a*) gives

$$T = GJ\phi' - EI_f \frac{s^2}{2} \phi'''$$

with the primer notation referring to differentiation with respect to *z*. Simplifying by inserting $c = GJ$ and $c_1 = EI_f s^2/2$ and rearranging leads to

$$\phi''' - k^2\phi' = -\frac{T}{c_1} \qquad k^2 = \frac{c}{c_1} \tag{d}$$

The general solution to Eq. (*d*) is

$$\phi = \frac{Tz}{c} + A_1 + A_2 \sinh kz + A_3 \cosh kz \tag{e}$$

For the segment $0 \le z \le d_1$, taking values of Eq. (*e*) and its derivatives at $z = 0$ gives

$$\phi(0) = A_1 + A_3 \qquad \phi'(0) = \frac{T_0}{c} + kA_2 \qquad \phi''(0) = k^2 A_3$$

with T_0 used because this segment restrains only that part of applied torsion T (Fig. 7.8*c*). The equation for the first segment is then obtained by making substitutions for the constants of integration, A_1, A_2, A_3, in Eq. (e) in terms of the function at $z = 0$

$$\phi_1 = \frac{T_0 z}{c} + \phi(0) - \frac{\phi''(0)}{k^2} + \left[\frac{\phi'(0)}{k} - \frac{T_0}{ck}\right] \sinh kz + \frac{\phi''(0)}{k^2} \cosh kz$$

$$= \frac{T_0}{ck}(kz - \sinh kz) + \phi(0) + \phi'(0)\frac{\sinh kz}{k} + \frac{\phi''(0)}{k^2}(\cosh kz - 1)$$

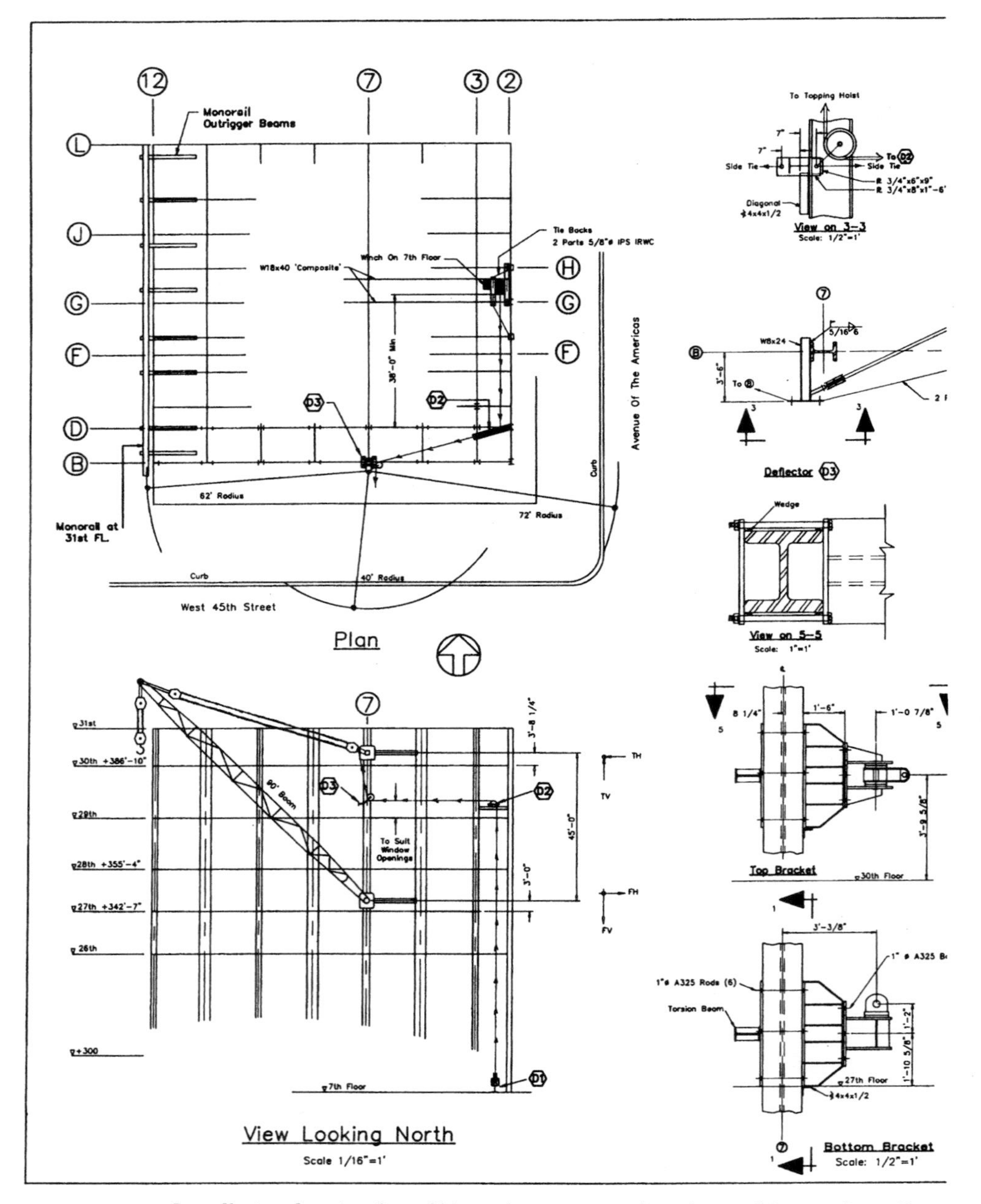

Figure 7.5 Installation drawing for a Chicago boom mounted on the steel frame of an office tower under construction. At both the top and the bottom brackets, beams have been used to relieve the columns of torsion loading. Although the boom foot is just above the twenty-seventh floor, the winch has been set on the seventh floor and provisions to carry the lead lines are shown. The derrick was used to erect precast facing panels and to feed panels to a monorail placing system.

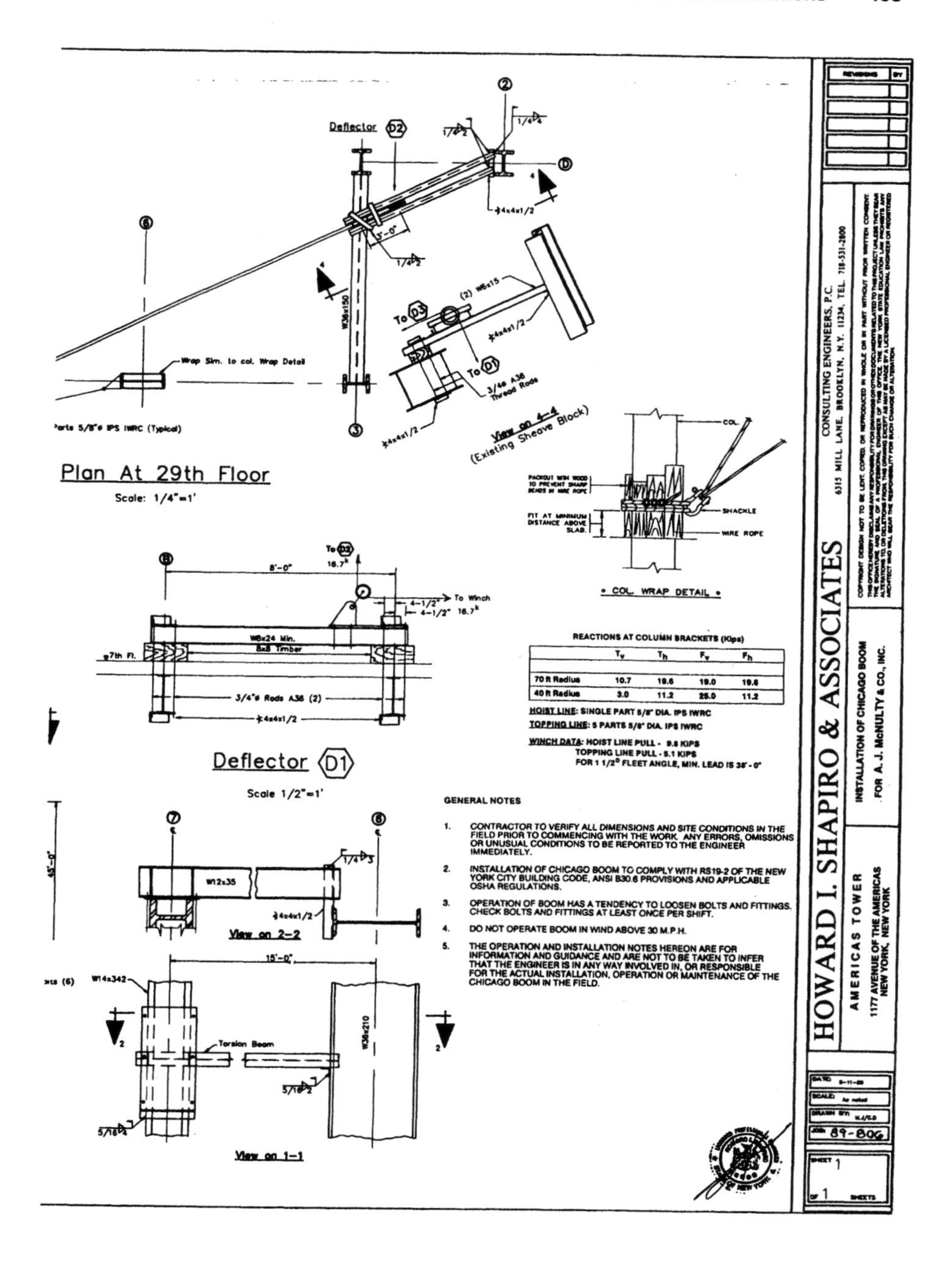

Deflector D2
1/4
1/4
3'-0"
W36x150
(2) W6x15
To D3
To D1
3/4ø A36 Thread Rods
4x4x1/2
Wrap Sim. to col. Wrap Detail
Parts 5/8"ø IPS IWRC (Typical)
View on 4-4 (Existing Sheave Block)
Plan At 29th Floor
Scale: 1/4"=1'
COL.
PACKOUT WITH WOOD TO PREVENT SHARP BENDS IN WIRE ROPE
FIT AT MINIMUM DISTANCE ABOVE SLAB.
SHACKLE
WIRE ROPE
COL. WRAP DETAIL
8'-0"
To D2
16.7k
4-1/2"
4-1/2"
To Winch
16.7k
W8x24 Min.
8x8 Timber
27th Fl.
3/4"ø Rods A36 (2)
Deflector D1
Scale 1/2"=1'
REACTIONS AT COLUMN BRACKETS (Kips)
Tv Th Fv Fh
70 ft Radius 10.7 19.6 19.0 19.6
40 ft Radius 3.0 11.2 25.0 11.2
HOIST LINE: SINGLE PART 5/8" DIA. IPS IWRC
TOPPING LINE: 5 PARTS 5/8" DIA. IPS IWRC
WINCH DATA: HOIST LINE PULL - 9.8 KIPS
TOPPING LINE PULL - 5.1 KIPS
FOR 1 1/2° FLEET ANGLE, MIN. LEAD IS 38'-0"
GENERAL NOTES
1. CONTRACTOR TO VERIFY ALL DIMENSIONS AND SITE CONDITIONS IN THE FIELD PRIOR TO COMMENCING WITH THE WORK. ANY ERRORS, OMISSIONS OR UNUSUAL CONDITIONS TO BE REPORTED TO THE ENGINEER IMMEDIATELY.
2. INSTALLATION OF CHICAGO BOOM TO COMPLY WITH RS19-2 OF THE NEW YORK CITY BUILDING CODE, ANSI B30.6 PROVISIONS AND APPLICABLE OSHA REGULATIONS.
3. OPERATION OF BOOM HAS A TENDENCY TO LOOSEN BOLTS AND FITTINGS. CHECK BOLTS AND FITTINGS AT LEAST ONCE PER SHIFT.
4. DO NOT OPERATE BOOM IN WIND ABOVE 30 M.P.H.
5. THE OPERATION AND INSTALLATION NOTES HEREON ARE FOR INFORMATION AND GUIDANCE AND ARE NOT TO BE TAKEN TO INFER THAT THE ENGINEER IS IN ANY WAY INVOLVED IN, OR RESPONSIBLE FOR THE ACTUAL INSTALLATION, OPERATION OR MAINTENANCE OF THE CHICAGO BOOM IN THE FIELD.
45'-0"
W12x35
View on 2-2
15'-0"
W14x342
Torsion Beam
W36x210
5/16
View on 1-1
HOWARD I. SHAPIRO & ASSOCIATES
CONSULTING ENGINEERS, P.C.
6315 MILL LANE, BROOKLYN, N.Y. 11234, TEL. 718-531-2800
INSTALLATION OF CHICAGO BOOM
FOR A. J. McNULTY & CO., INC.
AMERICAS TOWER
1177 AVENUE OF THE AMERICAS
NEW YORK, NEW YORK
JOB: 89-806
SHEET 1
OF 1 SHEETS

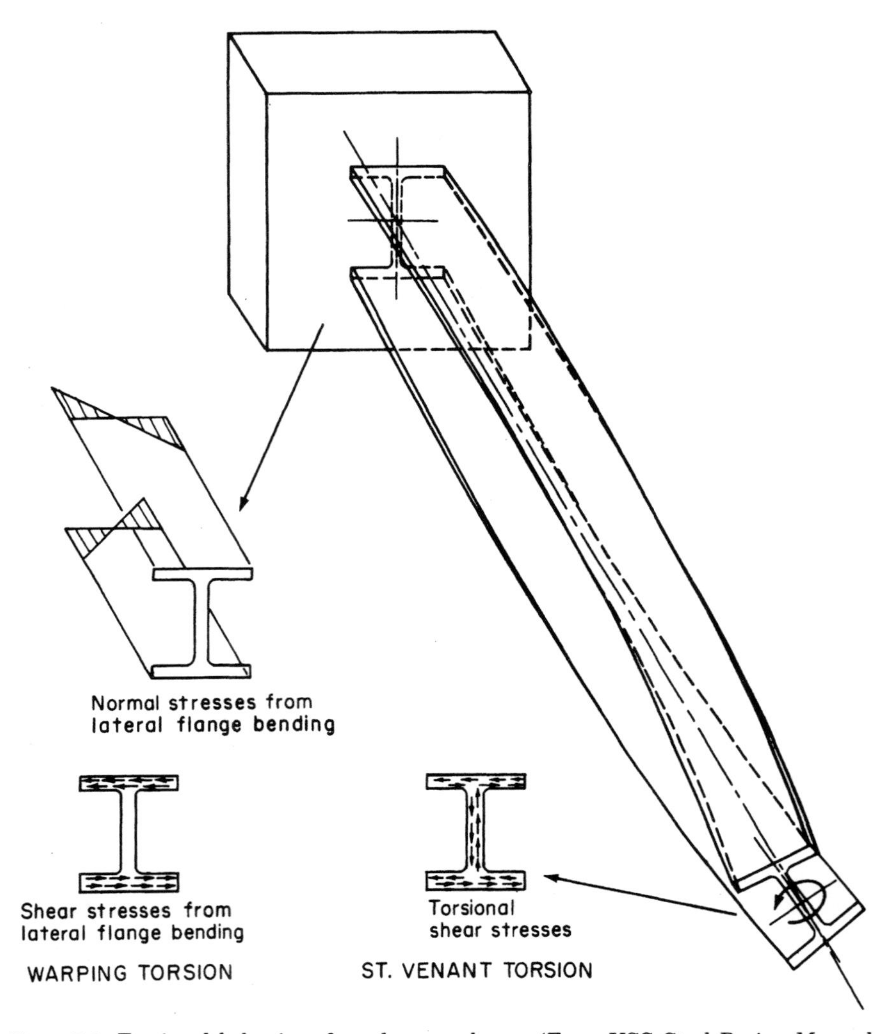

Figure 7.6 Torsional behavior of a column or beam. (*From USS Steel Design Manual ADUSS-27-3400-03, United States Steel Corp., May 1974.*)

Assuming full restraint at the ends of the column dictates that $\phi(0) = \phi'(0) = 0$, so that this expression becomes

$$\phi_1 = \frac{T_0}{ck}(kz - \sinh kz) + \frac{\phi''(0)}{k^2}(\cosh kz - 1) \qquad (f)$$

For the second segment, $d_1 \leq z \leq h$, the effect of applied moment T must be added, giving

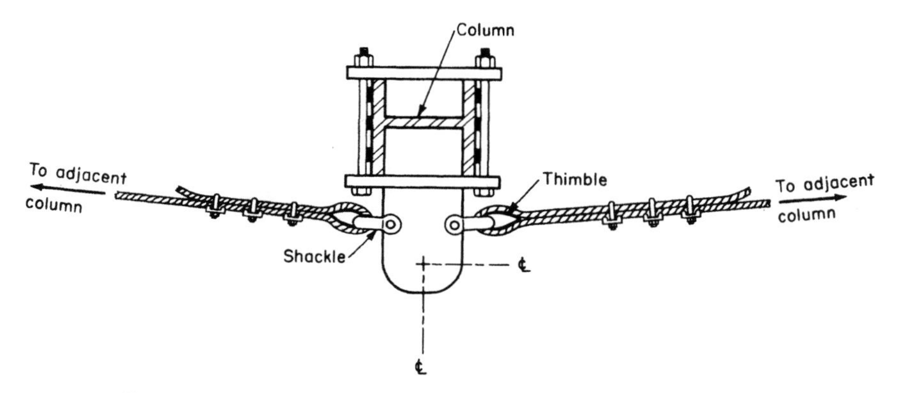

Figure 7.7 Torsion reducers on a Chicago boom column fitting.

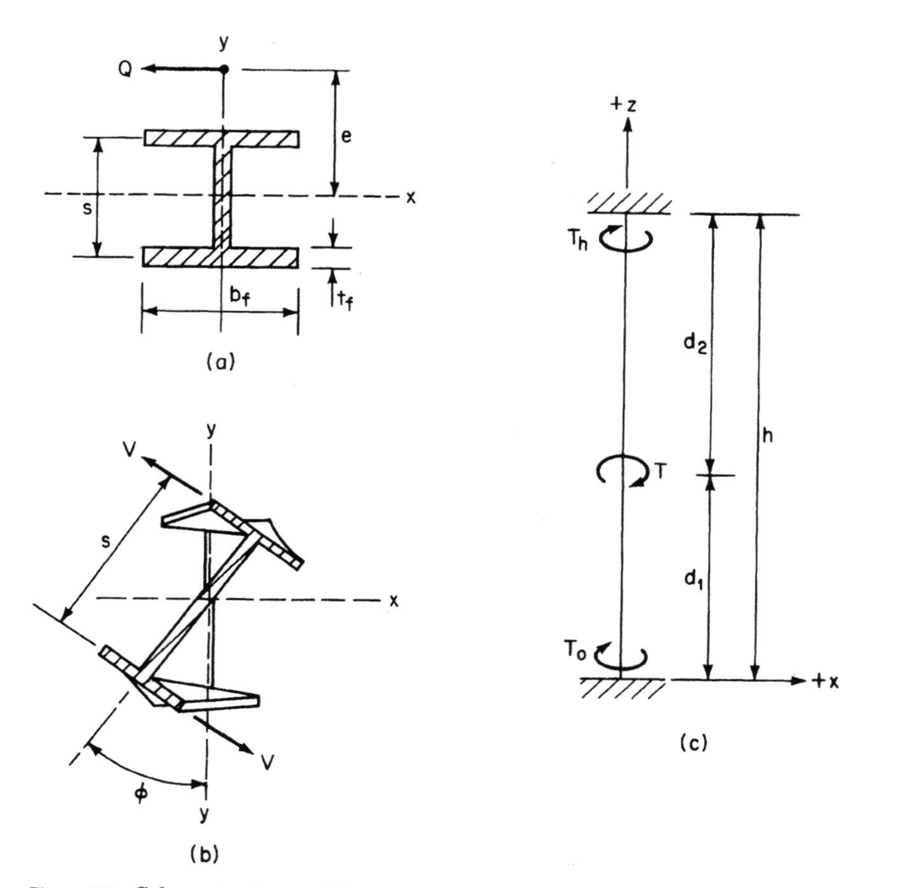

Figure 7.8 Column torsion model.

$$\phi_2 = \phi_1 + \frac{T}{ck}\,[k(z - d_1) - \sinh k(z - d_1)] \qquad (g)$$

Introducing the boundary condition at the upper end, $\phi_2(h) = 0$, and rearranging leads to

$$\phi''(0) = [T_0(\sinh kh - kh) + T(\sinh kd_2 - kd_2)]\,\frac{k/c}{\cosh kh - 1} \qquad (h)$$

The derivative of Eq. (*f*) is then inserted into the derivative of Eq. (*g*), and the boundary condition $\phi_2'(h) = 0$ is introduced. After rearranging this gives

$$\phi''(0) = [T_0(\cosh kh - 1) + T(\cosh kd_2 - 1)]\,\frac{k/c}{\sinh kh} \qquad (i)$$

Equating the two expressions for $\phi''(0)$ and solving for T_0, we have

$$T_0 = -uT$$

where
$$u = \frac{\cosh kd_2 - 1 + v(kd_2 - \sinh kd_2)}{\cosh kh - 1 + v(kh - \sinh kh)}$$

and
$$v = \frac{\sinh kh}{\cosh kh - 1} \qquad (7.1)$$

which, when substituted back into either Eq. (*h*) or (*i*), yields

$$\phi''(0) = \frac{Twk}{c}$$

with
$$w = \frac{\cosh kd_2 - 1}{\sinh kh} - \frac{u}{v} \qquad (7.2)$$

The entire torsional action has now been defined for a column with fixed ends, and the following expressions are readily drawn from Eqs. (*f*) and (*g*):

$0 \le z \le d_1$:

$$\phi_1 = \frac{T[w(\cosh kz - 1) - u(kz - \sinh kz)]}{ck}$$

$$\phi_1' = \frac{T[w \sinh kz - u(1 - \cosh kz)]}{c}$$

$$\phi_1'' = \frac{Tk(w \cosh kz + u \sinh kz)}{c}$$

$$\phi_1''' = \frac{Tk^2(w \sinh kz + u \cosh kz)}{c}$$

$d_1 \le z \le h$:

$$\phi_2 = \frac{T[w(\cosh kz - 1) - u(kz - \sinh kz) + k(z - d_1) - \sinh k(z - d_1)]}{ck}$$

$$\phi_2' = \frac{T[w \sinh kz - u(1 - \cosh kz) + 1 - \cosh k(z - d_1)]}{c}$$

$$\phi_2'' = \frac{Tk[w \cosh kz + u \sinh kz - \sinh k(z - d_1)]}{c}$$

$$\phi_2''' = \frac{Tk^2[w \sinh kz + u \cosh kz - \cosh k(z - d_1)]}{c}$$

(7.3)

Equations (7.3) are used to give values for each of the moments at any point within the column by referring back to the expressions given earlier

$$T_v = GJ\phi' \qquad M_f = \frac{c_1\phi''}{s} \qquad T_w = -c_1\phi'''$$

and, of course, ϕ is the angle of twist at any point. The stresses are then given by

$$\text{St. Venant shear stress} = f_v = \frac{T_v\delta t_f}{J} = G\phi'\delta t_f \tag{7.4}$$

$$\text{Flange bending stress} = f_t = \frac{M_f b_f}{2I_f} = \frac{Esb_f\phi''}{4} \tag{7.5}$$

$$\text{Nonuniform warping shear} = f_w = \frac{Vq}{I_f t_f} = \frac{Esb_f^2\phi'''}{16} \tag{7.6}$$

where δ is a stress coefficient given in either Fig. 7.9 or 7.10 and the shear modulus G can be taken as 11,000 kips/in^2 (75.8 GN/m^2). The notation q refers to the static moment of half of one flange area about the flange neutral axis, while J, the torsional constant, is approximately equal to the sum of the $bt^3/3$ values for each of the rectangular elements in an open section such as an *H*, *I*, channel, or angle member (when $b/t \ge 6$), but more accurate values for this parameter are given in the AISC manual† together with the warping constant $c_w = c_1/E$.

When the column ends are unrestrained, there will be no nonuniform warping and therefore no flange bending. The unrestrained case is ideal, however, and in actuality column ends will be partially to fully restrained.

---

†"Manual of Steel Construction," 9th ed., American Institute of Steel Construction, Inc., Chicago, 1989.

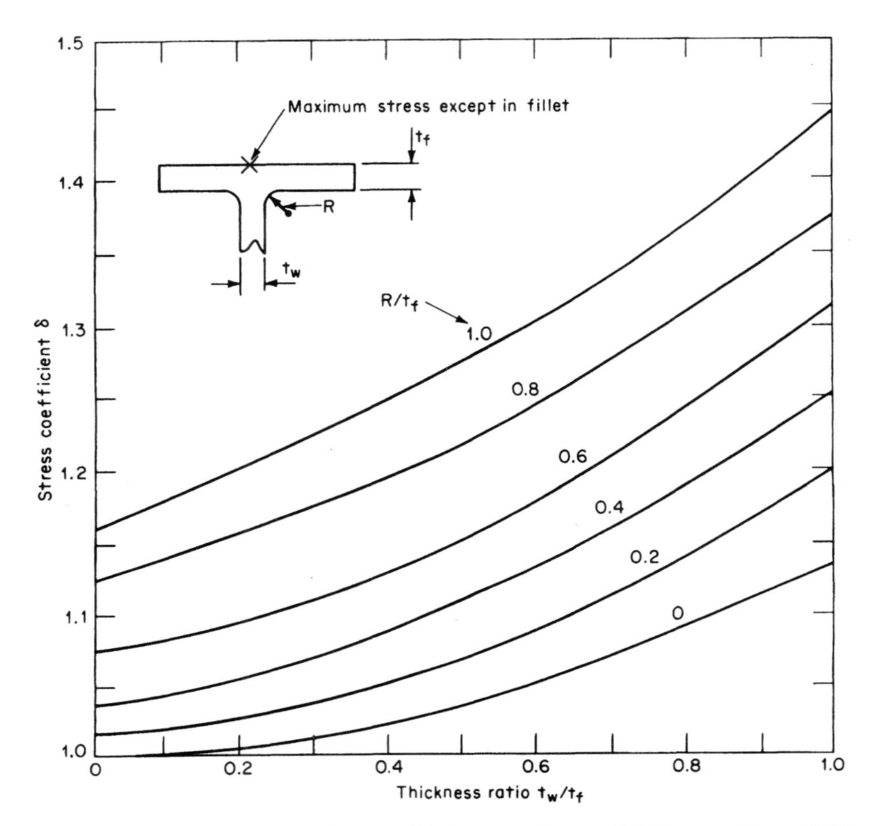

Figure 7.9 Stress coefficients for T or W shapes with parallel flanges. (*From USS Steel Design Manual ADUSS-27-3400-03, United States Steel Corp., May 1974.*)

In addition to the ordinary beam-column stress-interaction evaluation, it is also necesssary to sum the flange tip stresses at the ends of the column segment carrying the boom (and topping lift) mounting fittings and to monitor the strength margin against yield at those points.

At a restrained end, the St. Venant shear stress f_v is zero while the nonuniform warping shear stress f_w is maximum. As distance from the support increases f_w decreased and f_v rises.

If rod or rope torsion reducers are used, they will carry only a portion of the torsional load, the column taking the balance. The distribution will be such as to create consistent strains. Let Q_c be that portion of the torsion-inducing load Q that is carried by the column. Then Q_r will be the load induced in the reducer located at eccentricity r from the column center. The change in length of the loaded reducer is then

$$\Delta L = \frac{Q_r L_r}{A_r E_r}$$

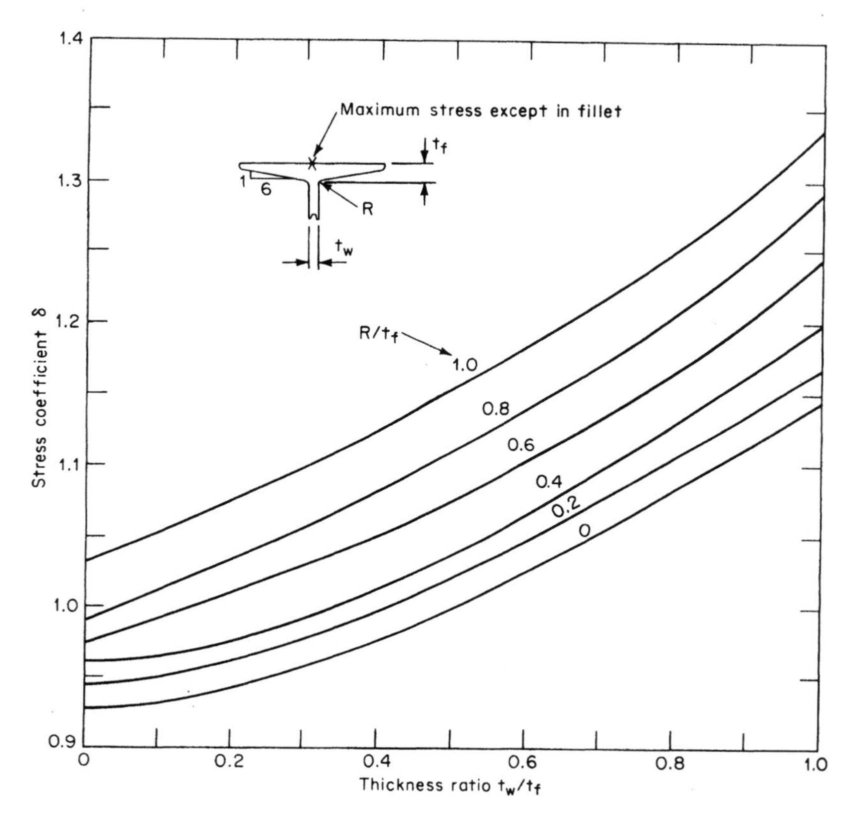

Figure 7.10 Stress coefficients for T or S shapes with sloping inner flanges. (*USS Steel Design Manual ADUSS-27-3400-03, United States Steel Corp., May 1974.*)

where L_r = length of reducer
A_r = metallic cross-sectional area
E_r = modulus of elasticity

To make the strains consistent

$$\Delta L = r\phi$$

but ϕ is the angle of rotation induced by $Q_c e$. For clarity, let $\phi_1(d_1)/T = m$; then

$$\Delta L = rQ_c em$$

The torsional moments will be in equilibrium if

$$Qe = Q_c e + Q_r r \qquad Q_c = Q - \frac{Q_r r}{e}$$

Substituting this expression for Q_c into the ΔL relationship gives

$$\Delta L = \left(Q - \frac{Q_r r}{e}\right) emr = \frac{Q_r L_r}{A_r E_r}$$

which when rearranged becomes

$$Q_r = \frac{Q}{L_r / emrA_r E_r + r/e} \tag{7.7}$$

The absolute value of parameter m is to be used in this equation.

With Q_r known, the rope or rod reducer can be checked for stress and Q_c can be found so that the column stresses can be evaluated. If column stresses are still excessive, A_r can be increased and the process repeated.

Example 7.1 (*a*) A W14 × 95 column section loaded in torsion has the following parameters (Fig. 7.8):

b_f = 14.545 in (369 mm)
s = 13.37 in (340 mm)
Q = 20 kips (89.0 kN)
d_1 = 24 in (610 mm)
J = 4.74 in^4 (197.3 cm^4)
E = 30,000 kips/in^2 (206.8 GN/m^2)
t_f = 0.748 in (19.0 mm)
e = 22.06 in (560 mm)
h = 150 in (3.81 m)
c_w = 17,200 in^6 (4.62 dm^6)
G = 11,000 kips/in^2 (75.8 GN/m^2)

Assuming fully restrained ends, calculate the maximum values for St. Venant and warping shear stresses and the flange bending stress induced by nonuniform warping.

First, a series of intermediate values are needed

$$T = Qe = 20(22.06) = 441.2 \text{ kip} \cdot \text{in } (49.85 \text{ kN} \cdot \text{m})$$
$$c_1 = c_w E = 17{,}200(30{,}000) = 5.16 \times 10^8 \text{ kip} \cdot \text{in}^4 \ (955.4 \text{ kN} \cdot \text{m}^4)$$
$$c = GJ = 11{,}000(4.74) = 52{,}140 \text{ kip} \cdot \text{in}^2 \ (149.6 \text{ kN} \cdot \text{m}^2)$$
$$k = \sqrt{\frac{c}{c_1}} = 0.01005 \text{ rad/in } (0.3958 \text{ rad/m})$$

$$\sinh kd_1 = 0.2436 \qquad \cosh kd_1 = 1.0292$$
$$\sinh kd_2 = 1.6334 \qquad \cosh kd_2 = 1.9152$$
$$\sinh kh = 2.1478 \qquad \cosh kh = 2.3692$$

From Eq. (7.1) $\quad v = 1.5687 \quad u = 0.9302$

From Eq. (7.2) $\quad w = -0.1669$

The St. Venant shear will be greatest at $z = d_1$ and is given by Eq. (7.4), but a value for ϕ' is needed first, and this is given by Eq. (7.3)

$$\phi_1'(d_1) = \frac{441.2[-0.1669(0.2436) - 0.9302(1 - 1.0292)]}{52{,}140}$$

$$= -\frac{5.9540}{52{,}140} \text{ rad/in} \left(-\frac{0.673}{149.6} \text{ rad/m}\right)$$

The stress coefficient δ is found from Fig. 7.9 after reference to additional section properties. For this W14 column section, $R = 0.60$ in (15.2 mm), so that $R/t_f = 0.80$ and $t_w = 0.465$ in (11.8 mm) for a t_w/t_f ratio of 0.62. δ is then read from the curves as 1.25. Now

$$f_v = 11{,}000\left(-\frac{5.954}{52{,}140}\right)(1.25)(0.748)$$

$$= -2.07 \text{ kips/in}^2 \ (-143 \text{ MN/m}^2) \qquad \textit{Ans.}$$

For the maximum flange bending stress, given by Eq. (7.5), ϕ'' must be evaluated at $z = 0$, or, by examination of Eq. (7.3),

$$\phi_1''(0) = \frac{Tkw}{c}$$

$$f_t = \frac{Esb_fTkw}{4c} = \frac{30{,}000(13.37)(14.545)(441.2)(0.01005)(-0.1669)}{4(52{,}140)}$$

$$= -20.7 \text{ kips/in}^2 \ (-143 \text{ MN/m}^2) \qquad \textit{Ans.}$$

Likewise, for warping shear ϕ''' needs to be evaluated at $z = 0$ before Eq. (7.6) can be solved. Again examining Eq. (7.3), we see that

$$\phi_1'''(0) = \frac{Tk^2u}{c}$$

$$f_w = \frac{Esb_f^2Tk^2u}{16c} = \frac{30{,}000(13.37)(14.545^2)(441.2)(0.01005^2)(0.9302)}{16(52{,}140)}$$

$$= 4.2 \text{ kips/in}^2 \ (29.1 \text{ MN/m}^2) \qquad \textit{Ans.}$$

(*b*) In order to reduce the somewhat high flange bending stress induced by the torsional load, torsion reducers are installed consisting of angle irons 6 by 6 by ½ in (152 by 152 by 12.7 mm) with the eccentricity of the torsion reducer from column centerline $r = 8.24$ in (209 mm), $L_r = 240$ in (6.1 m), $A_r = 5.75$ in^2 (37.10 cm^2) and, of course, $E_r = E$. What will be the tensile stress in the reducer and the new value of flange bending stress?

The parameter m in Eq. (7.7) is the absolute value of $\phi_1(d_1)/T$, or

$$m = \left| \frac{w(\cosh kd_1 - 1) - u(kd_1 - \sinh kd_1)}{ck} \right|$$

$$= \frac{0.2628}{52{,}140} \text{ kip} \cdot \text{in} \left(\frac{29.69}{149.6} \text{ N} \cdot \text{m} \right)$$

$$Q_r = \frac{20}{\dfrac{240(52{,}140)}{22.06(0.2628)(8.24)(5.75)(30{,}000)} + \dfrac{8.24}{22.06}}$$

$$= 10.57 \text{ kips } (47.0 \text{ kN})$$

The tensile stress in the reducer is then

$$f = \frac{10.57}{5.75} = 1.8 \text{ kips/in}^2 \ (12.7 \text{ MN/m}^2) \qquad \textit{Ans.}$$

$$Q_c = 20 - \frac{10.57(8.24)}{22.06} = 16.05 \text{ kips } (71.4 \text{ kN})$$

The remaining torsional flange bending stress is then

$$f_t = -\frac{20.7(16.05)}{20} = -16.6 \text{ kips/in}^2 \ (-114.5 \text{ MN/m}^2) \qquad \textit{Ans.}$$

(*c*) If r were increased to match the load eccentricity, e, using the same angle-iron reducer, what will be the effect?

$$Q_r = \frac{20}{\dfrac{240(52{,}140)}{22.06(0.2628)(22.06)(5.75)(30{,}000)} + 1}$$

$$= 12.76 \text{ kips } (56.8 \text{ kN})$$

$$f_t = \frac{20.7(20 - 12.76)}{20} = -7.5 \text{ kips/in}^2 \ (-51.7 \text{ MN/m}^2) \qquad \textit{Ans.}$$

Remedial measures

If lateral-torsion reducer rods, angles, or ropes are not practical, excessive effects of torsion can be materially reduced by adding plates to the column, making it a two-cell tubular section. Relatively thin plates are ordinarily all that will be required, but they must be welded into place and, depending upon architectural or other needs, may have to be removed later. Another method entails installation of a horizontal beam to transfer the torsional moment into a horizontal reaction at an adjacent column (Fig. 7.5). When this is done, the beam is positioned on the side of the derrick opposite to the load landing side so that it will not interfere with load handling.

Excessive column bending induced by a force parallel to the building face can be ameliorated by use of tension bars from the load point of the boom (or topping lift) carrying column to the ends of columns in adjacent bays. Judicious placement of the tension bars can provide the added benefit of having them act as torsion reducers at the same time.

If the bars will interfere with the landing of loads, a short steel compression member can be used instead. It is placed at a vertical angle between the column and a floor beam framing into it. The floor beam, of course, must be checked and may itself need reinforcing for the added loading from the diagonal strut. Similar means can be utilized when an excessive column moment occurs with the boom perpendicular to the building.

When the beam-to-column connections are not sufficiently rigid (or have not been completely made up) to restrain the couple formed by the horizontal reactions at boom foot and topping lift, a system of rod or wire-rope cross braces can be installed (Fig. 7.11). The braces must be designed for the loads in the couple, but if wire ropes are used, the effects of rope stretch on the distortion of the building frame must be considered. With either ropes or rods, turnbuckles are needed to remove slack from the system.

The attachment of load-carrying ropes to steel framing members must be arranged so that the sharp corners of the members do not

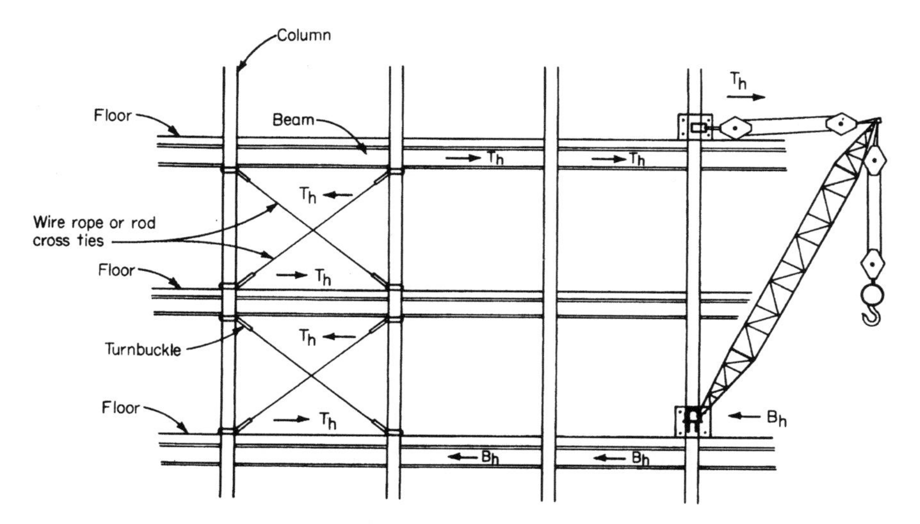

Figure 7.11 Wire-rope or steel-rod crossties can be used to stabilize a frame and prevent distortions from Chicago boom operations. The ties will carry all or a portion of the horizontal force component parallel to the building face, depending on the moment-transfer capabilities of the steel connections.

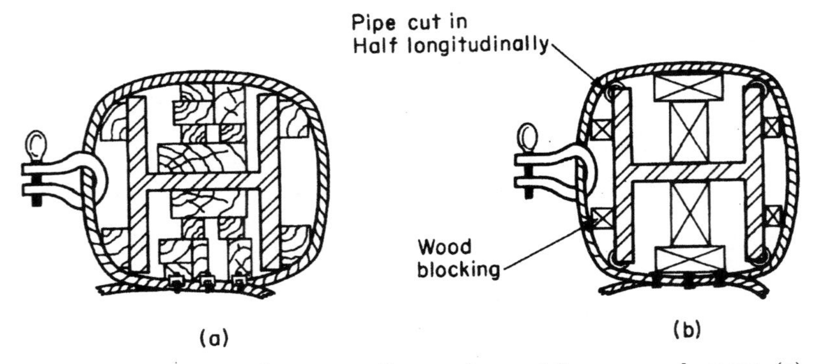

Figure 7.12 Beam or column rope ties can be used for rope anchorages. (*a*) Wood blocking is driven between steel and rope until the rope is tight and rope bends have been "softened." (*b*) An alternative means for corner "softening." When the first method is used, the spaces must be packed out fully both to tighten the rope and to soften the bends. With the second method, the packing is used only for tightening the rope.

shear the ropes. This can be accomplished with wood blocking, pipe segments, or similar material. When a wrap-and-shackle arrangement is used, as shown in Fig. 7.12, the angle that the rope wraps make with the shackle indicates a need to design the wraps for about twice the load in the line secured to the shackle.

Fitting attachment bolts

The bolts used to attach boom mounting fittings to a column have a tendency to work loose during the course of operations. It is essential that the bolts be checked and tightened each day before work begins. Nevertheless, loosening can cause the bolts to bend or distort, particularly if a fitting seat (Fig. 7.13*a*) is not provided. The bending will be a consequence of local overstresses resulting when the fitting plates slide on the column.

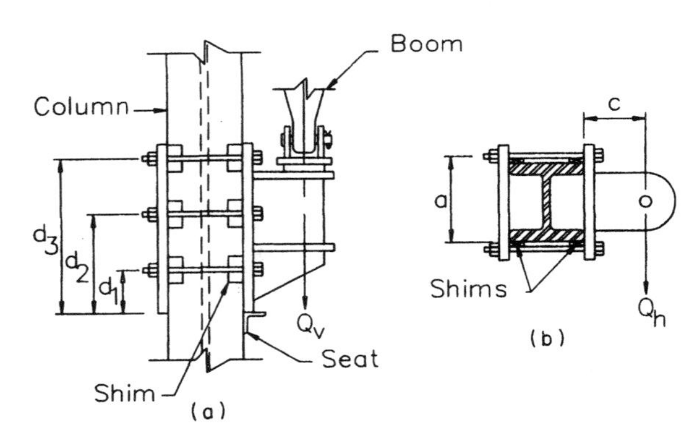

Figure 7.13

Using high-strength bolts, such as the ASTM A325 type (SAE grade 5 or ISO 8.8), can eliminate the problem when the bolts are installed properly. High-strength bolts are torqued to 70% of bolt tensile strength, but no danger to the bolts will result from higher preloads (see Bolting in Chap. 6).

To illustrate this bolted connection we assign the following values to the parameters in Fig. 7.13:

$$Q_v = Q_h = 40 \text{ kips } (18.1\ t) \qquad a = c = 15 \text{ in } (381 \text{ mm})$$

$$d_1 = 6 \text{ in } (152.4 \text{ mm}) \ d_2 = 12 \text{ in } (304.8 \text{ mm}) \ d_3 = 18 \text{ in } (457.2 \text{ mm})$$

From the vertical load moment, each bolt in the vertical lines of bolts can be assumed to be loaded in tension in proportion to bolt distance above the seat. This will make their strains consistent. If F_v is the load thus induced in each of the topmost bolts, then

$$2\left(F_v d_3 + \frac{F_v d_2^2}{d_3} + \frac{F_v d_1^2}{d_3}\right) = Q_v c$$

$$F_v = \frac{1}{2}\left(\frac{40(15)}{18 + 12^2/18 + 6^2/18}\right) = 10.71 \text{ kips } (47.6 \text{ kN})$$

The horizontal moment $Q_h c$ induces tension in each of the left side bolts of

$$F_h = \frac{Q_h c}{3a} = 40\,\frac{15}{3 \times 15} = 13.33 \text{ kips } (59.3 \text{ kN})$$

The upper left-hand bolt will then be the most heavily loaded, carrying tension of

$$13.33 + 10.71 = 24.04 \text{ kips } (106.9 \text{ kN})$$

The allowable working tensile stress on A325 bolts is 40,000 lb/in^2 (275 MN/m^2) on the nominal diameter of the bolt, while the permitted shear stress is 22,000 lb/in^2 (150 MN/m^2). The shear load will be the resultant of Q_v and Q_h, 40 $\sqrt{2}/6$ = 9.43 kips (41.9 kN) for each of the six bolts.

The bolt area A can now be found by using a stress-interaction relationship

$$\frac{9.43}{22A} + \frac{24.04}{40A} \leq 1 \qquad A \geq 1.03 \text{ in}^2 \ (664 \text{ mm}^2)$$

Thus, 1¼-in-diameter (32-mm) bolts are needed.

The proper preload torque for an A325 bolt of this size will induce initial bolt tension of at least 71,000 lb (316 kN) for each of the bolts. For the entire fitting, the net force of initial tension less the applied bolt loads, which vary with each bolt according to its position, will then be 343 kips (1526 kN). Assuming that the steel-to-steel coefficient of static friction will be 0.15, a conservative figure for building steelwork, the friction force developed will be about 51.5 kips (229 kN), which is just less than the resultant shear of $40\sqrt{2} = 56.6$ kips (252 kN). The seat and bolt wedges must then take that small excess. A much smaller preload would be applied to common bolts, and the excess of applied load over static-friction resistance would be very much greater.

The high preloads necessary when using A325 and similar bolts impose additional effects at Chicago boom anchorages. The plates comprising those anchorage fittings are subjected to bending moments at the column flange tips that are induced by the bolt forces. The plates must be thick enough to endure those bending stresses. Further, when those plates are mounted parallel to the column flanges, the plates together with the flanges must be stiff enough to prevent excessive bending about the column web. Excessive bending may be prevented, however, by wedging wood blocks between the tips of the flanges; stiffener plates would be better yet.

7.3 Guy Derricks

To an installation designer, the two most important aspects of a guy derrick are support of the mast loads and the selection and anchorage of the guys. The guys must be designed first, as their loadings contribute to the mast load. The positioning and anchoring of the guys are an important consideration in selecting guy derrick location in buildings under construction.

A guy derrick mast is seated on a base, or *footblock,* with a pivot arrangement that permits the mast to swing with the boom. The pivot is usually of the ball-and-socket type, which is necessary in order to allow the mast to lean freely in the direction of loading. It is the freedom of the mast to lean that makes the guying system for this type of derrick far less critical, in a design sense, than the mast guys of tower cranes, discussed previously. The mast of a guy derrick does not support any portion of the overturning forces by means of bending resistance. It is therefore not necessary to employ guy-force-measuring devices, but erection crews should still try to plumb the mast as closely as possible.

Guying systems

For most practical guy derrick installations, the number of guys used and their placement is based solely on site and operating conditions (Fig. 7.14). Although six or eight guys are usual, fewer guys can be used whenever necessary provided the design adequately reflects the actual conditions, stability is maintained and field personnel understand and implement operating limitations.

For any operating position of the boom, guys that lie within 90° of either side of the boom in a horizontal circle are called *dead guys,* as they do not contribute to the support of applied loads. Guys within 75° of either side of the boom are taken, however, as imposing dead-load moment to the system.

Guys in the remaining 180° of arc are called *live guys,* and those lying within the 150° zone opposite the boom are considered to be effective, i.e., participating, in the support of the mast (Fig. 7.15).

When the derrick is loaded, a horizontal force is induced at the mast top at the topping lift connection. The live guys will then become loaded in reaction to this force. The radiating guys will distribute the reaction among themselves in such manner that the transverse forces will be in equilibrium and the geometric displacements will be consis-

Figure 7.14 One of the guy derricks shown is mounted on a mobile-crane base, and the other is a standard derrick form. The congested site has made it necessary to place the guy anchorages far from the bases, some across the river. Flat guy angles, however, permit swinging at a large radius without hitting the guys.

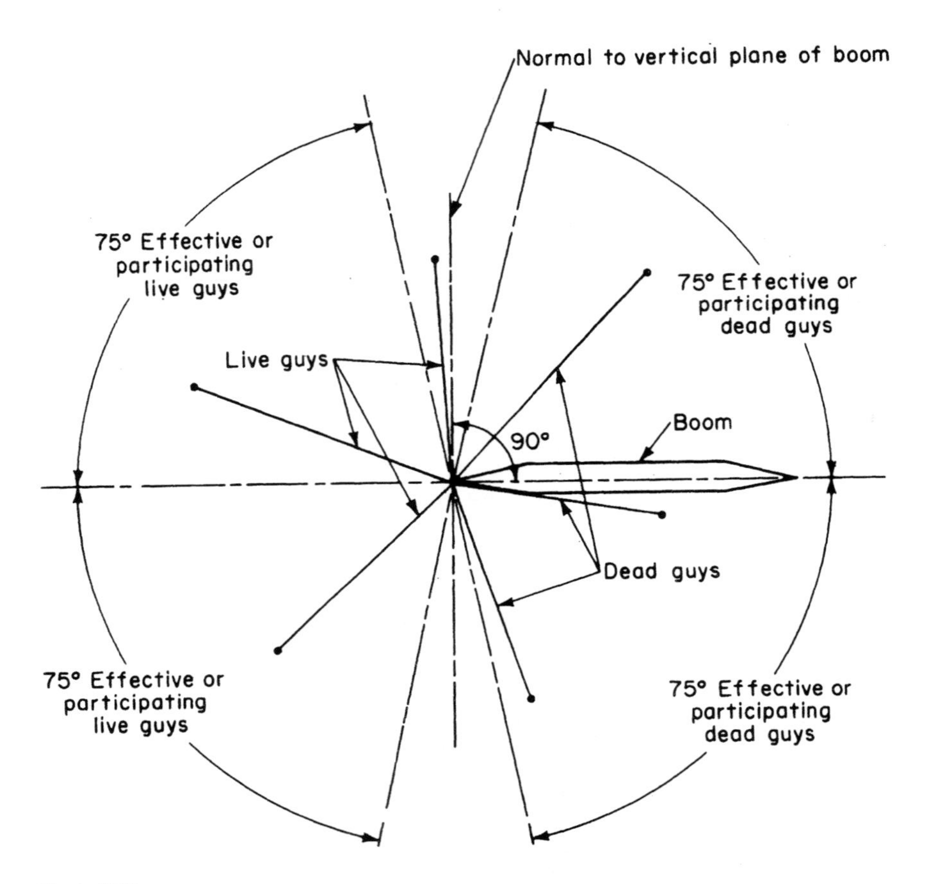

Figure 7.15

tent with rope strains and catenary sag changes. In doing this, there is not assurance that the mast top will displace directly toward the boom. On the contrary, it can be expected that some transverse displacement will also occur.

A transverse displacement of the mast top will be accompanied by a mast dead-load moment in that direction and some live-load effects as well. With small displacement (and *in practice* displacements will be small), the induced moment will not attain an important magnitude, and consequently the loads added to the opposite guys will not be great.

We do not intend to imply that design of the guys should be loosely treated, but it is a fact that this type of guying system is not very sensitive. Therefore, the practice in the United States is to use the

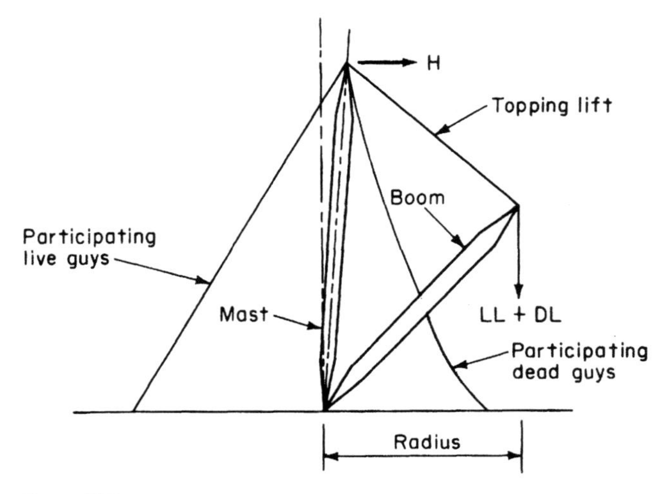

Figure 7.16

approximate AISC† method, which has proved to be adequate. In this method, the guys themselves are designed with a design factor of 3 against breaking load with the derrick assumed to be exposed to *static loading* only.

For derricks with tall masts, guy loadings should be explored under live plus impact loading in conjunction with wind at 30 mi/h (13.4 m/s). There are no rope design factor criteria for this case (in United States codes), so designers are left to use their own judgement. Guy anchorages, however, should be designed to withstand guy loads including wind and impact, and normal design criteria should be used.

The load applied to the participating live guys is a horizontal force H at the guy connection level of the mast (Fig. 7.16). The force is taken in the direction of the boom and comprises two parts. The first is induced by the dead loads of the boom and topping lift, along with the lifted-load and hoist-rope forces, while the second is an empirical quantity reflecting the effect of the participating dead guys. This dead load component of guy force can be taken as

$$H_g = \begin{cases} 0.006N(L_a - 12) & \text{rope diam} \leq 1 \text{ in} \\ 0.087N(L_a - 3.66) & \text{rope diam} \leq 25 \text{ mm} \\ 0.012N(L_a - 12) & \text{rope diam} > 1 \text{ in} \\ 0.0175N(L_a - 3.66) & \text{rope diam} > 25 \text{ mm} \end{cases} \tag{7.8}$$

†"Guide for the Analysis of Guy and Stiffleg Derricks," American Institute of Steel Construction, Inc., New York, 1974, p. 15.

where N is the number of effective dead guys and L_a is the average horizontal projection of the effective dead-guy lengths in a direction parallel to the boom in feet or meters. The resulting value for H_g will be in kips or kilonewtons in the respective formulas for inch and millimeter size ropes.

The load H must then be amplified to account for the small moment that could be added by a transverse lean of the mast, as mentioned earlier. The AISC suggests an increase of 5%, or a multiplier of 1.05, for derricks used in steel erection service. That same factor can be realistically taken for most installations.

If each of the live-guy loads, taken in the direction of the guy-rope centerlines, is denoted G_1, G_2, G_3, and G_4 and the guy slope lengths (actual length) as L_1, L_2, L_3, and L_4 with nominal rope cross-sectional areas A_1, A_2, A_3, and A_4, the following expressions can be applied when from one to four guys are effective:

With only one guy in effective live-guy area:

$$G_1 = \frac{HL_1}{x_1}$$

With two guys in effective live-guy area:

$$G_1 = \frac{1.05H}{x_1/L_1 + (L_1^2/A_1x_1)(A_2x_2^2/L_2^3)}$$

$$G_2 = \frac{1.05H}{x_2/L_2 + (L_2^2/A_2x_2)(A_1x_1^2/L_1^3)}$$

With three guys in effective live-guy area:

$$G_1 = \frac{1.05H}{x_1/L_1 + (L_1^2/A_1x_1)(A_2x_2^2/L_2^3 + A_3x_3^2/L_3^3)}$$

$$G_2 = \frac{1.05H}{x_2/L_2 + (L_2^2/A_2x_2)(A_1x_1^2/L_1^3 + A_3x_3^2/L_3^3)}$$

$$G_3 = \frac{1.05H}{x_3/L_3 + (L_3^2/A_3x_3)(A_1x_1^2/L_1^3 + A_2x_2^2/L_2^3)}$$

With four guys in effective live-guy area:

$$G_1 = \frac{1.05H}{x_1/L_1 + (L_1^2/A_1x_1)(A_2x_2^2/L_2^3 + A_3x_3^2/L_3^3) + A_4x_4^2/L_4^3)}$$

$$G_2 = \frac{1.05H}{x_2/L_2 + (L_2^2/A_2x_2)(A_1x_1^2/L_1^3 + A_3x_3^2/L_3^3 + A_4x_4^2/L_4^3)} \tag{7.9}$$

$$G_3 = \frac{1.05H}{x_3/L_3 + (L_3^2/A_3x_3)(A_1x_1^2/L_1^3 + A_2x_2^2/L_2^3 + A_4x_4^2/L_4^3)}$$

$$G_4 = \frac{1.05H}{x_4/L_4 + (L_4^2/A_4x_4)(A_1x_1^2/L_1^3 + A_2x_2^2/L_2^3 + A_3x_3^2/L_3^3)}$$

where x_1, x_2, x_3, and x_4 are the horizontal projections of the guys in a direction parallel to the boom (Fig. 7.17). Inasmuch as Eq. (7.9) contains only ratios of areas, nominal as opposed to actual rope metallic cross-sectional areas can be used.

When the AISC method is used, it will generally be necessary to evaluate guy loadings under several conditions. Normally, as is usual

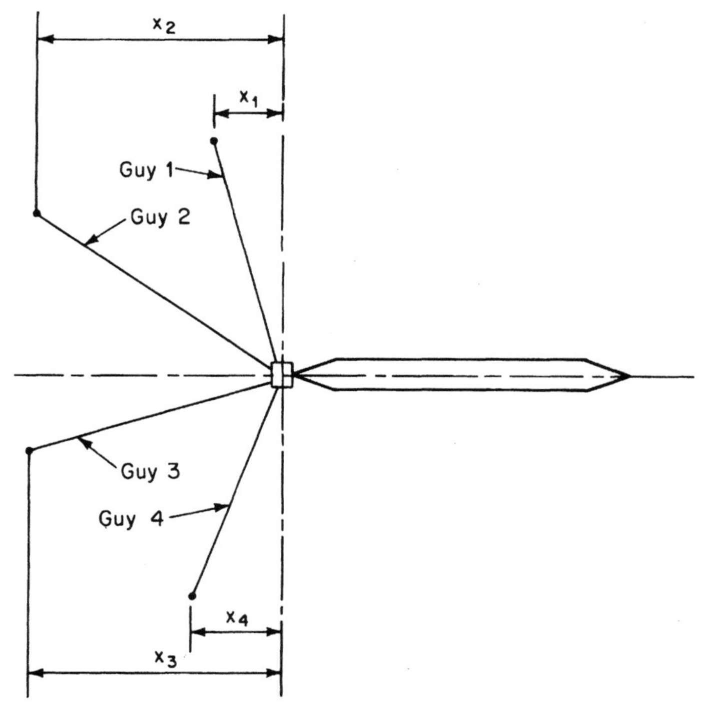

Figure 7.17

in steel erection work, the most severe loading will take place during initial lifting of loads from ground level truck unloading points to the working level. Each active guy must be evaluated for this case. Next, after booming in, the load will be swung to a landing position. The swing radius during operation must be rather small to permit the boom to pass under the guys, so it is unlikely that critical or governing loads will develop here, but if they do, each guy is checked when the boom is opposite to the guy in question. Finally, maximum conditions during load placement may well induce maximum loading on particular guys. Each guy is then designed for its greatest calculated load.

As a start, it is convenient to assume that each guy in the set will be of the same cross-sectional area; this permits the area parameter to drop out in Eq. (7.9). Once loads have been determined on that basis, guy ropes can be tentatively selected and the calculations repeated to check the ropes chosen.

The increase in loading due to impact must be allowed for in the design; the value chosen will be a matter for the designer's judgment after considering the type of work to be performed and the working conditions. For steel erection work, the AISC suggests 20% with a properly matched winch. Higher impact forces can occur when the winch is capable of much greater line pull than needed. Impact is used in the design of the guy anchorages and of the mast support but not of the guy ropes themselves, unless unusually high impact forces are expected.

Example 7.2 With the boom in the position shown in the guying system of Fig. 7.18, guys 2 and 3 receive their heaviest loading. In this position the horizontal reaction at the mast top induced by the boom and load is 55 kips (244.7 kN). Assuming that all dead guys will be greater than 1 in (25 mm) in diameter, design guys 2 and 3 using 6 × 19 class IWRC IPS wire rope. Also assume that guy 1 will be 1⅛ in (29 mm) in diameter.

Guys 4, 5, and 6 are each participating dead guys, so that parameter L_a of Eq. 7.8) is

$$L_a = \frac{32.4 + 141.0 + 70.7}{3} = 81.4 \text{ ft (24.8 m)}$$

The dead-load effect of the dead guys is then

$$H_g = 0.012[81.4 - 12)] = 2.5 \text{ kips (11.1 kN)}$$

making the total horizontal reaction at mast top

$$H = 55.0 + 2.5 = 57.5 \text{ kips (255.8 kN)}$$

For a first trial it will be assumed that all participating live guys will be

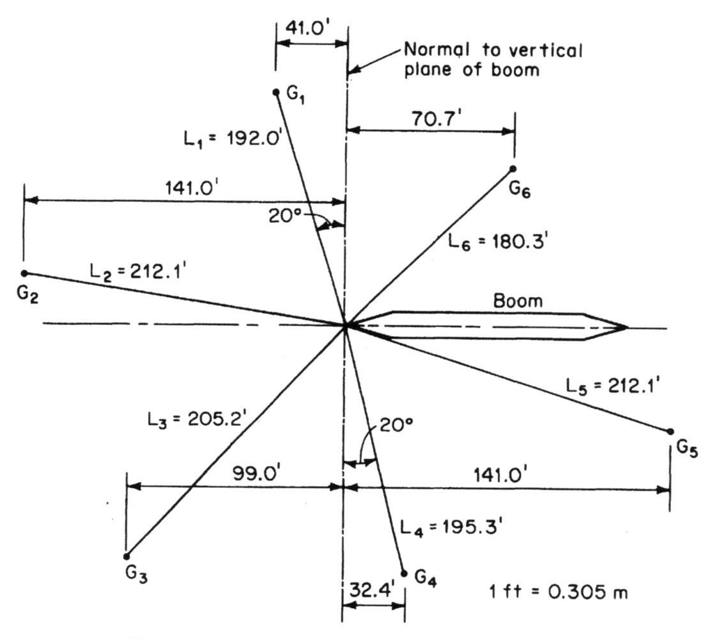

Figure 7.18 Example 7.2.

1⅛ in (29 mm) in diameter in order to simplify Eq. (7.9). Using only the needed expressions for three effective guys, we have

$$G_2 = \frac{1.05(57.5)}{\dfrac{141.0}{121.1} + \dfrac{212.1^2}{141.0}\left(\dfrac{41.0^2}{192.0^3} + \dfrac{99.0^2}{205.2^3}\right)} = 54.8 \text{ kips (244 kN)}$$

$$G_3 = \frac{1.05(57.5)}{\dfrac{99.0}{205.2} + \dfrac{205.2^2}{99.0}\left(\dfrac{41.0^2}{192.0^3} + \dfrac{141.0^2}{212.1^3}\right)} = 41.1 \text{ kips (183 kN)}$$

The break strengths required for ropes 2 and 3 are then

$$BS(2) = 54.8(3/2) = 82.2 \text{ tons (74.6 t)}$$

$$BS(3) = 41.1(3/2) = 61.7 \text{ tons (56.0 t)}$$

For guy 2 we tentatively choose a rope of 1⅜ in (35 mm) diameter and break strength of 87.1 tons (79.0 t), while for number 3 a 1¼-in-diameter (32-mm) rope of 70.4 tons (63.9 t) break strength will be tried. Therefore, A_2 = 1.49 in^2 (961 mm^2), A_3 = 1.23 in^2 (794 mm^2), and guy 1 was given as 1⅛ in (29 mm) diameter, or A_1 = 0.99 in^2 (639 mm^2) and break strength

57.0 tons (51.7 t). Inserting these values into Eq. (7.9) for the two ropes to be designed gives

$$G_2 = \frac{60.4}{\dfrac{141.0}{212.1} + \dfrac{212.1^2}{1.49(141.0)}\left[\dfrac{0.99(41.0^2)}{192.0^3} + \dfrac{1.23(99.0^2)}{205.2^3}\right]}$$

$$= 59.6 \text{ kips (265 kN)}$$

$$G_3 = \frac{60.4}{\dfrac{99.0}{205.2} + \dfrac{205.2^2}{1.23(99.0)}\left[\dfrac{0.99(41.0^2)}{192.0^3} + \dfrac{1.49(141.0^2)}{212.1^3}\right]}$$

$$= 36.9 \text{ kips (164 kN)}$$

For the revised loads, the required break strengths are

$$\text{BS}(2) = 59.6(\tfrac{3}{2}) = 89.4 \text{ tons (81.1 t)}$$

$$\text{BS}(3) = 36.9(\tfrac{3}{2}) = 55.4 \text{ tons (50.3 t)}$$

Guy 2 must be changed, as the rope previously selected lacks sufficient strength. When we substitute a rope of 1½ in (38 mm) diameter with break strength of 103 tons (93.4 t), A_2 becomes 1.77 in^2 (1140 mm^2). But guy 3 has excessive strength, and a smaller rope, 1⅛ in (29 mm) diameter, can be tried. Rechecking in Eq. (7.9), we get

$$G_2 = 66.4 \text{ kips (295 kN)} \qquad \text{BS(2)} = 99.6 \text{ tons (90.4 t)}$$

$$G_3 = 27.9 \text{ kips (124 kN)} \qquad \text{BS(3)} = 41.9 \text{ tons (38.0 t)}$$

Guy 2 will be 1½ in (38 mm) in diameter, and guy 3 will be 1⅛ in (29 mm) in Diameter. *Ans.*

Figure 7.19*b* shows that selecting guy positions can be quite difficult in the real world of the jobsite.

Footblock supports

Footblock loading consists mainly of the vertical reaction placed on it by the base of the mast. This reaction will include total derrick dead and live load, including impact, plus the vertical components of the forces in the guys (with impact added).

Because of the eccentricities of the connections of the guys and the topping lift to the mast as well as the offset boom foot mounting, there will be bending moments in the mast during most conditions of operation. This implies a horizontal reaction at the ball-and-socket connection with the base. Thus a base moment is present and must be transmitted to the host structure along with the horizontal shear and vertical force. Additionally, when the hoist and topping leads lines run out of the footblock to the winch at base level, another increment is added to the horizontal force at the footblock fastenings and a small moment as well.

Finally, the derrick can be slewed by means of a swinger winch and bull wheel, a manually operated crank and gear set, or other such mechanism. In either case, in calm air only friction forces and the effects of mast lean under load need to be overcome. But when wind is present, the swing force materially increases and may add enough horizontal torque to require the added shear forces at the anchorage bolts to be considered. When a winch is used for slewing, the line pull is an added horizontal force.

With all the footblock loads determined, the design of the anchorage and supports follows ordinary structural procedures. As shown in Fig. 7.19*a*, it is sometimes necessary to shore down several floors in order to distribute mast loads to within framing member capacity. In addition to stress analysis, elastic deflections in the supports may need to be checked. Excessive deflections can cause the derrick to behave as though the guys were too slack inducing the mast to experience added lean, with added guy loading and increased swing forces as a consequence.

Guy derricks that are used for steel erection are raised, or *jumped*, as the framework of the structure rises. The derrick can be used to raise itself unassisted, taking advantage of the fact that the boom length is limited so that it can pass under the guys and permit full-circle slewing.

First, the boom is brought to a vertical position and held by the topping lift while the boom foot pin is removed. The boom foot is then swung clear of the mast base, rotated 180°, and lowered onto a footplate that has been anchored to the structure in advance. A few temporary guys are made fast to the boom head, and the boom is leaned in toward the mast. By rotating the boom, the load fall has been placed on the mast side and the boom has been converted into a gin pole.

A lashing is placed about the mast at a point above its CG (including the footblock), and the hook of the gin pole is then attached. When the mast guys have been released, the mast is raised, together with its footblock, to the next level, and support beams are slid into place. The mast guys are then reset at the new level.

Finally, using the topping lift to support the boom head, the boom guys are released and the boom is rotated and then raised. With the boom foot pin reinstalled, the guy derrick is ready for service at the top of the steel.

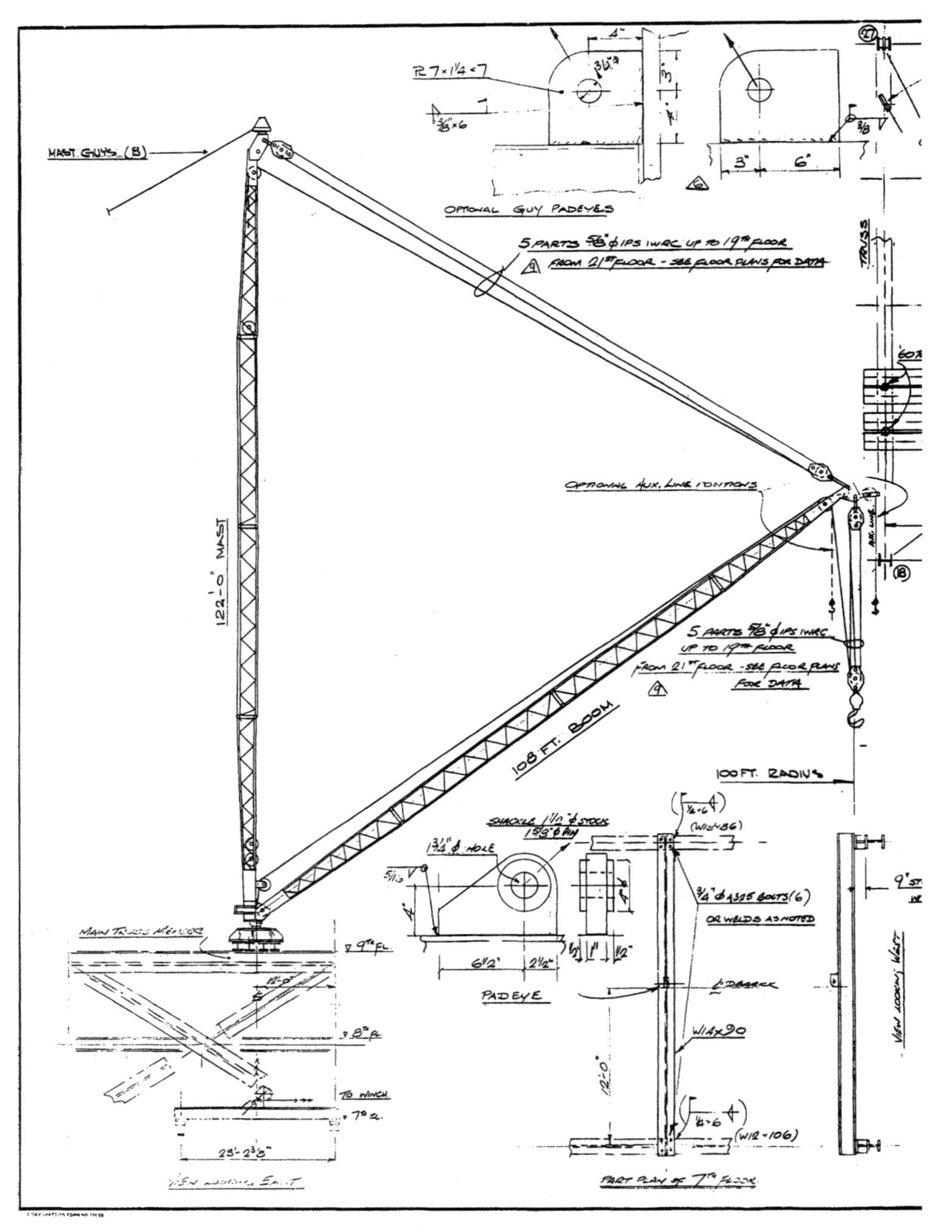

Figure 7.19 (*a*): One of set of drawings for two guy derricks used to erect the steel frame for an office building. Note the shoring needed to under the jumping beams supporting the mast.

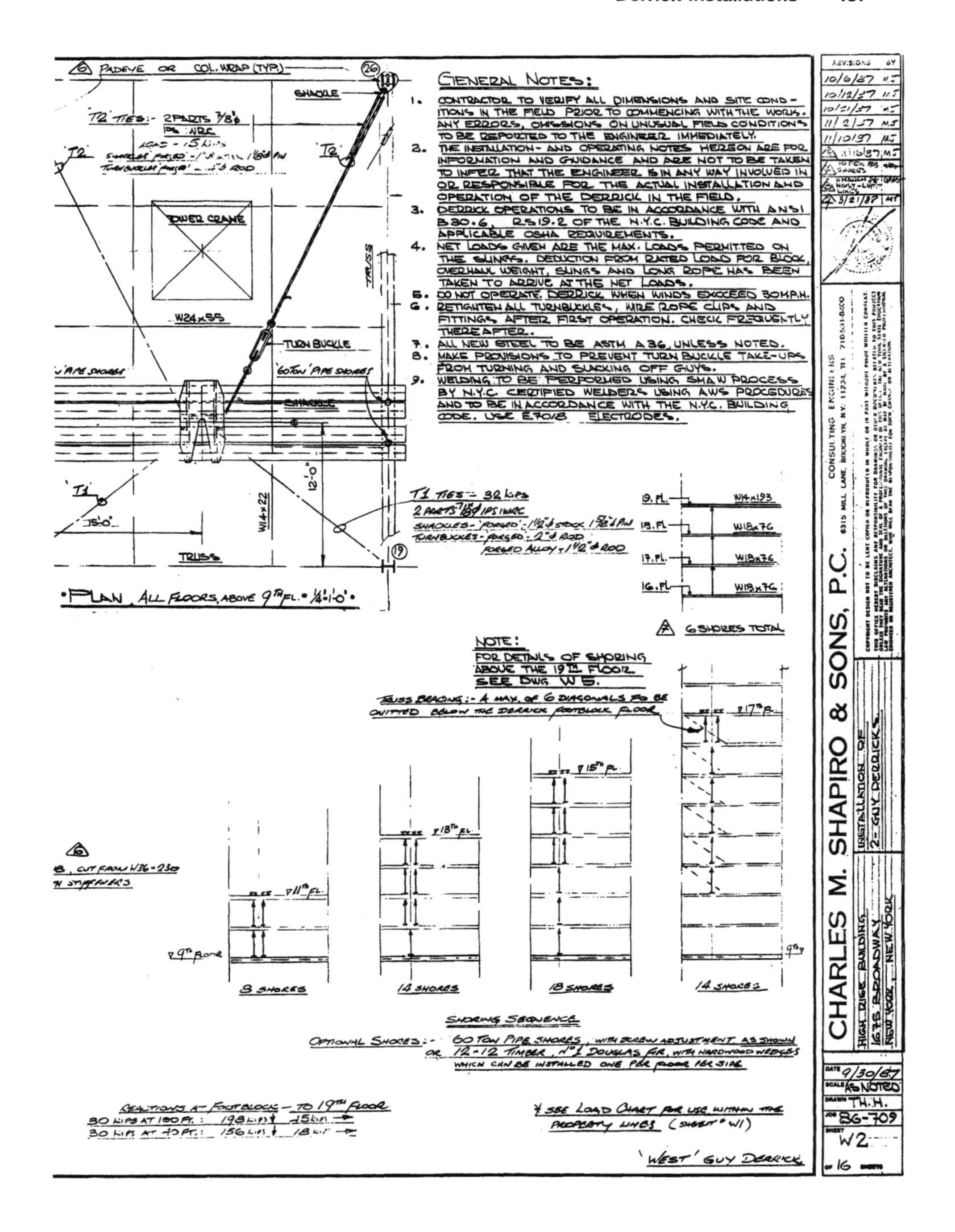
GENERAL NOTES:
1. CONTRACTOR TO VERIFY ALL DIMENSIONS AND SITE CONDITIONS IN THE FIELD PRIOR TO COMMENCING WITH THE WORK. ANY ERRORS, OMISSIONS OR UNUSUAL FIELD CONDITIONS TO BE REPORTED TO THE ENGINEER IMMEDIATELY.
2. THE INSTALLATION- AND OPERATING NOTES HEREON ARE FOR INFORMATION AND GUIDANCE AND ARE NOT TO BE TAKEN TO INFER THAT THE ENGINEER IS IN ANY WAY INVOLVED IN OR RESPONSIBLE FOR THE ACTUAL INSTALLATION AND OPERATION OF THE DERRICK IN THE FIELD.
3. DERRICK OPERATIONS TO BE IN ACCORDANCE WITH ANSI B30.6, RS19.2 OF THE N.Y.C. BUILDING CODE AND APPLICABLE OSHA REQUIREMENTS.
4. NET LOADS GIVEN ARE THE MAX. LOADS PERMITTED ON THE SLINGS. DEDUCTION FROM RATED LOAD FOR BLOCK, OVERHAUL WEIGHT, SLINGS AND LONG ROPE HAS BEEN TAKEN TO ARRIVE AT THE NET LOADS.
5. DO NOT OPERATE DERRICK WHEN WINDS EXCEED 30M.P.H.
6. RETIGHTEN ALL TURNBUCKLES, WIRE ROPE CLIPS AND FITTINGS AFTER FIRST OPERATION. CHECK FREQUENTLY THEREAFTER.
7. ALL NEW STEEL TO BE ASTM A36, UNLESS NOTED.
8. MAKE PROVISIONS TO PREVENT TURNBUCKLE TAKE-UPS FROM TURNING AND SLACKING OFF GUYS.
9. WELDING TO BE PERFORMED USING SMAW PROCESS BY N.Y.C. CERTIFIED WELDERS USING AWS PROCEDURES AND TO BE IN ACCORDANCE WITH THE N.Y.C. BUILDING CODE. USE E70I8 ELECTRODES.
PADEYE OR COL. WRAP (TYP.)
SHACKLE
T2 TIES: 2 PARTS 7/8"φ IPS IWRC
T2
TOWER CRANE
TRUSS
W24x55
TURNBUCKLE
60TON PIPE SHORES
SHACKLE
12'-0"
W14x22
T1
15'-0"
TRUSS
T1 TIES = 32 KIPS
2 PARTS 1 1/8"φ IPS IWRC
FORGED ALLOY 1 1/2"φ ROD
19.FL.
18.FL.
17.FL.
16.FL.
W14x193
W18x76
W18x76
W18x76
PLAN. ALL FLOORS ABOVE 9TH FL. = 1/4"=1'-0"
6 SHORES TOTAL
NOTE:
FOR DETAILS OF SHORING ABOVE THE 19TH FLOOR SEE DWG W5.
TRUSS BRACING: A MAX. OF 6 DIAGONALS TO BE OMITTED BELOW THE DERRICK FOOTBLOCK FLOOR
17TH FL.
15TH FL.
13TH FL.
11TH FL.
9TH FLOOR
8 SHORES
14 SHORES
18 SHORES
14 SHORES
SHORING SEQUENCE
OPTIONAL SHORES: - 60 TON PIPE SHORES, WITH SCREW ADJUSTMENT AS SHOWN OR 12x12 TIMBER, NO1 DOUGLAS FIR, WITH HARDWOOD WEDGES WHICH CAN BE INSTALLED ONE PER FLOOR PER SIDE
REACTIONS AT FOOTBLOCK - TO 19TH FLOOR
30 KIPS AT 100 FT.: 198 KIPS 15 KIPS
30 KIPS AT 70 FT.: 156 KIPS 18 KIPS
* SEE LOAD CHART FOR USE WITHIN THE PROPERTY LINES (SHEET W1)
'WEST' GUY DERRICK
CHARLES M. SHAPIRO & SONS, P.C.
CONSULTING ENGINEERS
6315 MILL LANE, BROOKLYN, NY 11234
HIGH RISE BUILDING
1675 BROADWAY
NEW YORK NEW YORK
INSTALLATION OF 2-GUY DERRICKS
REVISIONS BY
DATE 9/30/87
SCALE AS NOTED
DRAWN T.H.H.
JOB 86-709
SHEET W2
OF 16 SHEETS

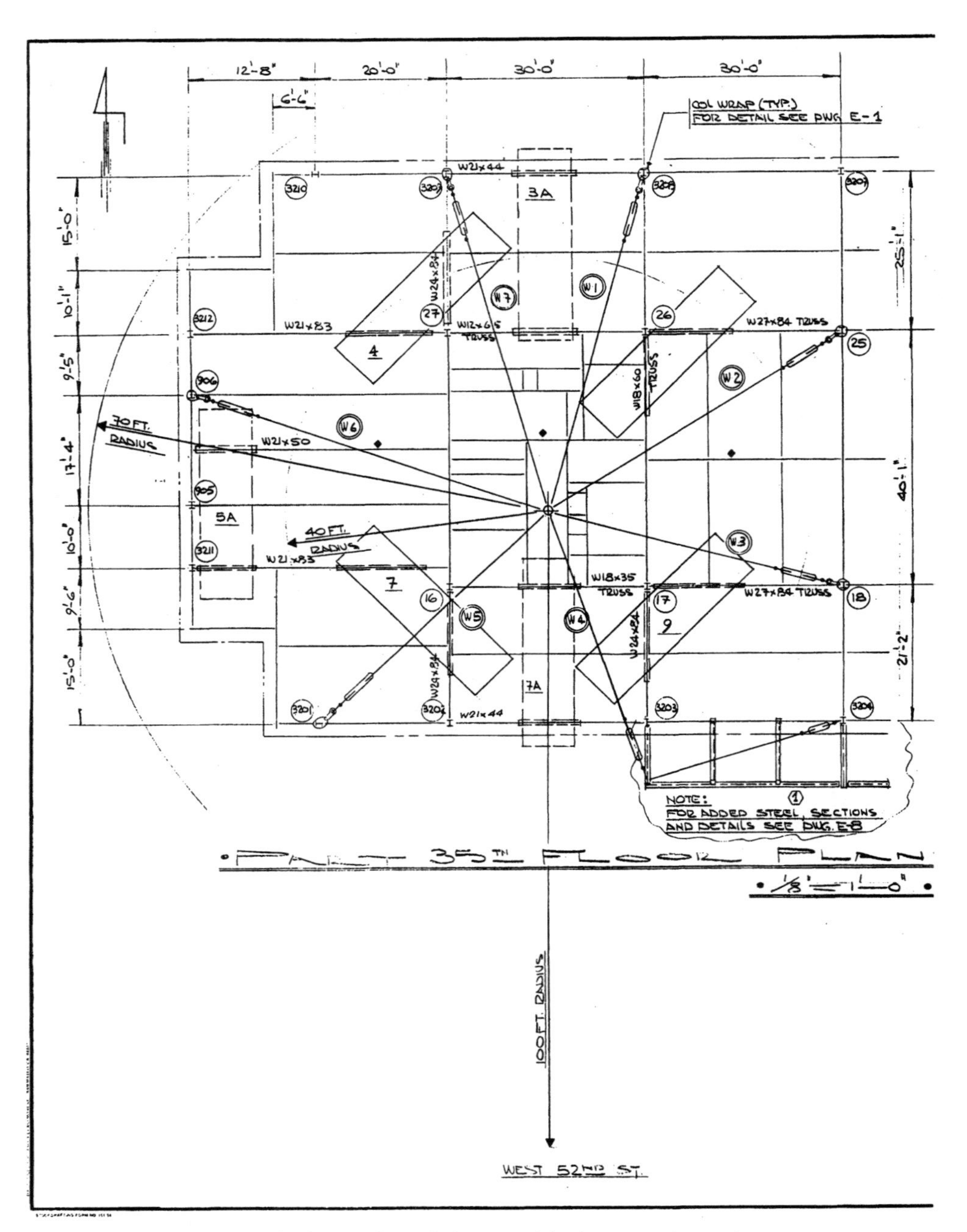

Figure 7.19 *(continued)* (*b*) Another of the set, this drawing shows the guy arrangement and loadings and positions where the derrick is permitted to land drafts of steel. An outrigger structure had to be added to improve the front guy angle at the 35th floor, because of a setback below

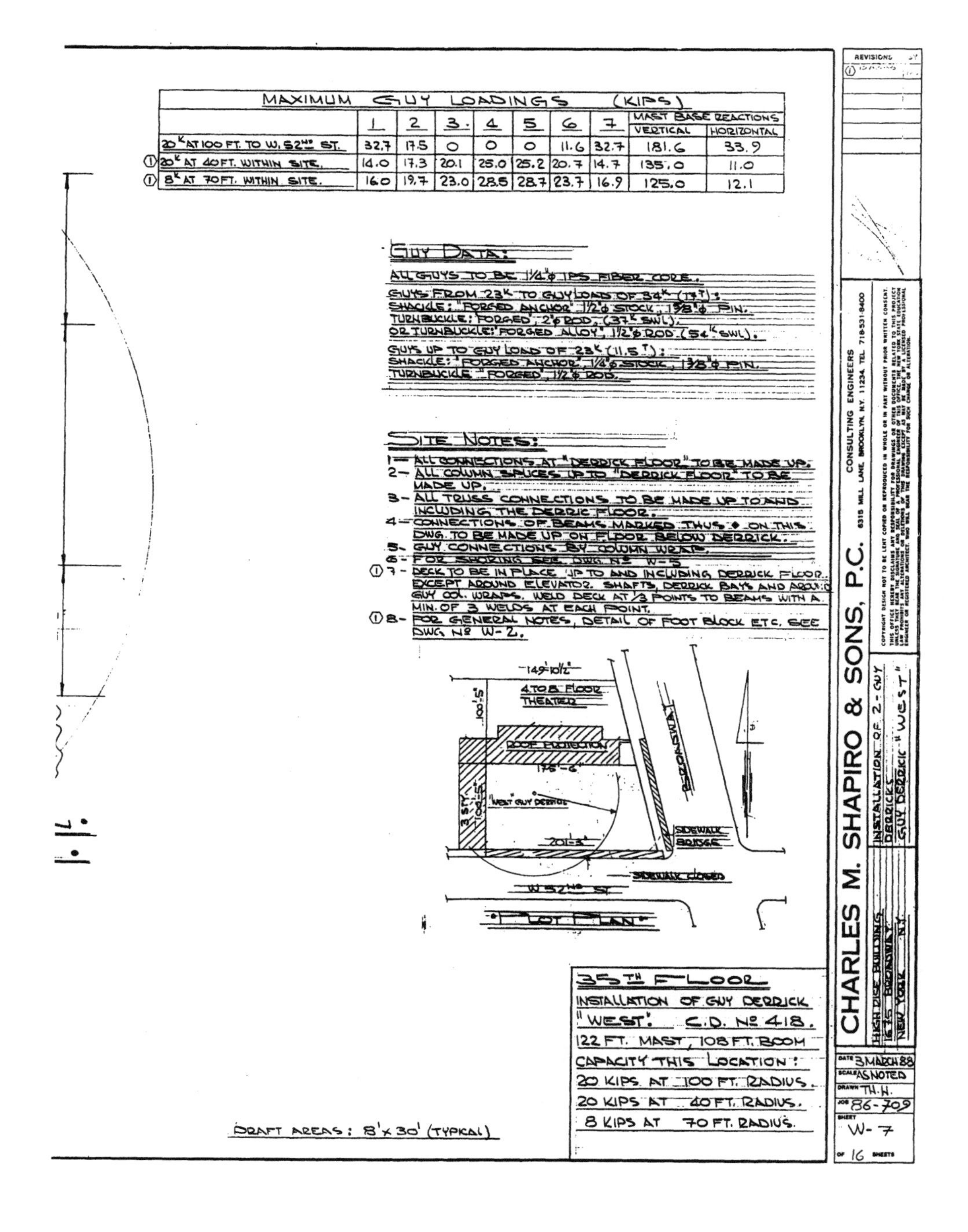

MAXIMUM GUY LOADINGS (KIPS)									
	1	2	3	4	5	6	7	MAST BASE REACTIONS VERTICAL	MAST BASE REACTIONS HORIZONTAL
20K AT 100 FT. TO W. 52ND ST.	32.7	17.5	0	0	0	11.6	32.7	181.6	33.9
(1) 20K AT 40 FT. WITHIN SITE.	14.0	17.3	20.1	25.0	25.2	20.7	14.7	135.0	11.0
(1) 8K AT 70 FT. WITHIN SITE.	16.0	19.7	23.0	28.5	28.7	23.7	16.9	125.0	12.1

Guy derrick adaptations

For added capacity, two guy derricks can be mounted side by side. The derricks can then be operated in tandem with two active hooks or with the load blocks made fast to a lifting beam or similar load-handling device. This setup can perform only lifting and luffing functions, and it is essential that procedures be established to assure that the motions are reasonably well synchronized.

Some large crawler-mounted mobile cranes can be equipped with optional guy derrick front end attachments, but when so outfitted, they are not longer mobile. The boom hoist is used to erect a mast to vertical, and the guys are then made fast. The derrick boom is raised by using a line from one of the hoist drums, which then serves to provide the luffing motion. A pair of guy derrick-equipped cranes can also be used in tandem or a fixed derrick can be paired with a crane-mounted rig (Fig. 7.20).

Another adaptation utilizes the mast and boom of a guy derrick but does not require the use of guys. This variant, called a *column derrick*

Figure 7.20 A mobile-crane-mounted guy derrick is paired with a fixed guy derrick for this tandem lift.

(Fig. 7.21), can be used on building setbacks or at any location where an adjacent structure is taller than the derrick mast; it has even been mounted on a frame cantilevered from the side of a building. The guys are replaced by structural member ties to the adjacent structure that are made fast to the derrick spider or to the gudgeon pin. When used this way, the mast must be held plumb throughout its limited swing arc. A column derrick can often be installed in lieu of a Chicago boom when long reaches and high-capacity lifts must be accommodated.

7.4 Gin Pole Derricks

A heavy-duty gin pole is nothing more than a guyed mast or boom rigged with a load fall. A single pole is set near vertical, usually not more than 10° from plumb, and is used for lifting without radius change. To assure adequate clearance between the load and the mast, the derrick height must often be considerably greater than lift height.

Six or more guys are arranged about the pole exactly as for a guy derrick mast, and guy design uses Eq. (7.9) as well as following the procedures previously described.

Guys on gin poles should be more snugly prestressed than on guy derricks because gin poles are more prone to radius increase when the load is picked. The greater the preloading, the less the guys will permit the radius to drift, but excessive guy tensioning can cause the mast itself to be overloaded in compression. Preload and radius increase can be calculated with reasonable accuracy using the concepts given under Exact Analysis of a Guyed Mast in Sec. 6.4.

Alternatively, field procedures can be followed that will preclude the need for exact mathematical analysis and permit the guys to be designed using the AISC method given earlier. The calculated value of H should be increased by about 15%, instead of the 5% suggested for guy derricks, to allow for expected increases in guy forces. Also, the preloading in the dead guys must be allowed for as an amplification of the dead load reaction H_g; two times H_g can be used.

The operation starts with adjustment of the guys to support the mast at a radius somewhat less than the desired value. The load is positioned at the planned radius, however. With the load falls made fast to the load, a strain is taken on the load line while the radius of the upper load block is monitored with a surveyor's transit. If the mast leans out too far, the strain is released and the participating live guys are given more preload. But if the load is about to lift off and the mast has not yet come to the required radius, the load line must be released and some slack given to the live guys. After successive adjustments have brought the boom to the proper loaded radius, the derrick is ready to lift.

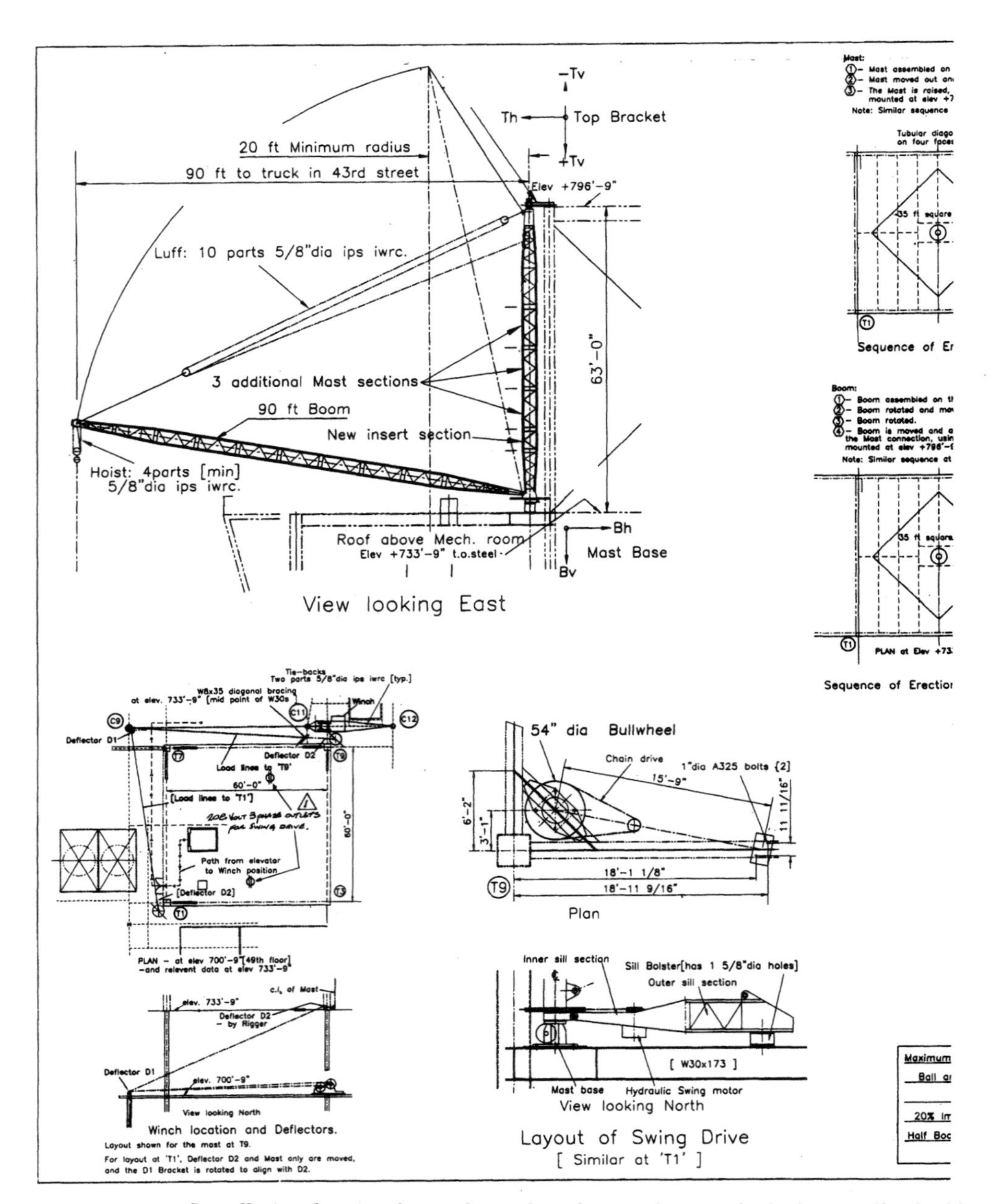

Figure 7.21 Installation drawing for a column derrick near the top of a high-rise office building in New York City. The derrick is intended for future use in changing out mechanical equipment. The derrick itself will not be left in place, but all of the arrangements for it will be permanent fixtures. In the center of the drawing are instructions for erecting the mast and the boom. Further details are given on a second sheet, not shown.

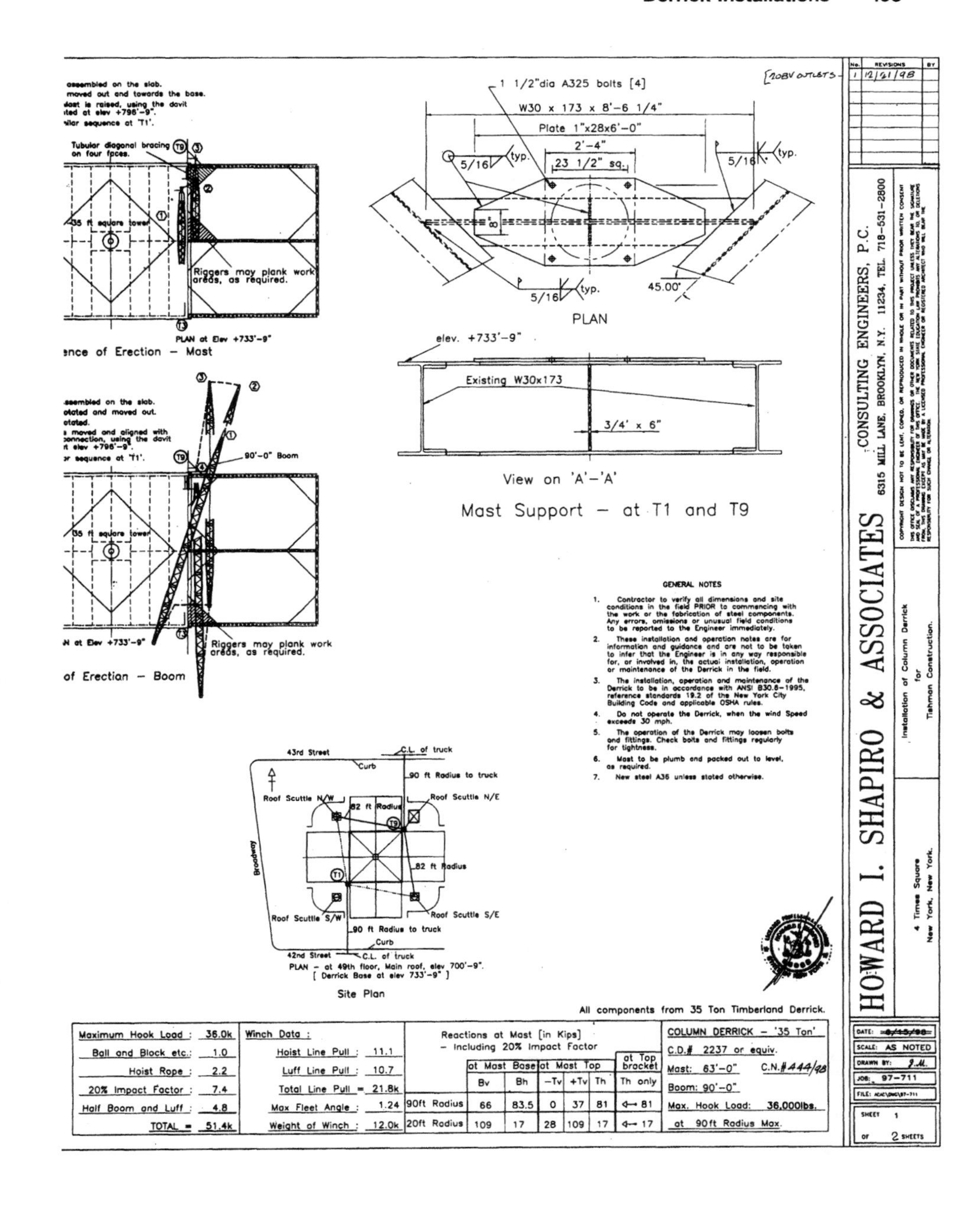

1 1/2"dia A325 bolts [4]
W30 x 173 x 8'-6 1/4"
Plate 1"x28x6'-0"
2'-4"
23 1/2" sq.
5/16 typ.
45.00°
PLAN
elev. +733'-9"
Existing W30x173
3/4" x 6"
View on 'A'-'A'
Mast Support – at T1 and T9
Tubular diagonal bracing on four faces.
35 ft square tower
Riggers may plank work areas, as required.
PLAN at Elev +733'-9"
ence of Erection – Mast
90'-0" Boom
of Erection – Boom
GENERAL NOTES
1. Contractor to verify all dimensions and site conditions in the field PRIOR to commencing with the work or the fabrication of steel components. Any errors, omissions or unusual field conditions to be reported to the Engineer immediately.
2. These installation and operation notes are for information and guidance and are not to be taken to infer that the Engineer is in any way responsible for, or involved in, the actual installation, operation or maintenance of the Derrick in the field.
3. The installation, operation and maintenance of the Derrick to be in accordance with ANSI B30.6-1995, reference standards 19.2 of the New York City Building Code and applicable OSHA rules.
4. Do not operate the Derrick, when the wind Speed exceeds 30 mph.
5. The operation of the Derrick may loosen bolts and fittings. Check bolts and fittings regularly for tightness.
6. Mast to be plumb and packed out to level, as required.
7. New steel A36 unless stated otherwise.
43rd Street
C.L. of truck
Curb
90 ft Radius to truck
Roof Scuttle N/W
Roof Scuttle N/E
82 ft Radius
Broadway
Roof Scuttle S/W
Roof Scuttle S/E
42nd Street
PLAN – at 49th floor, Main roof, elev 700'-9". [Derrick Base at elev 733'-9"]
Site Plan
All components from 35 Ton Timberland Derrick.
Maximum Hook Load : 36.0k
Ball and Block etc.: 1.0
Hoist Rope : 2.2
20% Impact Factor : 7.4
Half Boom and Luff : 4.8
TOTAL = 51.4k
Winch Data :
Hoist Line Pull : 11.1
Luff Line Pull : 10.7
Total Line Pull = 21.8k
Max Fleet Angle : 1.24
Weight of Winch : 12.0k
Reactions at Mast [in Kips] – Including 20% Impact Factor
at Mast Base: Bv, Bh; at Mast Top: –Tv, +Tv, Th; at Top bracket: Th only
90ft Radius: 66, 83.5, 0, 37, 81, 81
20ft Radius: 109, 17, 28, 109, 17, 17
COLUMN DERRICK – '35 Ton'
C.D.# 2237 or equiv.
C.N.#444/98
Mast: 63'-0"
Boom: 90'-0"
Max. Hook Load: 36,000lbs.
at 90ft Radius Max.
HOWARD I. SHAPIRO & ASSOCIATES
CONSULTING ENGINEERS, P.C.
6315 MILL LANE, BROOKLYN, N.Y. 11234, TEL 718-531-2800
Installation of Column Derrick for Tishman Construction.
4 Times Square New York, New York.
SCALE: AS NOTED
JOB: 97-711
SHEET 1 OF 2 SHEETS

Tandem gin poles

A pair of gin poles erected plumb are often used for erecting tall vessels at refineries or other such facilities; the vessel can be considerably taller than the poles (Fig. 7.22). Note that the guys must be positioned to allow clear space for the top end of the vessel as it rises.

The poles are initially erected with a slight lean away from each other. Guy preloading is then adjusted as previously described until the poles come into plumb when loaded. In addition, it is necessary to provide a tailing crane for the bottom of the vessel that can operate beneath the guys; otherwise the guys must be arranged to provide a clear area for that part of the operation.

With tandem gin poles, heavy awkwardly proportioned loads can be erected in rather confined working spaces.

Gin poles are simple, rugged, basic, low-tech hoisting devices, but as time goes by, fewer and fewer gin poles remain available for use.

Figure 7.22 A pair of heavy gin poles has been used to erect vessels considerably taller than the poles themselves in this incredibly tight work area. (*American Hoist and Derrick Co.*)

Sad to say, very few rigging firms now own them and there are not many people who are experienced any longer in their use. They are well adapted, low-cost and practical for a limited realm of lifting assignments, but their limitations are leading to extinction. They are being replaced by more expensive, more complex, but more versatile machines.

Because they are low tech, it is easy to conduct condition checks and to repair them in the field; they use standard fittings and accessories which are readily available. A tandem gin pole installation implies a three-hook lift, where the third hook is the tailing device. The mobile crane used to erect the poles is often used for tailing, making for an efficient operation. The side loading that can be expected to accompany tailing can be allowed for when the guying system is designed. As in any three-hook operation, tailing must be closely coordinated with hoisting, and the winches powering each pole must be coordinated with each other.

The guys supporting tandem gin poles fall into two categories, "back guys" which react to load weight, and "fore-and-aft" guys which stabilize the poles and react to side loading. Back guy forces can be calculated using the method described for guy derricks, but the side loading of the fore-and-aft guys depends on control procedures used in the field, as described under "Tailing Operations" in Chap. 5. The adequacy of each guy, however, depends on its anchorage, or *deadman*. The vertical component of guy force can be balanced by the dead weight of the deadman (with excess weight for a margin of safety), but the horizontal component is resisted by friction against the earth or in combination with lateral resistance of the soil when the deadman is buried. Those parameters are not easily determined.

Light-duty gin poles

Great flexibility in application is possible with light--duty gin poles, chiefly because their small loadings make guy anchorages and foot-block supports rather simple (Fig. 7.23). For this same reason, lean angles need not be restricted to near vertical as in the heavy-duty variety, and poles with a live topping lift are sometimes used as well.

When gin poles are to be used at large angles of lean, the entire load should be taken by no more than two participating live guys, but one single live guy is also acceptable. Guys are not preloaded, so that using more than two guys can result in excessive lateral displacement under load and too much uncertainty in distributing the load between the guys. The dead guys, two or more, are then placed on either side of the pole and serve only as steadiers, giving the system lateral stability.

Figure 7.23 The very light duty gin pole in the foreground is used to erect and dismantle the stiffleg derrick seen in part. For the small loads involved, the pole is seated on a piece of planking located above a roof beam, and wire/rope ties are used to stabilize the pole foot. (*Gendelman Rigging and Trucking, Inc.*)

When two effective live guys are used, they must be arranged symmetrically with respect to pole lean. Then in the two-guy expressions of Eq. (7.9) $L_1 = L_2$, $A_1 = A_2$, and $x_1 = x_2$, giving

$$G_1 = G_2 = \frac{1.05HL}{2x} = \frac{0.525HL}{x}$$

assuming that a 5% misalignment factor is appropriate for the installation conditions at hand.

Side guys for light-duty gin poles are often made up of several parts of manila rope, which permits easy pole alignment by hand (Fig. 7.24). The side guys should be preloaded to improve stability, and this can be done manually as well. Under ideal conditions, the side guys carry only a small preload, but in reality the guys must withstand wind on the gin pole and the load as well as forces induced by inadvertent load swing. In designing manila-rope side guys, it is more important to be concerned about rope stretch than strength, so a few more parts should be used than strength criteria dictate.

When the gin pole is arranged for radius change while under load, the locations of the side-guy anchorages become critical. With the anchorages placed exactly on the horizontal axis that is the extension of the pole foot pivot pin, luffing can be easily performed because guy length will remain constant throughout the luffing motion. But if the

Figure 7.24 The same gin pole shown in Fig. 7.23. The manila rope side guys are taken up tight to stabilize the pole, and a chainfall is used for handling the loads. (*Gendelman Rigging and Trucking, Inc.*)

side-guy anchorages are placed on the load side of the this line (or below it), the side guys will have to increase in length as radius decreases (that is, they will be subjected to tensile load and must stretch or be let out in order to permit the radius to decrease). Conversely, with increase in radius the lines will become slack and must be taken in.

On the other hand, with the side guys anchored on the opposite side of that line (or above it), the effect of radius change on the guys will be reversed. In either case, the derrick can be operated successfully only if the side guys are attended and continually adjusted during luffing. There is an element of risk, however, requiring the close supervision of an experienced field foreman.

The operation will be hazardous if the guy anchorages are located on a line that is not parallel to the pivot axis line. In this case, the rate of change in tautness of the side guys is different for each guy as radius changes. For example, if the forward guy is slacked off too slowly during luffing in, the pole will tend to lay over and transfer the topping load to the other side guy. This is usually accompanied by bending failure of the pole foot at the pivot unless the overloaded opposite side guy fails first. The same action can be initiated by a wind gust, but the possibility of topping load transfer can be eliminated

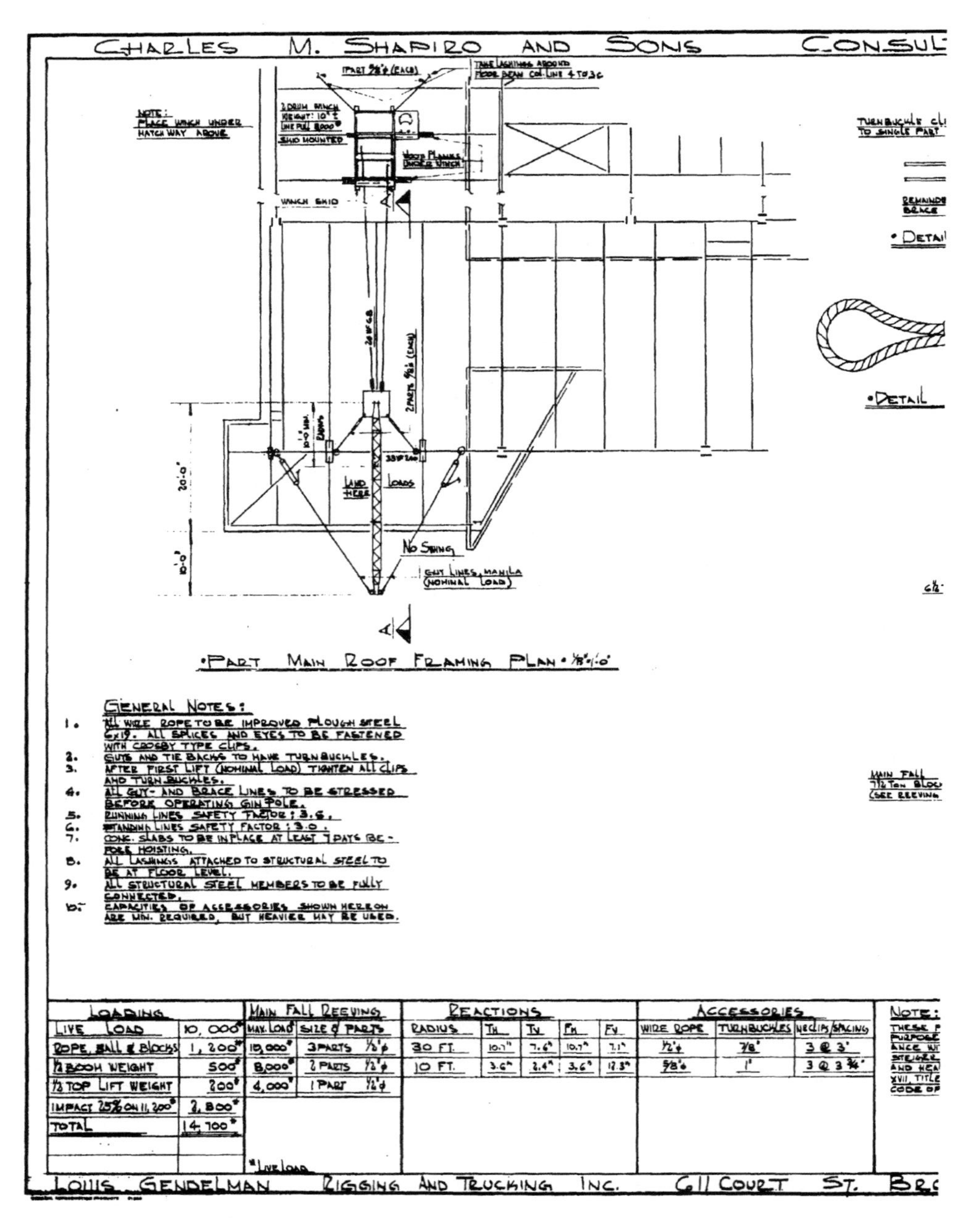

Figure 7.25 Installation drawing for a 5-ton (4.5-t) luffing gin pole to be placed on the roof of an office tower. The high topping lift gives the pole stability and permits anchorage of the side guys forward of the foot pivot pin.

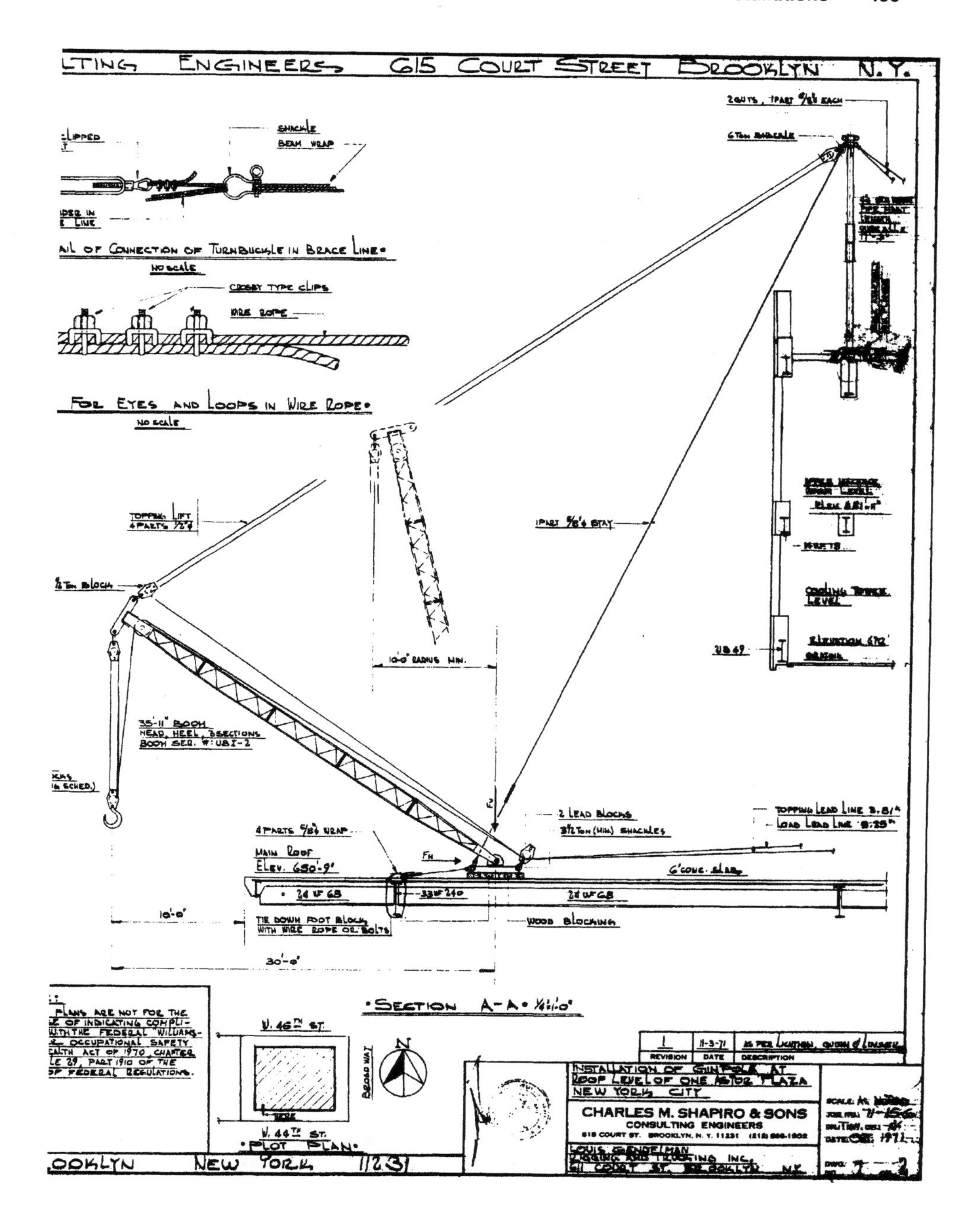
ENGINEERS 615 COURT STREET BROOKLYN N.Y.
SHACKLE
BEAM WRAP
CONNECTION OF TURNBUCKLE IN BRACE LINE
NO SCALE
CROSBY TYPE CLIPS
WIRE ROPE
FOR EYES AND LOOPS IN WIRE ROPE
NO SCALE
TOPPING LIFT
35'-11" BOOM
HEAD, HEEL, 3 SECTIONS
10'-0" RADIUS MIN.
2 LEAD BLOCKS
TOPPING LEAD LINE
LOAD LEAD LINE
4 PARTS 5/8" WRAP
MAIN ROOF
ELEV. 650'-9"
FH
6" CONC. SLAB
24 WF 68
TIE DOWN FOOT BLOCK WITH WIRE ROPE OR BOLTS
WOOD BLOCKING
10'-0"
30'-0"
SECTION A-A
COOLING TOWER LEVEL
PLANS ARE NOT FOR THE PURPOSE OF INDICATING COMPLIANCE WITH THE FEDERAL WILLIAMS-STEIGER OCCUPATIONAL SAFETY AND HEALTH ACT OF 1970, CHAPTER XVII, TITLE 29, PART 1910 OF THE CODE OF FEDERAL REGULATIONS.
W. 45TH ST.
W. 44TH ST.
BROADWAY
PLOT PLAN
N
REVISION
DATE
DESCRIPTION
INSTALLATION OF GIN POLE AT ROOF LEVEL OF ONE ASTOR PLAZA NEW YORK CITY
CHARLES M. SHAPIRO & SONS
CONSULTING ENGINEERS
615 COURT ST. BROOKLYN, N.Y. 11231
LOUIS GRENDELMAN RIGGING AND TRUCKING INC.
BROOKLYN NEW YORK 11231

altogether when the topping lift is made fast to a point higher than the gin pole head (Fig. 7.25).

The footbock must be anchored to prevent lateral movement from the horizontal component of pole axial load and from the lead-line forces. For light-duty poles, this can often be accomplished by using suitably anchored holdback ropes (Fig. 7.25).

7.5 Stiffleg Derricks

The first step in the design of a stiffleg derrick installation is, of necessity, determining the position in which the derrick is to be mounted. For a derrick situated on the ground, this is rarely difficult and involves only the usual considerations of where the loads must be taken from and where they must be placed. When the derrick is to be installed on a movable portal frame or tower, the additional feature of travel must be evaluated, but still the placement problem is usually not difficult.

Mounting a stiffleg derrick atop a structure, whether existing or under construction, is another matter altogether. Here the designer does not have the freedom to place the derrick in the position most favorable for load handling, as other constraints can have an overriding influence. The installation cost for any one derrick position can be markedly different from that for another. Installation cost can be so important that it may control selection of the derrick to be used independent of ordinary capacity and boom-length considerations.

A stiffleg derrick passes its loadings to the host structure through vertical reactions at the mast base and sill ends (Fig. 7.26). There are horizontal reactions as well, from hoist and topping lead lines, in addition to slewing and wind forces, but they are relatively small and can be accommodated without much difficulty. The most important single phase of stiffleg derrick installation design, on a structure, is therefore the positioning of the mast and sill anchorage points. Each of the three points can be subjected to both uplift and compressive loadings as the derrick operates, any single reaction often far exceeding the combined weight of the derrick and the load.

Few structures can support such a trio of loads randomly placed without significant reinforcement. The designer must then find that position where the necessary derrick operations can be performed efficiently and the structure can be made to accommodate the loads. Such accommodation may require reinforcement, shoring, ties, or a combination of these. The most efficient attachment results when the mast is positioned directly above a structure column as in Fig. 7.27. Whenever possible, one or both sill ends should be set above or close to a column as well.

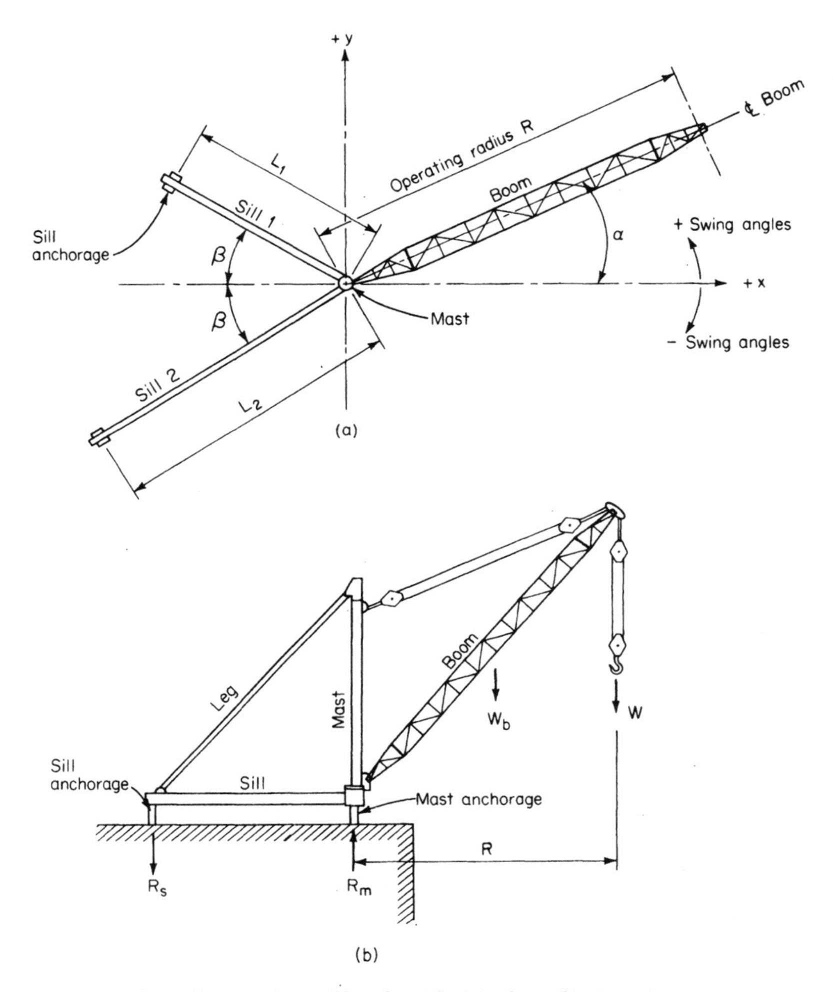

Figure 7.26 Installation of a stiffleg derrick: (*a*) plan; (*b*) elevation.

Sill lengths will not match column-to-column spacing except by chance, and some derricks are made with sills of different lengths, just as many buildings have different column spacing in each direction (Fig. 7.28).

The cost and difficulty associated with mast and sill anchorage, often make it necessary to opt for a derrick other than the one originally chosen, the aim being either to reduce reactions or to place them where they can be more readily accommodated. This rethinking may change original concepts relating to required operating radii or swing arcs relative to the stiffleg frame. In any event, the options cannot be adequately evaluated until reaction values have been calculated.

Figure 7.27 A 50-ton (45-t) stiffleg derrick mounted on a setback roof of a power plant structure. The derrick mast has been located above a building column, but the sill ends are supported on reinforced roof beams at this tight installation. Design was by Malcolm Johnston of the authors' firm. (*Photo by Jay P. Shapiro.*)

From Fig. 7.26, derrick reactions can be determined for any one boom position from the three equations of equilibrium, $\Sigma V = \Sigma M_x = \Sigma M_y = 0$. For the vertical forces, with W_b representing boom dead weight, and W for the weight of the load

$$R_{s1} + R_{s2} + R_m + W + W_b = 0$$

where a positive reaction value is used when the anchorage point experiences tension. Negative values for the solved reactions will therefore indicate compression at the anchorage. If the load moment is taken as positive, for moments about the y axis

Figure 7.28 Sill end attachment and leg connection of the stiffleg derrick in Fig. 2.4. The column stub, to which the sill is bolted, is welded to the roof framing steel. The bolt-down weldment straddles the wide-flange sill beam, permitting the weldment (anchorage) to be placed at any point from 23 to 30 ft (7 to 9 m) from the mast base. The sill, of course, has been designed for the bending moments that occur in any permitted position. Design by the authors' firm. (*Photo by Lawrence K. Shapiro.*)

$$M \cos \alpha - (R_{s1}L_1 + R_{s2}L_2) \cos \beta = 0$$

where $M = WR + M_b$ when M_b is the boom dead-load moment about the center of the mast. For the x axis

$$M \sin \alpha + (R_{s1}L_1 - R_{s2}L_2) \sin \beta = 0$$

The three equations define discrete solutions for the three unknown reactions. In matrix form, after rearranging, these equations become

$$\begin{bmatrix} 1 & 1 & 1 \\ L_1 & L_2 & 0 \\ -L_1 & L_2 & 0 \end{bmatrix} \begin{bmatrix} R_{s1} \\ R_{s2} \\ R_m \end{bmatrix} = \begin{bmatrix} -W - W_b \\ \dfrac{M \cos \alpha}{\cos \beta} \\ \dfrac{M \sin \alpha}{\sin \beta} \end{bmatrix}$$

Aftern solving by Kramer's rule and simplifying, this gives

$$R_{s1} = \frac{M}{L_1} \frac{\sin(\beta - \alpha)}{\sin 2\beta}$$

$$R_{s2} = \frac{M}{L_2} \frac{\sin(\beta + \alpha)}{\sin 2\beta} \tag{7.10}$$

$$R_m = \frac{-M}{\sin 2\beta} \left[\frac{\sin(\beta - \alpha)}{L_1} + \frac{\sin(\beta + \alpha)}{L_2} \right] - W - W_b$$

or

$$R_m = -(R_{s1} + R_{s2} + W + W_b)$$

From each of the reactions given by Eqs. (7.10), the appropriate dead-load reaction of the derrick frame must be subtracted (they are compressive and hence negative reactions).

Most stiffleg derricks have their sills set at 90° to each other, permitting the reaction equations to be simplified because $\sin 2\beta = 1$. When $|\alpha| < \beta$, both sill supports will experience uplift and the mast support will carry compressive loading. When $|\alpha| = \beta$, the reaction at the sill closest to the boom will be zero.

Figure 7.29 is a set of two drawings for the installation of a stiffleg derrick for removal of a large tower crane and for completion of steel erection. The general layout is shown on Fig. 7.29*a*, along with details of the somewhat complicated arrangements needed to lead the hoisting and derricking lines from the derrick to the winch. The winch is mounted some 35 ft (10.7 m) below the derrick base. This drawing sheet also includes the tower crane dismantling information. Details of the derrick mast and sill accommodations, support steelwork bracing, and building steel reinforcements are given on Fig. 7.29*b*. Interestingly, site restrictions made it necessary for the sills of the derrick to be spread to 94°.

In order to design the derrick anchorages, the maximum tensile and compressive reactions at each support point are needed, although on occasion it is necessary to know the values of the other reactions that will occur in conjunctions with them. At the sills, for any one load and radius condition, during swinging the maximum sill-uplift reaction will take place with the boom set normal to the other sill. The maximum compressive reaction will occur when the boom has reached its closest swing position to the sill under consideration.

For any trial derrick position, the path of each load at lift, swing, and placement must be plotted. From this layout, the load positions giving significant support reactions will be noted and calculations will reveal maximum values. Figure 7.29*a* includes a diagrammatic load-handling scheme and a table of maximum derrick reactions.

At the mast, the maximum compressive reaction will be produced when the boom is centered opposite the sills, $\alpha = 0°$, if both sills are of the same length. If not, the position inducing extreme loading can be found by differentiating the mast-reaction equation with respect to α. This new expression gives the rate of change in mast reaction as α changes, so that equating it to zero and solving will give the maximum-value equation

$$\frac{dR_m}{d\alpha} = 0$$

$$= \frac{M}{\sin 2\beta}\left[\frac{1}{L_1}(\sin\beta\sin\alpha + \cos\beta\cos\alpha) - \frac{1}{L_2}(\cos\beta\cos\alpha - \sin\beta\sin\alpha)\right]$$

which simplifies to

$$\alpha = \tan^{-1}\left(\frac{L_1 - L_2}{L_1 + L_2}\cot\beta\right) \tag{7.11}$$

The mast base will not always experience an uplift reaction, but when it does, it will be produced when $\alpha > 90°$.

When swing provisions are not built into the derrick itself (Fig. 7.30), the installation designer must give consideration to the means for providing the swing functions and the horizontal support reactions resulting therefrom.

Example 7.3 A stiffleg derrick will be sued to lift loads of 25 kips (11.3 t) at a radius of 55 ft (16.8 m) with the boom at $\alpha = 90°$. After raising the load and booming in to 25 ft (7.6 m) radius, the loads will be swung to $\alpha = -120°$ and set in place. The sills are each 20 ft (6.1 m) long and have dead-load reactions of 2 kips (8.9 kN) each, while the mast dead-load reaction is 3.5 kips (15.6 kN). The boom dead weight can be taken as 3.2 kips (1.5 t) acting at one-half the radius. Ignoring wind effects and impact for this preliminary study, find the maximum anchorage reactions when $\beta = 30°$.

The boom will not swing from the 55-ft-radius (16.8-m) position, so that each support reaction must be calculated for this condition ($\alpha = 90°$). The moment will be

$$M = WR + M_b = (25 \text{ kips})(55 \text{ ft}) + (3.2 \text{ kips})(\tfrac{1}{2})(55 \text{ ft})$$

$$= 1463 \text{ kip} \cdot \text{ft } (1984 \text{ kN} \cdot \text{m})$$

Using eq. (7.10) and subtracting the dead-load reactions gives

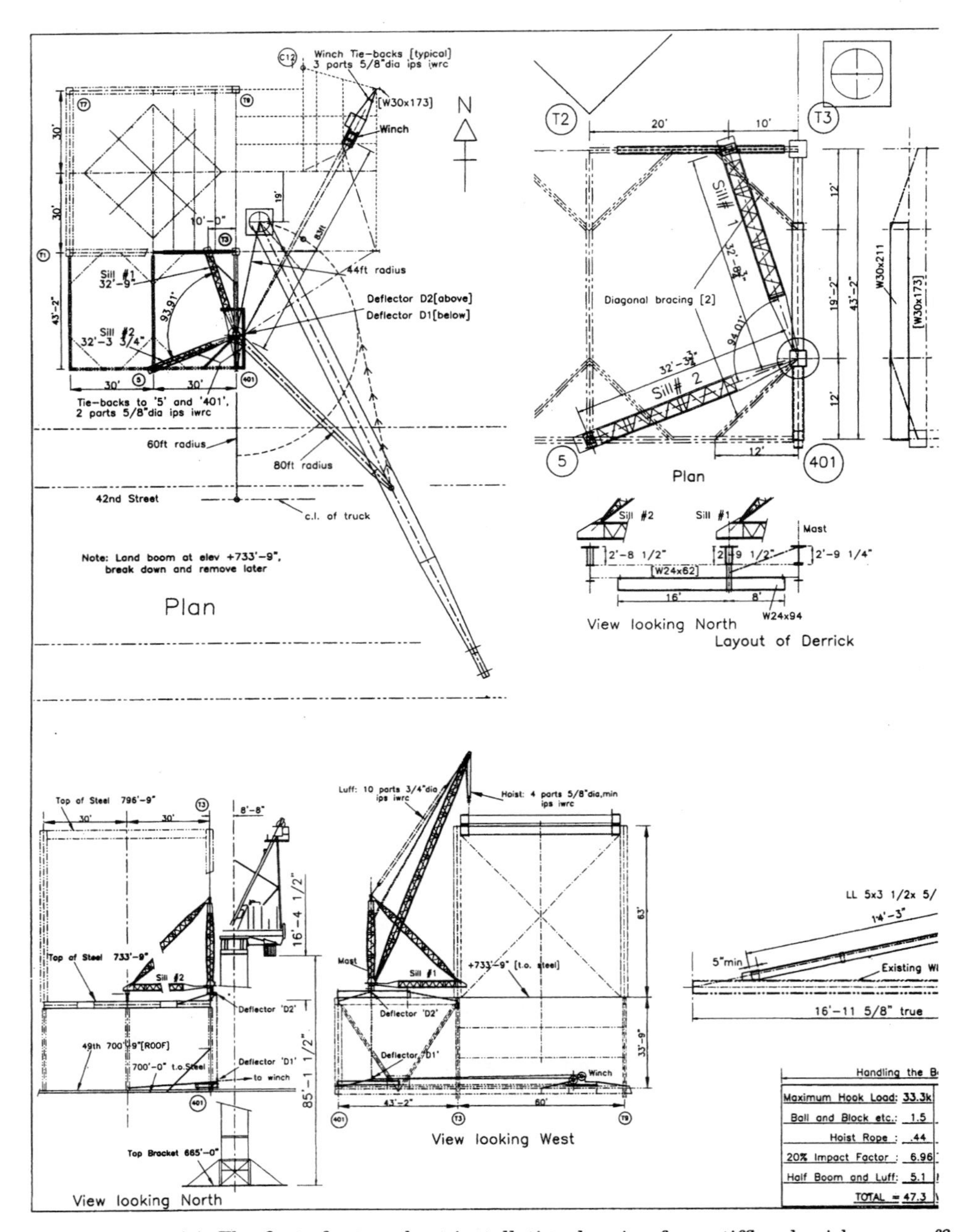

Figure 7.29 (*a*): The first of a two-sheet installation drawing for a stiffleg derrick on an office building roof (the same building as in Fig. 7.21). Tower crane removal data as well as installation details are shown. The derrick is to remove a tower crane and erect the last steel on this high-rise at New York City's Times Square.

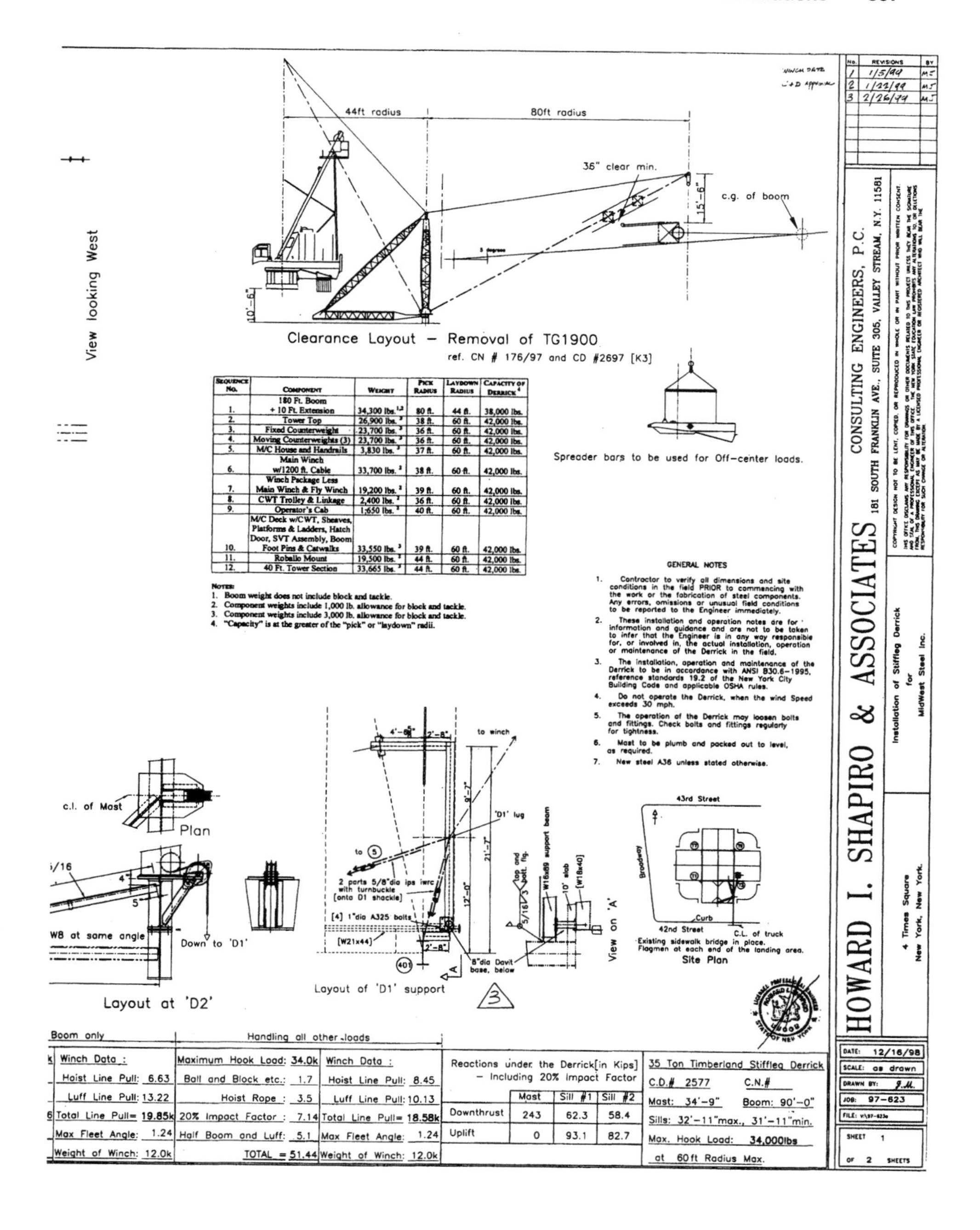

Sequence No.	Component	Weight	Pick Radius	Laydown Radius	Capacity of Derrick[4]
1.	180 Ft. Boom + 10 Ft. Extension	34,300 lbs.[1,2]	80 ft.	44 ft.	38,000 lbs.
2.	Tower Top	26,900 lbs.[3]	38 ft.	60 ft.	42,000 lbs.
3.	Fixed Counterweight	23,700 lbs.[3]	36 ft.	60 ft.	42,000 lbs.
4.	Moving Counterweights (3)	23,700 lbs.[3]	36 ft.	60 ft.	42,000 lbs.
5.	M/C House and Handrails	3,830 lbs.[3]	37 ft.	60 ft.	42,000 lbs.
6.	Main Winch w/1200 ft. Cable	33,700 lbs.[2]	38 ft.	60 ft.	42,000 lbs.
7.	Winch Package Less Main Winch & Fly Winch	19,200 lbs.[2]	39 ft.	60 ft.	42,000 lbs.
8.	CWT Trolley & Linkage	2,400 lbs.[2]	36 ft.	60 ft.	42,000 lbs.
9.	Operator's Cab	1,650 lbs.[2]	40 ft.	60 ft.	42,000 lbs.
10.	M/C Deck w/CWT, Sheaves, Platforms & Ladders, Hatch Door, SVT Assembly, Boom Foot Pins & Catwalks	33,550 lbs.[3]	39 ft.	60 ft.	42,000 lbs.
11.	Roballo Mount	19,500 lbs.[2]	44 ft.	60 ft.	42,000 lbs.
12.	40 Ft. Tower Section	33,665 lbs.[3]	44 ft.	60 ft.	42,000 lbs.

Notes:
1. Boom weight does not include block and tackle.
2. Component weights include 1,000 lb. allowance for block and tackle.
3. Component weights include 3,000 lb. allowance for block and tackle.
4. "Capacity" is at the greater of the "pick" or "laydown" radii.

GENERAL NOTES

1. Contractor to verify all dimensions and site conditions in the field PRIOR to commencing with the work or the fabrication of steel components. Any errors, omissions or unusual field conditions to be reported to the Engineer immediately.
2. These installation and operation notes are for information and guidance and are not to be taken to infer that the Engineer is in any way responsible for, or involved in, the actual installation, operation or maintenance of the Derrick in the field.
3. The installation, operation and maintenance of the Derrick to be in accordance with ANSI B30.6–1995, reference standards 19.2 of the New York City Building Code and applicable OSHA rules.
4. Do not operate the Derrick, when the wind Speed exceeds 30 mph.
5. The operation of the Derrick may loosen bolts and fittings. Check bolts and fittings regularly for tightness.
6. Mast to be plumb and packed out to level, as required.
7. New steel A36 unless stated otherwise.

Boom only — Winch Data:

Hoist Line Pull:	6.63
Luff Line Pull:	13.22
Total Line Pull=	19.85k
Max Fleet Angle:	1.24
Weight of Winch:	12.0k

Handling all other loads:

Maximum Hook Load:	34.0k
Ball and Block etc.:	1.7
Hoist Rope:	3.5
20% Impact Factor:	7.14
Half Boom and Luff:	5.1
TOTAL =	51.44

Winch Data:

Hoist Line Pull:	8.45
Luff Line Pull:	10.13
Total Line Pull=	18.58k
Max Fleet Angle:	1.24
Weight of Winch:	12.0k

Reactions under the Derrick [in Kips] – Including 20% Impact Factor

	Mast	Sill #1	Sill #2
Downthrust	243	62.3	58.4
Uplift	0	93.1	82.7

35 Ton Timberland Stiffleg Derrick
C.D.# 2577 C.N.#
Mast: 34'–9" Boom: 90'–0"
Sills: 32'–11"max., 31'–11"min.
Max. Hook Load: 34,000lbs at 60ft Radius Max.

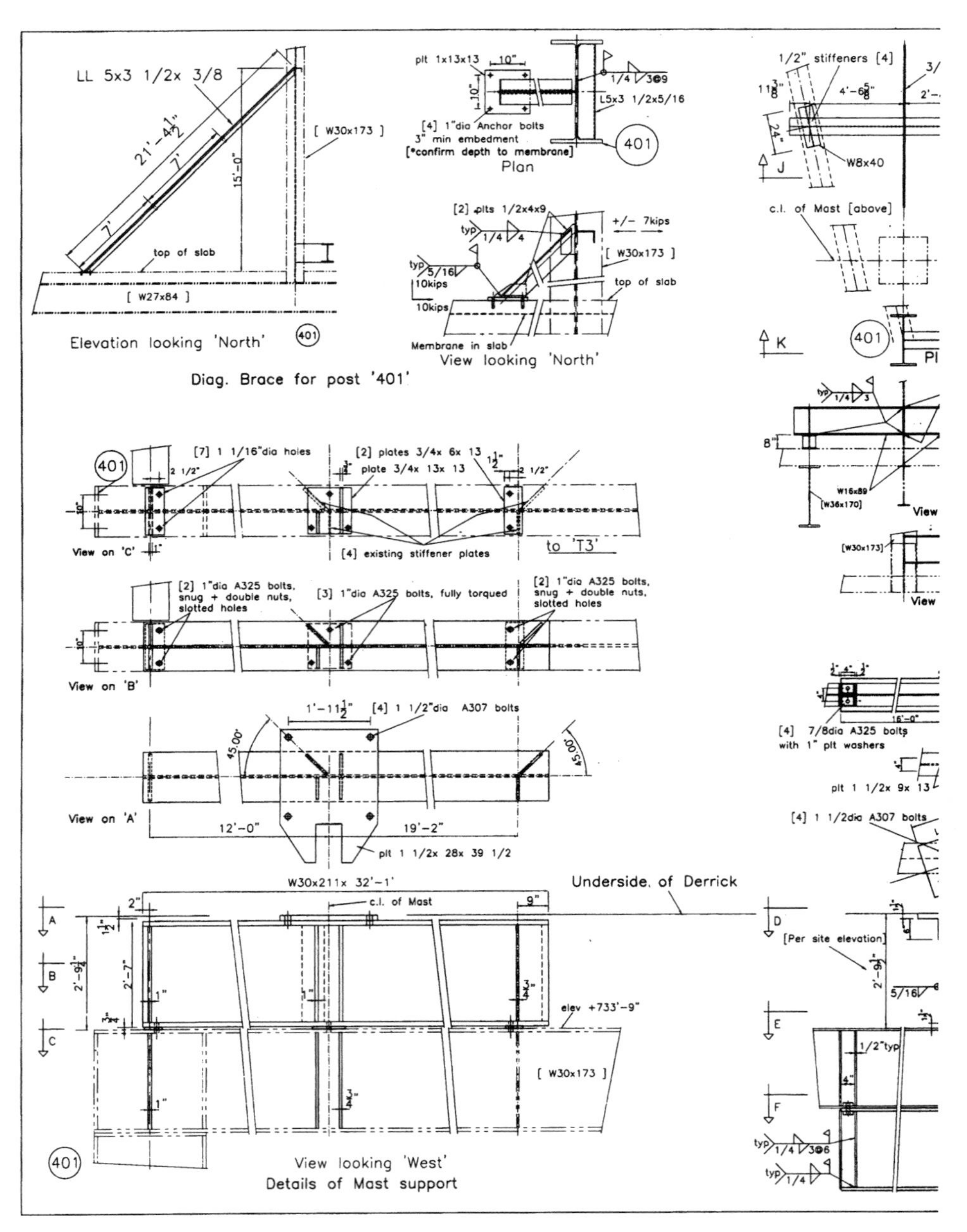

Figure 7.29 *(continued)* (*b*): This sheet shows the building reinforcing and derrick support steelwork details including arrangements for deflector sheaves to lead the hoist and luffing lines from the derrick to the winch.

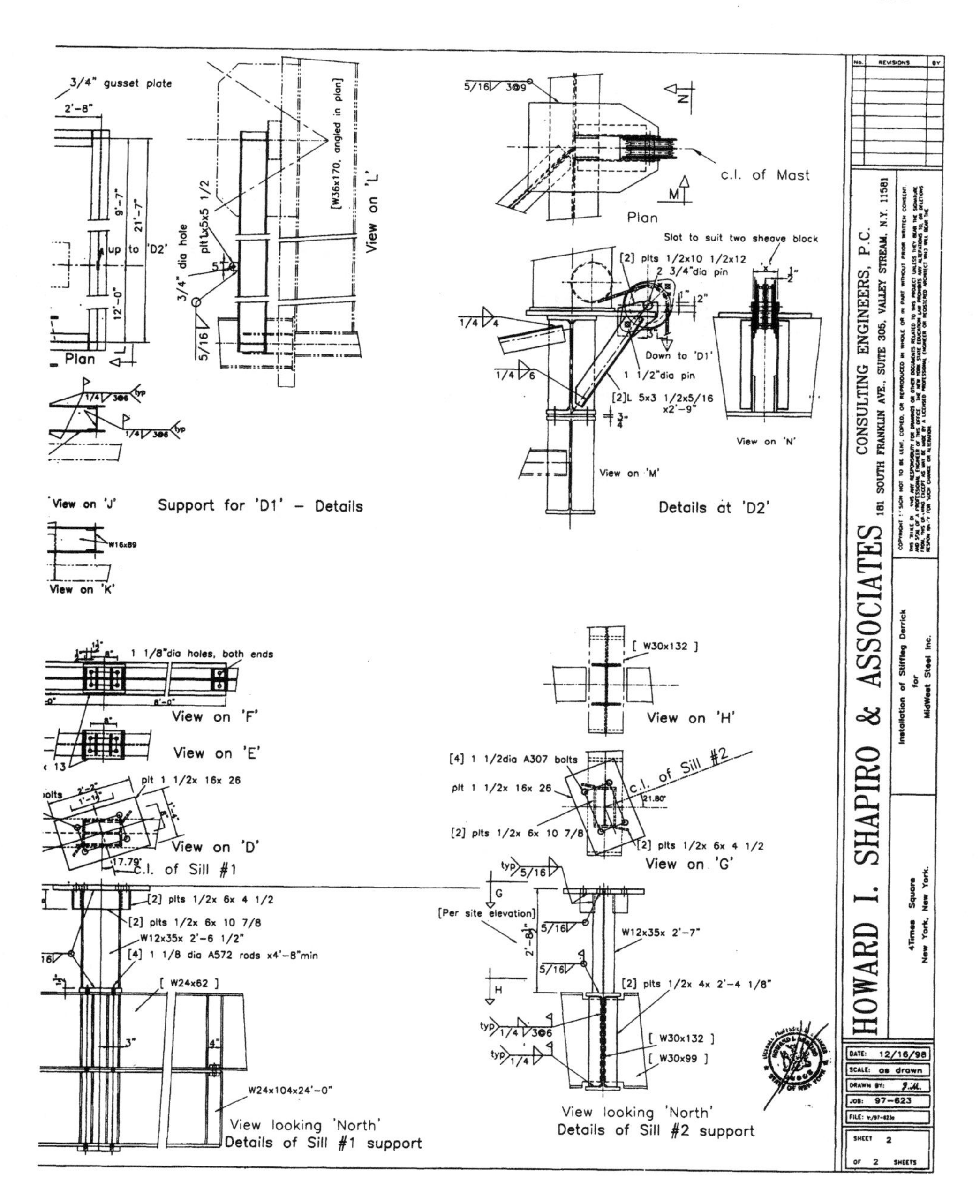

3/4" gusset plate
Plan
View on 'J'
View on 'K'
Support for 'D1' - Details
View on 'L'
[W36x170, angled in plan]
c.l. of Mast
Plan
Slot to suit two sheave block
[2] plts 1/2x10 1/2x12
2 3/4"dia pin
Down to 'D1'
1 1/2"dia pin
[2]L 5x3 1/2x5/16 x2'-9"
View on 'M'
View on 'N'
Details at 'D2'
1 1/8"dia holes, both ends
View on 'F'
View on 'E'
plt 1 1/2x 16x 26
View on 'D'
c.l. of Sill #1
[2] plts 1/2x 6x 4 1/2
[2] plts 1/2x 6x 10 7/8
W12x35x 2'-6 1/2"
[4] 1 1/8 dia A572 rods x4'-8"min
[W24x62]
W24x104x24'-0"
View looking 'North'
Details of Sill #1 support
[W30x132]
View on 'H'
[4] 1 1/2dia A307 bolts
plt 1 1/2x 16x 26
c.l. of Sill #2
[2] plts 1/2x 6x 10 7/8
[2] plts 1/2x 6x 4 1/2
View on 'G'
[Per site elevation]
W12x35x 2'-7"
[2] plts 1/2x 4x 2'-4 1/8"
[W30x132]
[W30x99]
View looking 'North'
Details of Sill #2 support
HOWARD I. SHAPIRO & ASSOCIATES
CONSULTING ENGINEERS, P.C.
181 SOUTH FRANKLIN AVE., SUITE 305, VALLEY STREAM, N.Y. 11581
Installation of Stiffleg Derrick for MidWest Steel Inc.
4Times Square New York, New York.
DATE: 12/16/98
SCALE: as drawn
JOB: 97-623
SHEET 2 OF 2 SHEETS

Figure 7.30 The bull wheel used to swing the stiffleg derrick of Fig. 7.27. A wire rope is anchored to the wheel, then passed around its circumference. The rope is then run to a capstan on the winch, around which a few turns are taken, and then back around the wheel in the opposite direction to an anchorage. The derrick swings when the capstan is rotated; reversing capstan rotation will reverse swing direction. (*Photo by Jay P. Shapiro.*)

$$R_{s1} = \frac{1463}{20}\,\frac{\sin(30° - 90°)}{\sin[2(30°)]} - 2.0 = -75.15 \text{ kips } (-334.3 \text{ kN})$$

$$R_{s2} = \frac{1463}{20}\,\frac{\sin(30° + 90°)}{\sin 60°} - 2.0 = 71.15 \text{ kips } (316.5 \text{ kN})$$

$$R_m = -(-75.15 + 71.15 + 25 + 3.2) - 3.5 - 2.0 - 2.0$$

$$= -31.70 \text{ kips } (-141.0 \text{ kN})$$

After booming in to 25 ft (7.6 m) radius, the maximum sill uplift will occur with the boom in two positions, $\alpha = 60°, -60°$, so that both sill supports must be designed for this force. The moment now becomes

$$M = 25(25) + 3.2(1/2)(25) = 665 \text{ kip} \cdot \text{ft } (902 \text{ kN} \cdot \text{m})$$

$$R_{s1} = R_{s2} = \frac{665}{20}\,\frac{\sin(30° + 60°)}{\sin 60°} - 2.0 = 36.39 \text{ kips } (161.9 \text{ kN})$$

The maximum compressive force at sill 2 will develop when the boom comes closest to that support, or when $\alpha = -120°$

$$R_{s2} = \frac{665}{20} \frac{\sin (30^\circ - 120^\circ)}{\sin 60^\circ} - 2.0 = -40.39 \text{ kips } (-179.7 \text{ kN})$$

The first position checked indicated the maximum compressive condition for sill 1.

Using Eq. (7.11), we find the horizontal boom angle at which the mast receives maximum compressive loading

$$\alpha = \tan^{-1} \left(\frac{20 - 20}{20 + 20} \cot 30^\circ \right) = 0^\circ$$

which will always be the case with sills of identical length. Then, from Eq. (7.10) we have

$$R_m = \frac{-665}{\sin 60^\circ} \left[\frac{\sin (30^\circ - 0^\circ)}{20} + \frac{\sin (30^\circ + 0^\circ)}{20} \right] - 25 - 3.2 - 7.5$$

$$= -74.09 \text{ kips } (-329.6 \text{ kN})$$

After swinging to $\alpha = -120^\circ$, the boom will be at its maximum incursion on the sill side of the y axis. The mast base may then be in tension; we find

$$R_m = - \left[\frac{665}{20} \frac{\sin (30^\circ + 120^\circ)}{\sin 60^\circ} - 40.39 + 25 + 3.2 \right] - 7.5$$

$$= -14.51 \text{ kips } (-64.5 \text{ kN})$$

which is a compressive force after all.

The design reactions are then

Unit	R_{s1}	R_{s2}	R_m
kips	−75.15, +36.39	−40.39, +71.15	−74.09
kN	−334.3, +161.9	−179.7, +316.5	−329.6

7.6 Other Derrick Forms and Details

The most basic form of lifting device, commoly used in manufacturing plants, service shops, and construction sites, is a self-contained hoisting unit which is available in capacities of from about ½ to 40 tons (0.45 to 36 t) or more. These hoists are powered electrically or pneumatically or made for manual operation by means of a hand chain; the manual units are often referred to as chainfalls (Fig. 2.1).

When used for vertical lifting, the units are merely suspended from any structurally adequate overhead support. The load on the support will include only rated load and the weight of the hoist for manual units because lifting motion is too slow to make dynamic effects significant. On the other hand, powered hoist supports should include an impact allowance of 25% or more if the manufacturer so advises.

Similar lifting functions can be performed by a set of blocks reeved to a winch (Fig. 7.31), a job-made device that can be arranged for any capacity. The overhead support will carry the same loads as for a powered hoist, but when the lead line runs out of the upper block, its effect must be added.

Figure 7.32 shows the support arrangement used for the hoist in Fig. 7.31. Note that the long hoist-support beams will elastically deflect when loaded and for this reason were blocked up clear of the concrete floor. Without this precaution, the lightly framed office building floor probably would have cracked.

When a lead line runs off at an angle to the block, it introduces a lateral force component at the block that could cause lateral load displacement when a few parts of line are being used. Likewise, the line

Figure 7.31 One of the most basic lifting arrangements, a shaft hoist. The load was rigged under the hook by rolling it over the timbers. With the load raised, the timbers will be moved aside and the load lowered to a subcellar several stories below the street. (*Photo by Lawrence K. Shapiro.*)

Figure 7.32 Support arrangement for the shaft hoist in Fig. 7.31. The long beams are blocked up at their ends so that they can deflect freely under load. At the center, they are tied in against timber separators for lateral torsional stability. The load block is suspended from the cross-beam. (*Photo by Lawrence K. Shapiro.*)

directions at a snatch block define the alignment of the anchorage reaction (Fig. 7.33) which is needed for design. With the lead-line directions defined at the snatch block, the lead-line-angle effect at the block can be determined with sufficient accuracy.

Using the notation of Fig. 7.34, the vertical load on the pivot point will be $W + P \cos \theta$ if W is taken to include the weight of the rigging. The upper block will be displaced laterally through distance Δ by the lead-line effect, and this will introduce a clockwise moment about the pivot point of

$$M = \Delta(W + P \cos \theta)$$

Equilibrium will be maintained by the counterclockwise moment

$$M = dP \sin \theta$$

Solving for the displacement, we get

$$\Delta = \frac{dP \sin \theta}{W + P \cos \theta} \tag{7.12}$$

The same effect takes place when the lead line runs up from the lower block, but the condition is exacerbated as the load is lowered

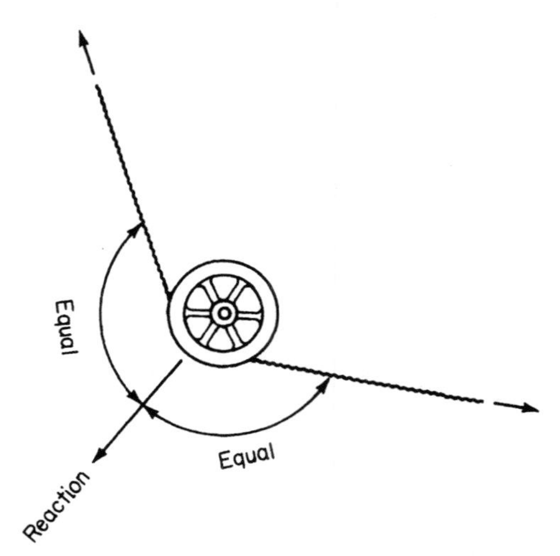

Figure 7.33 The line of action of the reaction at a block anchorage is a function of the loaded-rope directions.

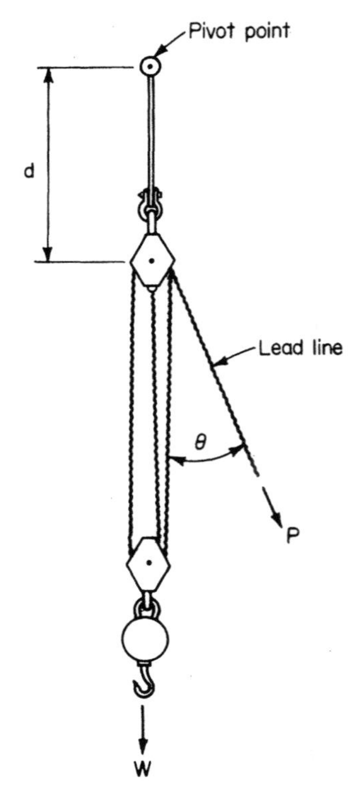

Figure 7.34

and the distance from the pivot point increases. Unless many parts of line are used, the load will severely displace and could cause clearance problems unless the lead angle is kept suitably small.

Example 7.4 (*a*) A load of 45 kips (20 t), including the weight of the rigging, is to be hoisted on six parts of line. The sheave shaft of the upper load block is 82 in (2.08 m) below the pivot at the suspension point, and the lead line runs from the upper block at 30° from the vertical. What distance will the load be displaced laterally from the pivot position?

If friction is neglected, the line pull will be 45/6 = 7.5 kips (33.4 kN). Using Eq. (7.12), we have

$$\Delta = \frac{82(7.5 \sin 30°)}{45 + 7.5 \cos 30°} = 6.0 \text{ in } (152 \text{ mm}) \qquad \textit{Ans.}$$

(*b*) Given the same load and rigging but with the lead running from the lower block at 30° (and included as one of the parts), what offset will have developed when the lower block has reached 30 ft (9.1 m) below the pivot?

Again neglecting friction, with the lead as one of the parts the lead-line load will be given by

$$5P + P \cos \theta = W$$

$$P = \frac{45}{5 + \cos 30°} = 7.67 \text{ kips } (34.1 \text{ kN})$$

which when inserted into Eq. (7.12) gives the displacement

$$\Delta = \frac{30(12)(7.67 \sin 30°)}{45 + 7.67 \cos 30°} = 26.7 \text{ in } (679 \text{ mm}) \qquad \textit{Ans.}$$

Of course, the lateral displacement will reduce the lead-line angle somewhat, making the calculated value an overstatement, the actual displacement will be about 2½ in (65 mm) less.

Further, the lead-line angle will change as the lower block is raised or lowered, making it necessary to do the calculation, using applicable lead angles, at each critical clearance point.

To reduce load displacement when the lead is from the lower block it is necessary to mount the snatch block close to the pivot, in other words, make the lead line near vertical. But when the lead must come from the top block, an ample lead angle must be established for the lead line to clear the load. If displacement is critical, or if the block is suspended well below the pivot, the problem can be overcome by fixing the block in place with stabilizing lines run laterally from the block bail.

Catheads

Another common basic form of lifting device is the cathead, which is nothing more than a cantilevered beam supporting an upper block at its outer end. When catheads are installed at construction sites, the beam is made to project well into the building to make erection and anchorage easier. The beam is locked into place either by clamping onto floor beams below or by wedging posts between the cathead and floor beams above. Another but less satisfactory method uses weights placed on the cathead beam. This makes the capacity depend upon stability imparted by the weights, with a great deal of uncertainty introduced because of the unknown value of the impact accompanying lifting and unloading.

The anchorage of the cathead must then be a two-point support, one at or near the building edge and the other inside (Fig. 7.35). Each support must carry on increment of vertical loading from the lifted load and cathead weight plus the effect of load moment. Taking support uplift as positive, we find the reactions R from

$$R = W\left(-\frac{1}{2} \pm \frac{b}{d}\right) \tag{7.13}$$

where the positive value is used for R_i and the negative for R_o.

As the lifted load and span increase, the size of the cathead beam and the anchorage reactions increase to a limit where a more economical arrangement is needed. A simple inexpensive adaptation can be designed with pendant ropes supporting the outboard end of the cat-

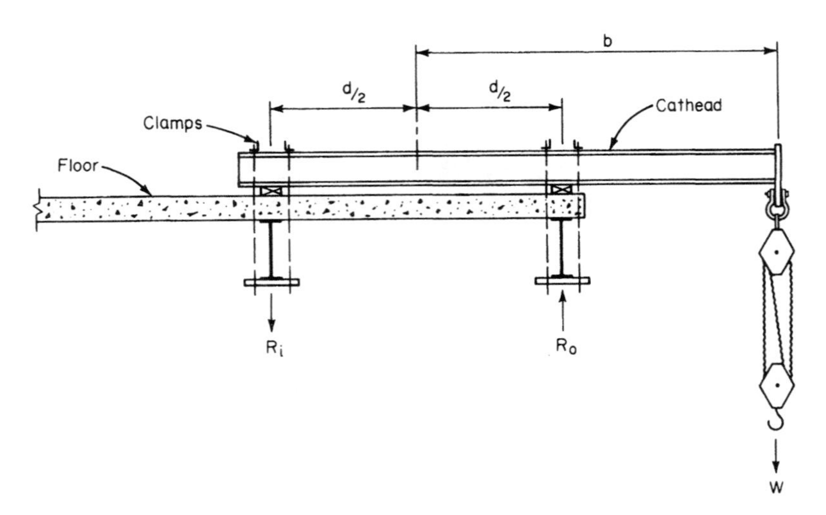

Figure 7.35 Basic cathead arrangement.

head (Fig. 7.36). The purpose of this arrangement is to relieve the cantilever of most of its bending load, but the portion of vertical load carried in the pendant introduces horizontal thrust into the cathead.

The cathead and the pendant share vertical load in a proportion that makes their elastic distortions consistent. For rope load P_r the vertical component of rope stretch (Fig. 7.36) will be

$$\Delta = \frac{P_r l \cos \theta}{A_r E_r} \tag{a}$$

and for cantilever-beam load P_c the tip vertical deflection will be

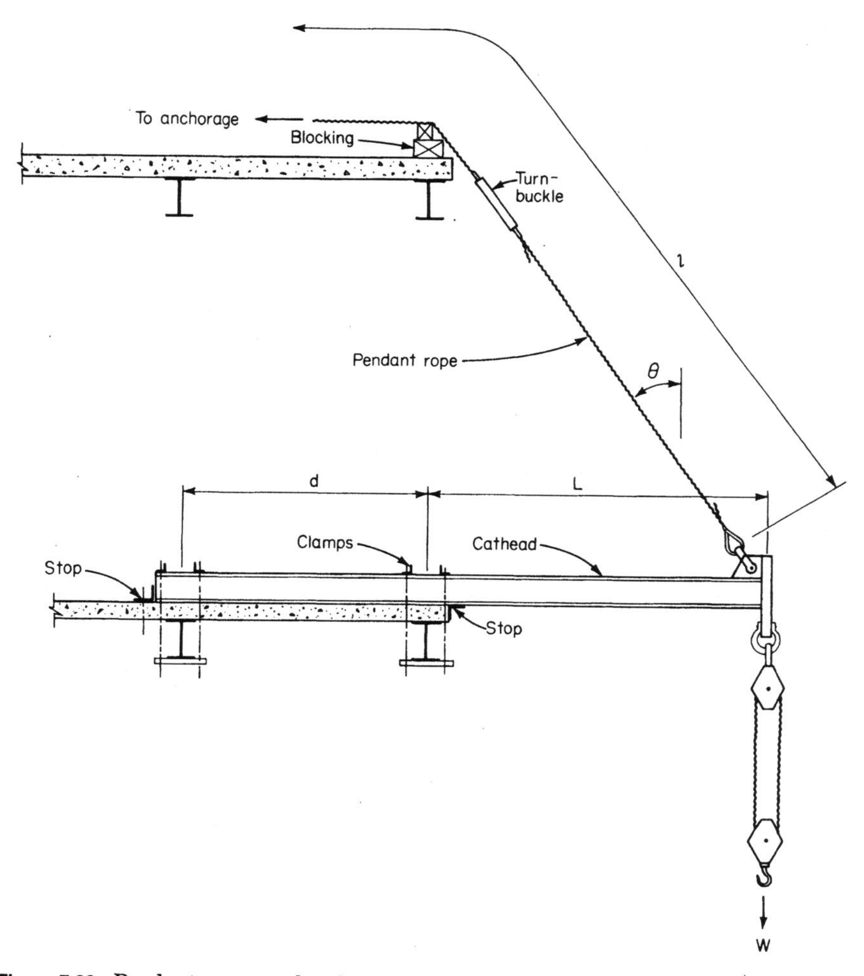

Figure 7.36 Pendant-supported cathead.

$$\Delta = \frac{P_c L^2}{3EI}(d + L) \qquad (b)$$

where I = moment of inertia of beam cross section
E = modulus of elasticity of beam material
A_r = metallic cross-sectional area of rope
E_r = rope modulus of elasticity

The vertical loads must be in equilibrium, as expressed by

$$P_r \cos\theta + P_c = W$$

Equating the deflection relationships (a) and (b), we have a second equation in the two unknowns. In matrix form

$$\begin{bmatrix} \cos\theta & 1 \\ \dfrac{l\cos\theta}{A_r E_r} & \dfrac{L^2(d + L)}{-3EI} \end{bmatrix} \begin{bmatrix} P_r \\ P_c \end{bmatrix} = \begin{bmatrix} W \\ 0 \end{bmatrix}$$

After solving by Kramer's rule and simplifying we get

$$P_r = \frac{W}{(1 + 1/e)\cos\theta} \qquad P_c = \frac{W}{1 + e} \qquad (7.14)$$

where

$$e = \frac{A_r E_r L^2(d + L)}{3EIl}$$

As the beam moment of inertia is reduced, the parameter e approaches infinity in the limit and the rope load becomes $W/\cos\theta$. This indicates that the cathead can be designed as a beam-column loaded with the thrust induced by the inclined pendant and a secondary moment that will be limited to the product of the thrust and the vertical displacement due to rope stretch. This moment will be

$$M = \Delta_{\text{rope}} P_r \sin\theta = \frac{P_r^2 l \sin\theta\cos\theta}{A_r E_r} = \frac{W^2 l \tan\theta}{A_r E_r} \qquad (7.15)$$

It is assumed, of course, that the design has included the provision that the centerline of the pendant intersect the line of action of the vertical load at the centerline of the cathead. If this is not the case, an additional end moment will be induced by the eccentricity.

Flying strut

The next logical evolutionary step to follow the pendant-supported cathead is the flying strut. In the previous form of cathead the clamp attachments to the floor served only to stabilize the device and hold it in place. With the load-transfer function of the clamps done away with, the floor is no longer an essential element in the installation.

The flying strut is a pendant-supported cathead held suspended in space (Fig. 7.37). It is held in position by the load fall, the pendant(s), and several wire-rope thrust-restraining and preventer lines. The number and positions of the lines are tailored to the individual installation and are arranged to transfer the strut thrust, derived from the inclined pendant, to the building frame and to assure that the strut will remain stable through any expected maneuvering of the load. It can easily be arranged to accommodate drifting of the load (out-of-plumb load falls) in any direction.

The flying strut is a pure pin-ended column subjected to axial loading plus some end moment introduced by imbalance in thrust-line loadings or tip eccentricity. In addition, side drifting of loads will impose torsional loading. Tube or pipe sections therefore make the most efficient strut members.

When adjustment capabilities are built into the pendant and other wire-rope lines, the strut can be "floated" laterally, vertically, or forward and backward to secure the most advantageous position (Fig. 7.38). With suitable arrangements, this can even be done under load.

7.7 Derrick Loads

Derricks can be mounted on the ground, in which case lift heights are small, or on a high-rise roof top, in which case lift heights can be great

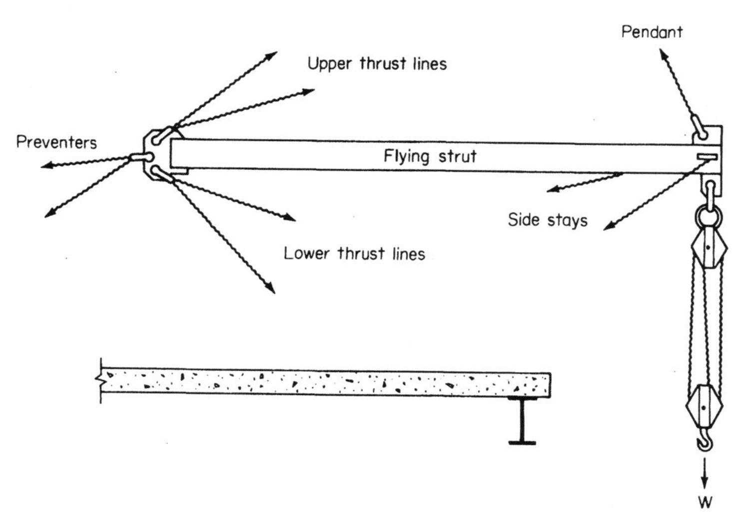

Figure 7.37 Flying strut.

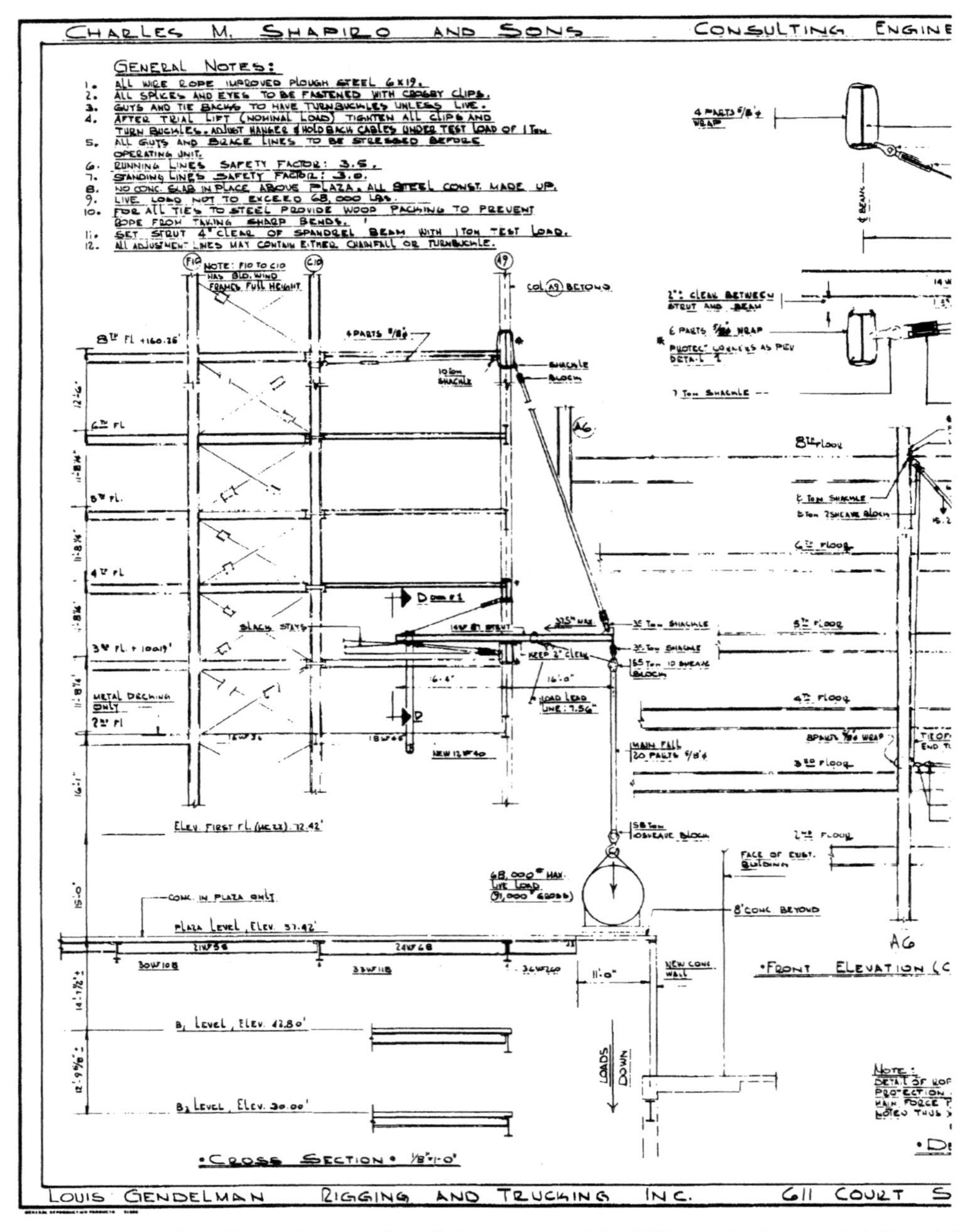

Figure 7.38 Installation drawing for a flying strut used for drifting a load some 18 ft (5.5 m) from the pickup point to its freely suspended lowering positions.

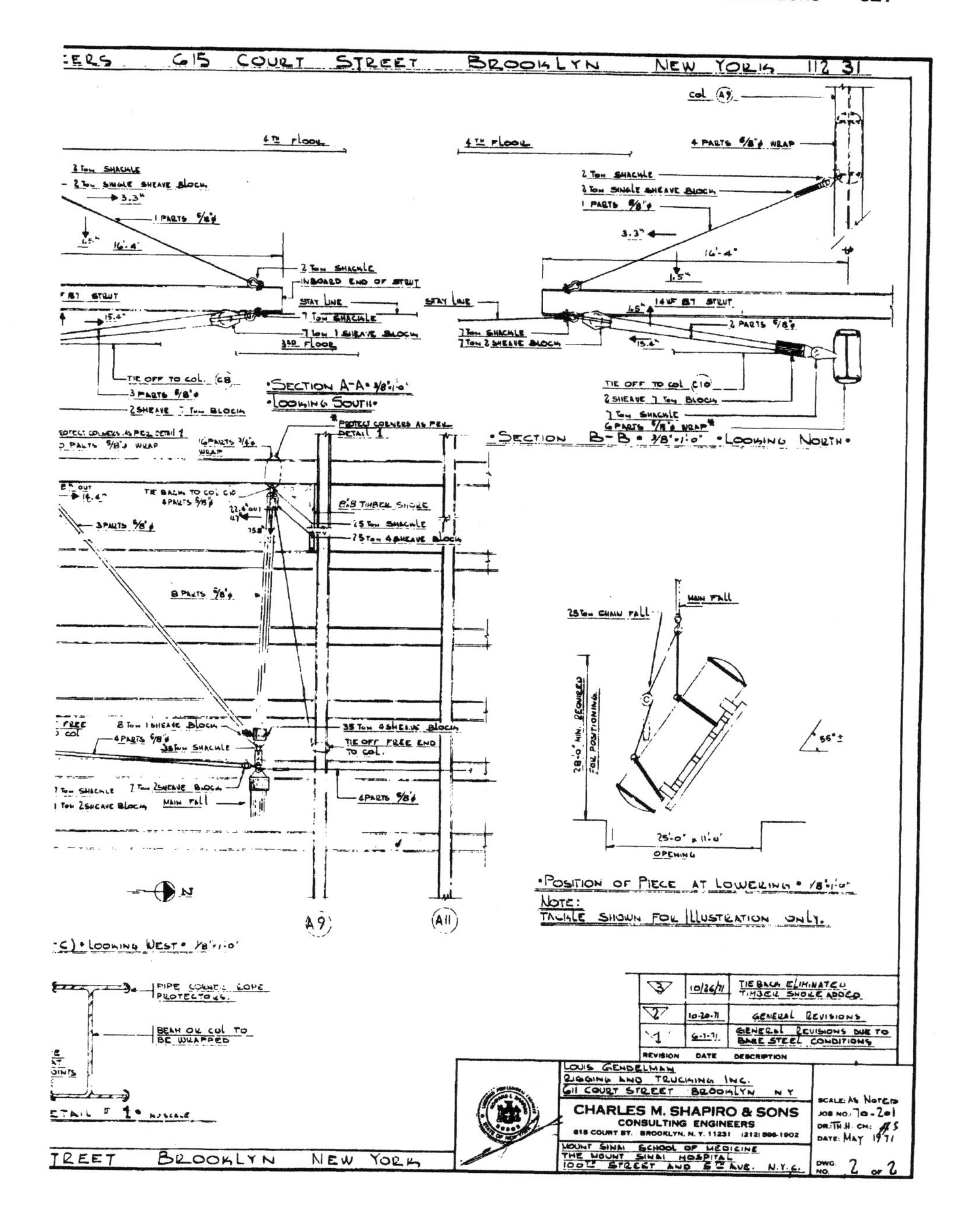

615 COURT STREET BROOKLYN NEW YORK 11231
4TH FLOOR
2 TON SHACKLE
2 TON SINGLE SHEAVE BLOCK
1 PARTS 5/8"ϕ
16'-4"
INBOARD END OF STRUT
STAY LINE
7 TON SHACKLE
7 TON 1 SHEAVE BLOCK
3RD FLOOR
TIE OFF TO COL. C8
3 PARTS 5/8"ϕ
2 SHEAVE 7 TON BLOCK
•SECTION A-A• 3/8"=1'-0"
•LOOKING SOUTH•
4 PARTS 5/8"ϕ WRAP
14WF 87 STRUT
2 PARTS 5/8"ϕ
7 TON SHACKLE
7 TON 2 SHEAVE BLOCK
TIE OFF TO COL C10
2 SHEAVE 7 TON BLOCK
6 PARTS 5/8"ϕ WRAP
•SECTION B-B • 3/8"=1'-0" •LOOKING NORTH•
PROTECT CORNERS AS PER DETAIL 1
8 PARTS 5/8"ϕ
25 TON SHACKLE
25 TON 4 SHEAVE BLOCK
35 TON 4 SHEAVE BLOCK
35 TON SHACKLE
TIE OFF FREE END TO COL.
MAIN FALL
25 TON CHAIN FALL
28'-0" MIN. REQUIRED FOR POSITIONING
25'-0" x 11'-0" OPENING
•POSITION OF PIECE AT LOWERING• 1/8"=1'-0"
NOTE: TACKLE SHOWN FOR ILLUSTRATION ONLY.
•LOOKING WEST• 1/8"=1'-0"
PIPE CORNER ROPE PROTECTORS.
BEAM OR COL TO BE WRAPPED
DETAIL # 1• NO SCALE
A9
A11
3 10/26/71 TIEBACK ELIMINATED TIMBER SHORE ADDED
2 10-20-71 GENERAL REVISIONS
1 6-1-71 GENERAL REVISIONS DUE TO BARE STEEL CONDITIONS
REVISION DATE DESCRIPTION
LOUIS GENDELMAN RIGGING AND TRUCKING INC. 611 COURT STREET BROOKLYN N.Y.
CHARLES M. SHAPIRO & SONS
CONSULTING ENGINEERS
615 COURT ST. BROOKLYN, N.Y. 11231
MOUNT SINAI SCHOOL OF MEDICINE THE MOUNT SINAI HOSPITAL 100TH STREET AND 5TH AVE. N.Y.C.
SCALE: AS NOTED
JOB NO: 70-201
DATE: MAY 1971
DWG. NO. 2 OF 2
STREET BROOKLYN NEW YORK

enough to make the weight of even a single part of line structurally significant. For this reason the hoist ropes below the upper block should be considered as part of the load.

During job planning it is often necessary to work out the load-handling process and determine the characteristics and criteria for special load-handling devices. These matters are outside the scope of this work and are mentioned only to indicate that additional derrick loads may derive from these devices. Likewise, winch location (elevation) controls overhauling characteristics and the size of the overhauling weight needed—winches for guy derricks that climb and for Chicago booms are often well below the derrick base.

Additionally, and this may be significant for high lifts on low-capacity derricks, the use of taglines on the load adds not only the weight of the lines but holdback forces as well. With wind and loads of appreciable wind sail area the taglines may have to be snubbed on a capstan or winch head (or on the bumper of a truck) to be effective; the holdback forces this implies can be considerable.

Just how much impact loading is appropriate to add for a particular installation is a matter for judgment. For steel erection, the AISC suggests 20%. In addition, the AISC notes (and this is true for all derrick applications) that impact will increase in proportion as winch-line pull capacity exceeds needed line pull and as rated load exceeds lifted load. The change in impact so noted is not necessarily in direct proportion.

For general lifting service, the FEM impact suggestions are incorporated in the expressions of Eq. (3.5). There are occasions when particular job-site conditions, at initial lift-off or on transferring the load into the building, can induce unusually high impact situations. These cases will be discovered and reflected in the design only when the planning work includes a thorough step-by-step consideration of each stage of moving the load into place.

It is difficult to offer advice on how important wind will be for a particular derrick installation design. The easiest advice to give, of course, is that wind effects should always be included. But in most design situations the effect of wind on derrick anchorages is not significant and the design would be adequate had wind been ignored. For tall gin poles and guy derricks, long-boomed stiffleg derricks, and gusty locations, wind can be an important loading in guys, bases, and swing lines both in and out of service. A derrick-installation designer will soon develop a "feel" or wind effects, however, and can then exercise judgment about their inclusion. Until that time, it is advisable at least to perform a rough evaluation of wind before excluding its effects.

Wind on the lifted load may be another matter, however, and operational planning had best include a check on wind action. This is particularly true when selecting boom tip side guys and swing or tagline arrangements. It is of crucial importance in planning procedures for loads that must be accurately spotted; otherwise, in many instances, work may have to be suspended at wind velocities well below the 30 mi/h (13.4 m/s) maximum normally permitted for derrick operations.

7.8 Winch Installation

Winches commonly used with construction derricks weigh about 10,000 to 20,000 lb (4.5 to 9 t), which is of the same order as the loads on the lead lines coming off the drums. What that means is that the winches must be anchored, that friction between the winch skids and the ground cannot be sufficient to resist the lead forces. As the lead lines transverse back and forth across the drums, when two or more fall on the same side of their drums at the same time, they set up a twisting moment that could move the winch into misalignment. Furthermore, the line loads induce a moment that tends to lift the rear of the winch if the front of the winch is anchored.

All of the unwanted effects mentioned above must be countered by fastenings designed to hold the winch in position and alignment. One option is to bolt the winch skid to the concrete floor below it, or similarly, to weld it to steel framing members if the concrete floor is not yet in place. Very often, however, it is more practical to tie the winch at its rear and sides with wire ropes. The ropes are made fast to framing members or other anchor points with turnbuckles used to snug up the lines.

Spreader beams are often necessary beneath winches to spread the weight and permit the structure below to take the load. These supports must be designed, and either steel or timber beams may be used.

The direction that lead lines exit at derrick bases is defined by the derrick design; on steel erector's guy derricks, for example, the lines run straight down through the bottom of the base, but for most stiffleg derricks the lines run out horizontally, bisecting the angle between the derrick legs. Often, however, the winch cannot be placed to receive the lead lines directly. When this occurs, it is necessary to install deflector sheaves to direct the lead lines to the winch. There is no limit to the ways in which deflector sheaves can be mounted, but there are clear-cut requirements for any mounting arrangement.

Deflector sheaves should be positioned so that when the ropes are loaded, they lead cleanly into the sheave grooves. When the alignment of the lead rope changes, as when Chicago booms swing, the deflectors must be mounted in a way that permits them to follow the swing.

The anchorages for deflector sheaves must accommodate the lead-line forces with consideration for the incoming and outgoing rope directions. The mountings also must be designed to keep the rope on the sheaves and the sheaves in position when unloaded. Figure 7.29 *a* and *b* shows two sets of deflector sheaves and their mountings that were used to carry lead lines from a stiffleg derrick to a winch placed more than 35 ft (10.7 m) below. Figure 7.5 includes a deflector arrangement used with a Chicago boom.

Chapter

8

Controlling Risk

Construction and rigging organizations have to be "lean and mean" to be competitive. That means they are limited in the personnel, material, and financial assets they can bring to bear on any particular lifting project. Hence, this chapter will offer guidance in using available assets effectively, by helping management decide where attention and planning are needed most, to what extent they should be utilized, and when to seek outside help. The means for accomplishing this will be through assessment of risk.

As used here, risk assessment is not a mathematical exercise, but rather a subjective application of knowledge and experience to the realities of the worksite. It weighs the difficulty of the work and the potential consequences of an accident.

There has been no industry-wide consensus on the division of responsibilities among the parties involved in lifting operations. Unfortunately, for many this becomes clear only after an untoward incident. In part, the uncertainty is because lifting equipment is used for such an extensive range of tasks, and in such varied situations. But recent efforts are starting to clarify these issues, and they will be discussed in this chapter.

The chapter closes by offering a rational scheme for quantifying risk. The system is judgmental, but comprises a framework that can be adapted by individual companies or projects to fit their own needs. When used thoughtfully, this approach can guide management towards more efficient use of company assets.

8.1 Introduction

One of the authors stopped at a jobsite some time ago to see if our crane-installation design had been properly executed. The crane, a large truck crane, was to place near-rated loads over the side after pickup off the rear. The crane was placed on a pontoon deck with the load-side outrigger floats centered on a retaining wall; the operating position for each set of loads was specified.

As he approached, a load was being swung at the specified radius and then, stopping that motion, the operator started to boom out to the placement radius. One outrigger beam broke free in its housing and then lifted while the other started to show signs of "softness" before the setting radius was reached.

The load was brought hack in and landed, the operator came out of the cab, and all concerned started to try to find out what had gone wrong. The only respect in which the installation differed from the design was that an off-the-rear screw-jack supplementary outrigger was in use. This optional accessory was not needed for this work and should not have affected over-the-side stability in any event. But when the operator was queried about it, the cause of the problem was revealed.

The operator had thought that perhaps some of the trucks delivering the loads would not be able to approach as close as planned and that greater reach over the rear might therefore be needed. He had noted in the past that the screw-jack arrangement of the off-the-rear outrigger did not permit the float to be fully preloaded. After setting the jack as hard as possible manually, he had thus added load by easing off on the hydraulic jacks of the rear standard outriggers.

That procedure produced a firmer "feel" to the crane in the rear sector, but on swinging over the side, the balance of forces was changed. With boom and load over the side, the pressure of that extra jack counteracted part of the machine weight and reduced stability. Full resisting moment, and therefore the ability to handle rated loads, could not be restored until the machine tilted sufficiently to unload the jack.

The principle in operation here is the same dictating that a crane be raised on its outriggers until all tires are relieved of crane weight. Although the operator apparently understood the rationale of this principle, he did not appreciate its analogy to the action of the supplementary rear outrigger float. What started out as an appropriate measure to improve lift safety instead had the opposite result.

8.2 Sources of Risk in Lifting Operations

Crane and rigging management are responsible for getting the job done safely and on time—while making money in the process. Whether consciously applied or not, risk control is embedded in the methods employed by successful managers. However mindful managers and employees are of the risks, and diligent at eliminating them, some accidents are inevitable; that is reality.

A magazine article examining a particular aircraft disaster in extensive detail describes three classes of errors that underlie accidents involving a mix of people and technology.† The article's author lists those classes as procedural, design, and system. We will now consider how they apply to crane and rigging work.

Procedural errors are errors of implementation; avoidance strategies would incorporate practical plans clearly communicated to field supervisors who are competent to diligently guide their execution. The crews executing the work must therefore be experienced, trained to their tasks, and supervised during performance.

Design errors are usually thought of as the responsibility of manufacturers and therefore out of the control of equipment users, but the class also includes equipment and methods originating with the user. A defect in the design of the lifted load would be in this category too. Of course, diligence and competency of designers and planners are key, but another element of risk avoidance is feedback—from field to planning office and from planners to vendors of material and equipment. Without feedback, there is no opportunity to learn from mistakes (or to discover them). White collar managers and credentialed engineers should be encouraged to overcome aloofness in order to learn from their blue collar colleagues. Too often modifications are made to head office planning by field supervisors. Their companies had not sought feedback, leading to delay, unnecessary cost, and heightened risk.

System errors are an outgrowth of complexity itself—they are often characterized by the left hand not knowing what the right hand is doing. Modern crane and rigging operations might involve input and oversight from numerous parties, and too often one group has no comprehension—or even awareness—of the work of the other individual participants. The system, as used here, is the big picture. It includes everyone and everything involved, from the crane manufacturer to the laborer smoothing the ground.

Unavoidably, the preparations for crane and rigging operations must rely on all of the parties; those involved could include engineers—equipment designers, building or facility designers, and installation designers; managers—of personnel, equipment, maintenance, traffic, and operations; and planners, schedulers, vendors, and regulators. The execution of the work itself often calls for the coordination of several trades, and may encompass activities at more than one location. Errors are likely to be made during the course of exe-

---

† The Lessons of ValueJet 592, William Langewiesche, *Atlantic Monthly,* March 1998.

cuting these disparate events and activities. Although each individual error may by itself be minor and not likely to cause an accident, several small errors combined could interact to produce disaster.† The greater the complexity, the greater is the opportunity for such interactions, and the less likely errors will be discovered and corrected. Therefore, whenever possible, managers should employ strategies to simplify and separate activities. There are a number of ways to accomplished simplification including:

- Forethought during selection of equipment—when crews are not experienced with particular machines, their use adds complexity. Some heavy-lift rigs require elaborate assembly procedures that depend fully on the competence of one or two technicians;
- Selection of control procedures—surveying instruments, electronic sensors, TV monitors, and human observers can be employed to monitor alignments, forces, and movements, but those multiple inputs to supervisors add complexity to control and decision making;
- Separation of portions of the operation into different distinct independent procedures—when portions of the operation can be isolated, performed and checked independent of the rest of the procedures, the system becomes less complex;
- Project-specific training—individuals trained for the tasks they must perform in a planned operation will be less likely to react in unexpected ways; this will reduce the likelihood that wayward behavior will complicate the actual event; and
- Advance study of the risks and benefits associated with particular simplification schemes.

Dry runs can prove to be a very effective way to reduce systemic risk during complex operations because they can uncover contingencies that were overlooked during planning, identify tasks that crew members do not fully understand, and highlight coordination problems. These practice runs are typically conducted without load, and are best followed by a full-crew debriefing. That gives each team member the opportunity to contribute feedback.

Consideration of maintenance and training is beyond the scope of this book, but we must start our consideration of risk control by as-

---

†"In complex systems, a simple failure rarely leads to harm. Human beings are impressively good at adjusting when an error becomes apparent, and systems often have built-in defenses. . . . When things go wrong, it is usually because a series of failures conspire to produce disaster." From *When Doctors Make Mistakes* by Atul Gawande, The New Yorker, Feb.1, 1999. The authors' work in accident investigations have revealed the same paradigm.

suming that all equipment is in suitable condition, having been properly maintained, and that personnel are sufficiently trained for their duties. And further, we will take it as given that neither drugs nor alcohol are permitted on the jobsite. If those matters have not been addressed, there is little point in considering the risks associated with the use of the equipment to lift loads.

Among crane and rigging accidents, there is a small percentage that arise from causes that could not have been reasonably anticipated, nor could they have been prevented. Irrational acts, improbable convergences of events, and acts of God are within this percentage. No meaningful risk reduction would be achieved by focussing on such uncontrollable events.

Attainable improvements can follow, on the other hand, by identifying tangible problems and implementing specific remedies. Risk categories where affirmative steps can be taken include:

- Pressure from cost/time constraints
- Inexperienced management
- Lack of training/skill
- Inadequate planning
- Unreasonable demands of management
- Environmental conditions
- Unclear instructions
- Operator errors
- Changed circumstances

Some aspects of the last four categories, however, defy prediction. To those, adding misjudgments and irrational acts completes the list of risk categories for which preventive actions are not likely to be fruitful.

An analysis of accidents reveals that the majority are the result of more than one contributing cause. The design standards and practices governing hoisting equipment usually enable them to overcome one error or failing, but adding a second or third mistake can overcome their inherent resistance to mishap. Most errors posing lifting equipment risk can be anticipated, making protective actions feasible. Each of the risk categories therefore warrants discussion.

Pressure from cost/time constraints

Most people strive to perform according to expectations, but some aspire to "set the world on fire". The latter is an example of what could be called internal cost/time pressure. Supervision and project schedules—and a job behind schedule—can create an atmosphere of tension, tending to force people to proceed faster than their comfort level or circumstances allow. The "too low" bid can result in an example of this. Those situations, usually trickling down from managers, could be called external cost/time pressures. But whether external or internal, the effects of the pressure can be similar.

Lifting operations and rigging work do not take well to efforts to move them more quickly than their proper pace, which should be unhurried. Some observers may assert they are often glacial. Fast crane lifts, swings, and landings introduce dynamic, or inertial, forces which act on the massive spring-like characteristics of crane components, and on the load, to make load movements irregular and harder to control. This gives rise to risk to the crane, to nearby objects, to the rigging crew and to workers in the vicinity. With the possible exception of duty cycle work, crane operations are meant (by equipment designers) to be gradual and deliberate. Generally, more work can be produced in a given period of time when a crane operates at a moderate speed, because the operator can better control the load in space and while landing, and because the crew would not be so apprehensive about being hit or pinned by the load.

Field people should bear in mind that schedules and budgets have been created by fallible human beings, in an office setting, making estimates and using their judgement; these data do not necessarily reflect ultimate truth, and indeed may not even be near correct. Management should recognize that budgets and schedules are goals, not absolutes, and should not encourage their field people to think otherwise. Further, an imposed atmosphere of pressure and a culture of "getting it done in spite of all odds" is apt to ultimately bring grief.

Some companies offer rewards or bonuses for work brought in on time or below budget, or for meeting specified production goals. This may appear to be forward looking and enlightened means to motivate staff, and often is, but unless backed up with effective discipline could lead to cutting corners, slipshod work, and added risk. Cutting corners suggests doing less than what is required and compromising safety. One may frequently get away with this, but eventually tragedy will strike.

A recent example of an overturning accident, which caused one fatality and a serious injury, provides a good example of time pressure affecting risk. A construction contractor had his own crane and long-serving crane operator on site. As bad luck would have it, that crane broke down during the morning before a large and important concrete pour. The job was on a ten-hour, four-day per week schedule, and the pour was set for a Thursday, which would allow the concrete to cure over the long weekend. To permit work to go forward, a local rental crane was called in, but it did not arrive until mid-afternoon. The rental crane operator, under pressure due to the lateness of the start, did not take the time to install the latticed extension, which would have materially increased lift capacity at the pertinent radii. Other circumstances led to use of an oversized concrete bucket and eventually to an overfilled bucket and tragic results. The company crane operator later stated that he would have objected to the oversized concrete bucket and refused the overfilled load, and that it was well understood company policy to respect a crane operator's objections. The rental operator was not aware of this policy—he was aware only of the pressure of time.

Inexperienced management

Experience cannot be learned—it can only be accumulated, and people either have it or they do not. When an individual attempts to work beyond his or her experience level, risk is introduced because that person may not fathom the limits of personal knowledge. The result can be either deficient or faulty planning, or shortcomings in execution. The experience required for an activity depends on the complexity of the equipment being used, or of the operation, or both. Should there be uncertainty about a manager's experience relative to the task, consideration ought to be given to bringing a suitably experienced person on board. We do not mean to suggest that managers should be denied the opportunity to learn by engaging in activities that are new to them, but we do urge awareness that the quest for learning should not overshadow a healthy regard for life, limb, and property.

Lack of training/skill

Earlier in this Chapter we stated that training is beyond the scope of this work, and that we must assume personnel are sufficiently trained for their duties. However, not all members of a particular work crew will be equally trained or experienced, nor is this often necessary. But when crew assignments are made, consideration should be given to the training, skill and experience required for the particular duties.

Crew members rely on one another to take needed actions at appropriate times, and one member's failing to do so could cause injury to another. Working with suspended loads is not a fully predictable exercise, because load behavior relies on a number of factors, including the actions and interactions between the crane or derrick operator, crew members, and the wind; the reactions of the slings, hoist ropes, and other crane components, and of the load itself. Crew members well placed and alert to respond to, and check, an unexpected load movement could protect other crew members from harm. It is experience that conditions people to react appropriately to the unexpected, and to prevent an incident from becoming an accident.

Nuances of crane and rigging work are not easily explained and are best learned, hence the time-honored practice of the tradesman passing on experience to the apprentice. An effective apprenticeship program, and an appropriate mix of journeymen and apprentices, are essential to sustaining skill levels over time.

Often crane operators are supplied by a union hall or just arrive with a crane from a rental house. Management and site supervisors do not really have the opportunity to assess the operator's competence other than by observing the operator at work. The exception would be those few large projects where formal testing procedures are in place.

There is now another option. A new organization called the CCO† conducts written and practical tests for mobile crane operators in a controlled environment, and issues a certificate to those passing. Accredited by the National Commission for Certifying Agencies (NCCA), the CCO is a not-for-profit organization that is:

- National in scope,
- Operated by the private sector,
- Independent of labor relations policies, and
- Tailored to different types of mobile cranes.

By requiring CCO certified operators, management can be assured that mobile crane operators will have demonstrated they have the basic knowledge and skills to operate the equipment safely. However, using certified operators will neither guarantee safety nor competence, but only make both more likely.

Inadequate planning

The extent of planning necessary for a particular operation is site and circumstance specific, however, at least minimal planning is needed whenever a crane or derrick is to be used to lift loads.

At one extreme is a crane at an open field site handling a load that is well below rated load, lifting it from the bed of a truck and landing it nearby. To do this simple task requires an adequate number of people, slings and fittings, and perhaps some dunnage material. Planning, therefore, can be informal, and may consist of nothing more than conversation between some of the individuals involved. On some pro-

†National Commission for the Certification of Crane Operators, Fairfax, VA.

jects, however, planning may need to go a little further and could require a written work order listing the personnel and equipment needed and the means to procure them, and perhaps requiring costing codes for the various items.

At the other extreme is the major lift which may utilize one or more primary lifting cranes, assist cranes, load transport arrangements, custom made rigging accessories, an extensive crew including specialists, and coordination of activities between a number of unrelated companies. Here planning starts well before equipment and people arrive at the jobsite, and may include detailed drawings, calculations, materials lists and orders, written procedures, specifications, check lists, time or critical path scheduling, monitoring and control equipment and procedures, and a myriad of other details.

In between those extremes are the overwhelming majority of lift assignments, where the amount of planning needed may not at all be clear, and where decisions must be made about how much planning to do. The authors maintain that for these situations, considerations of risk offer a logical basis for making decisions about the planning efforts that would be most beneficial. At the end of this chapter, a rational system will be presented to assist jobsite and lifting task management in doing so.

When planning has been well done, the appropriate crane or derrick will be sent to the site and the accessory equipment and people needed for the work will be on hand. The site will have been scouted to verify, among other considerations, that the crane has enough room for assembly, travel, and work, and that there is no danger from overhead power lines or other obstructions. When planning has been lacking, something needed may be missing or inappropriate, and field supervision may feel pressured to make do with what they have. Experienced field foremen and their coworkers have been known to demonstrate remarkable ingenuity in overcoming such obstacles and successfully completing their tasks. But the other side to that coin has been avoidable accidents with deaths, injuries and property damage.

Unreasonable demands of management

There are a number of ways in which management with a deficient understanding of crane and rigging work can impose its will inappropriately bringing unnecessary risk to the jobsite. Unrealistic or unachievable cost allowances and schedules have already been mentioned. Some will argue, as a matter of fact, that management shortcomings can be imputed as underlying most of the risks mentioned in this discussion, because it is management's duty to create and maintain a safe work environment. We leave that argument for the lawyers, and merely observe that a company's operations should benefit from consideration of risks and the actions needed to ameliorate them.

Manager's attitudes can directly contribute to risk. In some organizations, management shies away from actual participation in field activities, instead they only order what needs to be done. In effect management says, "Just do it. We have our own problems, and don't want to know about yours." That stance often will challenge field people, bringing out the best in them by compelling them to think and to innovate, but then again not all field people are thinkers or innovators.

Lifting and rigging activities, particularly the more complex ones, possess a degree of unpredictability, often requiring more time to complete than even a generous estimate allows. Or, the nature of an operation may call for continuous work until completion, forcing crews to be kept on duty for ten, twelve, or even sixteen hours straight. By nature, some work must be fit into down-time or off-hours. Needless to say, an individual's reaction time becomes longer and judgement becomes less sure and less dependable as the day stretches far beyond normal working hours—more errors are made and improper conditions go uncorrected. Having investigated many accidents, the authors have noted a repeated theme of failures taking place after long hours of work. We have never been able to discover a direct link, but the inference remains that long hours were a contributing factor to the accident. Management should neither require nor demand excessive hours. As a simple expedient, when it becomes necessary to work long hours, consideration should be given to allowing rest for those individuals who, when fatigued, pose a materially heightened risk.

Environmental conditions

Elements in the environment that can adversely affect lifting operation risk include wind, precipitation, light, noise, site conditions, and adjacent activities.

Chapter 3 covers action of wind, and offers extensive means for calculating wind loads, while Chapter 5 deals with wind on mobile cranes and Chapter 6 on tower cranes. Those chapters provide information for job pre-planning—what you might call "wind on paper". However, no level of planning will immunize the jobsite to the hazards of wind. Low-level gusty winds can make loads move excessively and unpredictably, endangering the load landing crew and raising the possibility of hitting other objects. Thunderstorms, predicted to be "normal", can strike with unforeseen intensity, and even tornado effects. Field supervisors may be faced with decisions respecting measures to deal with the wind after it arrives. Sometimes events can overtake even the best judgement, and damage occur before any defensive steps can be taken. The vagaries of storms can create risks without remedies and demand difficult decisions on short notice. For example, a supervisor in the middle of a concrete pour when a storm strikes is in no enviable position—perhaps having to choose between a cold joint in the concrete or a perilous continuation of the crane operation. This is the type of occasion when the question of balancing risks against costs has a degree of poignancy that leaves no room for prolonged debate or contemplation.

Rain and snow restrict visibility. When work proceeds during precipitation, hand signals may need to be modified or abandoned in favor of other means of communication. Slip and fall dangers may inhibit—or at least slow down—some work.

On occasion, it becomes necessary to carry out crane or derrick operations at night. Planners and managers must recognize that artificial lighting can never be equivalent to daylight. Lighting should therefore be planned with the same care as the rest of the operation. Obstructions in the path of the load and boom should be lighted and the positions of lights selected to avoid glare. Operations must be anticipated to be slower than in daylight. These same considerations might also apply to indoor lifting work.

Job supervisors should be prepared for changing conditions of visibility. Sun glare may suddenly inhibit the operator's ability to see signals. A delay may prolong a lift operation into dusk.

One of the authors, as a young engineer, was in charge of dismantling a tower crane inside of a large hyperbolic cooling tower. Aware of how early darkness engulfed the interior of the tower, he instructed the dismantling crew to wait until morning to make the difficult boom pick. He was overruled, and although the pick was started during daylight, it was not concluded until the interior was shrouded in deep shadow. Neither operator or riggers could see the boom tip of the truck crane as it lowered the load, nor see it bend like a fishing pole because the swing was locked during boom lay down. Finally noticed by the operator, the swing was released producing an abrupt and violent reaction. Disaster was narrowly averted.

Noise can bring about risk when it distracts crew members, or startles them, and when it interferes with radio or voice communication. High ambient noise levels limit verbal communication, but are not generally unanticipated. Fortunately, hand signals are universally recognized so that special measures need only be planned when more than hand signal communication is needed.

The environment also includes the site. Physical aspects of site conditions that could pose risk include the suitability of the area selected for crane assembly and set-up, and load pick up and lay down zones.

The condition of the ground and soil under the crane are another concern, as are pits, trenches, cellars, and underground utilities. Above ground obstructions to the path of the moving boom, load, counterweight, and gantry may pose hazards, particularly power lines. Preplanning should address these items, but for short-term assignments, such as "taxi-crane" operations or short term rentals, planning is usually done at the site after the equipment has arrived, and it is the crane operator who is encumbered with the task. Risk arises because of the limited time, resources, capabilities, and authority available to the typical operator to resolve some of the issues.

Activities adjoining crane and derrick operations can impose risk on the lifting operations too. Noise has been discussed, but excessive dust can also affect visibility, or even divert the attention of an operator by making breathing difficult. Plumbing contractors seem to delight in turning up unannounced with a backhoe to dig a trench at the edge of a crane, undermining support. In urban areas, they may do this at night so as to avoid interfering with traffic, and surprise the crane people when they arrive in the morning. Lastly, the presence and movement of people and machines in the immediate vicinity of lifting work can distract both the crane operator and crew from close attention to their own work.

Unclear instructions

Instructions include both the written and the verbal variety, such as rating charts, operator's manuals, notes on drawings, project specifications, written procedures and check lists, face-to-face, telephonic, and radio orders and explanations, and even crane or derrick hand signals. When any of them are ambiguous or misunderstood, risk ensues. Hand signals, of course, are affected by ambient light and by distance; when the operator cannot clearly discern intent, operations should not be permitted until signal clarity has been assured. An intermediate signal person or radios may be wanted.

Written and spoken communications need to be plain, direct, concise, and explicit using basic language appropriate to the recipient. One way to verify understanding of verbal instructions is by asking the other party to repeat the instructions in their own words. Teams that have worked together for years sometimes need only sparse communications, often little more than simple gestures. For those in less familiar circumstances, critical communications should be confirmed more explicitly and regularly. Confusion and coordination failures can also result from verbal changes to written procedures; be aware of the hazard when that must be done.

Using a radio to transmit signals to crane or derrick operators is not at all as simple as first impressions may imply. Radios can be interrupted by interference or by battery failure at either end. How then does the operator know when to stop moving the load? How can the operator be sure that the load will be slowed in time to avoid sudden brake application and bouncing as the load destination approaches?

Whenever radios are used, a signal communication protocol needs to be established, explained to the appropriate people, and enforced. It is our opinion that the signal person should maintain continuous chatter during signaling, by repeating motion commands over and over, and then advising distances as the endpoint approaches. In this way the operator will know to stop when the chatter stops, because this implies loss of contact.

Operator errors

Many accidents attributed to operator error are nothing of the kind. Insufficient planning can put crane operators in the unenviable position of having to make judgements that are beyond what should reasonably be expected of them. An example of this occurs when an operator is required to evaluate the ground beneath the crane and to select support cribbing when conditions are unusual or are not typical for the vicinity. Civil engineers devote class time to studying soil characteristics and behavior—and will usually require tests before rendering a judgement. Crane operators study the practical side of soil mechanics, but only by virtue of experience on the job. When conditions do not match their experience, they are being unfairly burdened with a task for which they have not been prepared, and which they may not perform successfully. To make matters worse, it is unlikely that adequate cribbing materials will be found on site to deal with the sub-standard conditions. An unsure operator using deficient materials is not apt to produce suitable crane supports.

Wind presents another series of situations for which crane operators are often burdened with responsibilities beyond their calling. Most mobile crane rating charts include a note requiring the operator to reduce rated loads to account for ambient wind. Chapter 3 demonstrates how complex and mathematical such a task can be for an engineer. An operator is not blessed with the analytical tools of the engineer, but must make those decisions nonetheless, based only on experience and judgement. Is this a reasonable burden to place on the operator?

It is not unusual for an operator to arrive at a jobsite from the union hall to operate a particular crane model never seen before. Requiring that operator to go right to work, without some warm up or "familiarization" time, raises the risk of mishap, because each crane model has its own peculiarities, feel, and response.

After all of that, operator errors do occur. They occur because of human frailties, or because of misjudgments when an operator has to react instinctively to an unexpected event. One-day-rental operators often leave the yard at a very early hour and have near a full day in by lunchtime. When work runs late, the operator will be more fatigued than the crew, and therefore more prone to making errors.

Changed circumstances

Here is where even the best planning can be foiled. Some party at the site does the unexpected contrary to agreements or understandings. It includes the plumber making a "surprise" trench, the excavator or materialman piling something where it shouldn't be, the "hot" wires that were supposed to be de-energized, or not there at all, and the promised delivery that isn't made. These changes often introduce risk, because now equipment may not be well matched to the task, or procedures well thought out may require "on-the-spot" changes, or time allotted for planned events may be lost to overcoming the new circumstances.

Changed circumstances are less likely to be disastrous when planners select a crane and position it so as not to be pushing the limits. A little extra capacity, boom length, clearance, and reach build some flexibility into the set up. But that may or may not be enough to overcome the new circumstances.

Conclusion

Crane safety and productivity are intimately associated, but not in the sense often assumed—to gain productivity it is necessary to sacrifice safety. The truth is quite the contrary. Measures taken to ameliorate risks will enhance safety for sure, but those same measures will also promote maximum equipment utilization and productivity. When people have confidence in the equipment, their co-workers, and the plan of action, they will perform at their best.

However, given the most thorough planning, the best equipment, and the most highly skilled personnel, some accidents will occur nevertheless. There will be instructions that are misunderstood; errors made by well trained and experienced crane operators; changes in circumstances that cannot be dealt with, such as sudden, unexpected high winds; misjudgments by supervisors and crew members who are properly qualified and experienced; and irrational acts of individuals, acts that offer no explanation and that could not have been anticipated.

A convergence of errors

A steel-erection subcontractor, we'll call them the ABC Company, undertook to construct the framing for a large-footprint, very-tall, long-span, single-story building. Their low bid, by about half a million dollars, anticipated operating heavy crawler cranes on the ground floor concrete deck built some 20 ft (6.1 m) above the cellar floor. The deck was robust, having been designed for heavy loadings.

Their lifting plan was accepted subject to the deck being shored. Shoring would cost about ¼ million, which was neither expected nor included in ABC's bid estimate. ABC protested, produced their numbers, and asked the owner to pay for that work, in which case they would still be low bidder. The request was refused, and ABC was ordered to proceed with the job. *Management,* the owner, *was being unreasonable* in rejecting that proven, equitable claim.

Now, *circumstances had changed* and a carefully worked out plan was scrapped and was replaced with an entirely new crane operational set up. The bid was also based on preassembling full-bay segments of wall framing on the ground to permit faster, lower-cost erection. That feature of the plan had to be retained.

The new lift plan required the cranes to erect from outside the building, which necessitated long travel runs with load. It was a quickly put together plan with little consideration for travel paths or crane support—in that respect, *the planning was inadequate.* The soil in the travel areas was of low capacity, unevenly compacted, and in need of grading.

The first load, of about 150 tons (136 t) measured about 120 ft (36.6 m) high by about 50 ft (15.2 m) wide, and had to be traveled about 700 ft (213.4 m) from assembly area to placement. Two cranes were used to trip the panel, and then travel commenced with the load under hook. The crane bogged down in the soil after a short run, one track digging in deeply, and exposing the boom to a high side loading. With assistance from the second crane, recovery was accomplished and the boom lowered and inspected for damage. Now, steel plates were procured for distributing track loads, but only a small number were available.

Travel continued without further mishap until the travel path came to a jog where the crane had to cut sharply around a projection on an adjacent building. Before cutting, because of the width of the load, the crane had to swing the load full over the side. So far, except for questionable ground support, the operation was in full compliance with the crane manufacturer's instructions.

After cutting left and then right, the crane started up a short upgrade of about 2% just a little distance from the setting location. The crane was stopped after breaking over the crest onto level ground, so that measurements could be taken for final crane positioning. But when stopped, about 1/3 of the track bearing length, at the rear, was not in ground contact—it was in the air because of the grade. What is more, the track closest to the load, the more heavily loaded track, had been stopped with a hard point (the top of a concrete foundation wall corner) at the track center. Most of the weight of the crane and load was concentrated on only a portion of one track, supported on a hard point with soft soil alongside.

The crane operator reported that he had been sitting there with his hands off the controls, doing nothing for about 2 minutes when all of a sudden the load took off to his left and tore the boom off. Although that may sound like a tall tale, it is exactly what happened. The CG of the crane and load was situated to the rear of the hard point, over soft soil and a few inches from the edge of the slope. The hard point would not give, but the soil did, slowly, ever so slowly, making the crane lean backwards. With the lean, the CG moved to the rear, until it passed the break of the slope. Suddenly, the crane tilted back until the rear of the tracks made ground contact. The load, of course, moved with the crane with the disastrous results described. Although there was substantial property damage, remarkably no one was injured.

The *pressure of time and cost constraints* led to inadequate travel path preparation, an unacceptable grade, and too few plates being used for crane support. But, the final straw was the *misjudgment* of stopping the crane where the tracks were not fully and firmly supported and in the presence of a hardpoint.

Each of the five risk categories involved in this accident played a part in the final denouement. Had the owner been reasonable and paid for shoring, there would have been no changed circumstances and hence nothing for us to discuss. Had more time and money been allowed, the travel path would have been properly graded and sufficient track support furnished—the lack of shoring and the changed circumstances would have been overcome. Had preparations been fully made and the crane's final location laid out in advance, the stop would not have been needed, and the accident would not have occurred. But given all of that, had the foreman used good judgement, the crane would have been moved farther forward to an adequate support area before being stopped, an additional move of less than 10 ft (3 m). The foreman had been placed in the position where he could make a bad judgement only because the other elements were present. Absent any one of them, he would not have been called upon to make such a judgement.

8.3 Responsibilities

There may be a number of independent parties taking part in one lifting operation. Who those parties are, and the parts they play, will vary with the nature and complexity of the operations. Again, there are extremes, from the taxi-crane making a single lift for a local business to the major building or petrochemical project. For that reason, there can be no single set of guidelines to cover all situations. There are, however, some principles which may assist the parties in sorting out responsibilities.

To clarify the role of each party, an SC&RA† task force has done considerable work with the intention of developing an industry understanding about who is responsible for what. The material that follows has been drawn from the SC&RA‡ efforts but with some minor alterations by the authors.

The parties that may be involved in lifting operations are:

Project Management—the party controlling the overall operations which may include activities other than lifting work; the project owner is included.

Crane Owner/Supplier—the party furnishing the crane or derrick, who may not have involvement in the use of the equipment at the site.

Site Supervision—the party supervising the people at the site which may include people not involved with the lifting work; this party may also be the lift director.

Lift Director—the person directing the lifts and the personnel involved in the lifts.

Crane or Derrick Operator—the person at the controls of the crane or derrick.

The following is a general list of responsibilities that could be adapted to most crane and rigging operations:

Project management:

1. Review proposed lifting operations for complexity and/or risk and verify that appropriate planning has been undertaken by other par-

†Specialized Carriers & Rigging Association, Fairfax, VA, a trade association with more than 1000 member companies engaged in crane, rigging, millwrighting, heavy specialized transport and related activities.

‡SC&RA pamphlet, *Guidelines for Crane Operations.*

ties; obtain assistance from an independent party, when appropriate, to take part in the review.

2. Coordinate activities to avoid interferences between lifting and other operations.
3. Arrange for site conditions necessary for proper lifting operations, in conjunction with Site Supervision.
4. Cooperate with Site Supervision in measures concerning electrical hazards when there are power lines present on the site.
5. Advise Site Supervision of the presence of underground utilities and sub-surface construction, and of the nature of other site activities that may impact lifting operations.
6. Know which local, state, and federal rules and regulations and which industry standards are applicable to lifting operations.
7. Monitor Site Supervision for general compliance with safety laws, rules and regulations.
8. Provide training and upgrading programs for project management personnel to impart knowledge concerning lifting equipment use and safety.

Crane owner/supplier:

1. Furnish a crane or derrick in proper condition and properly maintained.
2. Furnish a crane or derrick including necessary parts and components to satisfy the configuration and capacity requirements given by Site Supervision.
3. Furnish a crane or derrick complete with applicable rating charts, operator's manual, control identification labels, hand-signal placard, electrical hazard warning placard, and warning labels as originally furnished by the manufacturer.
4. Have maintenance and repair records available, if needed.
5. On request, assist the other parties by furnishing crane or derrick data and technical information.
6. Know which local, state, and federal rules and regulations and which industry standards are applicable to lifting operations.
7. Furnish competent personnel to assist in assembly and disassembly if requested.

Site supervision:

1. Visit the site to obtain first hand knowledge of conditions; particularly note power lines, obstructions, limitations on access and operating areas, and if personnel may have to be lifted with the crane or derrick.
2. Evaluate the operation to the extent necessary to select a crane or derrick with sufficient capacity, height, reach, clearance and other characteristics to do the work safely; arrange for the special provisions that may be needed for work in proximity to power lines or lifting of personnel; when appropriate, outside assistance should be sought.
3. Review operations for the purpose of identifying elements of risk and reducing or controlling them; when appropriate, outside assistance should be sought.
4. Verify that crane or derrick operations have been coordinated with other activities at the site.
5. Check with Project Management for the existence of underground utilities and sub-surface construction.
6. Verify that the crane or derrick provided is in proper condition and has been maintained as required; ropes should be verified for condition, length, and suitability for the work.
7. Know which local, state, and federal rules and regulations and which industry standards are applicable to lifting operations.
8. Verify that Lift Director is aware of and informed concerning lifting equipment safety laws, rules, and standards.
9. Brief the Lift Director as applicable on the data and rationale used for selecting the lifting equipment; review the parameters of the work, limitations of equipment, and potential hazards and their remedies, particularly for operations in the vicinity of power lines and the lifting of personnel.
10. Monitor the Lift Director concerning training and competence of personnel and for adherence to safety laws, rules, and standards.
11. Brief the Lift Director on the plan of lifting operations.
12. Before the lift equipment arrives at the site,
    a. Verify that access is suitable with respect to width, length, overhead clearances, turns, grading, and surface condition.
    b. Verify that there is sufficient space to erect or assemble the boom and its attachments, and that blocking is available, if needed, for boom assembly.

c. Verify that the crane operating area is suitable with respect to levelness, surface condition (compaction), proximity to slopes, power lines, underground utilities, sub-surface construction, and obstructions to crane operation.

13. If necessary, in coordination with the Crane Owner/Supplier, or using outside assistance, assess the ability of the ground to support the crane in its working condition.
14. Verify that blocking or mats are or will be available as may be needed for crane or outrigger float support.
15. Verify that barricade material is or will be available as may be needed to keep personnel or the public away from counterweight swing or away from crane operations.
16. For equipment that will be at the site for long periods of time, verify that appropriate maintenance and inspection procedures are taking place and that appropriate records are being kept.
17. Be aware of the unique differences in lifting equipment operations when working under the following specific conditions:
    - Multi-crane lifts,
    - Hoisting personnel platforms,
    - Clamshell operations,
    - Pile driving and pulling sheeting,
    - Concrete operations,
    - Demolition operations,
    - Magnet operations, and
    - Operation of barge mounted lift equipment.
18. For operations in the vicinity of power lines, verify that all concerned parties are instructed in OSHA† and ASME/ANSI B30.5‡ requirements for operational safety; inform personnel of the dangers and limitations of protective measures against electrical hazards, such as:
    - Proximity to electric power lines,
    - Proximity to radio and microwave transmitters,
    - Grounding (earthing),
    - Proximity warning devices,
    - Insulated links, and
    - Boom cages.

---

† U.S. Department of Labor Occupational Safety and Health Act, 29 CFR 1926, Construction Industry Standards, or 29 CFR 1910, General Industry Standards, as applicable.

‡ Safety Standard for Mobile and Locomotive Cranes, An American National Standard, American Society of Mechanical Engineers, New York, NY.

19. If personnel are to be lifted, verify that a proper personnel basket is or will be available and that personnel involved are informed of OSHA and ASME/ANSI B30.5 requirements; monitor the operation for conformity with those requirements.
20. Monitor lifting operations for conformity with lift plans and general compliance with safety laws, rules, and regulations.
21. Verify that the crane or derrick operator is knowledgeable or experienced with the lift equipment and attachments to be used; if in doubt, arrange for an operator test, or for practice time in a non-critical location. To assist crane or derrick operators in meeting their responsibilities, it is recommended that jobsite policy permit operators to refuse to handle a load until satisfied it can be handled safely.

Lift director:

1. Preparations for lifting operations, to include:
   a. Review planned operations with Site Supervision including working height, boom length, working radius, quadrant of operation, load weight, load dimensions, center of gravity and required blocking.
   b. Locate and identify site hazards such as electric power lines and piping (above and below ground) together with Site Supervision.
   c. Know which local, state, and federal rules and regulations and which industry standards are applicable to lifting operations.
   d. Know limitations of protective measures against electrical hazards, such as:
      - Proximity to electric power lines,
      - Proximity to radio and microwave transmitters,
      - Grounding (earthing),
      - Proximity warning devices,
      - Insulated links, and
      - Boom cages.
   e. Be aware of the unique differences in lifting equipment operations when working under the following specific conditions:
      - Multi-crane lifts,
      - Hoisting personnel platforms,
      - Clamshell operations,
      - Pile driving and pulling sheeting,
      - Concrete operations,
      - Demolition operations,

- Magnet operations, and
- Operation of barge mounted lift equipment.

f. Verify that the lifting equipment has been maintained and inspected and is in suitable condition for the work.

2. Preparation of personnel to include:
   a. Verifing that the crane or derrick operator is knowledgeable or experienced with the lifting equipment and attachments to be used.
   b. Reviewing operations with the crane or derrick operator and lift crew; discuss limitations of the equipment, potential hazards and measures that are to be taken to minimize risk. To assist crane or derrick operators in meeting their responsibilities, it is recommended that jobsite policy permit operators to refuse to handle a load until satisfied it can be handled safely, and that personnel be so informed.
   c. Verifing that the load riggers are experienced in slinging loads and in selecting suitable slings and fittings, when appropriate.
   d. When operations will be in the vicinity of power lines, see to it that all concerned parties are instructed in OSHA and ASME/ANSI B30.5 requirements for operational safety.
   e. When personnel are to be lifted by crane or derrick, verify that a proper personnel basket is or will be available, and see to it that personnel are informed of OSHA and ASME/ANSI B30.5 requirements.
   f. Be aware, and make sure that the crew is aware, as applicable, of the unique differences in operations when working under the following specific conditions:
      - Multi-crane lifts,
      - Hoisting personnel platforms,
      - Clamshell operations,
      - Pile driving and pulling sheeting,
      - Concrete operations,
      - Demolition operations,
      - Magnet operations, and
      - Operation of barge mounted lift equipment
3. Supervision of the work, to include:
   a. All work concerning lifting equipment.
   b. Checking that personnel are present and fit for work, including required protective clothing and tools.
   c. Verifying that the crane or derrick operator is performing all required daily maintenance and inspection tasks.
   d. Verifying the crane or derrick operator is aware of his or her responsibilities.

e. Establishing proper communications between the operator, crew and signal person, such as the use of hand signals, radios, etc.
f. If needed, assigning competent signal person(s), experienced in crane or derrick operations and in hand signals, and verifying that they are positioned where most effective.
g. Checking that proper slings, fittings, and lift accessories are at hand and that they are in acceptable condition.
h. Supervising the installation of the rigging on the load and hook.
i. Verifying that the crane is level and blocked, if appropriate, and positioned in the proper location.
j. Determining or verifying the weights of the loads to be lifted and making certain that this information is given to the load riggers and the crane or derrick operator.
k. Checking that loads are well secured and balanced in the slings.
l. Checking that the lift and swing path is clear of obstructions.
m. Checking that loads are free to be lifted, for example, that loads are not restrained by mud suction, frozen to the ground, or caught on adjacent materials.
n. Avoiding carrying loads over people.
o. Restricting access to work areas by unauthorized personnel; verifying that barricade material is or will be available as needed to keep people away from counterweight swing and from the potential impact area if an accident were to occur.
p. Checking for wind and take needed precautions to maintain stability of the load, and to avoid wind bringing the crane to a tipping condition, or exerting excessive side load on the boom, or overpowering personnel on tag lines.
q. Taking note of environmental conditions, such as temperature, lighting, and adjacent activities, and adjust operations accordingly so as not to endanger workers, the public, or property.
r. When personnel are to be lifted by crane or derrick, verifying that personnel are following OSHA and ASME/ANSI B30.5 requirements.

Crane operator:

1. Be in physical, mental and emotional condition to have full control of the crane or derrick. Know the contents of the Qualifications for Operators, Conduct of Operators and Operating Practices sections of ASME/ANSI B30.5.
2. Be familiar with and understand the contents of the crane or derrick's operating manual.

3. Know the particular crane or derrick model and configuration well enough to safely perform the work, to understand its functions and limitations as well as its operating characteristics.
4. Know how to use the crane or derrick's load chart including the meaning and application of its notes and warnings, and be able to calculate or determine the actual net capacity.
5. Inspect the crane or derrick and perform routine maintenance regularly, as prescribed by the owner and the manufacturer; keep appropriate records of inspections, maintenance and work done on the crane or derrick in the field.
6. Inform supervision and/or the Owner of any problems, or needed maintenance, or repairs, in writing, or in the equipment's log book; when the machine is turned over to another operator, inform that operator of observed problems.
7. Supervise the oiler or apprentice in their duties.
8. Be aware of any site condition that could affect crane or derrick operation and check that the site is adequately prepared for the equipment.
9. Be aware of the special requirements when lifting personnel.
10. Be aware of the presence of power lines or other electrical hazards and refuse to operate if the crane or derrick boom, hoist rope, slings or load will enter the prohibited zone specified in ASME/ANSI B30.5.
11. Review the planned operation and requirements with the Lift Director.
12. Check the load chart to verify that the crane or derrick has sufficient net capacity for the lift.
13. Know how to determine the load and rigging weight, where the load is to be placed, and verify operating radii with appropriate accuracy (the operator is not responsible for calculating the weight of the item to be lifted).
14. Know how to determine the number of parts of hoist line required; verify that the crane or derrick is properly reeved and that the rope and reeving accessories are adequate and in good condition.
15. Assist supervision in selecting (from the manufacturer's information) the appropriate boom, jib and crane configuration to suit the load, site and lift conditions.
16. Know the procedure and techniques for proper assembly, set up and rigging the crane in accordance with the manufacturer's instructions.

17. Consider the factors that might reduce crane or derrick capacity and inform supervision of the need to make appropriate adjustments.
18. Understand load rigging procedures and advise the responsible person if a doubt exists as to the adequacy of the rigging (when the load is visible to the operator).
19. Coordinate and communicate with the designated signal person, but obey a stop signal from whatever source.
20. Operate the crane or derrick in a smooth, controlled, and safe manner.
21. Not engage in any practice that will divert the operator's attention while operating the crane or derrick.
22. Know how to move the crane safely under its own power.
23. Shut down and secure the crane or derrick when it is to be unattended.
24. Understand the unique differences in operations when working under the following specific conditions:
    - Multi-crane lifts,
    - Hoisting personnel platforms,
    - Clamshell operations,
    - Pile driving and pulling sheeting,
    - Concrete operations,
    - Demolition operations,
    - Crane mounted on barge, and
    - Magnet operations.
25. Stay current in the skills and knowledge necessary to safely operate the crane or derrick.

Assigning responsibilities

The preceding lists apportion responsibilities for a major project with extensive crane and rigging work. The minor project with incidental crane use must also be considered. For example, using the taxi-crane case, the party filling the shoes of Project Management is often the owner of a local business who knows little or nothing about cranes or their use; it is not reasonable to expect that party to play more than a minimal role. Likewise, the Crane Owner/Supplier and Crane Operation Management will be one entity—the taxi-crane rental house.

The scope of such work makes it usual for management tasks to become abbreviated and less formal. A site visit might be replaced by telephoned questions and answers, and the Site Supervision role will be covered by the Crane Operator. Upon arrival, the operator most often becomes the person on site with the greatest crane and lifting expertise. In effect, the crane operator will wear three hats, including that of the Lift Director. Needless to say, these situations place a heavy burden of responsibility on the crane operator.

Typically, cranes on a construction project are peripheral to the principal project activities. More often than not, full responsibility is given to the subcontractors who hire and use the equipment, and Project Management takes a back seat. Responsibilities may be delegated to specialists or to vendor or subcontractor employees "wearing multiple hats".

On many small construction projects, Project Management, Crane Operation Management, Site Supervision, and even the Lift Director can be one person. Real planning may begin when the crane arrives, and it is exceptional for personnel on small sites to have extensive crane and rigging knowledge. The crane operator is counted upon, by the renter, to fill that void. Without being informed of this, the operator is expected to perform an entire range of additional responsibilities. The wide gap between the user's and the operator's perceptions of the latter's responsibilities are only revealed in a courtroom after an accident has occurred.

The lists furnished cover responsibilities for all parties potentially involved in lifting operations. Knowledgeable people will peruse those lists, and nod in appreciation. But what do people do with the lists, people involved in the typical small operation utilizing cranes? What use are those lists to them? General contractors, rigging companies, crane rental houses, and steel erectors, organizations very much involved in crane use, need to examine the lists and winnow them down to those items appropriate to their particular operations. Clearly the lists need to be abridged for the scale of the operation, but they also need to be disseminated to be beneficial. Education is the flow of knowledge from those who posses it to those who do not. Therefore, responsibilities should be discussed, incorporated in contract language and rental agreements, and become part of discussions about crane hire.

In reworking the lists, the worksite is the only focus. The goal is to provide:

- A competent operator,
- The right lifting equipment for the task,
- Sufficient numbers of personnel on site,
- The needed tools, rigging equipment, blocking, dunnage, and hardware,

- Adequate space for assembly, operation, and load handling,
- Proper crane support, and
- Coordination at the site to avoid interferences and hazards.

These basic requirements are applicable to every lifting operation.

8.4 Accident Statistics

Some accident statistics have become available since the earlier editions of this book, but it is difficult to draw accurate conclusions from them. On the one hand, each organization compiling data uses its own breakdown of categories with little overlap from one to another. On the other hand, some statistics are based on the nature or description of the accident while others are based on the cause of the accident. Compilations also differ in the time periods they cover and in the geographical area. Thus, New York City data lists no power line contact events, whereas OSHA reports that about 50% of all crane related fatalities derive from that cause, and Ontario, Canada statistics show 43%†.

The net result of these differences is that the newly available statistics do not offer much guidance for risk avoidance, other than to highlight the importance of power line contacts, overturnings, erection and dismantling incidents, wire rope breakage, rigging failures, and support deficiencies as significant hazards in North American crane use.

The authors find that the statistical data used in earlier editions is still pertinent in controlling the risks of crane operation and use. The data originated in a series of magazine articles‡ written by an engineer with the Building Research Establishment in Great Britain, which reported on the results of investigating 472 accidents involving construction-type cranes used in various activities as well as construction. Since the study was made from insurance company files of claims it includes some accidents that are not related to operations and others that are mechanical failures leading to localized damage, rather than true accidents in the sense of our discussion. Furthermore, the Butler data are more than 20 years old at the time of this writing. Stability-sensitive and operationally more complex telescopic cranes

† See *Crane Safety on Construction Sites*, Construction Division Task Committee on Crane Safety, American Society of Civil Engineers, Reston, VA, 1998, p. 4.

‡ A. J. Butler, Crane Accidents: Their Causes and Repair Costs, *Cranes Today*, Edgeware, Middlesex, England, no. 62, March 1978, p. 24; no. 63, April 1978, p. 28; and no. 65, June 1978, p. 27.

now represent a greater proportion of the crane population, therefore today's statistics will no doubt differ from those given below. However, because the data is from Great Britain where lines are below ground, no reports of electrocution from power line contact appear.

When mechanical items resulting in insurance claims, but not an accident, and some accidents not really related to crane use (such as another vehicle's striking a parked crane) are eliminated, the results for mobile cranes are as shown in Table 8.1. In total number of incidents, injuries, and fatalities, overturning accidents are by far the most serious. Furthermore, the articles give data showing that money damages associated with overturnings are disproportionately great. Of the three general types of mobile crane—crawler, latticed boom truck, and telescopic—the overturning statistics vary little except that damage costs for telescopic cranes are particularly high.

The category called "Structural damage, human error" in Table 8.1 was so called in order to separate that group of accidents from those brought on by machine defects and resulting in similar physical damage. In this category are included side loading of the boom, machine hit by load, overloads not leading to overturning, and assorted operator errors. An important cause of accidents, responsible for some 13½% of the total and 10% of injuries, this group can be effectively dealt with through risk control measures already mentioned. But also included in this class of mishaps are those which it must be admitted

TABLE 8.1 Mobile-Crane Accidents

Type of accident	Number	Percentage of total	Fatalities	Injuries	Percentage of total
Overturnings[a]	86	48.9	2	20	71.0
Structural damage, human error	24	13.6	....	3	9.7
Boom over cab	24	13.6[b]	....	1	3.2
Rope failures	22	12.5	....	3	9.7
Wind	11	6.3			
Structural damage, machine defect	8[c]	4.5	....	1	3.2
Miscellaneous	1[d]	0.6	1	....	3.2
	176	100.0	3	28	100.0

[a]Not including wind, which accounted for seven more overturnings.
[b]On the basis of only latticed boom cranes, 19.2%.
[c]Includes seven occurrences in which fatigue failure was suspected.
[d]A workman was crushed between the crane counterweight and an obstruction.
SOURCE: Data from A. J. Butler, Crane Accidents: Their Causes and Repair Costs, *Cranes Today,* Edgeware Middlesex, England, no. 62, March 1978.

are virtually unavoidable-accidents resulting from human failings that could occur in spite of all precautions.

The boom-over-cab occurrences, although common, are not responsible for a proportionate share of either injuries or costs. One-third of these incidents would likely have been prevented had a boom-hoist cutout device been operating.

Rope failures, causing 12½% of the total accidents and some 10% of the injuries, are confined to derricking lines and hoist lines (sling accidents have been removed from the data in Table 8.1). On cranes equipped with rope boom suspensions, the likelihood of failure in such systems is indicated as half again that for hoist lines.

Most of the wind accidents resulted in overturning, either directly or by blowing the load to greater radius. Apparently the wind accidents were either a slow process, were not catastrophic, or they occurred while the cranes were out of service, for there is no report of injury in this group.

Machine defects, including defective materials, workmanship, and component design, caused only 4½% of occurrences and 3% of injuries. It is likely that, with a number of suspected fatigue failures reported, cranes are being used for duty cycle work at load ratings set for construction work rather than at duty cycle ratings, which are lower due to the severity of that service.

The overturning, boom-over-cab, and rope-failure accidents reported are sufficiently numerous and severe to merit discussion in the following sections.

Of the tower crane accidents analyzed by Butler, about 36% occurred during erection or dismantling, 18% were wind-related, 10% were rope failures, 49% were due to human errors, and 5% were the result of machine defects. The total is greater than 100% because wind-related incidents are included in both the human-error and the erection-dismantling categories. Tower cranes will be discussed further following mobile crane matters.

8.5 Avoiding Accidents

Under each of the categories of accidents in the Butler data, a brief discussion of the nature of the events, the characteristics of the equipment, and the practicalities of the jobsite should offer insights into how risks can be controlled. Planners can then focus on the means to avoid situations that create hazards or institute controls to minimize risk. An understanding of the most common causes of accidents will allow supervisors to sharpen their focus too, and therefore devote time to those matters introducing the greatest risk.

Before going to the Butler categories, however, it is appropriate to address a cause of accidents of great importance in North America, but absent from that European data source.

Contact with power lines

The most reliable way to avoid power line contact accidents is to have the lines de-energized. Unfortunately, that option is not often available. Therefore, when cranes or derricks must be operated within a boom's length of an energized line, defensive measures must be undertaken, and it is management's obligation to see to it that adequate procedures are in place. There is an entire section within the ASME/ANSI B30.5 volume setting out minimum standards of good practice in this regard.

A prerequisite to developing adequate procedures is an understanding of the problem. What few people realize is that all overhead power transmission lines are bare wires; there is no insulation to offer protection in case of contact or to prevent arcing as a wire is approached. A boom, load or pendant line, slings, or the load itself, are all conductors which, in effect, make a crane a "mobile short circuit device". Thus, contact or near contact could allow the electrical current to take an unintended path from the power line through the crane and into the ground. But no harm is done unless the current flows, as evidenced by birds sitting unharmed on lines of all voltages. The greatest danger, in case of contact or near contact, is to the ground crew. When tires interrupt the current, a person touching the crane, load, slings, or even sometimes tag lines will complete the circuit and be electrocuted. Although crane tires act as insulation, steel belted tires have been known to catch fire when exposed to high voltages. The crane operator who remains in the cab, like the bird, will not be harmed.

Power lines are generally furnished with circuit protection devices, either fuses or reclosure devices that make a few attempts to restore power, within seconds of the short circuit, before shutting down the line. Those devices are intended to provide a reliable electric supply, one that is not constantly interrupted by rain or morning dew. The intent is to protect the power grid—none of them work quickly enough or at a low enough level of energy to offset the danger to people.

Electrical current can only harm crane or derrick crews if a part of the equipment either contacts or comes close to contacting an energized line. The minimum safe distance specified by ASME/ANSI B30.5 is 10 ft (3 m) for lines carrying 50,000 volts (50 kV) or less, with greater distances given for lines of higher voltage. Common experience informs us that current will not arc over such great distances. Why then are such distances required?

The distance limits are arbitrary numbers, but very practical nonetheless. They not only take into account electrical behavior, but they allow for how cranes and people behave as well. We should not lose sight of the fact that these distances have been mandated by OSHA for more than 25 years and included in ASME/ANSI B30.5 even longer, and yet crane operations electrocutions remain the single greatest cause of crane related fatalities. Based on that, one might argue that the 10 foot (3 m) minimum is not enough.

There are a number of elements that must be added together to explain the minimum distances specified. Wind, even light winds, will swing power lines and cause booms to wiggle as well. Additional boom movements come from the dynamics of hoisting and swinging loads, and these are also relatively small movements. But crane and derrick operators are required to focus their attention on either the load or a signal person. Therefore they cannot, at the same time, be judging distance from power lines. Should the operator's attention be focused on the power line, people may be endangered by the load, or by contact with some other obstruction; the operator may fail to see and respond to a signal.

Assume that a special signal person is on hand whose only duty is to observe proximity to the power line, which is a good practice. The special signal person will not be the main focus of the operator's attention; the regular signal person will be. Therefore, there will be a time lapse between the power line warning signal being given and when it is perceived by the operator. Reaction time and proper stopping time add to the delay. The potential movements during those time delays comprise most of the minimum clearance distances mentioned. That is why the distances specified are practical minimums, and actually offer only a small margin of safety.

A number of devices are on the market to assist crane operators in maintaining safe distances from power lines, such as boom cages, proximity warning devices, and insulated links. They each offer a somewhat different but limited type of protection, and although proven in the laboratory, the typical jobsite is a far different environment. Jobsite dirt or dust could impinge on their effectiveness. That is why both OSHA and ASME/ANSI B30.5 stipulate that even when those devices are used, the minimum distance and other requirements must be adhered to.

Reduction of overturning accidents

Table 8.2 was made after breaking down and reorganizing the Butler data on mobile-crane overturning accidents; it serves to help focus on the overturning problem. It is not surprising that overloading heads

TABLE 8.2 Mobile-Crane Overturning Accidents

Category of accident	Occurrences	Percentage of total
Overloading[a]	14	23.34
Slewing-related	12	20.00
Travel-related[b]	12	20.00
Trapped or caught loads	9	15.00
Wind	6	10.00
Operations on rubber[c]	3	5.00
Outrigger-related[d]	2	3.33
Unknown	2	3.33
	60	100.00

[a]Of the cranes involved in these accidents 11, or 78.6%, were equipped with load-indicating devices.

[b]Four, or 33.33%, of these accidents were related to soft ground and a like number to sloped ground.

[c]In all three cases, the cranes were without load at the time of the accident.

[d]With no blocking under outrigger floats in both cases.

SOURCE: Data from A. J. Butler, Crane Accidents: Their Causes and Repair Costs, *Cranes Today*, Edgeware, Middlesex, England, no. 62, March 1978.

the list, but that it is so closely followed by slewing- and travel-related incidents may be a surprise. Together, those three categories include almost two-thirds of all overturnings.

In the United States and Canada mobile cranes are rated under ideal operating conditions, that is, on a firm supporting surface, machine leveled to within 1% [1 in/100 in (2.5 cm/250 cm)], load freely suspended, a knowledgeable and experienced operator, slow operating speeds, and a virtual absence of wind. Therefore, crane ratings must be reduced by the operator when operating conditions are other than ideal.

Although the Butler statistics indicate that about one-quarter of the overturning accidents were caused by overloading, nearly 80% of the cranes in that sample were equipped with load or load-moment indicators. Unfortunately, the study did not go into that matter further. It should be noted, however, that those devices have improved significantly in the 20 years since the data was compiled. Basically, overloading is responsible for a relatively small portion of mobile-crane accidents simply because a very small proportion of lifted loads are at or near rated loads.

In concept, load and load-moment indicators are ideal means to assure that cranes will not be overloaded. In practice, they fall short of the ideal. The reasons are many and can only be briefly alluded to here.

Cranes furnished with load or load-moment indicators can be overloaded if (1) the device has been turned off or is down due to malfunction, (2) the device is out of calibration, or (3) operating conditions (wind or operating speeds or out of level) are so far from ideal that the published ratings can lead to failure. The mounting of a device is itself no assurance that operations will be safe. Just like oil pressure or temperature gauges, those devices are not safety devices; they are indicators that advise a competent operator of load parameters as an aid in making operating judgments.

Some authorities overstress the value of, or need for, load or load-moment indicators. There is no doubt that there are operating situations that require a device of that type, but on the other hand, in certain situations they offer mixed blessings. It has been demonstrated that there is a tendency for some operators to become overly reliant on the devices and to use them in place of judgment. This can lead to accidents when conditions are not ideal. An inadequately trained or inexperienced operator may use the device as a prop and as a substitute for knowing the machine, the load, and the rating chart. Operators who do not fully understand the meaning of the values on the rating chart, and who do not understand the limitations of the crane and its ratings, may operate imprudently or permit supervisors also lacking in knowledge to pressure them into doing so. The number of operators who do not understand rating charts is surprising. The number of supervisors who know little or nothing about cranes is shocking.

Typically, managers and supervisors assume that an operator is qualified because of proficiency at the controls or years of service. The truth is that many operators do not know how to read or interpret rating charts and most do not understand the strength or stability implications that underlie the ratings. To determine the useful load that can be lifted, rating chart values must be adjusted. In addition to reductions to account for the load block and for slings or other lifting attachments, reductions must be made for such items as jibs (either mounted or stored at the base of the boom), auxiliary or whip lines, optional head sheave arrangements, boom tip lights or pile lead adaptors, and the effects of operating conditions that differ from ideal. Each such item not taken into account reduces the margin of stability and makes the crane more prone to accident. All too often, these required capacity reductions are ignored.

At the time of this writing, standards committees in the United States are in the process of making load or load-moment indicators required equipment for mobile cranes. The decision has been a difficult one because there are persuasive and compelling arguments supporting both sides of this issue. A survey recently taken on this subject by a trade group† elicited an interesting response. Although respondants reported reliability and accuracy problems and considerable downtime, they favored continued use of these devices. They were strongly opposed, however, to mandating their use.

If loads that are close to rated loads at any planned working radius or operating arc are to be lifted, to achieve a safe operation it is necessary that the weight of the load be known with reasonable accuracy. To proceed otherwise requires seat-of-the-pants operation, bravado, or culpable negligence. There are many ways to determine load weight before the load is taken to a critical outreach. The following section deals with that subject in some detail.

Preventing overloads

What constitutes an overload is a relative matter. Technically and legally, any load in excess of that given in the rating chart for the crane configuration in use is an overload. As a practical matter, any load that causes the crane to overturn or to collapse is an overload; site, operating, or environmental conditions may override the rating chart and make it necessary to limit lifted load to less than chart values. Technical and legal overloads can be prevented by use of load or load-moment indicators or by adequate preplanning. These measures also help in preventing "practical" overloads, but well-trained and experienced staff are the main line of defense.

When load-indicating devices are relied upon to verify load weights or are depended upon to prevent overloading, it is not sufficient to just furnish a device in the cab. They are somewhat sensitive electronic or electro-mechanical instruments that require regular maintenance and periodic verification and recalibration. Further, crane operators must be specifically trained in their use.

For many construction operations load-indicating devices are not needed to control lift weights, although they can be effective as a checking means. When the rigging crew is aware of crane capacity and capable of determining load weights, lifting cycle time is minimized and efficient operations are maintained. The crew will make up and sling the loads, and the indicator will allow the operator to check on them. If an indicator is used as the primary means for determining load weight, loads will be slung, tested, unslung, and adjusted until a proper load is ready for lifting; that procedure can be intolerably slow and inefficient.

---

† Specialized Carriers and Rigging Association, Fairfax, Va., May 1999.

As part of the preplanning process, the procedures to be used to control load weights should be established. If manufactured items are to be lifted, weights can be secured in advance from the producer and made available to the jobsite. One person in the crane-operations crew should be assigned the task of load determination. This person must be competent in arithmetic, be comfortable with weights and measures, and understand the rating chart.

Before each lift the load controller should determine load weight, including the weights of lifting accessories; the operator can then be informed either verbally or by means of a sign placed on the load. For miscellaneous loads, the weights must be calculated by this designated person, and aids such as Table 8.3 or handbooks of steel-pipe, beam, or reinforcing-rod weights must be made available for this purpose.

Once when an operator who had been involved in an accident was asked why he had not referred to the rating chart before making the lift, he replied that he did not have the time to sort out the chart numbers while a crew of men was standing by waiting for him. A

TABLE 8.3 Weights of Common Materials

Material	Weight	
	lb/ft^3	t/m^3
Aluminum	165	2.64
Brick, hard	140	2.24
Common	120	1.92
Cement	90	1.44
Cereals, corn, rye, etc.	50	0.80
Cinders	45	0.72
Clay, dry	65	1.04
Damp	110	1.76
Coal, anthracite, loose	60	0.96
Bituminous, loose	55	0.88
Concrete, reinforced	150	2.40
Earth, dry, loose	80	1.28
Wet	110	1.76
Gasoline	42	0.67
Glass	160	2.56
Iron, cast	480	7.69
Paper	60	0.96
Petroleum, crude	55	0.88
Sand, gravel, dry	100	1.60
Wet	120	1.92
Steel	490	7.85
Stone, building	175	2.80
Crushed, loose	95	1.52
Timber, softwood	30	0.48
Hardwood	55	0.88

conscientious operator is sometimes torn between the pressure to proceed and his or her responsibilities toward safe operation. Planners, supervisors, and construction executives should be aware of this very human problem and should take the steps necessary to promote safety while maintaining productivity. For even when load weights are known, overloading and accidents can occur unless the known weights are related to the rating chart and to prevailing conditions.

One simple measure that can be taken is to provide each crane with a roll of masking tape and a marking pen. When the crane is being assembled, the operator can then run a vertical line of tape on either side of the rating section for the boom length to be used and for the jib length and offset as well. In this way the pertinent set of numbers becomes segregated from all the numbers that do not bear on the work at hand. Furthermore, the operator can mark radius numbers or any other needed data on the tape margins thus created.

When machinery or fabrications are to be trucked into a site, and the planners were only able to obtain approximate weights, the loads can be checked at certified scales or even at the site using portable wheel scales. Another option is to have a portable hook scale available to weigh loads during unloading; they can usually be lifted at a short radius.

Perhaps the strongest measure that can be taken to reduce overloading accidents is forethought in the office before the crane is sent to the site. Once the crane is on the site with a particular boom length mounted and minimum radius limited by site conditions, field crews will usually make every effort to place the loads given to them even if this means knowingly overloading the machine or bypassing indicator warnings. The pressures of time and money make this almost inevitable. Overloading accidents can therefore be prevented in advance by executing realistic and practical preplanning work. A small investment at this time will be more than reclaimed by improved productivity, not to mention accident prevention.

Slewing

The data on slewing-related overturnings are not as clear as we would like. They seem to imply that most overturnings are occasions of load swing-out accompanying excessive slewing speeds, but included in this category may well be incidents where the load was proper for the sector in which the load was lifted, but it was too great after swinging to another less stable sector. Some crane chassis deflect when swinging loads from over rear to the side. This will unintentionally increase load radius and may even tip the crane. There may also be cases where outrigger supports failed as the boom passed over.

All other conditions being ideal, lifting full-rated loads should not be attempted unless slewing speeds will be kept very low, and, of course, it is the least rating in the sectors of operation that will govern. Whenever an operation requires repeated crane cycles or a set production rate, higher than normal swing speeds can be expected; load rating reductions are necessary. Some rating charts specify reductions of 20 to 25% for such duty cycle work, although there may be an element of fatigue-failure protection included in such derates.

Most crane operators know that cranes can lift greater loads over the corner than they can at other positions, but rating charts provide no such information. This has resulted in accidents from attempts to "stretch" crane capacity by lifting over the corner. It is foolhardy to attempt this and supervisors should never permit it.

Few truck cranes are rated for lifting over the front. There is greatly reduced resistance to tipping when the boom is in the front quadrant, unless a front outrigger float or other stabilizing means has been provided. If there are no values in the rating chart labeled as applying over 360°, operation over the front is not permitted.

Travel

Although many truck cranes and nearly all rough terrain and all terrain cranes are equipped with off-road construction-type tires, this should not be construed to mean that they are able to roam freely about the countryside. On the contrary, cranes of all types, including crawlers, are too stability-sensitive to allow this, especially with long booms. While travelling assembled, with boom and counterweight, cranes must be kept near level in the side-to-side direction. When traveling without load, speed must be a function of travel-path smoothness. When traveling with load, speed must be held to rates recommended by the manufacturer and rated-load reductions must be followed. See Operations on Rubber (later in this section) and Sect. 5.10 Pick and Carry for more information on traveling with load.

Travel movements should be anticipated during the preplanning work, and arrangements can then be made in advance to assure that travel paths will be prepared to give adequate support (see Table 5.1) and be suitably smooth. Wherever grades must be traversed, and indeed for all travel movements, instructions should be prepared stating the direction the superstructure must face and the boom angle to be maintained.

Several precautionary measures can be taken to avoid danger during travel moves. When the ability of the ground surface to support the axle loads is in doubt, the heaviest available truck should be driven over the path first. Examination of the tire tracks may reveal soft spots and side-to-side irregularities that need be corrected before crane travel.

Moves should not be made with long booms in the air when windy or gusty conditions prevail; however, when truck cranes with long booms are moved in light winds, they can be protected from overturning. The outriggers should be fully extended with the floats held about 2 in (50 mm) clear of the ground surface on a prepared path. The move is then made at very slow speed, about 1 mi/h (1.6 km/h), with watchers posted at each outrigger to warn of hangups.

Trapped or caught loads

When a load is frozen to the ground or embedded in mud, considerable extra force may be required to pull it free. As this force is applied, the hoist lines and boom suspension system stretch elastically. As the load comes free, the elastic stretch induces an immediate rebound. Thus, regardless of the care taken, the load will lift while the boom shifts radius slightly, launching the load into a pendulumlike swing. Unless damped or returned to ground, the load could overturn the crane.

During investigation of an accident that occurred while lifting a furnace section in a building with a crane placed outside, it was discovered that a small projection on the load had become caught on a steel beam. As the lift continued, the load was made to tilt a little and then it broke free. The tilt caused the load to start swinging and in the end carried the crane over as well.

The descriptions above underscore the meaning of a phrase common to all domestic mobile-crane rating-chart notes, that the *rated loads apply to freely suspended loads*. Any other condition of the load implies side loading of the boom, the introduction of dynamic effects, load pendulum action, or a combination of these.

The only measures that can be taken to guard against this sort of accident are ground-crew and operator training and having alert and knowledgeable supervisors. In addition, when a load is being lifted and there is a possibility that it may become caught on an obstruction, watchers must be posted and there must be adequate light for clear observation of the load throughout its travel; these steps should be discussed during preplanning.

Wind conditions

Jobsite anemometers are not often seen even though they are relatively low in cost. The fact remains that few construction executives or supervisors are aware of crane sensitivity to wind and few if any realize that mobile-crane rated loads (where stability governs) are established without taking wind into account at all.

Wind-induced overturnings, although infrequent, could be nearly eliminated if anemometers were installed on jobsites. Other accidents, not easily recognized as wind related, could be reduced in number if supervisors and operators suitably adjusted crane ratings for ambient wind. However, without benefit of an anemometer wind speed can be approximated either by checking with the local airfield weather station or by using the indicators given in Table 8.4. In any case, when wind exceeds 10 mi/h (4.3 m/s), it is advisable that rating reductions be considered. For winds of 20 mi/h (8.9 m/s) or more, reductions in the strength-governed ratings must be made as well as in stability-controlled ratings. When winds exceed 30 mi/h (13.4 m/s), it is usually prudent to cease operations altogether. However, wind limitations given in an operator's manual supercede advice given here.

Mobile cranes with tower attachments are especially vulnerable to intermediate level winds, that is, roughly in the range of 30 to 60 mi/h (13.4 to 26.8 m/s). Some localities are prone to squalls of this intensity that arrive quickly with little warning. Mobile-crane users in those areas must be prepared at all times; those with tower attachments must be ready to fold down the boom or, if winds of 50 mi/h (22.4 m/s) or so are expected, to lower the tower to the ground. The actual measures to be taken depend on the particular manufacturer's instructions for securing the crane against wind.

The principal vulnerability of these machines is to frontal winds, especially when the boom is raised to a high angle. Long-boom mobile cranes are also vulnerable to squalls. With either a tower attachment or long boom, contingency plans should be made for laying the equip-

TABLE 8.4 Estimating Wind Speed

Approximate wind speed			
mi/h	m/s	Description	Effects
0–1	0–0.5	Calm	Smoke rises straight up
1–3	0.5–1.5	Light air	Smoke drifts
4–7	2–3	Slight breeze	Leaves rustle
8–12	3.5–5.5	Gentle breeze	Leaves and small twigs move
13–18	6–8	Moderate breeze	Dust and papers fly, small branches move
19–24	8.5–10.5	Fresh breeze	Small trees sway
25–31	11–14	Strong breeze	Large branches move
32–38	14.5–17	High wind	Walking difficult, trunks of trees hend
39–46	17.5–20.5	Gale	Twigs break off
47–54	21–24	Strong gale	Shingles can be carried away
55–63	24.5–28	Whole gale	Trees may be uprooted

ment on the ground. The area for doing this must be chosen in advance and steps taken to keep crane access to that area clear. At urban sites it will probably be necessary to coordinate with the local police, as streets will no doubt have to be closed to traffic during lowering and perhaps afterward as well.

Derates for wind can be approximated after a value for wind velocity has been determined and with a few basic dimensions at hand. For wind on a lifted load with wind-exposure area A and force coefficient C_f related to the shape of the object (see Sect. 3.4), the wind force will be

$$F = \begin{cases} \dfrac{v^2}{400} C_f A & \text{U.S. customary units} \\ \dfrac{3v^2}{5} C_f A & \text{SI units} \end{cases} \tag{a}$$

For the U.S. customary units, force will be in pounds when velocity is in miles per hour and area A is in square feet, while for the SI units the force will be given in newtons when velocity is in meters per second and the area is in square meters.

When wind blows from behind the operator, the wind force on the load acts horizontally and takes effect at the boom tip. The hook-load equivalent D_t to this force is then

$$D_t = F\frac{H}{R} = F\frac{\sqrt{L^2 - R^2}}{R} = F\sqrt{\frac{L^2}{R^2} - 1} \tag{b}$$

where L is the boom length and R is the operating radius in feet (meters). D_t is the rated-load deduct to account for wind on the lifted load.

Similarly, wind from the same direction acts on the boom itself, creating an overturning moment that can be represented as an equivalent hook load. Here, in determining the wind force on the boom we will use a configuration coefficient C instead of the force coefficient C_f used before (see Chap. 3) and will further define C below. This will make the boom derate D_b take the simplified form

$$D_b = \frac{FL^2}{2R} \tag{c}$$

The force coefficient for wind blowing normal to a latticed boom can be conservatively approximated as two-thirds when the total width of boom is used rather than the actual wind area. For each unit of boom length, then, $C_f A \approx 2b/3$ if b is the boom overall cross-sectional width in ft (m). For telescopic cranes $C_f A$ can be taken as 3 (0.9 in SI units).

But Eq. (3.27) tells us that $C = C_f \sin^2 \theta$ for wind blowing at angle θ to an object and F acting in the direction of the wind. Then

$$CA = \frac{2b}{3}\left(\frac{\sqrt{L^2 - R^2}}{L}\right) = \frac{2b}{3}\left(1 - \frac{R^2}{L^2}\right) \qquad \text{for latticed booms}$$

$$= 3\left(1 - \frac{R^2}{L^2}\right) \qquad \text{for telescopic booms} \qquad (d)$$

The full wind derate can now be stated as

$$\text{Derate} = D_t + D_b \tag{8.1}$$

D_t is given, after combining Eqs. (a) and (b) and taking $C_f = 1.33$, by the expression

$$D_t = \begin{cases} \dfrac{v^2}{300} A \sqrt{\dfrac{L^2}{R^2} - 1} & \text{U.S. customary units} \\[2ex] \dfrac{4v^2}{5} A \sqrt{\dfrac{L^2}{R^2} - 1} & \text{SI units} \end{cases} \tag{8.2}$$

After Eqs. (a), (c), and (d) are combined, D_b is stated in U.S. customary units by the formula

$$D_b = \begin{cases} \dfrac{v^2 b}{1200}\left(\dfrac{L^2 - R^2}{R}\right) & \text{for latticed booms} \\[2ex] \dfrac{3v^2}{800}\left(\dfrac{L^2 - R^2}{R}\right) & \text{for telescopic booms} \end{cases}$$

For SI units the formula is

$$D_b = \begin{cases} \dfrac{v^2 b}{5}\left(\dfrac{L^2 - R^2}{R}\right) & \text{for latticed booms} \\[2ex] \dfrac{9v^2}{10}\left(\dfrac{L^2 - R^2}{R}\right) & \text{for telescopic booms} \end{cases} \tag{8.3}$$

An additional check must be made to assure that when the wind comes from the side, boom strength will not be overtaxed. Mobile-crane booms are designed to withstand a 20 mi/h (8.9 m/s) side wind in conjunction with boom tip side loading at 2% of rated load and the rated load itself. The side loading is intended to allow for dynamic effects of slewing with some attendant load swing. We can use this side-load allowance as a measure of boom resistance to winds above 20 mi/h together with wind on the lifted load provided we do not fail to realize that in so doing the dynamic allowance no longer remains.

With the above in mind, boom strength is adequate for latticed boom cranes if

For US customary units and $v \geq 20$ mi/h

$$\left[\left(\frac{v^2}{400} - 1\right)\frac{d}{3}L + \frac{v^2}{400}A\right]50 \leq \text{rated load}$$

For SI units and $v \geq 8.9$ m/s

$$\left[\left(\frac{3v^2}{5} - 48\right)\frac{d}{3}L + \frac{3v^2}{5}A\right]50 \leq \text{rated load} \tag{8.4}$$

For telescopic booms, replace $d/3$ with 3/2 in U.S. customary and with 0.45 in SI equations. Rated load in Eq. (8.4) is the printed chart value without deductions while d is the depth of the boom cross section. If the left side of Eq. (8.4) is considerably less than rated load, some allowance for dynamic effects remains, but if the two sides of the equation are nearly equal, a reduction for dynamic effects is essential under any set of operating conditions.

The procedure to be used at the jobsite in preparation for lifts in the presence of wind is as follows:

1. Determine or estimate wind velocity (see Table 8.4) or select a value above which work will be stopped.
2. Determine by measurement the wind-exposure areas of the loads to be lifted.
3. Using Eq. (8.4), evaluate boom strength. For winds below 20 mi/h (8.9 m/s) only the load-area part of the equation need be used. If the left side of Eq. (8.4) is greater than rated load, reduce operating radius and/or boom length or postpone the lift until wind subsides. If the left side is more than half of rated load, reduce the rating by not less than 10% to allow for slewing effects.
4. Using Eq. (8.1), calculate the value to be deducted from rated load to account for the overturning effects of the wind. If the reduced rating is less than the load to be lifted, reduce the operating radius and recheck or postpone the lift until the wind subsides.

Example 8.1 A crane having 150 ft (45.7 m) of latticed boom mounted is to be used for placing precast concrete panels that measure 12.5 by 8 ft (3.81 by 2.44 m) and weigh 7500 lb (3400 kg). The boom is 72 in (1.83 m)

TABLE 8.5 Load Ratings for 150 ft (45.7 m) Boom, Side and Rear (for Example 8-1)

Operating radius		Rated load		Operating radius		Rated load	
ft	m	lb	kg	ft	m	lb	kg
40	12.2	36,400	16,500	100	30.5	8,400	3,800
45	13.7	30,400	13,800	110	33.5	7,050	3,200
50	15.2	25,900	11,750	120	36.6	5,950	2,700
60	18.3	19,500	8,850	130	39.6	5,050	2,300
70	21.3	15,400	7,000	140	42.7	4,250	1,900
80	24.4	12,400	5,600	150	45.7	3,650	1,650
90	27.4	10,100	4,600				

square in cross section, and the lower block and lifting accessories weigh 1200 lb (545 kg); Table 8.5 is the rating chart for the boom length mounted. What maximum operating radii can be utilized when ambient wind is at 15 and 30 mi/h (6.7 and 13.4 m/s):

Load area A is 12.5(8) = 100 ft^2 (9.29 m^2). When Eq. (8.4) is used, for wind at 15 mi/h (6.7 m/s) only the load area part of the equation applies

$$\frac{15^2}{400}(100)(50) = 2800 \text{ lb } (1275 \text{ kg}) \leqslant \text{rated load}$$

But rated load must be at least equal to 7500 + 1200 = 8700 lb (3950 kg); therefore, no strength limitation will govern, and there is certain to be a considerable side-load allowance remaining to accommodate dynamic effects associated with reasonable swing speeds.

For stability, the derate applying to wind on the load is given by Eq. (8.2)

$$D_t = \frac{15^2}{300} 100 \sqrt{\frac{150^2}{R^2} - 1} = 75 \sqrt{\frac{22{,}500}{R^2} - 1}$$

and the derate for wind on the boom is given by Eq. 8.3) as

$$D_b = \frac{15^2(72 \text{ in}/12)}{1200} \frac{150^2 - R^2}{R} = 1.125 \frac{22{,}500 - R^2}{R}$$

Then rated load must be greater than the lifted load plus the derate values, or

$$\text{RL} = 8700 + 75 \sqrt{\frac{22{,}500}{R^2} - 1} + 1.125 \frac{22{,}500 - R^2}{R}$$

from which the following table has been generated:

Radius		Minimum acceptable rated load		Radius		Minimum acceptable rated load	
ft	m	lb	kg	ft	m	lb	kg
40	12.2	9600	4350	70	21.3	9150	4150
45	13.7	9450	4300	80	24.4	9050	4100
50	15.2	9400	4250	90	27.4	9000	4075
60	18.3	9250	4200	100	30.5	8950	4050

From comparison with Table 8.5, it can be seen that an operating radius of 90 ft (27.4 m) cannot be safely exceeded when ambient wind is at 15 mi/h (6.7 m/s).

Likewise, for wind at 30 mi/h (13.4 m/s) Eq. (8.4) yields

$$\left[\left(\frac{30^2}{400} - 1\right)\frac{72/12}{3}\,150 + \frac{30^2}{400}\,100\right](50) = 30{,}000 \text{ lb } (13{,}600 \text{ kg}) \leq \text{RL}$$

Equation (8.2) gives

$$D_t = \frac{30^2}{300}\,100\sqrt{\frac{22{,}500}{R^2} - 1} = 300\sqrt{\frac{22{,}500}{R^2} - 1}$$

And Eq. (8.3) becomes

$$D_b = \frac{30^2(72/12)}{1200}\,\frac{22{,}500 - R^2}{R} = 4.50\,\frac{22{,}500 - R^2}{R}$$

Therefore the rated load must be at least

$$\text{RL} = 8700 + 300\sqrt{\frac{22{,}500}{R^2} - 1} + 4.50\,\frac{22{,}500 - R^2}{R}$$

but the minimum acceptable rated load must also satisfy the strength-rating limitation as well as stability.

Radius		Minimum stability-rated load	
ft	m	lb	kg
40	12.2	12,150	5500
45	13.7	11,700	5300
50	15.2	11,350	5150
60	18.3	10,800	4900
70	21.3	10,400	4750

Boom strength now clearly governs the selection of operating radius in the presence of an ambient wind of 30 mi/h (13.4 m/s). Thus operations can be conducted at radii not exceeding 45 ft (13.7 m) where rated load given by Table 8.5 is 30,400 lb (13,800 kg), which exceeds the minimum of 30,000 lb (13,600 kg) needed for lateral boom strength.

As a practical procedure for use at the jobsite, when loads of large wind area are to be handled, maximum operating radii can be calculated in advance for a series of wind speeds in increments between the limits used in the example problem. This will preclude the necessity for field calculations at the last minute while an operating crew stands by. It will also permit work to be safely performed with assurance when wind conditions might otherwise leave field crews uneasy about the crane or, worse yet, unwilling to proceed.

When the wind is gusty, it may be prudent to stop work even if the calculations show that adequate strength and stability are available at peak wind levels. Here the question is the stability of the load and whether it can be safely handled while being tossed about by the wind. The judgment of field supervisors must be relied upon, and their decisions will have to be made at the time and place of work. It is not often that work can be done safely when wind velocity is 30 mi/h (13.4 m/s) or higher.

Operations on rubber

Many truck, all terrain, and rough terrain cranes are rated for operations on rubber (tires), but those involved in such operations should be made aware of the greater sensitivity to overturning inherent in cranes in this configuration. Also, the limitations applicable to on-rubber ratings may vary from manufacturer to manufacturer. For example, ratings may apply within the entire rear quadrant, only for loads within the zone defined by extending imaginary lines through the centers of the tires, or within a defined angle, say ± 10°, from dead over the rear. Boom lengths may also be limited.

The cranes are capable of lifting their rated loads (of this there should be no doubt) since the ratings can be mathematically proved or verified by test. But a crane on rubber is on an elastic support; therefore any perturbation can cause the crane to lean out to a greater radius while the CG of the machine shifts closer to the tipping fulcrum at the same time. Few people realize that on-rubber ratings depend on a specific tire pressure that may be much higher than normal travel pressure. At the lower pressure, the crane cannot manage rated loads.

The danger from operations on tires derives from the heightened sensitivity of the machine to all things that are a danger for crane operations in general; in particular, they are susceptible to dynamic effects. If rated load reductions are recommended for a crane on crawlers or outriggers, they are imperative for a crane on rubber. On-rubber ratings are for static conditions.

Some cranes are rated to *pick and carry* (see Sect. 5.10). Those ratings apply for very slow travel speeds, as specified by the manufacturer, often on the order of 1 mi/h (1.6 km/h). And the ratings are typically for loads carried over front or rear, as specified.

Pick-and-carry operations require a firm level travel path. Small irregularities can start movements that amplify into serious side loads and radius changes if unchecked. Before executing a pick-and-carry movement with a meaningful load, the crane should be observed during an unloaded dry run. Unwanted load movements are minimized by minimizing travel speed; snubbing the load to the crane undercarriage can prevent it from swinging to greater radius, but beware of the shock load on the return if it is permitted to swing inward.

On urban and highway reconstruction projects, space for crane operation is often constrained, and this leads to cranes improperly set up with outrigger beams extended on one side only or partially extended on each side; some new cranes are rated for specific partial extension positions, however. Unless sanctioned by the manufacturer the crane should operate using on-rubber ratings with the outriggers considered as stabilizers that only reduce crane sensitivity. To do otherwise violates OSHA requirements and the manufacturer's operating instructions. (See Extension of Outriggers in Chap. 4.)

Prevention of boom-over-cab accidents

Many people mistakenly believe that boom stops will prevent boom-over-cab occurrences. They can, and were originally intended to do so but only when the boom is propelled toward the cab by a fairly low-level dynamic event. When a two-blocking situation continues, or when the boom hoist continues in engagement (overluffing), the boom stops do not have sufficient strength to withstand the rope forces or to cause the boom to buckle.

Some crane engineers feel that the philosophy here is due for change. When a boom goes over the cab at present, a portion of the superstructure housing can be crushed by the boom, posing a minor threat to the operator but a greater one to the machinery. Furthermore, the boom comes to the ground stretched out to nearly its full length. If the boom stops are strong enough to buckle the boom, it may arch over the housing, freeing that area from damage; then the boom would fall at a foreshortened length as well. An effectively shorter boom would pose less threat to people and property in the vicinity.

Be that as it may, all modern cranes are furnished with boom-hoist cutouts that stop the luffing motion as the boom contacts the boom stops. A functioning device of this type would preclude overluffing occurrences leading to boom-over-cab incidents (Table 8.6). The operator's daily machine functional checkout should therefore include verification that the boom-hoist cutout is in operating condition; every maintenance inspection should also include examination of the elements making up this device.

Boom-hoist cutouts will do nothing to prevent boom-over-cab accidents caused by two blocking, trapped or caught loads, rough travel paths, strong wind gusts or excessive dynamic bounce. Cranes should never be traveled with the boom at maximum angle, since uneven ground, wind gusts, ground slopes, a lurching start forward, or a combination of these can then launch the boom over the cab.

Crawler-crane travel is ideally performed with the boom at that angle which will cause the track pressure to be uniform through the length of the track; this implies that the moment of the crane about the axis of rotation will be zero. Setting Eq. (5.23) equal to zero and inserting Eq. (5.21), we find the expression for the optimum boom angle to be

$$\theta = \cos^{-1}\left[\frac{W_u d_u - (W + W_r)R}{W_b L_b} - \frac{t}{L_b}\right] - \theta_b \qquad (8.5a)$$

which can be simplified to give the approximate expression

TABLE 8.6 Boom-over-Cab Accidents

Category of accident	Occurrences	Percentage of total
Crane stationary	17	70.8
Crane traveling	7	29.2
Overluffing[a]	7	29.2
Maintenance, control self-engaged	4	16.7
Human error, inadvertent engagement[b]	4	16.7
Miscellaneous[c]	4	16.7
Hoist-line two blocking[d]	2	8.3
Travel, ground conditions	2	8.3
Trapped or caught loads[e]	1	4.1
	24	100.0

[a]In one instance it was a foreman, not an operator, in tbe cab.
[b]By the operator's knee in one case.
[c]Two cases of rope fouling, one of sling failure by slippage, and one of excessive boom "bounce."
[d]Crane run by other than operator in one case. No doubt two blocking in telescopic cranes represents a more significant source of accidents.
[e]Extracting piles.
SOURCE: Data from A. J. Butler, Crane Accidents: Their Causes and Their Repair Costs, *Cranes Today,* Edgeware, Middlesex, England, no.62, March 1978.

$$\theta \approx \cos^{-1} \frac{W_u d_u}{W_b L_b} \tag{8.5b}$$

resulting in satisfactory values for travel. When the arc cosine argument in either Eq. (8.5*a*) or (8.5*b*) is unity or greater, this means that the boom does not have sufficient weight to produce uniform track pressure even when extended out flat. The boom should then be carried at any reasonably low angle.

For truck cranes, the ideal travel mode will cause the axle loadings to be uniform (see Sect. 5.2), but this ideal should be relaxed if carrying the boom at or near the maximum angle is called for.

Avoidance of rope failures

Strength factors of 3.0 and 3.5 are used for standing and running lines, respectively, in mobile-crane and derrick practice (however, for running lines of rotation-resistant rope 5 is used). The factors apply against rope break strength, while most engineers are accustomed to thinking in terms of factors against yielding, as used in structural-steel design. Yield strength for wire rope is about 60% of the break strength, so that a factor of 3.0 implies working loads that are equal to the yield strength divided by 1.80, or only about 8% more than the 1.67 used for structural steel in tension. For running lines, the factor against yield, 2.10, is about 26% greater than 1.67.

Rope-strength factors stated in relation to break strength have misled many people into believing that rope is conservatively rated compared with ordinary structural steel. This is not so, and as the matter is looked into more deeply, it can be argued that rope is less conservatively rated.

All crane service rope is exposed to the weather, to alternate wetting and drying, changes in humidity, and variations in temperature. The rope itself is composed of a great many wires, giving it a large surface area compared with its metallic cross-sectional area. The corrosion susceptability which weather exposure gives it, coupled with the large relative surface area, makes rope a corrosion-critical item; a small amount of corrosion causes a disproportionately large loss of strength. What is more, surface pitting accompanying corrosion and the resulting increased stress (pitting is a stress raiser) is exacerbated by the exaggerated relationship between pit and wire size. Corrosion considerations alone should take up an important part of the strength factor and is reason to be attentive to proper lubrication of ropes.

The nature of loading in crane service is such that the rope is continuously exposed to varying levels of tensile force. Rope is sized by considering static load only, so that in use stresses may exceed the nominal design values and after many cycles of loading ultimate failure due to fatigue becomes most probable. In addition higher levels of tensile stress will be found in ropes that wind over sheaves, and the point of attachment of fittings on standing lines usually includes stress-raising situations.

But these are all fatigue-related conditions, and fatigue is a useful-life type of phenomenon rather than strictly a safety consideration. Are we talking about economy, then, rather than safety?

Fatigue life becomes strictly an economic factor only if the rope is removed from service before fatigue failure, or better yet with a reasonable margin of safety still remaining before fatigue failure. Without timely removal, the rope may fail without warning and with dire consequences. The criteria that have been evolved to warn of the time for removal are based on rope selected with the strength factors given above. This results in sufficient remaining strength at the time of removal to ensure that the needs of safety are adequately served. To use lower strength factors together with the same discard criteria would imply less than satisfactory strength at removal and a higher risk of failure just before removal.

The Butler data indicate that for latticed boom cranes the likelihood of rope failure in the derricking system is 50% greater than for the hoist lines. Our feeling is that instead of pointing up a weakness in design the data confirm what one would suspect, given human nature.

One tends to feel that the boom suspension is a rather sedentary system, as anyone familiar with crane operations is aware that derricking is performed far less often than hoisting. For this reason, and more particularly because the load-hoist drums of latticed boom cranes are within sight of the operator, load-hoist ropes are more closely observed. Signs of wear, broken wires, corrosion stains, kinks, birdcaging, and other damage can be noted by the operator from the work station as the rope spools in.

The boom hoist ropes can be observed only after climbing to the roof of the crane superstructure, but the pendants cannot be examined unless the boom has been lowered to the ground. The inaccessibility of the suspension ropes makes it less likely that they will be inspected often enough and closely enough.

It is only through inspection that ropes can be evaluated and the need for replacement determined. Telltale signs (discard criteria) reveal when fatigue damage has progressed to the farthest acceptable point, when wear has reached the limit that precludes further use, and when other abnormalities require rope replacement. Some rope inspection is required daily before operations begin, and a formal program of rope inspection, evaluation, and record keeping comprise an important part of all risk control programs.

Rope inspection and discard

Going back to the early days of mining and mine-shaft hoists, many groups have given consideration to the design of rope systems and to the criteria that define the point at which a rope must be removed from service. Usually it is not reasonable or economical to design for unlimited life, and for some aerial tramways the life of the rope was actually found to be increased by reducing the design strength factor.† Given the premise that rope will not last for the life of the structure or machine, it automatically follows that timely rope replacement is a requisite for safe operation.

For machines that continuously perform repetitious service, rope replacement records will provide the data necessary to predict when future rope replacements will be required, even though rope life is a statistically random variable. But for general-purpose machines, there is no way whatsoever to predict life reliably; rope inspection provides the only viable means for determining when replacement must be made.

To this end, the American Society of Mechanical Engineers (ASME) B30 Committee, Safety Standards for Cableways, Cranes, Derricks, Hoists, Hooks, Jacks and Slings, has offered recommendations for rope inspection and discard programs for various types of equipment. The following is a compendium including materials extracted from various works produced by that committee but is edited and includes our own thoughts as well. For committee consensus, the reader is referred to the ASME publications for each equipment type.

Inspection

1. Frequent inspection.
   a. For machines that are in service, all running ropes should be visually inspected each working day, the inspection to include observation of all rope that is expected to be in use during the work to be performed that day. This procedure is for the purpose of discovering gross damage which may be an immediate hazard, such as:
      (1) Distortion of the rope, i.e., kinking, crushing, unstranding, birdcaging, main-strand displacement, core protrusion, loss of rope diameter in a short length, or uneveness of outer strands, either of which provides evidence that the rope may need to be replaced.
      (2) General corrosion.

---

†J. Kogan, On the Calculation of Cables for Lifting Appliances, *Neve Int.*, Turin, Italy, Spring 1977, p. 50.

(3) Broken or cut strands.

(4) The number, distribution, and type of visible broken wires.

(5) Core failure in rotation-resistant rope. The discovery of such damage indicates either that the rope should be removed from service immediately or that a more detailed inspection and evaluation must be made.

b. Care must be taken when inspecting portions of the rope subjected to rapid deterioration such as *flange points, crossover points,* and that part of the rope which first contacts the drum during repetitive lifts.

c. Care must be taken when inspecting certain ropes such as:

(1) Rotation-resistant ropes because of their sensitivity during handling and their susceptibility to damage.

(2) Boom-hoist ropes because of the difficulty of inspection and the important nature of these ropes.

2. Periodic inspection

a. Inspection frequency should be determined by a person experienced with rope in practice, based on expected rope life, as indicated by past observations of this or similar installations, severity of the operating environment, relative frequency of capacity lifts, intensity of service, and exposure to impact loads. Inspections need not be at uniform calendar intervals but should occur more frequently as the rope nears the end of its useful life; the inspection should be made at least annually, however.

b. The person performing periodic inspections should be particularly qualified for this task. The inspection must cover the entire length of the rope, but only the surface wires of the rope should be checked, and no attempt should be made to open the rope. A determination must be made whether further use of the rope is permissible when deterioration resulting in appreciable loss of original strength is noted, as described below:

(1) Those points listed under frequent inspection.

(2) Reduction of rope diameter below the nominal diameter due to loss of core support, internal or external corrosion, or wear of the outside wires.

(3) Wires at end connections severely corroded or broken.

(4) Severely corroded, cracked, bent, worn, or improperly applied end connections.

c. Care must be taken in the inspection of portions of the rope subject to rapid deterioration such as the following, which are in addition to those mentioned under frequent inspection:

(1) Portions in contact with saddles, equalizer sheaves, or other sheaves where rope travel is limited.

(2) Portions of the rope at or near terminal ends, where corroded or broken wires may protrude.

Rope replacement

1. Continued use of rope that has been in service depends entirely on the strength remaining in the rope, which in turn implies that this is largely a matter that relies on the good judgment of the person inspecting and evaluating the rope. Since many variable factors are involved in such a determination, there are no precise rules that can be applied to fix the exact time for replacement.
2. Conditions such as the following are reason for questioning the continued use of the rope or for increasing the frequency of inspection:
   a. In running ropes, six randomly distributed broken wires in any one lay or three broken wires in one strand of one lay.
   b. One outer wire that has broken at its contact point with the core of the rope and protrudes or loops out from the rope body. Additional and closer inspection of this portion is required.
   c. Wear comprising one-third of the original diameter of outside individual wires.
   d. Kinking, crushing, birdcaging, or any other damage resulting in distortion of the rope structure.
   e. Evidence of heat damage from any cause.
   f. Reductions from the nominal rope diameter of more than the following:

Reduction		Nominal diameters			
		From		To	
in	mm	in	mm	in	mm
1/32	0.8	3/8	9.5	1/2	13
3/64	1.2	9/16	14.5	3/4	19
1/16	1.6	7/8	22	1 1/8	29
3/32	2.4	1 1/4	32	1 1/2	38

   g. In standing ropes, more than two broken wires in one lay in portions away from end connections or more than one broken wire at an end connection.
3. Replacement rope must have a strength rating at least as great as the original rope and any deviation from the original size, grade, or construction should only be made on the advice of the equipment manufacturer, a rope manufacturer, or a licensed professional engineer practicing in this field.

4. All rope installed on a crane that has been shut down or in storage for a month or more should be given an inspection equivalent to a periodic inspection before it is placed in service.
5. Rope that has been removed from service on a crane should not be used to make slings.

Rope safety program. The information presented above should make it quite clear to supervisors and executives that rope safety rests almost entirely on timely performance of inspections by qualified people. The daily (frequent) inspection made by the operator does not require record keeping other than a notation on the daily crane-inspection check sheet. Periodic inspections should be recorded on dated signed reports. The preparation of a report formalizes the procedure and enhances the importance of the inspections in the minds of those who would otherwise think of them as a routine thing that must be done. This does not refer to the inspector, for any inspector who does not appreciate the importance of the task should be assigned to other duties.

Inspection reports also serve another purpose in our litigious society. They are proof that can be entered as evidence in court that your firm has behaved responsibly in respect to the wire rope used on your cranes, assuming of course, that your rope maintenance and use practices are proper as well. Here we are getting to matters of work practices rather than engineering principles and so out of the scope of this book. Excellent material for field rope practices as well as photographs and drawings of the rope defects discussed, selection and use of slings and rigging hardware, and other practical information can be found in Dickie's "Rigging Manual."†

This book, together with its sister volumes, "Crane Handbook" and "Mobile Crane Manual," by the same author and publisher, should be part of the safety equipment available to both supervisors and field personnel of every firm that uses or owns cranes and derricks. They are practical, profusely illustrated sources for "hands on" data and necessary guidance in the use of the equipment. They portray both the pitfalls and the proper practices, not as lists of don'ts, but as guides as to how the work *can* be done.

Tower crane accidents

Overturning accidents, which so dominate the mobile-crane accident statistics, are almost unknown with tower cranes in North America.

†D. E. Dickie, compiler, "Rigging Manual," Construction Safety Association of Ontario, Toronto, 1975.

Except during erection, dismantling, and climbing, the vulnerability of tower cranes to accidents is relatively small. But serious accidents do occur from operator misjudgment, inadequate maintenance, tampering with controls and limits, wind, and initial defects.

We know of no adequate statistics of tower crane accidents, but our experience suggests that the leading category would be mishaps related to erection, dismantling, and climbing. Although not necessarily the most numerous, they tend to be the most destructive, exceeding the other categories in injuries, fatalities, and property damage.

Erection, dismantling, and climbing accidents

Erecting and dismantling tower cranes is a serious and sometimes difficult business. As discussed in Chap. 6, these activities must proceed according to a complete operational plan that includes sequencing, weights, erection and dismantling crane positions, component staging locations, and erection and dismantling clearances. The crew must be experienced in rigging and equipped with the necessary hardware and tools. At least one person must be present who knows how to assemble and disassemble the mechanical, electrical, and structural systems.

Erection-related accidents can occur at any time, even after the riggers have left the site. Mishandling or incorrect installation procedures can create conditions that lead to later failure. Poor rigging practice can cause damage, but damage initiated during storage or transport is more common. One can conclude from this that a final visual inspection of structural components before the crane is put to use should be part of the erection process.

If damage is found, an engineer's judgment maybe needed to decide on the course of action because many components of tower cranes are highly stressed and some are exposed to fatigue-inducing conditions. If field repairs are feasible, they must be made in accordance with the manufacturer's or an engineer's procedures; nondestructive testing may be advisable. Repair decisions are critical and should be neither hastily nor lightly made.

Bolting is a common source of installation errors. High-strength designations and markings of nuts and bolts are not widely understood, particularly those with ISO grading. And, few workers are aware that much greater harm can be done by undertightening bolts than by overtightening. Not surprisingly, the most common errors are undertightening and mismatching of nuts and bolts. There are several ways that mismatches can happen; common grade nuts are used on high-strength bolts, U.S. and metric components are mixed, or the threads on nuts and bolts do not match.

Dismantling presents other hazards. The greatest difficulties arise because a building is now present where earlier there was vacant space. Movements and clearances are often hampered by the new construction. Bolts and pins, after having been in place for months in all weathers, can be locked in place and hard to remove, requiring more time than anticipated and perhaps special tools. Chapter 6 discusses problems inherent in dismantling and particularly procedures required for jib removal. As with erection, planning is a prerequisite to a successful dismantling operation, but dismantling planning requires a higher level of practical expertise than erection.

During climbing operations, tower cranes are more vulnerable to mishap. This occurs because the equipment is relatively unsecured to allow for climbing movement; its means of support are continually shifting while the crane itself is moving in close proximity to fixed objects where hangups might occur. Conditions conducive to safe climbing are good lighting, ample working time, little or no wind, and an experienced, well-placed, and alert crew. The entire climbing system and the support elements, including shoring, should be inspected in advance of the climb.

Other causes of accidents

Causes of tower crane accidents are not always clear because accidents are often the result of a combination of undesirable conditions. But, if an accident requires more than one condition to occur, correction of only one of the undesirable conditions should be enough to prevent the mishap. This principal is both a reason for optimism and a key to accident prevention.

For example, all tower cranes are equipped with various automatic controls and limit switches to prevent overloading, overspeeding, and overtravel. A lazy and irresponsible operator will use these devices to control the crane; therefore if a device fails, an accident may ensue. The accident could be attributed to mechanical failure, but it is equally due to the failure of the operator to exercise control over the crane. Had the operator been performing properly, the accident would not have happened even though the mechanical failure occurred.

Another example concerns the prevention of fatigue failures. Some tower crane components are normally subjected to a high number of stress reversals—conditions conducive to fatigue. Crane makers take heed of this and do the best they can to avoid fatigue through careful detailing and fabrication. Fatigue potential, however, cannot be eradicated, and fatigue itself is a process where incremental damage accumulates over time. Thorough inspection is therefore important, especially as the crane ages. But even abused cranes can be saved from acute damage if the incipient cracks are discovered during inspection between jobs.

There is no better time to avoid fatigue and other ills than the period between jobs when the crane is idle. This is the opportunity to make a thorough visual inspection of the entire crane structure. Should indications of possible damage be found, more stringent examination procedures can then be used.

A wind-related accident can occur while the crane is working or when it is not in service. The magnitude of wind force that might occur while a crane is working is not sufficient by itself to cause collapse, but it can cause the operator to lose control. If the load has a large wind-catching area, damage can be done by either the boom or the load because the wind force may overcome the swing brakes and the plugging resistance of the swing motors. The remedy is to take the crane out of service when the wind starts to make load control tenuous. The difficulty here is that supervisors, from their ground level vantage point, will not feel the same wind intensity.

All tower cranes, when out of service, must be left free to weathervane unless the crane and installation have been specifically designed otherwise. This means that swing brakes and drives must be disengaged whenever the operator leaves the cab and shuts down the prime mover. For luffing cranes, the boom should be left at the lowest boom angle consistent with an unobstructed weathervaning path unless the site installation instructions specify a particular boom angle. As the boom angle decreases, wind effects on the crane decrease accordingly, hence the boom should be lowered when not working.

Out-of-service wind failures can be reduced to a small probability, but never eliminated, because the design wind is statistically determined and therefore can always be exceeded. The installation designer must choose a rational wind level that balances risks and economy of design (see Chap. 3).

If a hurricane-level storm is known to be approaching, defensive measures can be taken. The crane supplier or installation engineer should be consulted; the crane configuration at that particular time may not be storm-wind limited, or climbing cranes may perhaps be made safe by climbing down one or two increments. Needless to say, tower wedges, braces, or other attachments should be checked before the storm arrives.

Lifting personnel with cranes

Older ironworkers tell of riding up to their workplace while standing on a headache ball. In those days bravado went with an ironworker's job, OSHA did not exist, and managers did not have to worry much about liability.

More recently, workers would ride on a platform or a bosun's chair suspended from a crane hook. Though clearly better than the earlier practice, many serious accidents demonstrated that considerable risk was involved in using a crane as a personnel lift.

The federal government amended the OSHA personnel lifting standards during 1988. The new regulations are aimed at mitigating the dangers of personnel lifting without banning the practice outright. The new ASME/ANSI B30.23, Personnel Lifting Systems, covers the same subject in even more detail.

Casual personnel lifting is no longer permitted. Safeguards have been imposed and now there must be a demonstrable reason for using a crane or derrick to lift people; it must be shown that either it is the least hazardous means for access or that there is no reasonable alternative. A proposed operation satisfying one of those criteria will be permitted only with strict controls. Those controls are designed to directly improve safety, but they are also sufficiently onerous to dissuade people from lifting personnel with a crane unless it is really necessary.

Cranes and derricks are subjected to stricter limitations when used for lifting persons as compared to material handling. Rope design factors are doubled and lift capacity is halved. Freefalling is not permitted in either the boom hoist or the load hoist. The law requires two-block prevention equipment.

The lift platform can no longer be an ordinary skip box but must now be a specially designed platform used just for people and the tools and materials for the specific job.

Before the platform can be put to use at a jobsite, it is required to be proof loaded, followed by an inspection; the crane must be inspected as well. Each intended use has to be prefaced by a trial run through the anticipated motions using weights instead of people. Procedures are to be previewed in a meeting attended by the operator, lift supervisor, signalman, and the people who are to be lifted.

The comments given here are a summary. Personnel lifting operations should not he planned or implemented without first reviewing the full text of the regulations.

8.6 Codes and Standards

The standard design codes of the American Institute for Steel Construction and the American Concrete Institute, in addition to local or national building codes, provide the allowable stress and practice guidelines necessary for installation design. The practical information and procedures given throughout this book are in conformity with those codes (at the time of writing) or offer advice in areas not covered by code, except where specifically noted otherwise.

Basic guidance on inspection, maintenance, and operation of cranes and derricks is to be found in ASME/ANSI Standards,† which are nongovernmental consensus standards that have been adopted in large measure into the OSHA standards. The full series provides necessary information on the whole realm of equipment of interest here:

B30.1	Jacks
B30.2	Overhead and Gantry Cranes (Top Running Bridge, Single or Multiple Girder, Top Running Trolley Hoist)
B30.3	Construction Tower Cranes
B30.4	Portal, Tower and Pedestal Cranes
B30.5	Mobile and Locomotive Cranes
B30.6	Derricks
B30.7	Base Mounted Drum Hoists
B30.8	Floating Cranes and Floating Derricks
B30.9	Slings
B30.10	Hooks
B30.11	Monorails and Underhung Cranes
B30.12	Handling Loads Suspended from Rotorcraft
B30.13	Storage/Retrieval (S/R) Machines and Associated Equipment
B30.14	Side Boom Tractors
B30.16	Overhead Hoists (Underhung)
B30.17	Overhead and Gantry Cranes (Top Running Bridge, Single Girder, Underhung Hoist)
B30.18	Stacker Cranes (Top or Under Running Bridge, Multiple Girder With Top or Under Running Trolley Hoist)
B30.19	Cableways
B30.20	Below the Hook Lifting Devices
B30.21	Manually Lever Operated Hoists
B30.22	Articulating Boom Cranes
B30.23	Personnel Lifting Systems
B30.24	Container Cranes (not final at the time of this writing)
B30.25	Scrap and Material Handlers

The common introduction to the series includes a statement so basic to the use of cranes and derricks and so pertinent to the subject at hand that it should be quoted in full.

---

† Safety Standard for Cableways, Cranes, Derricks, Hoists, Hooks, Jacks and Slings, American Society of Mechanical Engineers, New York.

> The use of cableways, cranes, derricks, hoists, hooks, jacks and slings is subject to certain hazards that cannot be met by mechanical means, but only by the exercise of intelligence, care and common sense. It is therefore essential to have competent and careful personnel involved in the use and operation of the equipment, physically and mentally qualified, trained in the safe operation of the equipment and the handling of the loads.†

8.7 A Rational Means for Controlling Risk

The preceding discussion of the causes of lifting accidents was not presented to induce morbid fear. Rather, our purpose has been to offer insights, to foster thinking about risk reduction, and to introduce a rational, systematic method for lifting contractors, project managers, and field supervisors to confront risk. Clearly, not all crane or rigging operations need the same management attention. It is therefore necessary that managers establish policy guidelines to segregate operations that call for more attention, or deal with the issues on a case by case basis.

The authors have devised a general method for sorting out lifting operations, using four steps to separate those that need the least oversight from those that need the most. The method requires subjective judgments, and therefore is not foolproof. It should be tailored to the particular circumstances of each firm or jobsite. Used effectively, it should produce the dual benefits of risk reduction and directing management assets to where they can be most gainfully used.

Quantification of risk elements

The system requires that each of the items below must be assigned a value from 1 to 10, based on judgement. Allow a value of 1 if the item is of little significance or poses only minor problems, and 10 if of very high worth or presents very serious problems.
Items to be assigned values:

1. The money value of the load—replacement value if the load is destroyed including replacement shipping costs.
2. The potential for injury and/or property damage (other than to the load) should the load be dropped, or the crane or rigging equipment overturn, collapse or otherwise fail.

† Ibid., with permission of the American Society of Mechanical Engineers.

3. The potential for consequential damages; includes the costs of project delays due to loss or damage to the load, or while waiting for repairs or replacements, loss of business, and damage to the property of others.
4. The nature of the load: is it easy or difficult to handle, with well defined lift points or difficult attachment conditions; is it irregular or unsymmetrical, a large wind catcher, flexible and sloppy; are there a number of lift points requiring lift beams, or special accessories, or equalizers; how load weight relates to the rated capacity of the lift equipment.
5. The nature of site conditions: is it a clear, level, open site, or are there access problems, restrictions, obstructions, a constricted work area, or poor support conditions; will this be the only operation going on, or will coordination with other activities be needed; are there environmental restrictions or dangers, such as flammable gasses or liquids in adjacent pipes or vessels, or power lines; is this an open windy site or a protected one?

When assigning values, if the operation requires a multi-crane lift the "nature of load" value should be closer to 10; the lower end applies if a straight-forward unloading operation using two cranes and the higher end as complexity increases.

This evaluation procedure assumes that the equipment to be used will be in proper operating condition, and that OSHA rules, operator's manual instructions, and ASME/ANSI industry standards will be adhered to. After values have been assigned to each of the five listed items, the values are totaled for use in one of the Management Action tables below, either the Lifting Contractor table or the Project Management table. The tables are set up using value ranges which separate projects into four groups based on the level of management participation or attention required.

Lifting contractors:

Range 0 to 15 and no single value greater than 6; implies that the job is routine and that minimal management oversight is necessary; planning can be done by field personnel, but managers should nonetheless discuss plans and monitor the operation.

Range 16 to 25 but not more than one value of 6 or more; implies that management involvement is needed to review the operation, together with the Lift Director, to identify risks, and to plan measures that can be used to control or minimize them. Consideration should be given to the preparation of a formal plan of operations (lift plan) and/or having engineering participation in planning, otherwise engineering assistance may be advisable for reviewing plans, particularly when values are at the higher end of the range. Field operations should be monitored for compliance with plans and safety rules.

Range 26 to 35: implies the need for preparation of a formal written operations plan (lift plan), made with the assistance or under the supervision of an engineer. The plan should cover step-by-step procedures and details required to control risk. Depending upon circumstances and on the nature of the work, managers may elect to have the plan prepared by in-house staff and reviewed by an outside licensed engineer, but the extent of assistance needed depends on the level and kinds of risk. Field operations should be monitored for compliance with the operations plan and safety rules.

Range 36 to 50: when values fall in this range, full, complete, detailed, in-depth planning and review is necessary. If the project or company does not have access to a qualified in-house engineering staff, outside engineering help should be secured early in the process. The analysis and review should include an effort to identify alternative schemes for doing the job and to explore available means to control and to minimize risk. Field operations should be monitored for compliance with the operations plan and safety rules.

Project management:

Range 0 to 15 and no single value more than 6; implies that the job is routine and that only minimal oversight is necessary; for the most part, can be left to field personnel or to the Lifting Contractor. Management should monitor field operations for safety in general.

Range 16 to 25 but not more than one value of 6 or more; implies that management involvement is needed to review the operation, together with the Lifting Contractor, to identify risks, and to discuss measures that will be employed to control or minimize them. Consideration should be given to requiring the Lifting Contractor to submit a formal plan of operations (lift plan) and/or have an engineer participate in planning. Engineering participation may be advisable to assist in management review, particularly when values are at the higher end of the range. Field operations should be monitored for compliance with the operations plan and safety rules.

Range 26 to 35: implies that the Lifting Contractor should be required to prepare and submit a formal written operations plan (lift plan) with engineering participation. The plan should cover step-by-step procedures and details required to control risk. Depending upon circumstances, managers may accept the plan without comment or review the plan with the Lifting Contractor. If the plan is reviewed, engineering assistance should be sought, but the extent of assistance needed depends on the nature of the work. Field operations should be monitored for compliance with the operations plan and safety rules.

Range 36 to 50: when values fall in this range, full, complete, detailed, in-depth planning and review is necessary. Qualified in-house engineering staff or outside consulting engineers should be brought in early in the process. The analysis and review should include an effort to identify alternative schemes for doing the job and to explore available means to control and to minimize risk. Field operations should be monitored for compliance with the operations plan and safety rules.

This method was used to advantage on a project in Oakland, California where 16 separate crane set ups were needed for erecting steel girders on a viaduct leading to the Golden Gate Bridge. The lifts varied from 46 tons (41.7 t) by 91 ft (27.7 m) long to174 tons (158 t) by 231 ft (70.4 m) long. Some of the girders were curved, five of the positions required two cranes, and 13 involved erection over bridge access ramps or roadways.

Within the four value ranges, 0 to 15, 16 to 25, 26 to 35, and 36 to 50, seven operating positions were found to be in the 26 to 35 range while the remainder were between 16 and 25; the highest rating was 29 and the lowest 16. This quantification scheme identified the most critical parts of the work, focussing the attention of the planners and leading to changes in craneage, positioning, and procedures that improved the risk profile. Better yet, the risk evaluation work convinced state highway authorities to relax traffic control restrictions they had imposed. Those restrictions would have materially added to risk and the likelihood of an accident and a massive traffic tie-up.

Appendix

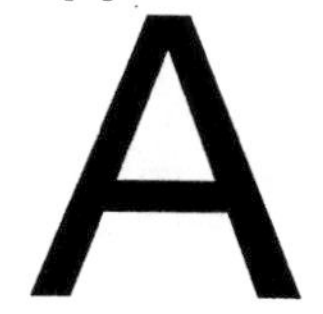

Conversions Between U.S. Customary, SI, and Other Metric Units

To convert from	*To*	*Multiply by*
ACCELERATION		
Feet per second^2 (ft/s^2)	Meters per second^2 (m/s^2)	0.30480
Free fall, standard	ft/s^2	32.174
	m/s^2	9.80665
AREA		
Square feet (ft^2)	Square meters (m^2)	0.092903
Square inches (in^2)	Square centimeters (cm^2)	6.4516
BENDING MOMENT, OR TORQUE		
Pound force–inches (lb · in)	Newton-meters (Nm)	0.11298
Pound force–feet (lb · ft)	Nm	1.3558
Kip force–inches (kip · in)	Nm	112.985
Kip force–feet (kip · ft)	Nm	1355.818
FORCE		
Kilograms force (kgf)	Newtons (N)	9.80665
Kips	N	4448.222
Pounds force (lb)	N	4.4482
LENGTH		
Feet (ft)	Meters (m)	0.30480
Inches (in)	m	0.02540
Miles, U.S. statute (mi)	m	1609.334
MASS		
kips	pounds	1000
Pounds mass (lb)	Kilograms (kg)	0.45359
Tons, long (= 2240 lb)	kg	1016.047

Conversions between U.S. Customary, SI, and Other Metric Units

To convert from	*To*	*Multiply by*
Tons or tonnes, metric (t)	kg	1000.00
Tons, short (= 2000 lb)	kg	907.185
POWER		
Foot-pounds force per second (ft · lb/s)	Watts (W)	1.3558
Horsepower (= 550 ft · lb/s)(hp)	W	745.70
Horsepower (metric)	W	735.499
PRESSURE OR STRESS		
Atmospheres, technical (at)	Kilopascals (kPa)	98.0665
Bars	kPa	100.000
Kilograms force per square meter (kgf/m^2)	Pascals (Pa)	9.80665
Pascals (Pa)	Newtons per square meter (N/m^2)	1.000
Pounds force per square foot (lb/ft^2)	Pa	47.8803
Pounds force per square inch (lb/in^2)	kPa	6.8947
UNIT WEIGHT		
Pounds per foot (lb/ft)	kg/m	1.4881
VELOCITY		
Feet per second (ft/s)	Meters per second (m/s)	0.30480
Kilometers per hour (km/h)	m/s	0.27778
Miles per hour (U.S.) (mi/h)	m/s	0.44704
	Kilometers per hour (km/h)	1.609344
VOLUME		
Cubic feet (ft^3)	Cubic meters (m^3)	0.028317
Gallons, U.S., liquid (gal)	m^3	0.0037854
Gallons, U.K. and Canada, liquid (gal)	m^3	0.0045461

Appendix

B

Glossary of Crane and Derrick Terms

A frame See Gantry and Jib mast. Also, a frame sometimes used with derricks, particularly barge-mounted, to support the boom foot and the topping lift.

Anti two-block device A mechanism which, when activated, disengages all crane functions whose movement can cause the lower load block to strike the upper load block or boom head sheaves.

Articulated jib A tower crane jib that in general has a pivot point somewhere in its middle area; also called pivoted luffing jib.

Axis of rotation The vertical line about which a crane or derrick swings; also called center of rotation (obsolete) and swing axis.

Back-hitch gantry See Gantry.

Ballast Weight added to a crane base to create additional stability; it does not rotate when the crane swings.

Barrel The lagging or body part of a rope drum in a drum hoist.

Base mounting The structure forming the lowest element of a crane or derrick; it transmits loads to the ground or other supporting surface. For mobile cranes this is synonymous with carrier or crawler mounting. For tower cranes, the term iucludes a travel base, knee frame base, or fixed base (footing).

Base section The lowermost section of a telescopic boom; it does not telescope but contains the boom foot pin mountings and the boom-hoist-cylinder upper end mountings.

Basic boom The minimum length of sectional latticed boom that can be mounted and operated, usually consisting of a boom base and tip section only.

Bogie An assembly of two or more axles arranged to permit both vertical wheel displacement and an equalization of loading on the wheels.

Boom A crane or derrick member used to project the upper end of the hoisting tackle in reach or in a combination of height and reach; also called jib (European).

Boom angle The angle between the horizontal and the longitudinal centerline of the boom, boom base, or base section.

Boom base The lowermost section of a sectional latticed boom having the attachment or boom foot pins mounted at its lower end; also called boom butt or butt section.

Boom butt See Boom base.

Boom foot mast A component of some mobile-crane boom suspensions. It consists of a frame hinged at or near the boom foot that serves to increase the height of the inboard end of the fixed-boom suspension ropes, thereby increasing the angle those ropes make with the boom while being itself controlled by the boom-hoist ropes. Its purpose is to decrease the axial compressive force on the boom; also called hi-light gantry.

Boom guy line See Pendant.

Boom head The portion of a boom that houses the upper load sheaves.

Boom hoist The rope drum(s), drive(s), and reeving controlling the derricking motion of the boom.

Boom-hoist cylinder Hydraulic ram used instead of a rope boom suspension; the most common means of derricking telescopic booms.

Booming in (out) See Derricking.

Boom inserts Center sections of a sectional latticed boom usually having all four chords parallel.

Boom point See Boom tip section.

Boom stay See Pendant.

Boom stop A device intended to limit the maximum angle to which the boom should be derricked.

Boom suspension A system of ropes and fittings, either fixed or variable in length, that supports the boom and controls the boom angle.

Boom tip section The uppermost section of a sectional latticed boom, which usually includes the weldment mounting the upper load sheaves as an integral part; also called boom point, head section, or tapered tip.

Bridle See Floating harness.

Bull pole A pole, generally of steel pipe, mounted to project laterally from the base of a derrick mast. It is used to swing the derrick manually.

Bull wheel A horizontally mounted circular frame fixed to the base of a derrick mast to receive and guide the ropes used for swinging.

Butt section See Boom base.

Carbody That part of a crawler crane base mounting which carries the superstructure and to which the crawler side frames are attached.

Carrier A wheeled chassis that is the base mounting for mobile truck and rough terrain cranes.

Center of rotation (obsolete) See Axis of rotation.

Cheek weights Overhauling weights attached to the side plates of a lower load block.

Climbing frame A supplemental structure placed on, and forming a sleeve around, tower crane mast; it is used in top climbing the crane; also called a climbing section.

Climbing ladder A steel member with crossbars (used in pairs) suspended from a frame and used as jacking support points when some tower cranes climb.

Climbing schedule A diagram or chart giving information for coordinating the periodic raising (climbing or jumping) of a tower crane with the increasing height of the building structure as the work progresses.

Counter jib A horizontal member of a tower crane on which the counterweights and usually the hoisting machinery are mounted; also called counterweight jib.

Counterweight Weights added to a crane superstructure to create additional stability. They rotate with the crane as it swings.

Counterweight jib See Counter jib.

Crawler frames Part of the base mounting of a crawler crane attached to the carbody and supporting the crawler treads, the track rollers, and the drive and idler sprockets. Crawler frames transmit crane weight and operational loadings to the ground; also called side frames.

Cribbing Timber mats, steel plates, or structural members placed under mobile-crane tracks or outrigger floats to reduce the unit bearing pressure on the supporting surface below.

Crossover points Points of rope contact where one layer of rope on a rope drum crosses over the previous layer.

Dead end The point of fastening of one rope end in a running rope system, the other (live) end being fastened at the rope drum.

Deadman An object or structure, either existing or built for the purpose, used as anchorage for a guy rope.

Derricking Changing boom angle by varying the length of the boom suspension ropes; also called luffing, booming in (out), or topping.

Dog A pawl used in conjunction with a ratchet built into one flange of a rope drum to lock the drum from rotation in the spooling-out direction; also, one of a set of projecting lugs that support the weight of a tower crane.

Dogged off The condition of a rope drum when its dog is engaged.

Drift The vertical clearance between the top of a lifted load and the lifting crane hook when in its highest position; a measure of the gap available for slings and other rigging or for manipulating the load.

Drifting Pulling a suspended load laterally to change its horizontal position.

Drum hoist A hoisting mechanism incorporating one or more rope drums; also called hoist, winch, or hoisting engine.

Duty cycle work Steady work at a fairly constant short cycle time with fairly constant loading levels for one or more daily shifts.

EOT crane Electric overhead traveling crane.

Expendable base For static-mounted tower cranes, a style of bottom mast section that is cast into the concrete footing block. All or part of this mast section is lost to future installations.

Extension cylinder Hydraulic ram used to extend a section of a telescopic boom; the most common but not the only means for power-extending boom sections.

Fall See Parts of line.

Flange point The point of contact between the rope and drum flange where the rope changes layers on a rope drum.

Fleet angle The angle the rope leading onto a rope drum makes with the line perpendicular to the drum rotating axis when the lead rope is making a wrap against a flange.

Fleeting sheave Sheave mounted on a shaft parallel to the rope-drum shaft and arranged so that it can slide laterally as the rope spools, permitting close sheave placement without excessive fleet angle.

Float An outrigger pan that distributes the load from the outrigger to the supporting surface or to cribbing placed beneath it; part of the crane's outrigger support system.

Floating harness A frame, forming part of the boom suspension, supporting sheaves for the live suspension ropes and attached to the fixed suspension ropes (pendants); also called bridle, spreader, upper spreader, live spreader, or spreader bar.

Fly section On a telescopic boom, the outermost powered telescoping section.

Footblock A steel weldment or assembly serving as the base mounting for a guy derrick, gin pole, or Chicago boom derrick.

Free fall Lowering the hook (load) or boom by gravity; the lowering speed is controlled only by a retarding device such as a brake.

Frequency of vibration Number of vibration cycles that will occur in a unit of time, usually expressed in hertz (cycles per second); also called frequency.

Front end attachments Optional load-supporting members for use on mobile cranes, e.g., taper tip boom, hammerhead boom, guy derrick attachment, and tower attachment.

Gantry A structure, fixed or adjustable in height, forming part of the superstructure of a crane, to which the lower spreader (carrying live boom-suspension ropes) is anchored; also called A-frame gantry, A-frame, or back-hitch gantry.

Gate block See Snatch block.

Gooseneck boom A boom with an upper section projecting at an angle to the longitudinal centerline of the lower section.

Gudgeon pin The pin at the top of a derrick mast forming a pivot for the spider or for the mast of a stiffleg derrick.

Guy rope A fixed-length supporting rope intended to maintain a nominally fixed distance between the two points of attachment; also called stay rope or pendant.

Hammerhead boom A boom tip arrangement in which both the boom suspension and the hoist ropes are greatly offset from the boom longitudinal centerline to provide increased load clearance.

Head section See Boom tip section.

Hog line See Pendant, Intermediate suspension.

Hoist To lift an object; the mechanism used for lifting.

Horse See Jib mast.

Impact Increase in load effect from dynamic causes.

In-service wind Wind encountered while a crane is working; usually used to define the maximum permissible level of wind pressure or velocity at the site before the crane must be taken out of service.

Intermediate suspension An additional set of boom-suspension lines attached to the boom at some point between the main suspension attachment and the boom foot. On mobile cranes it is used to reduce boom elastic deflection during erection; on horizontal jib tower cranes it is used as part of the primary support; also called midpoint suspension, midpoint hitch, intermediate hitch, or intermediate hog line.

Jib In American practice, an extension to the boom mounted at the boom tip, in line with the boom longitudinal axis or offset to it. It is equipped with its own suspension ropes made fast to a mast at the boom tip, which in turn is supported by guy ropes or a derricking system. (Europeans call this a *fly jib* and use *jib* to refer to a boom.)

Jib mast A short strut or frame mounted on the boom head to provide a means for attachment of the jib support ropes; also called jib strut, rooster, horse, or A-frame.

Jib strut See Jib mast.

Jumping Raising a tower crane or guy derrick from one operating elevation to the next in concert with the completion of additional floors of the building.

Lagging Removable shells (optional) for use on a rope drum to produce change in line pull or line speed.

Latticed boom A boom constructed of four longitudinal corner members, called *chords,* assembled with transverse and/or diagonal members, called *lacings,* to form a trusswork in two directions. The chords carry the axial boom forces and bending moments, while the lacings resist the shears.

Layer A series of wraps of wire rope around a rope drum barrel, extending full from flange to flange.

Level luffing An automatic arrangement whereby the crane or derrick hook does not significantly change elevation as the boom derricks.

Line pull The pulling force attainable in a rope leading off a rope drum or lagging at a particular pitch diameter (number of layers).

Line speed The speed attainable in a rope leading off a rope drum or lagging at a particular pitch diameter (number of layers).

Live mast See Boom foot mast.

Live spreader See Floating harness.

Loading An external agency that induces force in members of a structure. It may be in the form of a direct physical entity (e.g., weight superimposed on the structure or pressure applied by wind) or in the form of an abstract condition (e.g., inertial effects associated with motion).

Load jib See Saddle jib.

Load radius See Radius.

Lower spreader A frame, forming part of the boom suspension, supporting sheaves for the live suspension ropes and attached to the gantry or superstructure.

Luffing See Derricking.

Luffing jib A tower crane boom or mobile crane jib that is raised and lowered about a pivot to move the hook radially, that is, to change its working radius.

Lumped masses A concept used to simplify mathematical analysis, whereby distributed or discrete masses are replaced by aggregates of mass concentrated at a point (or points) where the resultant of the inertia forces associated with the actual masses seem to act.

Machine resisting moment The moment of the dead weight of the crane or derrick, less boom weight, about the tipping fulcrum; hence, the moment that resists over-turning; also called machine moment or stabilizing moment.

Manual insert An optional nonpowered section in a telescopic boom forming the outermost boom section when provided and usable either fully extended or fully retracted.

Mast An essentially vertical load-bearing component of a crane or derrick; the tower of a tower crane. See also Boom foot mast and Jib mast.

Mast cap See Spider.

Midpoint suspension See Intermediate suspension.

Midsection On a telescopic boom, the intermediate powered telescoping section(s) mounted between the base and fly sections.

OET crane See EOT crane.

Offset angle The angle between the longitudinal centerline of a jib and the longitudinal centerline of the boom on which it is mounted.

Operating radius See Radius.

Operating sectors Portions of a horizontal circle about the axis of rotation of a mobile crane providing the limits of zones where over-the-side, over-the-rear, and over-the-front ratings are applicable.

Out-of-service wind The wind speed or pressure that an upright inoperative crane is exposcd to; usually used to define the maximum level of wind that the inoperative crane is dcsigncd to safely sustain.

Outriggers Extensible arms attached to a crane base mounting, which include means for relieving the wheels (crawlers) of crane weight; used to increase stability.

Overhauling weight Weight added to a load fall to overcome resistance and permit unspooling at the rope drum when no live load is being supported; also called headache ball; see also Cheek weights.

Overturning moment The moment of the load plus the boom weight about the tipping fulcrum. Wind and dynamic effects can be included when appropriate.

Parking track For rail-mounted traveling cranes, a section of track supported so that it is capable of sustaining storm-induced bogie loads; it is provided with storm anchorages when required.

Parts of line A number of running ropes supporting a load or force; also called parts or falls.

Pawl See Dog.

Paying out Adding slack to a line or relieving load on a line by letting (spooling) out rope.

Pendant A fixed-length rope forming part of the boom-suspension system; also called boom guy line, hog line, boom stay, standing line, or stay rope.

Pitch diameter The diameter of a sheave or rope drum measured at the centerline of the rope; tread diameter plus rope diameter.

Pivoted luffing jib See Articulated jib.

Preventer In rigging practice, a means, usually comprising but not limited to a wire rope, for preventing an unwanted movement or occurrence or acting as a saving device when an anchorage or attachment fails.

Radius, load (operating) Nominally, the horizontal distance from the axis of rotation to the center of gravity of a lifted load. In mobile-crane practice, this is more specifically defined as the horizontal distance from the projection to the ground of the axis of rotation before loading to the center of a loaded but vertical hoist line.

Range diagram A diagram showing an elevation view of a crane with circular arcs marked off to show the luffing path of the tip for all boom and jib lengths and radial lines marking boom angles. A vertical scale indicates height above ground, while a horizontal scale is marked with operating radii. The diagram can be used to determine lift heights, clearance of the load from the boom, and clearances for lifts over obstructions.

Reach Distance from the axis of rotation of a crane or derrick; sometimes used synonymously with radius.

Recurrence period The interval of time between occurrences of repeating events; the statistically expected interval of time between occurrences, such as storms, of a given magnitude.

Reeving diagram A diagram showing the path of the rope through a system of sheaves (blocks).

Revolving superstructure See Superstructure.

Rooster Vernacular term for one or more struts at the top of a boom or mast, such as a jib strut, a tower-crane top tower, or the struts at the top of the mast of a mobile-crane tower attachment.

Root diameter See Tread diameter.

Rope drum That part of a drum hoist which consists of a rotating cylinder with side flanges on which hoisting rope is spooled in or out (wrapped).

Rotation-resistant rope A wire rope consisting of an inner layer of strand laid in one direction covered by a layer of strand laid in the opposite direction.

Running line A rope that moves over sheaves or drums.

Saddle jib The horizontal live-load-supporting member of a hammerhead-type tower crane having the load falls supported from a trolley that traverses the jib; also called load jib.

Sheave A wheel or pulley with a circumferential groove designed for a particular size of wire rope; used to change direction of a running rope.

Side frames See Crawler frames.

Side guys Ropes supporting the flanks of a boom or mast to prevent lateral motion or lateral instability.

Side loading A loading applied at any angle to the vertical plane of the boom.

Sill One of the horizontal stationary members of a stiffleg derrick, it secures the vertically diagonal stifflegs.

Slewing See Swing.

Snatch block A single- or double-sheave block arranged so that one or both cheek plates can be opened, permitting the block to be reeved without having to use a free rope end; also called gate block.

Spider A fitting mounted to a pivot (gudgeon pin) at the top of a derrick mast, providing attachment points for guy ropes; also called mast cap.

Spreader See Floating harness, Lower spreader.

Spreader bar See Floating harness.

Stabilizers Devices for increasing stability of a crane; they are attached to the crane base mounting but are incapable of relieving the wheels (crawlers) of crane weight.

Stabilizing moment See Machine resisting moment.

Standing line A fixed-length line that supports loads without being spooled on or off a drum; a line of which both ends are dead; also called guy line, stay rope, or pendant.

Static base Tower-crane support (base mounting) where the crane mast is set on or into a foundation with or without knee braces.

Stay rope See Guy rope, Pendant.

Strand A group of wires helically wound together forming all or part of a wire rope.

Strength factor Failure load (or stress) divided by allowable working load (or stress).

Structural competence The ability of the equipment and its components to support the stresses imposed by operating loads without the stresses exceeding specified limits.

Superstructure The entire rotating structure of a crane less the front end attachment; also called upper, or revolving superstructure.

Swing A crane or derrick function wherein the boom or load-supporting member rotates about a vertical axis (axis of rotation); also called slewing.

Swing axis See Axis of rotation.

Tackle An assembly of ropes and sheaves designed for pulling.

Tagline A rope (usually fiber) attached to the load and used for controlling load spin or alignment from the ground. Also, for clamshell operations, a wire rope used to retard rotation and pendulum action of the bucket.

Tailing crane In a multimachine operation in which a long object is erected from a horizontal starting position to a vertical final position, the crane controlling the base end of the object.

Taking up The process of removing slack from a line or drawing (spooling) in on a line; loading a line by drawing in on it.

Tapered tip See Boom tip section.

Telescoping A process whereby the height of a traveling or freestanding tower crane is increased by adding sections at the top of the outer tower and then raising the inner tower. There are cranes that are telescoped by adding to the inner tower from below.

Tipping fulcrum The horizontal line about which a crane or derrick will rotate should it overturn; the point(s) on which the entire weight of a crane or derrick will be imposed during tipping.

Tipping load The load for a particular operating radius that brings the crane or derrick to the point of incipient overturning.

Top climbing A method of raising a tower crane by adding mast sections; with the crane balanced, a climbing frame lifts the crane upperworks to permit insertion of a new tower section beneath the turntable. The process is repeated as required.

Topping See Derricking.

Topping lift See Boom hoist.

Top tower A tower mounted above the jibs of some tower cranes providing means for attachment of the pendants; also called rooster or towerhead.

Tower Tower crane mast; on some tower cranes there is a *top tower* used to support the jib and counterjib pendants.

Tower attachment A combination of vertical mast and luffing boom mounted to the front end of a mobile crane in place of a conventional boom.

Tower head See Top tower.

Transit Movement or transport of a crane from one jobsite to another.

Travel Movement of a mobile or wheel-mounted tower crane about a jobsite under its own power.

Travel base The base mounting for a wheel-mounted (traveling) tower crane.

Tread diameter The diameter of a sheave or grooved rope drum measured at the base of the groove; the diameter of a smooth barrel on a rope drum.

Trolley A carriage carrying the hook block for radial movement along the lower chords of a horizontally mounted tower crane jib; in some tower cranes, a *counterweight trolley* allows the counterweights to be moved radially to modulate their backward moment in proportion to load hook radius.

Two blocking The condition in which the lower load block or hook assembly comes in contact with the upper load block or boom point sheave assembly.

Upper See Superstructure.

Upper spreader See Floating harness.

Vangs (vang lines) Side lines reeved to a derrick boom and used to swing the boom.

Walking beam A bogie member whose long axis is nominally horizontal and parallel to the direction of travel; it is pivoted at its center and mounts a wheel, wheel pair, axle, or the center pivot of another walking beam at each end. Its purpose is to permit wheel oscillation, thus equalizing wheel loading during passage over travel-path irregularities.

Weathervane To swing with the wind when out of service so as to expose a minimal area to the wind.

Whip line A secondary or auxiliary hoist line; also called whip, runner, or auxiliary line.

Winch See Drum hoist.

Wrap One circumferential turn of wire rope around a rope-drum barrel.

Index

ABOUT THE AUTHORS

HOWARD I. SHAPIRO is president of Howard I. Shapiro & Associates, Consulting Engineers, P.C., of which JAY P. SHAPIRO and LAWRENCE K. SHAPIRO are vice presidents and principal engineers. Their firm has an international practice in crane and derrick application engineering, and also specializes in providing engineering assistance to high-rise construction contractors. For nearly 30 years the firm has played an active role in developing technical and safety standards for cranes and derricks under the auspices of the Society of Automotive Engineers, American Society of Mechanical Engineers, and the International Organization of Standardization.